Plant Polyphenols in the Food and Cosmetic Industries

Plant Polyphenols in the Food and Cosmetic Industries

Extraction and Applications

Edited by

José Manuel Lorenzo

ELSEVIER

ACADEMIC PRESS

An imprint of Elsevier

Academic Press is an imprint of Elsevier
125 London Wall, London EC2Y 5AS, United Kingdom
50 Hampshire Street, 5th Floor, Cambridge, MA 02139, United States

ISBN: 978-0-443-19137-4

For information on all Academic Press publications visit our website at https://www.elsevier.com/books-and-journals

Publisher: Mara Conner
Acquisitions Editor: Nicole Denis
Editorial Project Manager: Manvi Sharma
Production Project Manager: Sugumar Gomathi
Cover Designer: Raman Kumar

Typeset by TNQ Tech

Contents

SECTION 3 Cosmetic applications

Contributors

Mohamed Addi
Laboratoire d'Amélioration des Productions Agricoles, Biotechnologie et Ecophysiologie (LAPABE), Faculty of Sciences, Mohammed Premier University, Oujda, Morocco

Meryem Bakour
Laboratory of Biotechnology, Conservation, and Valorization of Bioresources (BCVB), Research Unit: Api-Phytotherapy, Physiology, Environment, and Health, Department of Biology, Faculty of Sciences Dhar Mehraz, Sidi Mohamed Ben Abdellah University, Fez, Morocco; The Higher Institute of Nursing Professions and Health Techniques (ISPITS), Fez, Morocco

Shi-Hui Cheng
School of Biological and Environmental Sciences, Unievrsity of Nottingham Malaysia, Semenyih, Selangor, Malaysia

Aurora Cittadini
Centro Tecnológico de la Carne de Galicia, Ourense, Spain

Beatriz Curros
Centro Tecnológico de la Carne de Galicia, Ourense, Spain

Noemí Echegaray
Centro Tecnolóxico da Carne, Ourense, Spain

Halina Ekiert
Department of Medicinal Plant and Mushroom Biotechnology, Jagiellonian University Medical College, Kraków, Poland

Amine Elbouzidi
Laboratoire d'Amélioration des Productions Agricoles, Biotechnologie et Ecophysiologie (LAPABE), Faculty of Sciences, Mohammed Premier University, Oujda, Morocco

Asmae El Ghouizi
Laboratory of Biotechnology, Conservation, and Valorization of Bioresources (BCVB), Research Unit: Api-Phytotherapy, Physiology, Environment, and Health, Department of Biology, Faculty of Sciences Dhar Mehraz, Sidi Mohamed Ben Abdellah University, Fez, Morocco; The Higher Institute of Nursing Professions and Health Techniques (ISPITS), Fez, Morocco

Mounir Haddou
Laboratory of Bioresources, Biotechnology, and Sustainable Development (L2BD), Faculty of Sciences, Mohammed Premier University, Oujda, Morocco

Karolina Jafernik
Department of Medical Cosmetology, Faculty of Medicine and Health Sciences, Andrzej Frycz Modrzewski Krakow University, Krakow, Poland

Marta Klimek-Szczykutowicz
Department of Pharmaceutical Sciences, Collegium Medicum, Jan Kochanowski University, Kielce, Poland

Aleksandra Cvetanović Kljakić
University of Novi Sad, Faculty of Technology, Novi Sad, Serbia

Hassan Laaroussi
Laboratory of Biotechnology, Conservation, and Valorization of Bioresources (BCVB), Research Unit: Api-Phytotherapy, Physiology, Environment, and Health, Department of Biology, Faculty of Sciences Dhar Mehraz, Sidi Mohamed Ben Abdellah University, Fez, Morocco; Centre of Biological Engineering (CEB), University of Minho (UMINHO), Braga, Portugal

José Manuel Lorenzo
Centro Tecnológico de la Carne de Galicia, Ourense, Spain; Área de Tecnoloxía dos Alimentos, Facultade de Ciencias, Universidade de Vigo, Ourense, Spain

Badiaa Lyoussi
Laboratory of Natural Substances, Pharmacology, Environment, Modeling, Health, and Quality of Life (SNAMOPEQ), Department of Biology, Faculty of Sciences Dhar Mehraz, Sidi Mohamed Ben Abdellah University, Fez, Morocco

Magdalena Anna Malinowska
Department of Organic Chemistry and Technology, Faculty of Chemical Engineering and Technology, Cracow University of Technology Cracow, Poland

Aleksandra Mišan
University of Novi Sad, Institute of Food Technology, Novi Sad, Serbia

Driss Ousaaid
Laboratory of Biotechnology, Conservation, and Valorization of Bioresources (BCVB), Research Unit: Api-Phytotherapy, Physiology, Environment, and Health, Department of Biology, Faculty of Sciences Dhar Mehraz, Sidi Mohamed Ben Abdellah University, Fez, Morocco; Laboratory of Drug Sciences, Faculty of Medicine, Pharmacy, and Dental Medicine, Sidi Mohamed Ben Abdellah University, Fez, Morocco

Branimir Pavlić
University of Novi Sad, Faculty of Technology, Novi Sad, Serbia

Milica Pojić
University of Novi Sad, Institute of Food Technology, Novi Sad, Serbia

Marta Sharafan
Cracow University of Technology, CUT Doctoral School, Faculty of Chemical Engineering and Technology, Department of Organic Chemistry and Technology, Cracow, Poland

Paulo Eduardo Sichetti Munekata
Centro Tecnológico de la Carne de Galicia, Ourense, Spain

Elżbieta Sikora
Department of Organic Chemistry and Technology, Faculty of Chemical Engineering and Technology, Cracow University of Technology Cracow, Poland

Alena Stupar
University of Novi Sad, Institute of Food Technology, Novi Sad, Serbia

Agnieszka Szopa
Department of Medicinal Plant and Mushroom Biotechnology, Jagiellonian University Medical College, Kraków, Poland

Mohamed Taibi
Laboratory of Bioresources, Biotechnology, and Sustainable Development (L2BD), Faculty of Sciences, Mohammed Premier University, Oujda, Morocco

Nemanja Teslić
University of Novi Sad, Institute of Food Technology, Novi Sad, Serbia

Soumaya Touzani
Laboratory of Mini-Invasive Surgery, Robotics, Artificial Intelligence and Educational Innovations, Faculty of Medicine Pharmacy and Dentistry, University Sidi Mohamed Ben Abdellah, Fez, Morocco

CHAPTER

1 Plant polyphenols in the food and cosmetic industries: Extraction and applications

José Manuel Lorenzo[1,2], Noemí Echegaray[1]

[1]*Centro Tecnolóxico da Carne, Ourense, Spain;* [2]*Área de Tecnoloxía dos Alimentos, Facultade de Ciencias, Universidade de Vigo, Ourense, Spain*

1. Introduction

In addition to being essential for human life, food plays an increasingly vital role in human health and well-being. Similarly, cosmetic products have become an integral part of our daily lives, contributing to various aspects of personal care and self-expression. These products, including creams, lotions, shampoos, and sunscreens, serve the purpose of providing protection, enhancing beauty, and fostering confidence in individuals worldwide (Rathee et al., 2023). The convergence of these two vital industries, food and cosmetics, underscores the growing awareness of holistic well-being. As consumers prioritize health-conscious choices, the demand for food products enriched with bioactive compounds and cosmetics featuring natural ingredients continues to rise. Both industries are fueled by innovation, seeking sustainable solutions that contribute to individual health and environmental preservation.

In the face of this dynamic landscape, the use of polyphenols as natural compounds is more prevalent than ever, with the aim of displacing other synthetic substances that may have negative effects on human health. Polyphenols belong to a class of chemical substances widely distributed in various plant sources, including fruits, vegetables, grains, herbs, and spices (Dias et al., 2022). These biocompounds exhibit diverse biological features, including antioxidant, antimicrobial, antiinflammatory, and photoprotective properties (Kumar & Goel, 2019), making them potential ingredients for both the food and cosmetic industries. Thus, the allure of polyphenols lies in their potential to bestow beneficial health effects while simultaneously enhancing product stability and desired attributes when incorporated into various food and cosmetic preparations (Ahmed et al., 2022; Khan et al., 2019; Saad et al., 2021).

In this chapter, we aim to examine the importance of polyphenolic compounds in the food and cosmetic industries, encompassing their extraction, utilization, and key

Plant Polyphenols in the Food and Cosmetic Industries. https://doi.org/10.1016/B978-0-443-19137-4.00004-1

characteristics that make them highly versatile ingredients aligning with current consumer demands.

2. Classification, chemical structure, and sources of plant polyphenols

Polyphenols are constituted of a broad group of natural metabolites synthesized only by plants (Singla et al., 2019). These natural substances are distinguished by their particular chemical structure, which includes at least two phenyl rings (with conjugated double bonds), as well as one or more hydroxyl substituents (-OH). This description encompasses a heterogeneous array of compounds with a repertoire of over 8000 structural variations (Han et al., 2007) distributed into five main subclasses, namely flavonoids, phenolic acids, stilbenes, lignans, and tannins. Understanding the classification and structure of polyphenols is essential for comprehending their biological activities and optimizing their extraction, purification, and formulation techniques for various applications in the food and cosmetic industries.

Flavonoids, with over 5000 compounds, are the largest and most well-known subclass of polyphenols (Rusconi & Conti, 2010). Their structure consists of two benzene rings connected by a heterocyclic pyran ring. They can be further divided into 13 groups, with the most notable ones being anthocyanins, flavonols, flavanols, flavones, flavanones, and isoflavones (Scalbert & Williamson, 2000). Flavonoids are commonly found in fruits, vegetables, tea, and red wine and are principally represented by catechin, fisetin, gallocatechin, kaempferol, myricetin, and quercetin (Han et al., 2007). After flavonoids, phenolic acids stand out within polyphenols due to their abundance in the plant kingdom. These compounds typically contain a single phenolic ring with one or more hydroxyl groups. Although phenolic acids technically do not meet the strict definition of having a minimum of two phenolic rings, they are still classified within the broader category of polyphenols due to their phenolic structure and involvement in similar chemical reactions (Belščak-Cvitanović et al., 2018; Khan et al., 2019). Phenolic acids can be categorized into two major groups, namely hydroxybenzoic acids (such as gallic acid, protocatechuic acid, and vanillic acid) and hydroxycinnamic acids (including caffeic acid, chlorogenic acid, ferulic acid, and p-coumaric acid). Some phenolic acids are particularly abundant in acidic fruits and are also present in coffee, certain berries, black radish, and onions (Kumar et al., 2014).

Stilbenes are a smaller subclass of polyphenols but are notable for their potent antioxidant properties. Their chemical structure consists of two phenolic rings connected by a two-carbon methylene bridge (Han et al., 2007). Resveratrol is the most well-known stilbene and is found in grapes, berries, and peanuts (Tian & Liu, 2020). For their part, lignans are polyphenols characterized by their complex and diverse structures. However, their common feature is that they possess two or more phenylpropanoid units. Many lignans have been identified in the plant kingdom, including matairesinol, niranthin, secoisolariciresinol, sesamin, and terameprocol. The most abundant dietary source of lignans is flaxseed, with secoisolariciresinol being the most prominent representative (Cui et al., 2020). Finally, tannins also belong to the group of polyphenolics. They are known to possess multiple phenolic subunits that are connected through different linkages. They can be classified as hydrolyzable tannins or nonhydrolyzable tannins, also known as condensed tannins or proanthocyanidins. Hydrolyzable tannins are composed of gallic or ellagic acid units, whereas condensed tannins consist of flavan-3-ols and/or flavan-3,4-diols units (such as catechin and epicatechin). Some common sources of tannins include red fruits, grapes, pomegranates, nuts, red wine, and tea (Belščak-Cvitanović et al., 2018; Khanbabaee & van Ree, 2001).

3. Key characteristics of plant polyphenols

The unique structure of plant polyphenols is responsible for their diverse biological activities, enabling them to manifest a wide array of features (Fig. 1.1). These include antioxidant, antimicrobial, health-promoting, and photoprotective properties (Leri et al., 2020). In the following sections, we will explore these fundamental biological activities of plant polyphenols, which play a crucial role in their extensive application in the food and cosmetic industries.

3.1 Antioxidant properties of plant polyphenols

Plant polyphenols are renowned for their remarkable antioxidant activity, which has been extensively studied and well-documented (Dienaitė et al., 2019; Dou et al.,

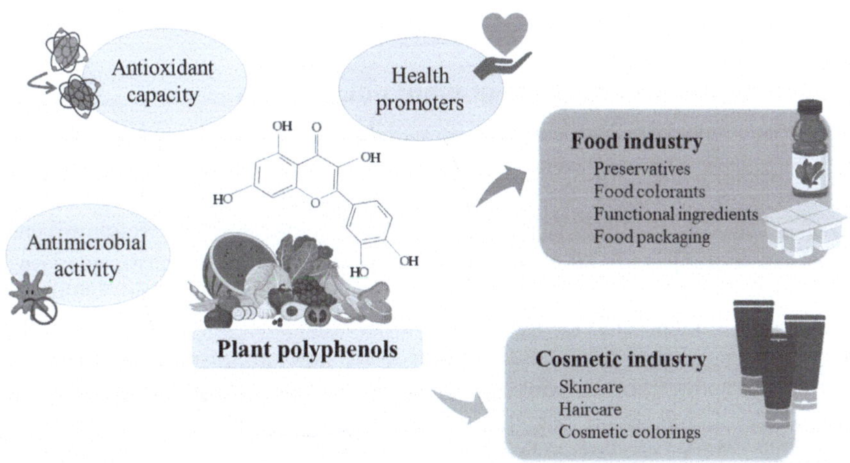

FIGURE 1.1

Key properties and versatile applications of plant polyphenols in the food and cosmetic industry.

2023; Li et al., 2020; Lučić et al., 2023; Radajewska et al., 2023; Zheng et al., 2020). These bioactive compounds exhibit a wide range of mechanisms underlying their antioxidant action, including the direct scavenging of reactive free radicals, chelation of transition metal ions implicated in free radical formation, inhibition of enzymes involved in free radical production, and the restoration of membrane-bound antioxidants, which can occur simultaneously (Belščak-Cvitanović et al., 2018). Importantly, the presence of hydroxyl groups in plant polyphenols, with their specific chemical structure (multiple phenolic rings, conjugated double bonds, and potential presence of other functional groups), contributes significantly to their antioxidant potency. The hydroxyl substituents act as electron donors, allowing polyphenols to effectively neutralize free radicals by hydrogen-atom transfer (HAT) or single-electron transfer (SET) mechanisms. In the first case, polyphenols act as chain-breaking antioxidants by donating an electron, transforming themselves into a highly stable free radical; meanwhile, in the SET mechanism, after electron donation, the antioxidant is converted into a stable radical cation (Di Meo et al., 2013). Furthermore, polyphenols can also neutralize free radicals through the mechanism of sequential proton loss electron transfer (SPLET), whereby they first donate a proton and subsequently transfer an electron to form stable molecules (Zeb, 2020).

On the contrary, polyphenols have the ability to chelate transition metals (such as Cu, Mg, and Co), forming stable compounds. As a result, these antioxidants can inhibit the reduction in Fe^{3+} and thereby diminish the formation of reactive hydroxyl free radicals ($OH^{\bullet}$) from the Fenton reaction, thereby mitigating oxidative processes (Wijeratne et al., 2006). Moreover, the ability of certain plant polyphenols to inhibit oxidase enzymes such as cyclooxygenase (CO), myeloperoxidase (MPO), lipoxygenase (LO), reduced nicotinamide adenine dinucleotide phosphate (NADPH) oxidase, and xanthine oxidase (XO) is associated with their antioxidant activity, as it helps reduce the formation of excessive amounts of reactive oxygen species (ROS) and organic hydroperoxides in vivo (Belščak-Cvitanović et al., 2018).

3.2 Antimicrobial properties of plant polyphenols

Plant polyphenols possess exceptional antimicrobial features against a wide range of microorganisms, including bacteria, fungi, and viruses. Due to their diverse chemical structures, polyphenols possess a multitude of potential mechanisms for antimicrobial activity capable of impacting the integrity (cell membrane or cytoplasm) and function of microbial cells (Rathod et al., 2023). The precise mechanism underlying the antimicrobial effects of polyphenols remains uncertain and warrants additional research (Wu & Zhou, 2021). However, it is known that plant polyphenols can damage the cell membrane by generating pores in its structure, disrupting membrane proteins, disrupting the cell wall, and/or inhibiting cell wall synthesis. This can also modify the membrane potential, ultimately leading to cell death (Ben Arfa et al., 2006; Wu et al., 2016). In addition, polyphenols can induce alterations in the cytoplasm, resulting in the release of cellular components, acidification of the cytoplasm, and potential chelation of metal ions (Martínez-Arámburu et al.,

2015). Furthermore, the activity of microbial cells can be compromised as polyphenols can disrupt the synthesis of both DNA and RNA, as well as proteins (Rathod et al., 2023). Polyphenols can also inhibit enzymatic activity, inactivate toxins produced by pathogenic bacteria, and hinder the adhesive abilities and biofilm formation capabilities of microorganisms (Rathod et al., 2023; Tapia-Rodriguez et al., 2017).

Just like with antioxidant capacity, microbial activity is strongly related to the chemical structure of polyphenols, where the number of hydroxyl groups plays a key role in determining the antimicrobial activity of a polyphenolic compound (Chibane et al., 2019).

3.3 Health benefits of plant polyphenols

Including their antioxidant and antimicrobial activities, polyphenols exhibit diverse attributes that have considerable health-promoting effects. These compounds have shown great potential in various areas of health. Firstly, their antioxidant properties help protect cells from damage caused by free radicals, reducing oxidative stress and its association with chronic diseases such as cardiovascular diseases, cancer, diabetes, and neurodegenerative conditions (Dong et al., 2021). Secondly, polyphenols have an antimicrobial effect that can contribute to improving the microbial balance in the host's gut, promoting intestinal health (Rodríguez-Daza & de Vos, 2023; Scarano et al., 2023). Moreover, polyphenols have demonstrated antihypertensive effects by exerting a beneficial impact on vascular endothelial functions (Oak et al., 2018). They can help prevent the activation of the angiotensin system and produce a vasorelaxant effect, contributing to the management of hypertension (Kolodziej, 2017). In addition, extensive research has shown that polyphenols have immunomodulatory properties. They can regulate various components of the immune system, including immune cells like macrophages, dendritic cells, and lymphocytes, as well as cytokines and signaling pathways involved in immune responses (Yahfoufi et al., 2018). Apart from that, polyphenols have also been shown to be effective in improving inflammatory stress in both the short and long term (Joseph et al., 2016).

Polyphenols have also emerged as promising for cancer treatment due to their low toxicity and minimal side effects. Regular consumption of polyphenols has been associated with a reduced risk of various types of cancer. These bioactive compounds have the ability to influence multiple cell signaling pathways, exerting an anticancer effect (Rathod et al., 2023; Yi et al., 2019). Furthermore, polyphenols have been found to provide benefits to the aging brain. They have displayed enhancements in cognition, learning, and memory, as well as a decreased risk of dementia and prevention of neurodegenerative diseases (Annunziata et al., 2021). In terms of lipid metabolism, polyphenols play a role in preventing the impairment of lipid metabolism associated with atherosclerosis. They reduce cholesterol synthesis, increase the expression of low-density lipoprotein receptors, and inhibit cholesterol absorption from micelles (Ziółkiewicz et al., 2023).

Lastly, polyphenols have been recognized for their role in the prevention and management of Type 2 diabetes mellitus. They act through both insulin-dependent and insulin-independent mechanisms, such as protecting pancreatic islet β-cells, reducing β-cell apoptosis, promoting β-cell proliferation, alleviating oxidative stress, activating insulin signaling pathways, inhibiting glucose absorption, suppressing digestive enzymes, modulating intestinal microbiota, regulating inflammation response, and inhibiting the formation of advanced glycation end products (Sun et al., 2020).

It is noteworthy that despite the well-established health benefits of polyphenols, ongoing research endeavors continue to delve into elucidating their underlying mechanisms of action and examining their potential implications in the realms of disease prevention and treatment.

3.4 Photoprotective properties of plant polyphenols

Plant polyphenols have emerged as promising candidates for providing photoprotective benefits due to their unique characteristics. Specifically, the ability of polyphenols to act as screens against solar radiation is based on their chemical structure (presence of chromophore molecules in their structure) and photophysical properties (Khan et al., 2019). When exposed to ultraviolet (UV) radiation (which includes ultraviolet A (UVA) and ultraviolet B (UVB) radiation), polyphenols can absorb energy from light in a specific region of the electromagnetic spectrum (Bose et al., 2017). This energy absorption excites electrons in the phenolic groups, causing them to transition to higher energy states. However, these excited electrons are unstable and tend to return to their lower energy state, releasing the absorbed energy in the form of heat or fluorescent light. This capacity to absorb and dissipate UV radiation energy makes polyphenols excellent natural sunscreens (Pizano-Andrade et al., 2022). By absorbing UV radiation, polyphenols prevent harmful light from reaching the deeper layers of the skin, where it can cause cellular damage and trigger oxidative processes. In addition, by dissipating energy, polyphenols avoid energy from accumulating as free radicals and causing oxidative damage (Cefali et al., 2020).

It is essential to consider that the photoprotective efficacy of polyphenols may vary depending on their specific chemical structure and concentration in the product or material in which they are used.

4. Innovations in the extraction of plant polyphenols

The polyphenols present in plants are commonly stored in plant tissues at cellular (bound to the cell wall) and subcellular (in the vacuoles) levels. Therefore, they are rarely found in their free form. Instead, they are often conjugated with other chemical substances (Treutter, 2005). In addition, polyphenols are found in a dispersed form and in relatively low quantities. To maximize their beneficial properties, it is necessary to efficiently extract and concentrate these compounds (Fig. 1.2). This allows for the production of extracts with high purity and activity, which is essential for the food and cosmetic industries.

In light of these facts, recent years have witnessed remarkable advancements in the extraction of plant polyphenols, driven by the growing demand for natural bioactive compounds in various industries. Traditional extraction methods, such as maceration and heat-assisted extraction, have often been complemented or even replaced by innovative techniques or their combinations, offering improved efficiency, selectivity, sustainability, and high-purity and high-quality extracts (Pal & Jadeja, 2020). This section explores the latest innovations in plant polyphenol extraction (Table 1.1), highlighting emerging extraction techniques like US extraction, MW extraction, and PEF extraction. It also showcases cutting-edge methodologies such as SCF extraction and NADES extraction. Through these innovations, researchers and industry professionals are unlocking new possibilities for harnessing the full potential of plant polyphenols in several applications.

4.1 Ultrasound (US) extraction for the recovery of polyphenols

The extraction of polyphenols using US utilizes high-frequency US waves (20–100 kHz) to facilitate the retrieval of desired compounds from plants. The application of US induces a hydrodynamic cavitation phenomenon, causing the rapid formation and collapse of vapor bubbles in the extraction medium. This bubble collapse disrupts the cell wall of the plant material, leading to an increased diffusion coefficient and ultimately facilitating higher extraction rates of polyphenol compounds compared with conventional techniques (Martínez-Olivo et al., 2023). Thus, the

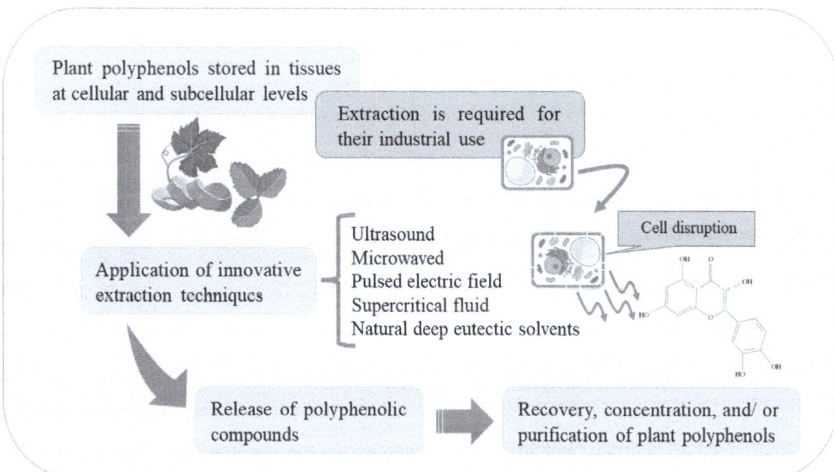

FIGURE 1.2

Approach to extraction and recovery of plant polyphenols.

Table 1.1 Representative cases of the application of innovative extraction techniques for the recovery of different polyphenols from distinct plant matrices.

Plant source	Target compounds	Treatment conditions	Main findings	References
Ultrasound (US) extraction				
Hibiscus sabdariffa calyces	Anthocyanins (delphinidin-3-O-sambubioside and cyanidin-3-O-sambubioside)	100–500 W; 20 kHz; 0%–100% ethanol; 30–35°C; 3–36.5 min	- Optimal conditions: 296.6 W; ethanol 39.1%; 26.1 min. - Higher yields compared with heat-induced extraction. - Increasing the solid-to-solvent ratio enhances the US extraction yield. - The US was presented as an efficient extraction method.	Pinela et al. (2019)
Fresh and freeze-dried *Vitis vinifera* cv *Tannat* grape pomace	Phenolic compounds (including monomeric anthocyanins)	50–100 W; 39 kHz; 50% ethanol; 15–60 °C; 5–50 min	- The US exhibited improved extraction efficiency for both fresh and freeze-dried pomace samples. - The freeze-dried pretreatment improves US extraction. - US provided extracts with higher antioxidant capacity.	González et al. (2020)
Defatted canola, flax, and hemp seed cakes	Phenolic compounds (including flavonoids)	200 W; methanol: acetone: water mixture; 40–70 °C; 20–40 min	- Optimal conditions: 5: 50 solid-to-solvent ratio; 70 °C; and 20 min. - The US extraction under optimal conditions yielded higher levels of polyphenols compared with maceration extraction. - Polyphenol-rich extracts obtained by US showed twice the antioxidant capacity compared with maceration extraction.	Teh and Birch (2014)
Clitoria ternatea flowers	Phenolic compounds (including flavonoids)	250–650 W; 25–65 °C; 60% ethanol; 10–50 min	- Optimal conditions: 359 W, 45 °C, and 30 min. - US provided extracts with higher antioxidant capacity and antimicrobial activity compared with maceration and reflux techniques.	Santos and Martins (2023)

Microwave (MW) extraction				
Tomato peel waste	Phenolic compounds (including flavonoids and phenolic acids)	0–500 W; 25–90 °C; different solvents; 5–10 min	- Extraction time did not influence the extraction yield. - Lower concentrations of organic solvent (50% vs. 70%) improved the extraction. - Increasing temperature increased the extraction yield.	Bakić et al. (2019)
Thymene *Thymus serpyllum* residue	Phenolic compounds (including apigenin glucuronide, luteolin 7-O-glucuronide, and rosmarinic acid)	Unspecified W; 200°C; 0%–60% ethanol; 10–190 s	- Optimal conditions: 200 °C; 48% ethanol; 86 s. - Lyophilized polyphenol-rich extracts exhibited excellent antioxidant activities (under optimal MW extraction conditions). - Lyophilized polyphenol-rich extracts showed good effectiveness against Gram-positive bacteria (under optimal MW extraction conditions).	Jovanović et al. (2022)
Coffee pulp residue	Phenolic compounds	400–1000 W; 20%–80% ethanol; 55–85 min	- Optimal conditions: 1000 W; 80 °C; 42.5% ethanol; 85 min. - Solid-to-solvent ratio affected the extraction process. - MW was conducted on extracts with excellent antioxidant capacity.	Tran et al. (2022)
Mango peel waste	Phenolic compounds and carotenoids	225–275 W; 50%–80% ethanol; 30–180 s	- Optimal conditions for the recovery of polar compounds: 675 W; 50% ethanol; 105 s. - Extraction time did not influence the extraction yield. - Polyphenol-rich extracts exhibited excellent antioxidant activity.	Vélez-Erazo et al. (2021)
Pulsed electric field (PEF) extraction				
Calendula officinalis flowers, *Castanea sativa* leaves, and *Rosa canina* fruits	Phenolic compounds	1.2–2.0 kV/cm; pulses of 10 μs, 24 °C; 20 min (1 pulse/ms)	- Electric field intensities higher than 1.4 kV/cm increase the recovery of individual and total phenolic compounds compared with control extraction. - *Rosa canina* fruits were the highest source of phenolic compounds.	Lakka et al. (2021)

Continued

Table 1.1 Representative cases of the application of innovative extraction techniques for the recovery of different polyphenols from distinct plant matrices.—*cont'd*

Plant source	Target compounds	Treatment conditions	Main findings	References
Lemon (*citrus limon*) residues	Phenolic compounds (including flavonones, hesperidin, and eriocitrin)	3.0–9.0 kV/cm; 0–100 pulses of 3 μs; room temperature; 0–100 μs	- PEF improved polyphenolic compound extraction by 300% compared with untreated samples. - The electric field of 7 kV/cm exhibited the highest polyphenol recoveries. - Antioxidant capacity of extracts was significantly correlated with total phenolic content.	Peiró et al. (2019)
Brewers spent grain	Phenolic compounds (including flavan-3-ols, flavonoids, phenolic acid derivatives, and total free and bound phenols)	0.5–2.5 kV/cm; 50–150 Hz; pulses of 10 μs; 5–15 s	- Optimal conditions: 2.5 kV/cm; 50 Hz; 14.5 s. - Under optimal conditions, PEF enhanced the recovery of total free and bound polyphenols by 2.7-fold and 1.7-fold, respectively, in comparison with untreated samples.	Martín-García et al. (2020)
Supercritical fluid (SCF) extraction				
Fresh frozen tea (*Camellia assamica* L.) leaves	(–)-Epigallocatechin-3-gallate (EGCG); (–)-epicatechin-3-gallate (ECG); (–)-epigallocatechin; and epicatechin	SC-CO_2; 100–200 solid-to-solvent ratio; 40–60 °C; 15–25 MPa; 1.2%–3.6% ethanol modifier	- Improved the extraction of catechins compared with the Soxtec extraction. - Optimal conditions: 200 solvent-to-material ratio; 200 °C; and 25 MPa. - Experimental yield of 4.20% for optimal conditions. - Allowed obtaining extracts rich in EGCG (723–848 mg EGCG/g vs. 55 EGCG/g in conventional extraction).	Gadkari et al. (2015)
Maackia amurensis heartwood	Biological compounds, including a variety of polyphenol compounds	SC-CO_2; 31–70 °C; 5–40 MPa; 2% ethanol modifier	- Enhanced the extraction of biological compounds (including polyphenols) compared with maceration extraction. - Recovery of several hydroxycoumarins, flavan-3-ols, flavanones, flavones, flavonols, phenolic acids, stilbenes, and lignans.	Razgonova et al. (2023)

Labisia pumila leaves	Phenolic acids (gallic acid, methyl gallate, and caffeic acid)	SC-CO_2; 60 °C; 20 MPa; 10% of different modifiers (water, 50%–100% ethanol and methanol); 30 min	- Optimal conditions: 55 °C; and 10 MPa. - Generated extracts with higher antioxidant activity under optimal extraction conditions. - The type and concentration of modifiers influenced the extraction yield, phenolic acids recovery, and antioxidant capacity of extracts. - 70% ethanol proved to be the most effective modifier for phenolic acids recovery. - SC-CO_2 significantly reduced the extraction time compared with traditional solvent extraction methods.	Li et al. (2020)
Natural deep eutectic solvents (NADESs) extraction				
Mulberry	Anthocyanins (cya- nidin-3-glucoside equivalents)	10 different NADESs; 40°C; 10 min	- 60% of the NADESs showed anthocyanin extraction yields higher than those of conventional acidified 80% ethanol. - NADESs with basic properties do not extract any anthocyanin. - Choline chloride: lactic acid stood out for its high recoveries of anthocyanins.	Bi et al. (2020)
Prunella vulgaris	Phenolic acids	12 different NADESs; 60–100 °C; 20–60 min	- Maximum yield of phenolic acids was achieved with choline chloride: Ethylene glycol at a 1: 4 mole ratio. - The increase in the proportion of hydrogen bond donors decreased the extraction yield of polyphenolic acids. - NADESs extraction demonstrated superior yields compared with conventional extraction methods.	Xia et al. (2015)

US extraction process involves placing the plant material in a suitable solvent in the presence of US waves (which can be transmitted through a bath or directly by a probe) (Pinton et al., 2021). After extraction, the mixture is separated using techniques such as centrifugation, filtration, or evaporation to obtain the polyphenol-enriched extracts (Roselló-Soto et al., 2015).

The frequency, wave amplitude, exposure time, and other variables (such as solvent nature and solid-to-solvent ratio) of the US process must be adjusted to successfully optimize the extraction of polyphenols (Xu et al., 2022). Existing studies show that high frequencies or powers can lead to the formation of hydroxyl radicals that can quickly react with these bioactive substances, degrading or oxidizing them. On the contrary, high wave amplitudes can promote the release of bioactive compounds such as polyphenols. Nevertheless, as mentioned earlier, these parameters need to be adjusted for each specific matrix as different behaviors may occur (Medina-Torres et al., 2017). In this field, several investigations have been conducted with different plant matrices, where US extraction has been optimized for the recovery of polyphenolic compounds and, in some cases, compared with other extraction methods. An example is the study carried out by Pinela et al. (2019), where they revealed that the optimization of US extraction parameters provided a more efficient process for obtaining anthocyanins (namely, delphinidin-3-O-sambubioside and cyanidin-3-O-sambubioside) from *Hibiscus sabdariffa* calyces compared with conventional heat-assisted extraction (in terms of yield). Specifically, the optimal conditions of US extraction (power of 296.6 W; time of 26.1 min; and 39.1% ethanol as solvent) provided nearly 3 times higher amounts of anthocyanins compared with heat-assisted extraction. Identically, US extraction applied to a red wine production by-product (skins of *Vitis vinifera* cv *Tannat*) significantly improved the recovery yield of polyphenols compared with conventional maceration extraction. The outcomes demonstrated that US extraction resulted in an increase in total polyphenol content and total monomeric anthocyanin content of 50% and 180% when the starting matrix was fresh or previously freeze-dried, respectively. Furthermore, the US process allowed to produce higher quality extracts, as evidenced by their increased antioxidant capacity. Other works have also demonstrated the enhanced antioxidant capacity of polyphenol-enriched extracts obtained through US extraction. Based on Teh and Birch (2014), the use of US for the recovery of phenolics and flavonoids from defatted canola, flax, and hemp seed cakes resulted in extracts with significantly higher antioxidant potentials compared with conventional extraction methods, exhibiting a two-fold increase. Santos and Martins (2023) also provided evidence of the positive impact of US extraction on the antioxidant capacity of polyphenol-rich extracts derived from *Clitoria ternatea* flowers compared with maceration and reflux techniques.

4.2 Microwave (MW) extraction for the recovery of polyphenols

MW refers to electromagnetic waves with nonionizing properties that fall within the frequency range of 300 MHz to 300 GHz, corresponding to wavelengths ranging from 1 mm to 1 m (Puligundla, 2013). The MW extraction technique utilizes MW energy to facilitate the recovery of polyphenolic compounds from plants since the application of this energy induces modifications in the treated plant material by interacting with dipoles of polar and polarizable materials (particularly water molecules). When exposed to electromagnetic beams, these molecules become energized and oscillate in response to the alternating electric field. This continuous realignment creates friction, resulting in localized heating (Ciriminna et al., 2016). In addition, ionic compounds present in the matrix migrate under the influence of the electric field. The intermolecular friction generated by these processes leads to rapid vaporization of water and an increase in internal cell pressure, potentially causing cell wall rupture. Furthermore, MW energy selectively heats the solvent and the plant material (Orsat et al., 2017). As a result, MW treatment facilitates mass transfer processes, allowing for the release of polyphenols trapped within the cells into the medium at an accelerated rate (Zin et al., 2020). The process of MW extraction involves placing the plant material in a suitable solvent and subjecting it to microwave irradiation in an MW oven or an MW reactor. After the extraction process, the resulting polyphenol-enriched solution is typically separated from the solid residue using similar techniques employed in US extraction, including centrifugation, filtration, and evaporation (Hu et al., 2022).

Optimization studies are necessary to determine the most effective MW parameters (such as solid-to-solvent ratio, power, time, temperature, etc.) for each particular case. The choice of solvent and its compatibility with MW irradiation should be considered to ensure efficient extraction and preservation of the target compounds. According to Bakić et al. (2019), the use of different solvents significantly influenced the recovery of total phenolic compounds, total flavonoids, and individual compounds (such as kaempferol-3-O-rutinoside, p-coumaric acid, and chlorogenic acid derivatives). Specifically, these authors demonstrated that binary solvent systems with low concentrations of an organic fraction generally resulted in higher extraction yields, as water has a higher dielectric constant compared with less polar solvents. Thus, the use of 50% methanol and ethanol exhibited better extraction compared to 70% methanol and ethanol. Similar ethanol concentrations (48% and 42.5%) were proposed as optimal for the extraction of polyphenols from thyme residue (Jovanović et al., 2022) and coffee pulp waste (Tran et al., 2022), respectively. Regarding temperature, it plays a crucial role, as it directly influences the diffusion rate and, consequently, the extraction ratio of polyphenolic compounds. Higher temperatures generally result in faster diffusion and extraction. For instance, several studies have demonstrated that elevated temperatures (up to 200°C) have a positive impact on polyphenol extraction without compromising their stability (Bakić et al., 2019; Jovanović et al., 2022). However, it is important to note that high temperatures can lead to the degradation of heat-sensitive compounds, including certain polyphenols, especially during prolonged extraction periods. Therefore, careful temperature control is necessary to achieve a balance between extraction efficiency and the preservation of heat-sensitive compounds (Orsat et al., 2017).

4.3 Pulsed electric field (PEF) extraction for the recovery of polyphenols

The PEF technology used in the extraction of polyphenols from plants employs short pulses of high-voltage electric fields, frequently between 10 and 80 kV/cm (Naliyadhara et al., 2022). These electric fields entail the creation of temporary (reversible) or permanent (irreversible) pores in the cell membranes, a process referred to as electroporation. Specifically, the high voltage causes the widening of the pores in plant cells, which can lead to cell rupture. The irreversible permeabilization of the cell membrane permits the extraction medium to penetrate and extract the desired substances, improving mass transfer and facilitating the release of polyphenols, as well as other compounds present, into the medium solvents (Cemazar et al., 2018; Koubaa et al., 2015). The PEF treatment of plants involves placing the solid plant material between two electrodes, typically immersed in a conductive liquid, which increases the contact with the electrodes and reduces the nonhomogeneity of the electric field (Wang et al., 2023). Distilled or tap water is the most commonly utilized solvent in this process (Bilge et al., 2022; Franco et al., 2020; Giancaterino et al., 2023; Ramon-Mascarell et al., 2022). Following the extraction process, the mixture is subjected to separation methods, including centrifugation, filtration, or evaporation, to obtain extracts enriched in polyphenol substances.

The characteristics of PEF technology are inherently associated with a multitude of parameters that can influence the extraction process. Key process parameters include electric field intensity, electric field strength, specific energy density, pulse waveform, number of pulses, exposure time, and the temperature alteration experienced by the treated sample (Arshad et al., 2020). A study conducted by Lakka et al. (2021) investigated the effect of electric field intensity (ranging from 1.2 to 2.0 kV/cm) on the extraction efficiency of polyphenols from *Calendula officinalis* flowers, *Castanea sativa* leaves, and *Rosa canina* fruits. The outcomes showed that PEF treatment significantly increased the extraction of polyphenols from all three plant materials. Concretely, the application of moderate to high electric field intensities significantly enhanced the recovery of both total and individual polyphenols. In the case of *Rosa canina* fruits, the recovery was particularly remarkable, as the extraction of total phenolic compounds increased by up to 63%, while the extraction of eriodictyol 7-O-rutinoside, an individual polyphenol, increased by up to 83%. Even more exceptional recovery values of polyphenolic compounds were reported by Peiró et al. (2019) during the use of PEF on lemon waste, as the extraction efficiency of these bioactive substances improved by 300% compared with untreated samples when the applied electric field intensity was 7 kV/cm. Furthermore, the PEF lemon extracts obtained exhibited significant antioxidant activity, which is significantly related to the total phenolic content. PEF technology has also been used as a pretreatment to enhance the extraction of polyphenols from grains. For instance, Martín-García et al. (2020) employed PEF before conducting a solid-liquid extraction with 80% ethanol, resulting in 2.7- and 1.7-times higher recoveries of total and bound phenolic compounds compared with untreated samples when the conditions employed were 2.5 kV/cm, 50 Hz, during 14.5 s.

4.4 Supercritical fluid (SCF) extraction for the recovery of polyphenols

The SCF technology utilizes a fluid in a supercritical state, which exists in an intermediate phase between a liquid and a gas, to extract the desired compounds from plant matrices (Agregán et al., 2023). The most commonly used supercritical fluid in polyphenol extraction is carbon dioxide (CO_2) in its supercritical state (SC–CO_2). This solvent is chosen due to its low cost, ability to selectively dissolve certain compounds, and nontoxic nature. Moreover, it also ensures the generation of minimal or no waste after its utilization (Soquetta et al., 2018). Nonetheless, in the extraction of polyphenolic compounds, CO_2 is typically supplemented with small organic molecules, such as acetone, ethanol, or even water itself, which function as secondary solvents or cosolvents (also referred to as modifiers or entrainers) (Da Porto et al., 2014; Gadkari et al., 2015; Tai et al., 2022). These additional compounds are employed in different concentrations to enhance the extraction of these polar substances, since CO_2 is of nonpolar nature, making the extraction process of polyphenols more efficient (Vinitha et al., 2022). The procedure of SCF extraction comprises several steps. First, the plant material is placed into a specialized extractor. Subsequently, the temperature and pressure conditions are adjusted to achieve the supercritical state of the fluid. The SCF is passed through the plant material, and as it traverses the matrix, it dissolves the polyphenols (and other bioactive compounds) present in the plant. Then, the resulting mixture is taken to an expansion chamber, where the pressure is rapidly reduced, causing the separation of the supercritical fluid and the extracted compounds. Finally, the SCF is recovered for reuse in future extractions, while the extracted polyphenols are obtained in a concentrated form and further purified through additional processes such as chromatographic separation (Ahmad et al., 2019). Thus, the main transport mechanism that enables mass transfer in SCF technology is the convection flow of the supercritical solvent phase, since the passage of SCF through the plant matrix cracks its tissues and opens its pores, altering the cell wall and organelles and facilitating the release of bioactive compounds (Gadkari et al., 2015).

It is important to note that the design and SCF extraction parameters may vary depending on the type of polyphenol and the plant source, so specific studies need to be conducted to optimize the results for each particular case. Gadkari et al. (2015) conducted an early study optimizing the extraction of catechins (namely, (−)-epigallocatechin-3-gallate (EGCG); (−)-epicatechin-3-gallate (ECG); (−)-epigallocatechin; and (−)-epicatechin) from fresh frozen tea (*Camellia assamica* L.) leaves using SC-CO_2 and compared this technique to conventional solvent extraction (Soxtec extraction with ethanol). They found that SC-CO_2 extraction using a solvent-to-material ratio of 200, a temperature of 50 °C, and a pressure of 25 MPa yielded the highest extractable solids, obtaining a percentage of 4.20% (weight/weight) at the experimental level for these conditions. Moreover, the use of SC-CO_2 allowed them to acquire extracts rich in EGCG (>95%), showcasing the capability of this technique to procure high-quality polyphenolic extracts.

Similarly, a recent study conducted on *Maackia amurensis* heartwood demonstrated the effectiveness of SC-CO_2 in the recovery of bioactive compounds, including numerous polyphenols as cirsimaritin, (epi)catechin-5-O-D-glycopyranoside; 6,4′-dimethoxyisoflavone-7-O-glucoside; formononetin; hydroxy hexamethoxyflavon; myricetin, quercetin; ferulic acid; and resveratrol, compared with ethanol and methanol maceration extraction (Razgonova et al., 2023). Specifically, this research revealed that the optimal conditions for the recovery of bioactive compounds were 55 °C and 10 MPa, resulting in yields of 4.3 mg/100 g, as well as providing extracts with higher antioxidant capacity. Under these conditions, several flavones, flavonols, flavan-3-ols, flavanones, phenolic acids, stilbenes, lignans, and hydroxycoumarins were recovered.

Not only do pressure and temperature conditions qualitatively and quantitatively modify the recovered polyphenols, but the nature of the modifiers used can also affect the extraction efficiency. This was confirmed by Radzali et al. (2020), who informed that the use of different types and concentrations of cosolvents (namely, 50%,-100% ethanol, 50%–100% methanol, and water) in SC-CO_2 extraction had an impact on the yield, composition, and antioxidant capacity of the extracts obtained from *Labisia pumila* leaves. The study suggests that 70% ethanol is the optimal modifier for maximizing the extraction yield, the recovery of phenolic acids (such as gallic acid, methyl gallate, and caffeic acid), and the antioxidant activity of the extracts obtained.

4.5 Natural deep eutectic solvents (NADESs) extraction for the recovery of polyphenols

Deep eutectic solvents (DESs) are a homogeneous mixture of two nontoxic components that exhibit a lower melting point than their individual constituents, primarily due to noncovalent interactions. These solvents typically consist of a hydrogen-bond acceptor (HBA) combined with a hydrogen-bond donor (HBD), which can be quaternary ammonium, tetraalkylammonium, or phosphonium for HBA and salts, acids, alcohols, amines, or carbohydrates for HBD (Zhang et al., 2012), with water forming part of their supramolecular structure (Ma et al., 2018). DESs possess several advantageous chemical properties, including low vapor pressure, a relatively wide liquid range, nonflammability, and water insensitivity. In addition, DESs are cost-effective, easily prepared without the need for purification steps, exhibit low toxicity or are nontoxic, and are biodegradable and recyclable (Palos-Hernández et al., 2022). Among DESs, a specific type known as natural deep eutectic solvents (NADESs) is gaining significance in the extraction of biocompounds. This term refers to a liquid composed of molecules abundantly found in nature. Typically, NADESs are composed of nontoxic quaternary ammonium salts acting as HBA, such as choline chloride, and naturally derived uncharged HBD, such as alcohols, amines, carboxylic acids, and sugars (Ozturk et al., 2018). The remarkable potential of NADESs in industrial applications stems from their cost-effectiveness, versatile physicochemical properties, simple and inexpensive preparation, and low cytotoxicity. They emerge to address the issues of commonly utilized solvents frequently employed in conventional extractions (such as ethanol, methanol, ethyl lactate, glycerol, or 2-methyltetrahydrofuran), as these, despite having some shared advantages with NADESs, are not as versatile as the latter and exhibit limited solvation capabilities. Nevertheless, the high viscosity and very low vapor pressure of NADESs present evident limitations (Bosiljkov et al., 2017). Extraction using NADESs involves the initial synthesis of the solvents themselves, which can be carried out through heating and stirring, evaporation, and to a lesser extent, freeze-drying methods (Ruesgas-Ramón et al., 2017). Subsequently, various extraction processes can be performed, which typically resemble conventional extractions (immersion of plant material and agitation), but with the use of these advanced solvents. However, other extraction technologies can also be combined with the use of NADESs, such as US (Bosiljkov et al., 2017) and MW (López-Linares et al., 2021) extractions.

The wide range of structural variations and adjustable physicochemical properties of NADESs present a vast potential for their application in the extraction of phenolic compounds. Nevertheless, the proper selection of NADES for extracting target compounds from the plant matrix is crucial due to their effects on diffusion, solubility, viscosity, surface tension, polarity, and physicochemical interactions during solid-liquid extraction (Palos-Hernández et al., 2022). In a study conducted by Bi et al. (2020), the potential of 10 different NADESs was investigated for the extraction of mulberry anthocyanins. The tested NADESs combined the use of choline chloride, choline bitartrate, or choline acetate (as HBA) with urea, glycerin, ethylene glycol, propylene glycol, malonic acid, lactic acid, and/or acetic acid (as HBD) in a molar ratio of 1:2, respectively (except for choline bitartrate: acetic acid, which had a molar ratio of 1:4). Out of the 10 NADESs investigated, six exhibited an anthocyanin extraction yield greater than 80% of acidified ethanol, with the combination of choline chloride: lactic acid (1:2) providing the highest yield. The study also indicated that the recovery of anthocyanins did not occur with eutectic mixtures characterized by having strong basic properties (such as the one containing urea), as these could degrade these polyphenols. Meanwhile, the use of choline chloride-based NADESs in combination with lactic acid provided improved half-life and extended storage times for extracted anthocyanins. Polyphenolic acids have also been successfully extracted using NADESs. For instance, rosmarinic acid and salviaflaside from *Prunella vulgaris* L. were extracted using six different alcohol-based NADESs with varying mole ratios fluctuating from 1:2 to 1:6 (Xia et al., 2015). The maximum yield was achieved with choline chloride: ethylene glycol at a 1:4 mole ratio, while augmenting the proportion of HBD led to a significant drop in polyphenolic acid extraction yields. After optimizing the extraction conditions for this NADES (including the water/NADES ratio), experimental extraction yields were obtained for rosmarinic acid and salviaflaside that were higher than those reported in conventional extractions. However, alcohol-based NADESs are not suitable for the recovery of flavonoids since these compounds can be considered a type of HBD, and thus, they can compete with alcohol-based HBDs. Therefore, once again, the importance of selecting appropriate

NADESs according to the target polyphenols is highlighted (Palos-Hernández et al., 2022).

5. Food applications of plant polyphenols

The versatile applications of polyphenolic compounds in the food industry are driven by their role as natural food preservatives, as they exhibit antioxidant and antimicrobial properties. These valuable characteristics offer a promising solution for enhancing food stability and prolonging shelf life. In addition, polyphenols serve as natural colorants, imparting vibrant hues to various food products. Furthermore, their incorporation as functional ingredients in food formulations supports potential health benefits, making them a valuable asset in promoting overall well-being. Examples of the diverse utilization of plant polyphenols in food applications are illustrated in Table 1.2.

5.1 Application of plant polyphenols as natural food preservatives (antioxidant and antimicrobial additives)

In the context of the food industry, the importance of antioxidants and antimicrobials cannot be underestimated, as oxidation reactions and microbial proliferation are the primary culprits responsible for food deterioration (Sun et al., 2022). Frequently, the initial alterations observed in food oxidation processes are related to the sensory attributes of the nourishment, encompassing variations in color, texture, and the development of off-flavors and off-odors (Domínguez et al., 2019). These changes directly impact consumer perception and acceptance, ultimately defining the food's shelf life based on the point at which consumers detect the oxidative-related modifications. In addition, oxidative processes result in the degradation of nutritional components as a consequence of various reactions leading to decreased levels of essential fatty acids, essential amino acids, vitamins, and other biocompounds. Moreover, the oxidation of food can present toxicological concerns due to the formation of substances with toxic properties and the generation of free radicals during these oxidative steps (Echegaray et al., 2021). Similarly, the deterioration caused by microbial growth plays a pivotal role in the food industry since microbial spoilage can imply various changes in food, including altered texture, off-flavors, and unpleasant odors. These alterations have a direct impact on consumer acceptance and influence the shelf life of the product. Furthermore, microbial activity can result in the degradation of essential nutrients and biocompounds, leading to a loss of nutritional value in the food. In addition, the presence of harmful microorganisms can pose significant food safety concerns, as they can produce toxic compounds and pose health risks to consumers (Karanth et al., 2023). Therefore, it is imperative to exercise control over factors that promote oxidation and microbial development in the food industry to mitigate economic losses on an industrial scale and address potential nutritional, toxicological, health, and safety issues.

Table 1.2 Application of plant polyphenols in food manufacturing.

Plant source	Main polyphenols	Food	Tested concentration	Main findings of polyphenol addition to foods	References
Date pit	Catechol, gallic, and protocatechuic acid, catechin, and quercetin	Beef burger	1.5%, 3%, and 6%	- Improved the preservation and shelf life of burgers. - Stabilized color, prevented oxidation, and inhibited microbial growth - Increased fiber and bioactive compound content. - Improved cooking properties and acceptability. - 3% concentration was found to be the most suitable.	Sayas-Barberá et al. (2020)
Tea	NS	Sea bass fillets	Immersed in 2 mg of polyphenols or 1 mg of polyphenols + 1.6 mg of kojic acid/mL	- Synergistic antibacterial effect of tea polyphenols with kojic acid. - A combination of polyphenols with kojic acid prevented lipid oxidation of the fillets. - Decreased total volatile basic nitrogen content compared with the control. - Stabilized the texture and color of the fillets in synergy with kojic acid. - Extended shelf life from 5 to 6 days.	Wang et al. (2021)
Clove	NS	Beef burger	2%, 4%, and 6%	- Enhanced the preservation and extended the burger shelf life.	Ahmed et al. (2022)

Continued

Table 1.2 Application of plant polyphenols in food manufacturing.—*cont'd*

Plant source	Main polyphenols	Food	Tested concentration	Main findings of polyphenol addition to foods	References
				- Enhanced antioxidant activity and microbial stability. - Improved burger cooking characteristics. - Sensory acceptability was higher for burgers containing 2% and 4%. - 2% and 4% concentration proved to be the optimal choice.	
Bana peal	NS	Yogurt	0.1%, 0.2%, 0.4%, 0.6%, 0.8%, and 1%	- Increased antioxidant activity and improved lipid stability. - Did not affect yogurt quality parameters (color, pH, acidity, and whey separation). - Had no impact on yogurt acceptability (optimal concentration of 0.6%).	Kabir et al. (2021)
Cinnamon, clove, ginger, and mint	Catechin, chlorogenic gallic acid, pyrogallol, apigenin-6-arabinose, and hesperidin	Cucumber juice	0.003%	- In general, extracts reduced the growth of bacteria and fungi. - Clove and mint extracts displayed the greatest antioxidant activity. - Clove extract had a positive influence on the overall juice taste and flavor. - Cinnamon and mint extracts positively influenced the juice color.	Saad et al. (2021)

Colorants					
Black rice bran	Anthocyanins (including cyanidin-3-O-glucoside)	Fermented Thai pork sausage	0.8%, and 0.1%	- Increased the presence of bioactive compounds. - Preserved C* compared with a control. - Helped partially replace the use of nitrites in sausages.	Loypimai et al. (2017)
Rubus ulmifolius Schott fruits	Anthocyanins	Donuts	0.86%	- Minimal effect on the chemical composition. - Imparted pink/lilac color to the product. - Color remained stable throughout storage.	Da Silva et al. (2019)
Black bean seed coat	Anthocyanins (together with saponins)	Maize tortillas and cookies	0.3% and 0.7%	- Did not influence flour rheology. - Did not influence the texture of tortillas and cookies. - Reduced the values of L*, a*, and b* and produced darker products. - Presence of cyanindin-3-glucoside equivalents, myricetin 3-O-glucosid, quercetin 3-O-glucoside, and kaempferol 3-O-glucoside. - After baking, 80% and 60% of bioactive compounds were retained in the tortillas and cookies, respectively.	Chávez-Santoscoy et al. (2016)
Blueberry	Anthocyanins	Beverage system	100 mg anthocyanin-loaded nanoparticles	- The stability of anthocyanins was enhanced in the anthocyanin-loaded chitosan nanoparticles. - Anthocyanin-loaded chitosan nanoparticles exhibited reduced degradation in simulated gastrointestinal fluid	He et al. (2017)

Continued

Table 1.2 Application of plant polyphenols in food manufacturing.—*cont'd*

Plant source	Main polyphenols	Food	Tested concentration	Main findings of polyphenol addition to foods	References
Euterpe edulis Martius	Anthocyanins	Dairy beverage	4%–8% powdered microencapsulated anthocyanin	- Acidity, pH, and anthocyanin content remained constant. - A slight overall color difference was detected between the analyzed beverages.	Lima et al. (2019)
Functional ingredients					
Persimmon by-product	Polyphenols (including gallic acid and flavanone glucosides)	Pork liver pâté	3%, and 6%	- Lipid-rich foods can serve as excellent carriers of bound polyphenols. - Polyphenols remained intact until reaching the colon, thus being available for utilization by the intestinal microbiota. - The addition of 6% persimmon flour favored the oxidation of PUFA.	Lucas-González et al. (2021)
Prickly pear peel	NS	Bread	5%, 10%, 15%, 20%, and 50%	- The 50% concentration was not viable for bread production. - The polyphenol-rich flour increased the fiber content and produced a functional bread. - Baking did not prevent the presence of a high polyphenol content in the bread (possible protective effect of the food matrix). - The formulation with 10% addition resulted in the dough with better characteristics and the bread with the highest scores in sensory evaluation.	Parafati et al. (2020)

Passion fruit peel	NS	Biscuit	5%, 10%, and 15%	- Improved fiber content. - Enhanced biscuit texture. - Negatively influenced biscuit color.	Weng et al. (2021)
Vitis labrusca L. solid by-product flourand/or grape	NS	Goat milk yogurt	20 g/100 mL	- Increased the content of polyphenolic compounds, leading to improved nutritional value in yogurt. - Improved sensory acceptance and consumer preference compared with yogurt without grape ingredients - Provided high antioxidant capacity. - Led to high *Lactobacillus acidophilus* La-05counts during storage. - The in vitro GI digestion of grape polyphenols increased the antioxidant capacity of yogurt at the stomach and intestinal levels.	Silva et al. (2022)

In contrast to synthetic additives, the use of plant polyphenols, whether isolated, in combination, or in extract form, has emerged as a hopeful approach to mitigate the aforementioned challenges. In meat products, the use of polyphenol-rich extracts has been employed in a promising manner (Munekata et al., 2021). For instance, extracts rich in phenolic acids (as catechol, gallic, and protocatechuic acid) and flavonoids (as catechin and quercetin), like clove powder (Ahmed et al., 2022) and date pit powder (Sayas-Barberá et al., 2020), have been employed in burgers to reduce oxidation reactions and microbial growth, thereby extending the shelf life of these perishable edibles. Furthermore, these polyphenol-rich ingredients enriched the meat product with fiber and enhanced cooking properties and acceptability when added at an optimal concentration. Similarly, polyphenols can help extend the shelf life of fish, which are highly perishable (Zeb, 2020). The use of natural preservatives such as tea-derived polyphenols, in synergy with kojic acid, significantly slowed down the growth of spoilage bacteria and lipid oxidation in sea bass fillets (Wang et al., 2021). In addition, the addition of these natural preservatives reduced the formation of total volatile basic nitrogen (TVB-N) while stabilizing the physical qualities (color and texture) of fish fillets, thereby extending the shelf life from 5 to 6 days.

Plant polyphenols have also been successfully used as preservatives in dairy products, such as yogurts, helping to improve oxidative stability during storage without affecting quality parameters like color, pH, acidity, and whey separation. Moreover, their use does not negatively influence the overall acceptability of these goods (Kabir et al., 2021). Likewise, their utilization in juices proved to be interesting as it reduced the growth of bacteria and fungi, while also positively influencing the flavor and color in the case of clove, mint, and cinnamon-rich polyphenol extracts, respectively (Saad et al., 2021).

5.2 Application of plant polyphenols as natural food colorants

Color represents a sensory characteristic that significantly impacts the consumer acceptance of food products. Thus, dyes are incorporated into these products to enhance their visual appeal. The demand for safe and ecofriendly alternatives to synthetic colorants has led to a growing interest in exploring the vibrant hues offered by several polyphenolic compounds derived from plants. By harnessing the rich pigmentation of these compounds, food manufacturers seek to enhance the visual appeal of their products while meeting consumer expectations for natural and wholesome ingredients. Among the polyphenolic compounds, anthocyanins stand out for their potential as natural dyes in various food sectors (Echegaray et al., 2023). In meat products, the utilization of pigmented anthocyanin sources has proven beneficial, not only for achieving desirable colorations but also for partially replacing the nitrite content. In fermented sausages, the application of natural colorant from black rice bran (rich in anthocyanins) increased bioactive compounds (specifically cyanidin-3-O-glucoside) while preserving chroma (C*) compared with a control (sausages with a nitrite content of 120 ppm) (Loypimai et al., 2017). For their part, in bakery products, extracts rich in anthocyanins have been frequently employed to enhance the colors obtained after baking, replacing synthetic dyes. The pigment rich in anthocyanins extracted from *Rubus ulmifolius* Schott fruits proved to be a promising colorant, imparting prominent lilac and pink hues to donuts, which remained stable throughout storage with minimal impact on the final baked product's chemical composition (Da Silva et al., 2019). In addition, the use of black bean seed coat has been applied in the preparation of gluten-free tortillas and cookies from nixtamalized maize flours (Chávez-Santoscoy et al., 2016). Its addition allowed the incorporation of various flavonoids such as myricetin 3-O-glucoside, quercetin 3-O-glucoside, and kaempferol 3-O-glucoside, resulting in darker edibles with lower lightness (L*) value as well as reduced redness (a*) and yellowness (b*) values.

Despite the potential use of anthocyanins as coloring agents, it should be taken into consideration that these pigments can be unstable in food systems. They may degrade over time during storage due to factors such as pH values, oxidation reactions, and the presence of metals and other reactive substances in the food. As a result, these compounds have been frequently applied after stabilization, such as through copigmentation, encapsulation, and the use of nanoparticles (Gençdağ et al., 2022; Ghosh et al., 2022). For instance, the use of edible nanoparticles of chitosan could be a strategy to enhance anthocyanin stability in beverages, as well as to delay their degradation in the gastrointestinal system (He et al., 2017). In dairy products as well, encapsulation has been shown to be a technique that helps maintain the stability of anthocyanins over time (da Silva Carvalho et al., 2016; Lima et al., 2019).

5.3 Application of plant polyphenols as healthy ingredients

The utilization of plant polyphenols extends beyond their role as natural food preservatives and colorants. Their remarkable health-promoting properties have garnered significant attention, making them a focal point in the development of functional food products. These bioactive compounds are increasingly sought after by consumers who prioritize nutrition and well-being. As researchers explore the potential health benefits of polyphenols, their incorporation (particularly as part of extracts, powders, flours, or other forms) into various food formulations aims to enhance nutritional value and promote overall health (Zeb, 2020). However, the full potential of polyphenols as healthy ingredients in diverse foods is still subject to further investigation. Significant progress has been made in understanding the role of dietary phenolics, specifically in terms of their absorption and chemical transformations within the body. The digestive process and absorption of polyphenols, where they are converted into aglycone form, occur during the passage of food through the digestive system and into the circulatory system, reaching the liver. Bioavailability plays a crucial role in ensuring that the most bioavailable phenolic compounds in the diet exert their optimal effects on the human body (Prabhu et al., 2021). To fully harness the benefits of polyphenols, it is essential to explore the influence of various nutrients (including carbohydrates, fiber, minerals, lipids, and proteins) and food processing operations on their bioaccessibility. More investigations are needed to elucidate the impact of polyphenols (both in their direct use and as part of other plant matrices) within the food matrix on human health. Conducting in vitro and in vivo assessments in human models could provide valuable insights for their potential applications in food and appropriate dosages (Rajha et al., 2022).

The use of polyphenols in the development of meat products such as pâté has recently been investigated in a study where persimmon by-product flour (rich in gallic acid and flavanone glucosides) was employed to enhance the polyphenolic profile and the health-promoting characteristics of the product (Lucas-González et al., 2021). This investigation also revealed that lipid-rich foods can serve as excellent carriers of bound polyphenols, which remain intact until reaching the colon and can be metabolized by the intestinal microbiota, exerting their beneficial effects. However, it also showed that higher concentrations (6%) of persimmon flour could induce the oxidation of polyunsaturated fatty acids. The use of polyphenol-rich flours in the production of bakery doughs has also demonstrated the possibility of obtaining functional products with elevated nutritional characteristics (such as fiber and mineral content) without negatively affecting the sensory properties of the prepared foods (Bhardwaj et al., 2022). For instance, prickly pear peel flour proved to be an effective bioactive and functional ingredient in breadmaking, providing bread with high fiber and polyphenol content, as well as significant antioxidant capacity without negatively impacting dough properties and sensorial bread evaluation (Parafati et al., 2020). Similarly, the incorporation of passion fruit flour (rich in biocompounds such as polyphenols) in biscuit formulation improved the product's fiber content while significantly enhancing its texture. However, it was also observed that this had a negative effect on the biscuits' color parameters (Weng et al., 2021).

The production of functional products through polyphenol enrichment has also been observed in dairy products. For example, the incorporation of *Vitis labrusca* L. solid by-product flour and/or grape ingredients into goat yogurt not only increased the content of polyphenolic compounds, resulting in improved nutritional value, but it also showcased significant health-enhancing properties (Silva et al., 2022). The enriched dairy product exhibited high antioxidant capacity, which can play a crucial role in neutralizing harmful free radicals and reducing oxidative stress in the body. Moreover, the presence of grape ingredients positively influenced the yogurt's microbial stability, as indicated by the high *Lactobacillus acidophilus* La-05 counts observed during storage. This suggests that the polyphenols present in the grape may have contributed to the preservation of probiotic microorganisms. Furthermore, the study exploring the in vitro gastrointestinal digestion of grape ingredients demonstrated an intriguing finding. The digestion process increased the antioxidant capacity of the yogurt not only in the stomach but also in the intestinal environment. This indicates that the polyphenols from the grape remain bioavailable and bioactive even after undergoing digestion, enhancing their potential health benefits upon consumption.

5.4 Application of plant polyphenols in food packaging

Although the food packaging industry is not directly part of food manufacturing, its significance is undeniable in the context of food preservation, protection, and promotion. In this field, the development of innovative food packaging has gained attention as a response to mounting environmental concerns, the increasingly challenging issue of managing plastic waste, the utilization of food industry by-products, and the growing consumer demand for natural, nutritious, and sustainable food solutions. This concept aims to address critical issues related to environmental impact and sustainability while meeting the evolving preferences of conscious consumers (Umaraw et al., 2020). In this context, the utilization of polyphenolic substances in the development of packaging can provide a highly significant ingredient for various reasons, as described in Table 1.3. Firstly, polyphenols are abundant natural compounds found in plants and fruits, making them a sustainable and renewable source of raw material for packaging manufacturing. By utilizing these compounds, the dependency on nonrenewable petroleum-derived materials, which have a significant environmental impact, can be reduced. Secondly, polyphenols possess antioxidant and antimicrobial properties, which enable their use in the development of active packaging with antimicrobial and antioxidant characteristics. This allows the polyphenols to be released from the polymeric matrix (which can be edible or nonedible), contributing to extending the shelf life of products and reducing food waste, thus addressing crucial issues related to food safety and sustainability. An example of this can be seen in the development of an active film made from gelatin cross-linked with dialdehyde quince seed gum and incorporated with apple polyphenols (Maroufi et al., 2022). Through the use of these biocompounds, the antioxidant activity of the original film (composed solely of gelatin and dialdehyde quince seed gum) was significantly improved up to 30 times, along with exhibiting antimicrobial properties against *Escherichia coli* and *Staphylococcus aureus*. These properties could extend the shelf life of food products and reduce food waste. Similarly, Bermúdez-Oria et al. (2019) observed that the incorporation of hydroxytyrosol and 3,4-dihydroxyphenylglycol into a pectin-fish gelatin edible film improved the antioxidant properties of the packaging, consequently aiding in the preservation of refrigerated beef meat.

At the same time that antioxidant properties are improved, certain polyphenolic compounds can also enhance other characteristics of packaging. For instance, substances such as tannins (Huang et al., 2022) and curcumin (Roy & Rhim, 2020) enhanced the barrier properties, mechanical strength, and UV light blocking of cellulose and gelatin films, respectively, exhibiting significant potential for their application in food packaging. These features play a crucial role in preserving the quality and freshness of packaged foods while preventing the loss of nutrients and flavors.

Another advantage of using polyphenolic substances in innovative packaging lies in their suitability for the production of intelligent packaging. Concretely, anthocyanins, among the natural colorants, have garnered significant attention as color indicators due to their remarkable color-changing response in the presence of

Table 1.3 Application of plant polyphenols in food packaging. clicking.

Plant source	Main polyphenols	Matrix packaging	Tested concentration	Main findings	References
Apple peal	NS	Gel/DAQSG film	0.5%, and 1%	- Increased the film's antioxidant capacity up to 30 times. - Enhanced the film's antimicrobial activity against *Escherichia coli* and *Staphylococcus aureus.*	Maroufi et al. (2022)
Olive by-product	Hydroxytyrosol and 3,4-dihydroxyphenylglycol	Pectin-fish gelatin film containing beeswax	0.1%, and 0.5%	- Improved antioxidant properties of the film. - Effective in delaying oxidation in refrigerated packaged beef.	Bermúdez-Oria et al. (2019)
Black wattle	Tannins	Cellulose film	10.1%, 14.3%, and 19.4%	- Improved the antioxidant properties of the film. - Enhanced hydrophobicity and water vapor barrier properties of the film. - Improved UV-blocking. - Exhibited slightly lower dry tensile strength but higher toughness.	Huang et al. (2022)
Turmeric	Curcumin	Gelatin film	0.25%, 0.5%, 1.0%, and 1.5%	- Increased UV blocking effect by over 99% (at a concentration of 1.5%). - Enhanced mechanical strength and water vapor barrier properties of the film (up to a concentration of 1%). - Displayed remarkable antimicrobial activity against *Escherichia coli* and *Listeria monocytogenes.*	Roy and Rhim (2020)

Purple tomato (*Solanum lycopersicum*)	Anthocyanins	Chitosan film	10%, 30%, and 50%	- Visual color change from pink to light yellow with the variation of pH value (3–11). - The 10% concentration was the one that best showed the color change. - Proven useful freshness indicator for milk and fish in intelligent packaging.	Li et al. (2021)
Blueberry	Anthocyanins	Protein/carbohydrate nanocomplex film	Anthocyanins loaded in ovalbumin-carboxymethyl cellulose nanocomplexes (NS %)	- Visual color change from purple to pink during the mushroom storage. - Useful freshness indicator for intelligent packaging.	Liu et al. (2022)
Turmeric	Curcumin	Chitosan and polyethylene nanofiber	2.5%	- Visual color change from right yellow to reddish color during the chicken breast storage. - Useful freshness indicator for intelligent packaging.	Yildiz et al. (2021)

different pH values (Echegaray et al., 2023). Thus, the indicators based on anthocyanins are an innovative and effective way to monitor the freshness and quality of food in real-time. Li et al. (2021) conducted a study where they utilized purple tomatoes (*Solanum lycopersicum*) as a novel source of anthocyanins to create an intelligent colorimetric indicator film based on a chitosan matrix. The research demonstrated that the film's color changed from pink to light yellow with an increasing pH range from 3 to 11. Furthermore, the application of this film proved useful in monitoring the deterioration of both milk and fish, as it also changed color (darkening) during the spoilage process of these packaged foods. In a similar manner, Liu et al. (2022) and Yildiz et al. (2021) developed intelligent indicators using anthocyanins for monitoring the freshness of mushrooms and chicken breast. Concretely, for the monitoring of mushrooms, Liu et al. (2022) utilized anthocyanins sourced from blueberries, which were loaded into a protein-carbohydrate nanocomplex film. In the case of chicken breast, Yildiz et al. (2021) employed a packaging composed of chitosan and polyethylene nanofibers loaded with turmeric anthocyanins. Both packaging solutions exhibited color changes (from purple to pink and yellow to red, respectively) as the perishable foods deteriorated, proving highly effective as freshness indicators in intelligent packaging.

6. Cosmetic applications of plant polyphenols

The significance of polyphenolic compounds in cosmetic products primarily lies in their antioxidant properties, since the utilization of antioxidants in these items helps reduce and prevent oxidative stress and damage, offering a viable option for therapy and prevention of premature aging. These compounds also provide photoprotective effects and aid in the treatment of sensitive or sun-damaged skin due to their antiinflammatory activity. These characteristics have also led to the use of polyphenols in haircare. In addition, antioxidant compounds are utilized to prevent or minimize oxidative degradation of active ingredients in cosmetics and to avoid the oxidation of oily components present in formulations. Table 1.4 presents illustrative cases showcasing the application of plant polyphenols in the cosmetic industry.

6.1 Applications of plant polyphenols in skincare

Skin aging is a complex biological process involving a combination of factors. It encompasses changes in DNA repair and stability, mitochondrial function, cell cycle and apoptosis, ubiquitin-induced proteolysis, activity of skin-aging-related enzymes (anticollagenase, antielastase, antityrosinase, and xanthine oxidase), and cellular metabolism. One of the primary contributors to skin aging is the decline in physiological hormones, which plays a vital role in maintaining skin health. In addition, the skin gradually loses its ability to repair itself, further exacerbating the signs of aging. Furthermore, external factors such as UV, visible (VIS), and infrared (IR) radiation, tobacco smoke, air pollutants, poor nutrition, sleep deprivation, and stress exacerbate skin aging through ROS-mediated signaling pathways. Together, these factors

Table 1.4 Application of plant polyphenols in cosmetic products.

Plant source	Main polyphenols	Biological activities	References
Ripe strawberries	Anthocyanins (including Cy- and Pg-glycosides)	- Reduced intracellular ROS levels. - Increased cell viability. - Reduced lipid membrane oxidation and decreased DNA damage. - Improved mitochondrial functionality. - Skin antiaging.	Giampieri et al. (2014)
Tomato and rosemary	Polyphenols in general, together with carotenoids and estradiol	- Protected dermal fibroblasts from ROS-induced damage. - Reduced MMP-1 secretion - Increased pro-collagen levels. - May improve skin health and appearance.	Darawsha et al. (2021)
Benitaka grape peel	Flavonoids	- No effect on cell viability - Skin-protecting activity against sun radiation. - Possessed antioxidant activity. - Potential for use as a sunscreen.	Cefali et al. (2020)
Grapevine leaf	NS	- Antiaging cellular protection. - Produced transcriptional and DNA methylation alterations.	Letsiou et al. (2020)
Malaxis acuminata leaf and stem	Phenolic acids, flavonoids, and tannins	- Exhibited promising photoprotective activity against UVA and UVB radiation. - Showed promising inhibitory activity against major skin aging-related enzymes and antiinflammatory potential.	Bose et al. (2017)

Continued

Table 1.4 Application of plant polyphenols in cosmetic products.—*cont'd*

Plant source	Main polyphenols	Biological activities	References
		- Significant antioxidant potential. - Potential rejuvenator and antiaging ingredient.	
Olive mill wastewater	NS	- Promoted hair growth. - Provided protection against oxidative stress-induced hair loss. - Facilitated the release of IGF-1 from dermal papilla cells. - Prevented the formation of free radicals in the scalp and hair follicles. - Promising ingredient in haircare formulations.	Sittek et al. (2021)
Green tea	Epigallocatechin-3-gallate	- Stimulation of hair growth in hair follicles. - Increased proliferation of dermal papilla cells. - Promoted cell proliferation and prevented apoptosis. - Beneficial compound for maintaining healthy hair and potentially prolonging the anagen stage.	Kwon et al. (2007)
Carthamus tinctorius L., *Clitoria ternatea* L., and *Papaver rhoeas* L.,	Caffeic, gallic and quinic acid; rutin, quercetin kaempferol-3-O-rutinoside, and kaempferol-3-O-glucoside	- Potential as cosmetic colorants (sources of yellow, orange, blue, and purple dyes). - Maintained their color during storage as ingredients in a model micellar cosmetic. - Provided benefits to the skin (hydration, UV protection, and antiaging action).	Bujak et al. (2022)

Table 1.4 Application of plant polyphenols in cosmetic products.—*cont'd*

Plant source	Main polyphenols	Biological activities	References
Cacao pod husk epicarp	Principally, myricetin, jaceosidin, phloretin, procyanidin B1, and quercetin	- Potential as cosmetic colorants (sources of yellow, red, and purple dyes). - High anthocyanin content. - UV absorption capability. - Strong dyeing capacity. - Noncytotoxicity.	Oñate-Gutiérrez et al. (2023)

contribute to the visible signs of aging, such as wrinkles, loss of elasticity, and uneven skin tone (Ferreira et al., 2021; Yaghoubi et al., 2019). In this field, plant polyphenol extracts are considered to be among the most effective functional raw materials for antiaging cosmetics due to their reported efficacy in mitigating the advancement of the aging process (de Lima Cherubim et al., 2020).

In an early study, it was observed that strawberry extract, rich in anthocyanins and other bioactive compounds, exhibited high antioxidant power, resulting in a protective effect on skin cells against oxidative stress-induced damage (Giampieri et al., 2014). The authors demonstrated that the anthocyanin-rich extract was effective in reducing intracellular ROS and protecting lipids, DNA, and mitochondrial functionality in primary human dermal fibroblasts (HuDe). A positive relationship was highlighted between Cy- and Pg-glycosides anthocyanins and the protective properties. Likewise, it has been suggested that polyphenols (together with carotenoids and estradiol) found in rosemary and tomato extracts exhibit protective effects on dermal fibroblasts by reducing levels of ROS and mitigating oxidative stress-induced damage (Darawsha et al., 2021). These bioactive compounds have been shown to reduce the secretion of matrix metalloproteinases (MMPs) (which are responsible for the degradation of collagen in the extracellular matrix) and increase the levels of procollagen, thereby potentially improving skin health and enhancing its appearance. Furthermore, grapevine (*Vitis vinifera*) leaf extracts have been associated with their potential to provide antioxidant and antiaging cellular protection against environmental stressors (Letsiou et al., 2020). This was demonstrated in research where the effects were observed on UV-stressed primary human dermal fibroblasts at the transcriptional and DNA methylation levels. The findings suggest that grapevine leaf extracts possess protective features that extend to cellular and genetic levels in response to environmental stressors. For their part, phenolic acids have additional benefits in the control of inflammation, immune system stimulation, and prevention of collagen fiber degradation through various mechanisms. These include facilitating the formation of natural collagen bonds, protecting against the harmful effects of free radicals, and improving blood circulation. Collectively, these factors play a significant role in contributing to antiaging effects in skin (Bjørklund et al., 2022).

Generally, phenolic compounds can absorb UV radiation and act as a barrier, preventing solar radiation from penetrating the skin. This property significantly enhances the sun protection provided by products containing phenolic compounds and effectively counteracts the detrimental effects of oxidative stress induced by sun exposure (Khan et al., 2019). A rich flavonoid extract derived from Benitaka grape (*Vitis vinifera*) peel was investigated for its sun protection and antioxidant activity, demonstrating promising results (Cefali et al., 2020). The study revealed that although the extract did not exhibit effects on cell viability, it displayed noteworthy properties as a sunscreen filter. Moreover, the inclusion of Benitaka grape peel extract resulted in oil-in-water cosmetic formulations with desirable characteristics for skincare. In their study, Bose et al. (2017) found that methanolic extracts of leaves and stems from *Malaxis acuminata* exhibited a promising protective effect against ultraviolet A (UVA) and ultraviolet B (UVB) radiation. This effect was attributed to the presence of various bioactive compounds, including several polyphenols such as phenolic acids, flavonoids, and tannins. In addition, the *Malaxis acuminata* methanolic extracts demonstrated inhibitory activity against aging-related enzymes and significant antiinflammatory and antioxidant potential. As a result, it has been proposed as a rejuvenating and antiaging ingredient in cosmetics.

6.2 Applications of plant polyphenols in haircare

Haircare is a critical aspect of personal grooming, and understanding the factors that influence it is essential for developing effective cosmetic solutions. In recent years, scientific research has increasingly turned its attention to the application of polyphenols in haircare, albeit to a lesser extent than in skincare. These natural compounds offer promising benefits for promoting hair growth, preventing hair loss, and maintaining a healthy scalp. For instance, Sittek et al. (2021) demonstrated how phenol-enriched olive mill wastewater extract emerges as a promising ingredient in haircare formulations, since it promotes hair growth while simultaneously providing protection against oxidative stress-induced loss. Specifically, they evidenced that the polyphenol-enriched olive mill wastewater extract facilitated the release of insulin-like growth factor-1 (IGF-1) from the dermal papilla cells of the hair follicle, in addition to preventing the formation of free radicals in the scalp and hair follicles. For their part, Kwon et al. (2007) explored the potential of EGCG for hair growth promotion. In vitro studies revealed that this catechin stimulated hair growth in hair follicles and increased the proliferation of dermal papilla cells. This effect was attributed to the dual actions of EGCG, which promotes cell proliferation and prevents apoptosis. These findings indicate the potential of EGCG as a beneficial compound for maintaining healthy hair and potentially prolonging the growth phase of the hair growth cycle (in other words, the anagen stage).

6.3 Application of plant polyphenols in cosmetic coloring

In the realm of cosmetics, the application of polyphenols in makeup and cosmetic coloring has gained increasing attention. These natural compounds, renowned for their antioxidant properties and versatile benefits, are now being harnessed to enhance the efficacy and safety of makeup products and coloring formulations. A recent study investigated different flower extracts (specifically, *Carthamus tinctorius* L., *Clitoria ternatea* L., *Gomphrena globosa* L., *Papaver rhoeas* L., and *Punica granatum* L.) rich in polyphenols such as caffeic, gallic, and quinic acid, kaempferol-3-O-rutinoside, kaempferol-3-O-glucoside, quercetin, and rutin (Bujak et al., 2022). Upon analyzing their diverse color characteristics and conducting antioxidant, antiinflammatory, antiaging, and cytotoxicity studies, Bujak et al. (2022) discovered that the extracts of *Carthamus tinctorius*, *Clitoria ternatea*, and *Papaver rhoeas* exhibited remarkable potential as cosmetic colorants. Notably, these extracts maintained their color stability throughout storage as ingredients in a model micellar cosmetic, while concurrently demonstrating advantageous properties for the skin, including hydration, protection against UV radiation, and antiaging action. Similarly, Oñate-Gutiérrez et al. (2023) investigated the cacao pod husk epicarp as a potential source for colorant extraction. Specifically, they obtained three different extracts (yellow, red, and purple), which mainly contained polyphenols such as myricetin, jaceosidin, phloretin, procyanidin B1, and quercetin. These extracts stood out not only for their high anthocyanin content but also for their ability to absorb UV radiation, their dyeing capacity, and their noncytotoxicity. Thus, cacao pod husk epicarp extracts also presented themselves as potential sources of flavonoids with coloring properties for the cosmetic industry.

7. Challenges in the incorporation of polyphenols within the industry

Plant polyphenols and plant polyphenols-rich extracts exhibit potent biological properties. However, several challenges still hinder their successful incorporation into food and cosmetic formulations.

7.1 Challenges in the incorporation of polyphenols in the food industry

One of the primary challenges in the food industry lies in the intense bitter taste and astringency that some polyphenols can impart to foods. In addition, during food processing and storage, polyphenols are susceptible to degradation caused by heat, light, and oxygen. This lack of stability may result in a reduction in their concentration in the final product, consequently diminishing their potential benefits in the food, as an additive, or in the human body, as a functional ingredient (Bazinet et al., 2010; Khan et al., 2019). Another limiting factor is the interaction of polyphenols with other components present in the food matrix, such as carbohydrates, lipids, or proteins (Martinengo et al., 2021). These interactions can affect the quality and stability of the product, leading to the formation of poorly soluble complexes, precipitation, or changes in the food's texture and appearance. Also, the rapid release of polyphenols, their low solubility in the food products, and their limited bioactivity could restrict their benefits in the food sector (Wu & Zhou, 2021).

Some of these limitations can be addressed through encapsulation techniques, as this process allows the protection of polyphenolic compounds within a matrix (which may be composed of arabic gum, inulin, modified starch, whey protein isolate, etc.), thus preventing both their degradation and the imparting of undesirable taste characteristics to the food (Gençdağ et al., 2022). In addition to encapsulation, other stabilization methodologies are under investigation, such as complexation with suitable nanocarriers that prevent the polyphenols from succumbing to environmental stress (Zhang et al., 2021). Furthermore, formulations incorporating phenolic compounds must be prepared optimally, enabling the full potential of these substances to be harnessed while avoiding interactions with other food components (Ahmed et al., 2022; Sayas-Barberá et al., 2020).

7.2 Challenges in the incorporation of polyphenols in the cosmetic industry

Despite the numerous advantages of incorporating polyphenols into topical formulations, the practical use of bioactive compounds in cosmetics is limited due to their low solubility in topical applications without the addition of potentially skin- or hair-sensitizing solvents. Nevertheless, the use of surfactants can enhance the inclusion of different polyphenols, such as flavonoids, in preparations without compromising the cosmetic stability (de Oliveira & Dario, 2019). In addition, the use of polyphenols in cosmetic formulations is constrained by their topical absorption, as their ultimate function must take place in external and localized areas (skin, hair, and scalp). Cutaneous barriers may hinder the absorption of polyphenols, thus minimizing the action of these active ingredients in cells (Bharadvaja et al., 2023). As a result, novel drug delivery systems (NDDS) have been developed, primarily focused on enhancing and delivering therapeutic biomolecules to specific sites (Ajazuddin & Saraf, 2010). In this field, nanotechnology has enabled the development of different nanocarriers (such as lipid nanoparticles, solid-lipid nanoparticles, liposomes, and liquid crystal nanoparticles) that are specifically designed to facilitate the transportation of polyphenols across the epithelium, enhance their distribution and pharmacokinetics, improve stability, reduce skin irritation, control the release of bioactive compounds, and consequently improve their efficacy at the cellular level, among others (Archana et al., 2015; Avadhani et al., 2017; Gupta & Dixit, 2011; Montenegro et al., 2017; Shtay et al., 2019).

8. Conclusions

The significance of plant polyphenols in the food and cosmetic industries is evident, as they offer a multitude of beneficial properties, including antioxidant, antimicrobial, antiinflammatory, and photoprotective properties. These biocompounds are widely distributed in plant sources, so various innovative extraction techniques, such as US, MW, PEF, SCF, and NADES technologies, are presented to recover polyphenols efficiently from different plant materials. These advancements open new possibilities for harnessing the full potential of these compounds in a wide array of applications within the food industry and cosmetic formulations. However, process optimization in extraction technologies is essential to obtain high-quality and versatile ingredients that are also aligned with consumer demand.

Incorporating polyphenols into food and cosmetic formulations offers a wide range of opportunities, enabling their use as antioxidants, antimicrobials, colorants, functional ingredients, photoprotective agents, and skincare and haircare components. Nevertheless, challenges exist in effectively integrating polyphenols into matrices and formulations, necessitating strict control over processing and storage to minimize degradation reactions.

To capitalize on the benefits of polyphenols, the food and cosmetic industries must continue to explore ways to optimize their extraction and posterior utilization while addressing challenges related to stability and desired attributes. Embracing these natural substances as key ingredients can lead to enhanced product quality, promoting holistic well-being and sustainable practices in both sectors. Future research and innovation will be pivotal in maximizing the potential of polyphenols and elevating the offerings of the industry to meet the demands of health-conscious consumers.

Acknowledgments

N. Echegaray acknowledges support from the Axencia Galega de Innovación (GAIN), Xunta de Galicia, for her postdoctoral fellowship extension (grant IN606C-2025/001). N. Echegaray and J.M. Lorenzo are members of the Competitive Reference Group "FunFood" (Functional Food), funded by GAIN (Nº. IN607A2023/01).

References

Agregán, R., Bangar, S. P., Hassoun, A., Hano, C., Pateiro, M., & Lorenzo, J. M. (2023). Green technologies for sustainable food production and preservation: Supercritical fluids. *Reference Module in Food Science, 2023*, Article 780. https://doi.org/10.1016/B978-0-12-823960-5.00078-0

Ahmad, T., Masoodi, F. A. A., Rather, S., Wani, S. M., & Gull, A. (2019). Supercritical fluid extraction: A review. *Journal of Biological and Chemical Chronicles, 5*(1), 114–122. https://doi.org/10.33980/jbcc.2019.v05i01.019

Ahmed, I. A. M., Babiker, E. E., Al-Juhaimi, F. Y., & Bekhit, A. E. D. A. (2022). Clove polyphenolic compounds improve the microbiological status, lipid stability, and sensory attributes of beef burgers during cold storage. *Antioxidants, 11*(7), 1354. https://doi.org/10.3390/antiox11071354

Ajazuddin, & Saraf, S. (2010). Applications of novel drug delivery system for herbal formulations. *Fitoterapia, 81*(7), 680–689. https://doi.org/10.1016/j.fitote.2010.05.001

Annunziata, G., Sureda, A., Orhan, I. E., Battino, M., Arnone, A., Jiménez-García, M., Capó, X., Cabot, J., Sanadgol, N., Giampieri, F., Tenore, G. C., Kashani, H. R. K., Silva, A. S., Habtemariam, S., Nabavi, S. F., & Nabavi, S. M. (2021). The neuroprotective effects of polyphenols, their role in innate immunity and the interplay with the microbiota. *Neuroscience & Biobehavioral Reviews, 128*, 437–453. https://doi.org/10.1016/j.neubiorev.2021.07.004

Archana, A., Sri, K. V., Madhuri, M., & Kumar, C. H. A. (2015). Curcumin loaded nano cubosomal hydrogel: Preparation, in vitro characterization and antibacterial activity. *Chemical Science Transactions, 4*(1), 75–80. https://doi.org/10.7598/cst2015.913

Arshad, R. N., Abdul-Malek, Z., Munir, A., Buntat, Z., Ahmad, M. H., Jusoh, Y. M. M., Bekhit, A. E. D., Roobab, U., Manzoor, M. F., & Aadil, R. M. (2020). Electrical systems for pulsed electric field applications in the food industry: An engineering perspective. *Trends in Food Science and Technology, 104*, 1–13. https://doi.org/10.1016/j.tifs.2020.07.008

Avadhani, K. S., Manikkath, J., Tiwari, M., Chandrasekhar, M., Godavarthi, A., Vidya, S. M., Hariharapura, R. C., Kalthur, G., Udupa, N., & Mutalik, S. (2017). Skin delivery of epigallocatechin-3-gallate (EGCG) and hyaluronic acid loaded nano-transfersomes for antioxidant and anti-aging effects in UV radiation induced skin damage. *Drug Delivery, 24*(1), 61–74. https://doi.org/10.1080/10717544.2016.1228718

Bakić, M. T., Pedisić, S., Zorić, Z., Dragović-Uzelac, V., & Grassino, A. N. (2019). Effect of microwave-assisted extraction on polyphenols recovery from tomato peel waste. *Acta Chimica Slovenica, 66*(2), 367–377. https://doi.org/10.17344/acsi.2018.4866

Bazinet, L., Araya-Farias, M., Doyen, A., Trudel, D., & Têtu, B. (2010). Effect of process unit operations and long-term storage on catechin contents in EGCG-enriched tea drink. *Food Research International, 43*(6), 1692–1701. https://doi.org/10.1016/j.foodres.2010.05.015

Belščak-Cvitanović, A., Durgo, K., Huđek, A., Bačun-Družina, V., & Komes, D. (2018). Overview of polyphenols and their properties. In C. M. Galanakis (Ed.), *Polyphenols: Properties, recovery, and applications* (pp. 3–44). Woodhead Publishing. https://doi.org/10.1016/B978-0-12-813572-3.00001-4

Ben Arfa, A., Combes, S., Preziosi-Belloy, L., Gontard, N., & Chalier, P. (2006). Antimicrobial activity of carvacrol related to its chemical structure. *Letters in Applied Microbiology, 43*(2), 149–154. https://doi.org/10.1111/J.1472-765X.2006.01938.X

Bermúdez-Oria, A., Rodríguez-Gutiérrez, G., Rubio-Senent, F., Fernández-Prior, Á., & Fernández-Bolaños, J. (2019). Effect of edible pectin-fish gelatin films containing the olive antioxidants hydroxytyrosol and 3,4-dihydroxyphenylglycol on beef meat during refrigerated storage. *Meat Science, 148*, 213–218. https://doi.org/10.1016/j.meatsci.2018.07.003

Bharadvaja, N., Gautam, S., & Singh, H. (2023). Natural polyphenols: A promising bioactive compounds for skin care and cosmetics. *Molecular Biology Reports, 50*(2), 1817–1828. https://doi.org/10.1007/s11033-022-08156-9

Bhardwaj, K., Najda, A., Sharma, R., Nurzyńska-Wierdak, R., Dhanjal, D. S., Sharma, R., Manickam, S., Kabra, A., Kuča, K., & Bhardwaj, P. (2022). Fruit and vegetable peel-enriched functional foods: Potential avenues and health perspectives. *Evidence-Based Complementary and Alternative Medicine, 2022*, Article 8543881. https://doi.org/10.1155/2022/8543881

Bi, Y., Chi, X., Zhang, R., Lu, Y., Wang, Z., Dong, Q., Ding, C., Yang, R., & Jiang, L. (2020). Highly efficient extraction of mulberry anthocyanins in deep eutectic solvents: Insights of degradation kinetics and stability evaluation. *Innovative Food Science and Emerging Technologies, 66*, Article 102512. https://doi.org/10.1016/j.ifset.2020.102512

Bilge, G., Yurdakul, M., Buzrul, S., & Bulut, O. (2022). Evaluation of the effect of pulsed electric field on coffee Arabica beans. *Food and Bioprocess Technology, 15*(5), 1073–1081. https://doi.org/10.1007/s11947-022-02802-7

Bjørklund, G., Shanaida, M., Lysiuk, R., Butnariu, M., Peana, M., Sarac, I., Strus, O., Smetanina, K., & Chirumbolo, S. (2022). Natural compounds and products from an anti-aging perspective. *Molecules, 27*(20), 1–29.

Bose, B., Choudhury, H., Tandon, P., & Kumaria, S. (2017). Studies on secondary metabolite profiling, anti-inflammatory potential, in vitro photoprotective and skin-aging related enzyme inhibitory activities of *Malaxis acuminata*, a threatened orchid of nutraceutical importance. *Journal of Photochemistry and Photobiology B: Biology, 173*, 686–695. https://doi.org/10.1016/j.jphotobiol.2017.07.010

Bosiljkov, T., Dujmić, F., Cvjetko Bubalo, M., Hribar, J., Vidrih, R., Brnčić, M., Zlatic, E., Radojčić Redovniković, I., & Jokić, S. (2017). Natural deep eutectic solvents and ultrasound-assisted extraction: Green approaches for extraction of wine lees anthocyanins. *Food and Bioproducts Processing, 102*, 195–203. https://doi.org/10.1016/j.fbp.2016.12.005

Bujak, T., Zagórska-Dziok, M., Ziemlewska, A., Nizioł-łukaszewska, Z., Lal, K., Wasilewski, T., & Hordyjewicz-Baran, Z. (2022). Flower extracts as multifunctional dyes in the cosmetics industry. *Molecules, 27*(3), 922. https://doi.org/10.3390/molecules27030922

Cefali, L. C., Ataide, J. A., Sousa, I. M., de, O., Figueiredo, M. C., Ruiz, A. L. T. G., Foglio, M. A., & Mazzola, P. G. (2020). In vitro solar protection factor, antioxidant activity, and stability of a topical formulation containing Benitaka grape (*Vitis vinifera* L.) peel extract. *Natural Product Research, 34*(18), 2677–2682. https://doi.org/10.1080/14786419.2018.1550758

Cemazar, M., Sersa, G., Frey, W., Miklavcic, D., & Teissié, J. (2018). Recommendations and requirements for reporting on applications of electric pulse delivery for electroporation of biological samples. *Bioelectrochemistry, 122*, 69–76. https://doi.org/10.1016/J.BIOELECHEM.2018.03.005

Chávez-Santoscoy, R. A., Gutiérrez-Uribe, J. A., Serna-Saldivar, S. O., & Perez-Carrillo, E. (2016). Production of maize tortillas and cookies from nixtamalized flour enriched with anthocyanins, flavonoids and saponins extracted from black bean (*Phaseolus vulgaris*) seed coats. *Food Chemistry, 192*, 90–97. https://doi.org/10.1016/j.foodchem.2015.06.113

Chibane, L. C., Degraeve, P., Ferhout, H., Bouajila, J., & Oulahal, N. (2019). Plant antimicrobial polyphenols as potential natural food preservatives. *Journal of the Science of Food and Agriculture, 99*(4), 1457–1474. https://doi.org/10.1002/jsfa.9357

Ciriminna, R., Carnaroglio, D., Delisi, R., Arvati, S., Tamburino, A., & Pagliaro, M. (2016). Industrial feasibility of natural products extraction with microwave technology. *ChemistrySelect, 1*(3), 549–555. https://doi.org/10.1002/slct.201600075

Cui, Q., Du, R., Liu, M., & Rong, L. (2020). Lignans and their derivatives from plants as antivirals. *Molecules, 25*(1), 183. https://doi.org/10.3390/molecules25010183

Da Porto, C., Natolino, A., & Decorti, D. (2014). Extraction of proanthocyanidins from grape marc by supercritical fluid extraction using CO_2 as solvent and ethanol-water mixture as co-solvent. *Journal of Supercritical Fluids, 87*, 59–64. https://doi.org/10.1016/j.supflu.2013.12.013

da Silva Carvalho, A. G., da Costa Machado, M. T., da Silva, V. M., Sartoratto, A., Rodrigues, R. A. F., & Hubinger, M. D. (2016). Physical properties and morphology of spray dried microparticles containing anthocyanins of jussara (*Euterpe edulis* Martius) extract. *Powder Technology, 294*, 421–428. https://doi.org/10.1016/j.powtec.2016.03.007

Da Silva, L. P., Pereira, E., Prieto, M. A., Simal-Gandara, J., Pires, T. C. S. P., Alves, M. J., Calhelha, R., Barros, L., & Ferreira, I. C. F. R. (2019). *Rubus ulmifolius* Schott as a novel source of food colorant: Extraction optimization of coloring pigments and incorporation in a bakery product. *Molecules, 24*(11), 2181. https://doi.org/10.3390/molecules24112181

Darawsha, A., Trachtenberg, A., Levy, J., & Sharoni, Y. (2021). The protective effect of carotenoids, polyphenols, and estradiol on dermal fibroblasts under oxidative stress. *Antioxidants, 10*(12), 2023. https://doi.org/10.3390/antiox10122023

de Lima Cherubim, D. J. B., Martins, C. V., Oliveira Fariña, L., & da Silva de Lucca, R. A. (2020). Polyphenols as natural antioxidants in cosmetics applications. *Journal of Cosmetic Dermatology, 19*(1), 33–37. https://doi.org/10.1111/jocd.13093

de Oliveira, C. A., & Dario, M. F. (2019). Bioactive cosmetics. In L. M. T. Martínez, O. V. Kharissova, & B. I. Kharisov (Eds.), *Handbook of ecomaterials* (Vol. 5, pp. 1–23). Springer. https://doi.org/10.1007/978-3-319-68255-6_157

Di Meo, F., Lemaur, V., Cornil, J., Lazzaroni, R., Duroux, J. L., Olivier, Y., & Trouillas, P. (2013). Free radical scavenging by natural polyphenols: Atom versus electron transfer. *Journal of Physical Chemistry A, 117*(10), 2082–2092. https://doi.org/10.1021/jp3116319

Dias, M., Romaní-Pérez, M., Romaní, A., de la Cruz, A., Pastrana, L., Fuciños, P., & Amado, I. R. (2022). Recent technological advances in phenolic compounds recovery and applications: Source of nutraceuticals for the management of diabetes. *Applied Sciences, 12*(18), 9271. https://doi.org/10.3390/app12189271

Dienaitė, L., Pukalskienė, M., Pukalskas, A., Pereira, C. V., Matias, A. A., & Venskutonis, P. R. (2019). Isolation of strong antioxidants from *Paeonia officinalis* roots and leaves and evaluation of their bioactivities. *Antioxidants, 8*(8), 249. https://doi.org/10.3390/antiox8080249

Domínguez, R., Pateiro, M., Gagaoua, M., Barba, F. J., Zhang, W., & Lorenzo, J. M. (2019). A comprehensive review on lipid oxidation in meat and meat products. *Antioxidants, 8*, 429. https://doi.org/10.3390/antiox8100429

Dong, R., Yu, Q., Liao, W., Liu, S., He, Z., Hu, X., Chen, Y., Xie, J., Nie, S., & Xie, M. (2021). Composition of bound polyphenols from carrot dietary fiber and its in vivo and in vitro antioxidant activity. *Food Chemistry, 339*, 127879. https://doi.org/10.1016/j.foodchem.2020.127879

Dou, J. F., Wu, C. E., Fan, G. J., Li, T. T., Li, X. J., Zhou, D. D., Zhu, J. P., & Li, C. M. (2023). Insights into the pigment and non-pigment phenolic profile of polyphenol extracts of jujube peel and their antioxidant and lipid-lowering activities. *Food Bioscience, 52*, 102493. https://doi.org/10.1016/j.fbio.2023.102493

Echegaray, N., Guzel, N., Kumar, M., Guzel, M., Hassoun, A., & Lorenzo, J. M. (2023). Recent advancements in natural colorants and their application as coloring in food and in intelligent food packaging. *Food Chemistry, 404*, 134453. https://doi.org/10.1016/j.foodchem.2022.134453

Echegaray, N., Pateiro, M., Munekata, P. E. S., Lorenzo, J. M., Chabani, Z., Farag, M. A., & Domínguez, R. (2021). Measurement of antioxidant capacity of meat and meat products: Methods and applications. *Molecules, 26*, 3880. https://doi.org/10.3390/molecules26133880

Ferreira, M. S., Magalhães, M. C., Oliveira, R., Sousa-Lobo, J. M., & Almeida, I. F. (2021). Trends in the use of botanicals in anti-aging cosmetics. *Molecules, 26*(12), 1–18.

Franco, D., Munekata, P. E. S., Agregán, R., Bermúdez, R., López-Pedrouso, M., Pateiro, M., & Lorenzo, J. M. (2020). Application of pulsed electric fields for obtaining antioxidant extracts from fish residues. *Antioxidants, 9*(2), 90. https://doi.org/10.3390/antiox9020090

Gadkari, P. V., Balarman, M., & Kadimi, U. S. (2015). Polyphenols from fresh frozen tea leaves (*Camellia assamica* L.) by supercritical carbon dioxide extraction with ethanol entrainer - Application of response surface methodology. *Journal of Food Science and Technology, 52*(2), 720–730. https://doi.org/10.1007/s13197-013-1085-9

Gençdağ, E., Özdemir, E. E., Demirci, K., Görgüç, A., & Yılmaz, F. M. (2022). Copigmentation and stabilization of anthocyanins using organic molecules and encapsulation techniques. *Current Plant Biology, 29*, 100238. https://doi.org/10.1016/j.cpb.2022.100238

Ghosh, S., Sarkar, T., Das, A., & Chakraborty, R. (2022). Natural colorants from plant pigments and their encapsulation: An emerging window for the food industry. *Lebensmittel-Wissenschaft & Technologie, 153*, 112527. https://doi.org/10.1016/J.LWT.2021.112527

Giampieri, F., Alvarez-Suarez, J. M., Mazzoni, L., Forbes-Hernandez, T. Y., Gasparrini, M., Gonzàlez-Paramàs, A. M., Santos-Buelga, C., Quiles, J. L., Bompadre, S., Mezzetti, B., & Battino, M. (2014). Polyphenol-rich strawberry extract protects human dermal fibroblasts against hydrogen peroxide oxidative damage and improves mitochondrial functionality. *Molecules, 19*(6), 7798–7816. https://doi.org/10.3390/molecules19067798

Giancaterino, M., Fauster, T., Krottenthaler, A., & Jäger, H. (2023). Effect of pulsed electric field pre-treatment on the debittering process of cherry kernels. *Innovative Food Science and Emerging Technologies, 83*, 103234. https://doi.org/10.1016/j.ifset.2022.103234

González, M., Barrios, S., Budelli, E., Pérez, N., Lema, P., & Heinzen, H. (2020). Ultrasound assisted extraction of bioactive compounds in fresh and freeze-dried *Vitis vinifera* cv *Tannat* grape pomace. *Food and Bioproducts Processing, 124*, 378–386. https://doi.org/10.1016/j.fbp.2020.09.012

Gupta, N. K., & Dixit, V. K. (2011). Development and evaluation of vesicular system for curcumin delivery. *Archives of Dermatological Research, 303*(2), 89–101. https://doi.org/10.1007/s00403-010-1096-6

Han, X., Shen, T., & Lou, H. (2007). Dietary polyphenols and their biological significance. *International Journal of Molecular Sciences, 8*, 950–988.

He, B., Ge, J., Yue, P., Yue, X. Y., Fu, R., Liang, J., & Gao, X. (2017). Loading of anthocyanins on chitosan nanoparticles influences anthocyanin degradation in gastrointestinal fluids and stability in a beverage. *Food Chemistry, 221*, 1671–1677. https://doi.org/10.1016/j.foodchem.2016.10.120

Hu, Y., Yan, B., Chen, Z. S., Wang, L., Tang, W., & Huang, C. (2022). Recent technologies for the extraction and separation of polyphenols in different plants: A review. *Journal of Renewable Materials, 10*(6), 1471–1490. https://doi.org/10.32604/jrm.2022.018811

Huang, X., Ji, Y., Guo, L., Xu, Q., Jin, L., Fu, Y., & Wang, Y. (2022). Incorporating tannin onto regenerated cellulose film towards sustainable active packaging. *Industrial Crops and Products, 180*, Article 114710. https://doi.org/10.1016/j.indcrop.2022.114710

Joseph, S. V., Edirisinghe, I., & Burton-Freeman, B. M. (2016). Fruit polyphenols: A review of anti-inflammatory effects in humans. *Critical Reviews in Food Science and Nutrition, 56*(3), 419–444. https://doi.org/10.1080/10408398.2013.767221

Jovanović, A. A., Vajić, U. J. V., Mijin, D. Z., Zdunić, G. M., Šavikin, K. P., Branković, S., Kitić, D., & Bugarski, B. M. (2022). Polyphenol extraction in microwave reactor using by-product of *Thymus serpyllum* L. and biological potential of the extract. *Journal of Applied Research on Medicinal and Aromatic Plants, 31*, 100417. https://doi.org/10.1016/j.jarmap.2022.100417

Kabir, M. R., Hasan, M. M., Islam, M. R., Haque, A. R., & Hasan, S. M. K. (2021). Formulation of yogurt with banana peel extracts to enhance storability and bioactive properties. *Journal of Food Processing and Preservation, 45*(3), 1–10. https://doi.org/10.1111/jfpp.15191

Karanth, S., Feng, S., Patra, D., & Pradhan, A. K. (2023). Linking microbial contamination to food spoilage and food waste: The role of smart packaging, spoilage risk assessments, and date labeling. *Frontiers in Microbiology, 14*, 1198124. https://doi.org/10.3389/fmicb.2023.1198124

Khan, M. K., Paniwnyk, L., & Hassan, S. (2019). Polyphenols as natural antioxidants: Sources, extraction and applications in food, cosmetics and drugs. In Y. Li, & F. Chemat (Eds.), *Plant based "green chemistry 2.0". Moving from evolutionary to revolutionary* (pp. 197–235). Singapore: Springer. https://doi.org/10.1007/978-981-13-3810-6

Khanbabaee, K., & van Ree, T. (2001). Tannins: Classification and definition. *Natural Product Reports, 18*(6), 641–649. https://doi.org/10.1039/b1010611

Kolodziej, H. (2017). Beneficial vascular responses to proanthocyanidins: Critical assessment of plant-based test materials and insight into the signaling pathways. In K. Yoshida, V. Cheynier, & S. Quideau (Eds.), *Recent advances in polyphenol research* (pp. 226–258). John Wiley & Sons. https://doi.org/10.1002/9781118883303.ch11

Koubaa, M., Roselló-Soto, E., Šic Žlabur, J., Režek Jambrak, A., Brnčić, M., Grimi, N., Boussetta, N., & Barba, F. J. (2015). Current and new insights in the sustainable and green recovery of nutritionally valuable compounds from *Stevia rebaudiana* Bertoni. *Journal of Agricultural and Food Chemistry, 63*(31), 6835–6846. https://doi.org/10.1021/acs.jafc.5b01994

Kumar, H., Choudhary, N., Kumar, N., & Seth, R. (2014). Phenolic compounds and their health benefits : A review. *Journal of Food Research and Technology, 2*(2), 46–59.

Kumar, N., & Goel, N. (2019). Phenolic acids: Natural versatile molecules with promising therapeutic applications. *Biotechnology Reports, 24*, e00370. https://doi.org/10.1016/j.btre.2019.e00370

Kwon, O. S., Han, J. H., Yoo, H. G., Chung, J. H., Cho, K. H., Eun, H. C., & Kim, K. H. (2007). Human hair growth enhancement in vitro by green tea epigallocatechin-3-gallate (EGCG). *Phytomedicine, 14*, 551–555. https://doi.org/10.1016/j.phymed.2006.09.009

López-Linares, J. C., Campillo, V., Coca, M., Lucas, S., & García-Cubero, M. T. (2021). Microwave-assisted deep eutectic solvent extraction of phenolic compounds from brewer's spent grain. *Journal of Chemical Technology and Biotechnology, 96*(2), 481–490. https://doi.org/10.1002/jctb.6565

Lakka, A., Bozinou, E., Stavropoulos, G., Samanidis, I., Athanasiadis, V., Dourtoglou, V. G., Makris, D. P., & Lalas, S. I. (2021). Enhancement of polyphenols recovery from rosa canina, calendula officinalis and castanea sativa using pulsed electric field. *Beverages, 7*(3), 63. https://doi.org/10.3390/beverages7030063

Leri, M., Scuto, M., Ontario, M. L., Calabrese, V., Calabrese, E. J., Bucciantini, M., & Stefani, M. (2020). Healthy effects of plant polyphenols : Molecular mechanisms. *International Journal of Molecular Sciences, 21*, 1250. https://doi.org/10.3390/ijms21041250

Letsiou, S., Kapazoglou, A., & Tsaftaris, A. (2020). Transcriptional and epigenetic effects of *Vitis vinifera* L. leaf extract on UV-stressed human dermal fibroblasts. *Molecular Biology Reports, 47*(8), 5763–5772. https://doi.org/10.1007/s11033-020-05645-7

Li, Y., Wu, K., Wang, B., & Li, X. (2021). Colorimetric indicator based on purple tomato anthocyanins and chitosan for application in intelligent packaging. *International Journal of Biological Macromolecules, 174*, 370–376. https://doi.org/10.1016/j.ijbiomac.2021.01.182

Li, W., Zhang, X., He, Z., Chen, Y., Li, Z., Meng, T., Li, Y., & Cao, Y. (2020). In vitro and in vivo antioxidant activity of eucalyptus leaf polyphenols extract and its effect on chicken meat quality and cecum microbiota. *Food Research International, 136*, 109302. https://doi.org/10.1016/j.foodres.2020.109302

Lima, E. M. F., Madalão, M. C. M., dos Santos, W. C., Bernardes, P. C., Saraiva, S. H., & Silva, P. I. (2019). Spray-dried microcapsules of anthocyanin-rich extracts from *Euterpe edulis* M. as an alternative for maintaining color and bioactive compounds in dairy beverages. *Journal of Food Science and Technology, 56*(9), 4147–4157. https://doi.org/10.1007/s13197-019-03885-5

Liu, L., Wu, W., Zheng, L., Yu, J., Sun, P., & Shao, P. (2022). Intelligent packaging films incorporated with anthocyanins-loaded ovalbumin-carboxymethyl cellulose nanocomplexes for food freshness monitoring. *Food Chemistry, 387*, 132908. https://doi.org/10.1016/j.foodchem.2022.132908

Loypimai, P., Moongngarm, A., & Naksawat, S. (2017). Application of natural colorant from black rice bran for fermented Thai pork sausage - Sai Krok Isan. *International Food Research Journal, 24*(4), 1529–1537.

Lučić, D., Pavlović, I., Brkljačić, L., Bogdanović, S., Farkaš, V., Cedilak, A., Nanić, L., Rubelj, I., & Salopek-Sondi, B. (2023). Antioxidant and antiproliferative activities of kale (*Brassica oleracea* L. Var. *acephala* DC.) and wild cabbage (*Brassica incana* Ten.) polyphenolic extracts. *Molecules, 28*(4), 1840. https://doi.org/10.3390/molecules28041840

Lucas-González, R., Pérez-álvarez, J.Á., Viuda-Martos, M., & Fernández-López, J. (2021). Pork liver pâté enriched with persimmon coproducts: Effect of in vitro gastrointestinal digestion on its fatty acid and polyphenol profile stability. *Nutrients, 13*(4), 1332. https://doi.org/10.3390/nu13041332

Ma, C., Laaksonen, A., Liu, C., Lu, X., & Ji, X. (2018). The peculiar effect of water on ionic liquids and deep eutectic solvents. *Chemical Society Reviews, 47*(23), 8685–8720. https://doi.org/10.1039/c8cs00325d

Maroufi, L. Y., Shahabi, N., Ghanbarzadeh, M., dokht, & Ghorbani, M. (2022). Development of antimicrobial active food packaging film based on gelatin/dialdehyde quince seed gum incorporated with apple peel polyphenols. *Food and Bioprocess Technology, 15*(3), 693–705. https://doi.org/10.1007/s11947-022-02774-8

Martín-García, B., Tylewicz, U., Verardo, V., Pasini, F., Gómez-Caravaca, A. M., Caboni, M. F., & Dalla Rosa, M. (2020). Pulsed electric field (PEF) as pre-treatment to improve the phenolic compounds recovery from brewers' spent grains. *Innovative Food Science and Emerging Technologies, 64*, 110196. https://doi.org/10.1016/j.ifset.2020.102402

Martínez-Arámburu, D., González-Quijano, G. K., Dorantes-Alvarez, L., Aparicio-Ozores, G., & Lopez-Villegas, E. O. (2015). Changes in microstructure of *Salmonella typhimurium* and *Listeria monocytogenes* exposed to hydroxycinnamic salts. *Revista Mexicana de Ingenieria Quimica, 14*(2), 347–354.

Martínez-Olivo, A. O., Durán-Castañeda, A. C., López-Cárdenas, F. G., Rodríguez-Romero, J. de J., Sánchez-Burgos, J. A., Sáyago-Ayerdi, S. G., & Zamora-Gasga, V. M. (2023). Advances in ultrasound-assisted extraction of bioactive compounds (antioxidant compounds) from agrofood waste. In A. Basile, A. Cassano, & C. Conidi (Eds.), *Advanced technologies in wastewater treatment. Food processing industry* (pp. 217–246). Elsevier. https://doi.org/10.1016/B978-0-323-88510-2.00005-1

Martinengo, P., Arunachalam, K., & Shi, C. (2021). Polyphenolic antibacterials for food preservation: Review, challenges, and current applications. *Foods, 10*, 2469.

Medina-Torres, N., Ayora-Talavera, T., Espinosa-Andrews, H., Sánchez-Contreras, A., & Pacheco, N. (2017). Ultrasound assisted extraction for the recovery of phenolic compounds from vegetable sources. *Agronomy, 7*(3), 47. https://doi.org/10.3390/agronomy7030047

Montenegro, L., Parenti, C., Turnaturi, R., & Pasquinucci, L. (2017). Resveratrol-loaded lipid nanocarriers: Correlation between in vitro occlusion factor and in vivo skin hydrating effect. *Pharmaceutics, 9*(4), 58. https://doi.org/10.3390/pharmaceutics9040058

Munekata, P. E. S., Pateiro, M., Bellucci, E. R. B., Domínguez, R., da Silva Barretto, A. C., & Lorenzo, J. M. (2021). Strategies to increase the shelf life of meat and meat products with phenolic compounds. In F. Toldrá (Ed.), *Advances in food and nutrition research* (pp. 171–205). Academic Press. https://doi.org/10.1016/bs.afnr.2021.02.008

Naliyadhara, N., Kumar, A., Girisa, S., Daimary, U. D., Hegde, M., & Kunnumakkara, A. B. (2022). Pulsed electric field (PEF): Avant-garde extraction escalation technology in food industry. *Trends in Food Science and Technology, 122*, 238–255. https://doi.org/10.1016/j.tifs.2022.02.019

Oñate-Gutiérrez, J. A., Díaz-Sánchez, L. M., Urbina, D. L., Pinzón, J. R., Blanco-Tirado, C., & Combariza, M. Y. (2023). Exploring the chemical composition and coloring qualities of cacao fruit epicarp extracts. *RSC Advances, 13*(19), 12712–12722. https://doi.org/10.1039/d3ra01049j

Oak, M. H., Auger, C., Belcastro, E., Park, S. H., Lee, H. H., & Schini-Kerth, V. B. (2018). Potential mechanisms underlying cardiovascular protection by polyphenols: Role of the endothelium. *Free Radical Biology and Medicine, 122*, 161–170. https://doi.org/10.1016/j.freeradbiomed.2018.03.018

Orsat, V., Raghavan, G. S. V., & Krishnaswamy, K. (2017). Microwave technology for food processing: An overview of current and future applications. In M. Regier, K. Knoerzer, & H. Schubert (Eds.), *The microwave processing of foods: Second edition* (pp. 100–116). Woodhead Publishing. https://doi.org/10.1016/B978-0-08-100528-6.00005-X

Ozturk, B., Parkinson, C., & Gonzalez-Miquel, M. (2018). Extraction of polyphenolic antioxidants from orange peel waste using deep eutectic solvents. *Separation and Purification Technology, 206*, 1–13. https://doi.org/10.1016/j.seppur.2018.05.052

Pal, C. B. T., & Jadeja, G. C. (2020). Microwave-assisted extraction for recovery of polyphenolic antioxidants from ripe mango (*Mangifera indica* L.) peel using lactic acid/sodium acetate deep eutectic mixtures. *Food Science and Technology International, 26*(1), 78–92. https://doi.org/10.1177/1082013219870010

Palos-Hernández, A. G., Fernández, M. Y., Escuadra Burrieza, J., Pérez-Iglesias, J. L., & González-Paramás, A. M. (2022). Obtaining green extracts rich in phenolic compounds from underexploited food by-products using natural deep eutectic solvents. Opportunities and challenges. *Sustainable Chemistry and Pharmacy, 29*, 100773. https://doi.org/10.1016/j.scp.2022.100773

Parafati, L., Restuccia, C., Palmeri, R., Fallico, B., & Arena, E. (2020). Characterization of prickly pear peel flour as a bioactive and functional ingredient in bread preparation. *Foods, 9*(9), 1189. https://doi.org/10.3390/foods9091189

Peiró, S., Luengo, E., Segovia, F., Raso, J., & Almajano, M. P. (2019). Improving polyphenol extraction from lemon residues by pulsed electric fields. *Waste and Biomass Valorization, 10*(4), 889–897. https://doi.org/10.1007/s12649-017-0116-6

Pinela, J., Prieto, M. A., Pereira, E., Jabeur, I., Barreiro, M. F., Barros, L., & Ferreira, I. C. F. R. (2019). Optimization of heat- and ultrasound-assisted extraction of anthocyanins from *Hibiscus sabdariffa* calyces for natural food colorants. *Food Chemistry, 275*, 309–321. https://doi.org/10.1016/j.foodchem.2018.09.118

Pinton, M. B., dos Santos, B. A., Lorenzo, J. M., Cichoski, A. J., Boeira, C. P., & Campagnol, P. C. B. (2021). Green technologies as a strategy to reduce NaCl and phosphate in meat products: An overview. *Current Opinion in Food Science, 40*, 1–5. https://doi.org/10.1016/j.cofs.2020.03.011

Pizano-Andrade, J. C., Vargas-Guerrero, B., Gurrola-Díaz, C. M., Vargas-Radillo, J. J., & Ruiz-López, M. A. (2022). Natural products and their mechanisms in potential photoprotection of the skin. *Journal of Biosciences, 47*(4), 77. https://doi.org/10.1007/s12038-022-00314-2

Prabhu, S., Molath, A., Choksi, H., Kumar, S., & Mehra, R. (2021). Classifications of polyphenols and their potential application in human health and diseases. *International Journal of Physiology, Nutrition and Physical Education, 6*(1), 293–301. https://doi.org/10.22271/journalofsport.2021.v6.i1e.2236

Puligundla, P. (2013). Potentials of microwave heating technology for select food processing applications - A brief overview and update. *Journal of Food Processing & Technology, 4*, 11. https://doi.org/10.4172/2157-7110.1000278

Radajewska, A., Szyller, J., Niewiadomska, J., Noszczyk-Nowak, A., & Bil-Lula, I. (2023). Punica granatum L. polyphenolic extract as an antioxidant to prevent kidney injury in metabolic syndrome rats. *Oxidative Medicine and Cellular Longevity, 2023*, 4967. https://doi.org/10.1155/2023/6144967

Radzali, S. A., Markom, M., & Saleh, N. M. (2020). Co-solvent selection for supercritical fluid extraction (SFE) of phenolic compounds from *Labisia pumila*. *Molecules, 25*(24), 5859. https://doi.org/10.3390/molecules25245859

Rajha, H. N., Paule, A., Aragonès, G., Barbosa, M., Caddeo, C., Debs, E., Dinkova, R., Eckert, G. P., Fontana, A., Gebrayel, P., Maroun, R. G., Napolitano, A., Panzella, L., Pasinetti, G. M., Stevens, J. F., Schieber, A., & Edeas, M. (2022). Recent advances in research on polyphenols: Effects on microbiota, metabolism, and health. *Molecular Nutrition & Food Research, 66*(1), 2100670. https://doi.org/10.1002/mnfr.202100670

Ramon-Mascarell, F., Martí-Quijal, F. J., Castagnini, J. M., Phimolsiripol, Y., Ruksiriwanich, W., Rajoka, M. S. R., Mehwish, H. M., & Barba, F. J. (2022). Influence of geographical location of spirulina (*Arthrospira platensis*) on the recovery of bioactive compounds assisted by pulsed electric fields. *Separations, 9*(9), 257. https://doi.org/10.3390/separations9090257

Rathee, P., Sehrawat, R., Rathee, P., Khatkar, A., Akkol, E. K., Khatkar, S., Redhu, N., Türkcanoğlu, G., & Sobarzo-Sánchez, E. (2023). Polyphenols: Natural preservatives with promising applications in food, cosmetics and pharma industries; problems and toxicity associated with synthetic preservatives; impact of misleading advertisements; recent trends in preservation and legislation. *Materials, 16*(13), 4793.

Rathod, N. B., Elabed, N., Punia, S., Ozogul, F., Kim, S.-K., & Rocha, J. M. (2023). Recent developments in polyphenol applications on human health: A review with current knowledge. *Plants, 12*(6), 1217. https://doi.org/10.3390/plants12061217

Razgonova, M. P., Cherevach, E. I., Tekutyeva, L. A., Fedoreyev, S. A., Mishchenko, N. P., Tarbeeva, D. V., Demidova, E. N., Kirilenko, N. S., & Golokhvast, K. (2023). Maackia amurensis Rupr. et Maxim.: Supercritical CO2 extraction and mass spectrometric characterization of chemical constituents. *Molecules, 28*, 2026. https://doi.org/10.3390/molecules28052026

Rodríguez-Daza, M. C., & de Vos, W. M. (2023). Polyphenols as drivers of a homeostatic gut microecology and immuno-metabolic traits of *Akkermansia muciniphila*: From mouse to man. *International Journal of Molecular Sciences, 24*(1), 45. https://doi.org/10.3390/ijms24010045

Roselló-Soto, E., Galanakis, C. M., Brnčić, M., Orlien, V., Trujillo, F. J., Mawson, R., Knoerzer, K., Tiwari, B. K., & Barba, F. J. (2015). Clean recovery of antioxidant compounds from plant foods, by-products and algae assisted by ultrasounds processing. Modeling approaches to optimize processing conditions. *Trends in Food Science and Technology, 42*(2), 134–149. https://doi.org/10.1016/j.tifs.2015.01.002

Roy, S., & Rhim, J. W. (2020). Preparation of antimicrobial and antioxidant gelatin/curcumin composite films for active food packaging application. *Colloids and Surfaces B: Biointerfaces, 188*, Article 110761. https://doi.org/10.1016/j.colsurfb.2019.110761

Ruesgas-Ramón, M., Figueroa-Espinoza, M. C., & Durand, E. (2017). Application of deep eutectic solvents (DES) for phenolic compounds extraction: Overview, challenges, and ppportunities. *Journal of Agricultural and Food Chemistry, 65*(18), 3591–3601. https://doi.org/10.1021/acs.jafc.7b01054

Rusconi, M., & Conti, A. (2010). *Theobroma cacao* L., the food of the gods: A scientific approach beyond myths and claims. *Pharmacological Research, 61*(1), 5–13. https://doi.org/10.1016/j.phrs.2009.08.008

Saad, A. M., Mohamed, A. S., El-Saadony, M. T., & Sitohy, M. Z. (2021). Palatable functional cucumber juices supplemented with polyphenols-rich herbal extracts. *Lebensmittel-Wissenschaft & Technologie, 148*, Article 111668. https://doi.org/10.1016/j.lwt.2021.111668

Santos, L. G., & Martins, V. G. (2023). Optimization of the green extraction of polyphenols from the edible flower *Clitoria ternatea* by high-power ultrasound: A comparative study with conventional extraction techniques. *Journal of Applied Research on Medicinal and Aromatic Plants, 2023*, Article 100458. https://doi.org/10.1016/j.jarmap.2023.100458

Sayas-Barberá, E., Mar, A., Cherif, S., Ben-abda, J., & Pérez-Álvarez, J.Á. (2020). Effect of Date (Phoenix dactylifera L.) pits on the shelf life of beef burgers. *Foods, 9*, Article 102. https://doi.org/10.3390/foods9010102

Scalbert, A., & Williamson, G. (2000). Dietary intake and bioavailability of polyphenols. *Journal of Nutrition, 130*(8), 2073S–2085S. https://doi.org/10.1093/jn/130.8.2073s

Scarano, A., Laddomada, B., Blando, F., De Santis, S., Verna, G., Chieppa, M., & Santino, A. (2023). The chelating ability of plant polyphenols can affect iron homeostasis and gut microbiota. *Antioxidants, 12*(3), Article 630. https://doi.org/10.3390/antiox12030630

Shtay, R., Keppler, J. K., Schrader, K., & Schwarz, K. (2019). Encapsulation of (—)-epigallocatechin-3-gallate (EGCG) in solid lipid nanoparticles for food applications. *Journal of Food Engineering, 244*, 91–100. https://doi.org/10.1016/j.jfoodeng.2018.09.008

Silva, F. A., Queiroga, R.de C. R. do E., de Souza, E. L., Voss, G. B., Borges, G. da S. C., Lima, M.dos S., Pintado, M. M. E., & Vasconcelos, M. A. da S. (2022). Incorporation of phenolic-rich ingredients from integral valorization of Isabel grape improves the nutritional, functional and sensory characteristics of probiotic goat milk yogurt. *Food Chemistry, 369*, Article 130957. https://doi.org/10.1016/j.foodchem.2021.130957

Singla, R. K., Dubey, A. K., Garg, A., Sharma, R. K., Fiorino, M., Ameen, S. M., Haddad, M. A., & Al-Hiary, M. (2019). Natural polyphenols: Chemical classification, definition of classes, subcategories, and structures. *Journal of AOAC International, 102* (5), 1397–1400. https://doi.org/10.5740/jaoacint.19-0133

Sittek, L.-M., Schmidts, T. M., & Schlupp, P. (2021). Polyphenol-rich olive mill wastewater extract and its potential use in hair care products. *Journal of Cosmetics, Dermatological Sciences and Applications, 11*(04), 356–370. https://doi.org/10.4236/jcdsa.2021.114029

Soquetta, M. B., Terra, L.de M., & Bastos, C. P. (2018). Green technologies for the extraction of bioactive compounds in fruits and vegetables. *CyTA – Journal of Food, 16*(1), 400–412. https://doi.org/10.1080/19476337.2017.1411978

Sun, X., Wang, J., Dong, M., Zhang, H., Li, L., & Wang, L. (2022). Food spoilage, bioactive food fresh-keeping films and functional edible coatings: Research status, existing problems and development trend. *Trends in Food Science and Technology, 119*, 122–132. https://doi.org/10.1016/j.tifs.2021.12.004

Sun, C., Zhao, C., Guven, E. C., Paoli, P., Simal-Gandara, J., Ramkumar, K. M., Wang, S., Buleu, F., Pah, A., Turi, V., Damian, G., Dragan, S., Tomas, M., Khan, W., Wang, M., Delmas, D., Portillo, M. P., Dar, P., Chen, L., & Xiao, J. (2020). Dietary polyphenols as antidiabetic agents: Advances and opportunities. *Food Frontiers, 1*(1), 18–44. https://doi.org/10.1002/fft2.15

Tai, H. P., Hong, C. T. T., Huu, T. N., & Thi, T. N. (2022). Extraction of custard apple (*Annona squamosal* L.) peel with supercritical CO_2 and ethanol as co-solvent. *Journal of Food Processing and Preservation, 46*(11), 1–10. https://doi.org/10.1111/jfpp.17040

Tapia-Rodriguez, M. R., Hernandez-Mendoza, A., Gonzalez-Aguilar, G. A., Martinez-Tellez, M. A., Martins, C. M., & Ayala-Zavala, J. F. (2017). Carvacrol as potential quorum sensing inhibitor of *Pseudomonas aeruginosa* and biofilm production on stainless steel surfaces. *Food Control, 75*, 255–261. https://doi.org/10.1016/j.foodcont.2016.12.014

Teh, S. S., & Birch, E. J. (2014). Effect of ultrasonic treatment on the polyphenol content and antioxidant capacity of extract from defatted hemp, flax and canola seed cakes. *Ultrasonics Sonochemistry, 21*(1), 346–353. https://doi.org/10.1016/j.ultsonch.2013.08.002

Tian, B., & Liu, J. (2020). Resveratrol: A review of plant sources, synthesis, stability, modification and food application. *Journal of the Science of Food and Agriculture, 100*(4), 1392–1404. https://doi.org/10.1002/jsfa.10152

Tran, T. M. K., Akanbi, T., Kirkman, T., Nguyen, M. H., & Vuong, Q. Van (2022). Maximising recovery of bioactive compounds from coffee pulp waste using microwave-assisted extraction. *European Journal of Engineering and Technology Research, 7*(3), 1–6. https://doi.org/10.24018/ejeng.2022.7.3.2812

Treutter, D. (2005). Significance of flavonoids in plant resistance and enhancement of their biosynthesis. *Plant Biology, 7*(6), 581–591. https://doi.org/10.1055/s-2005-873009

Umaraw, P., Munekata, P. E. S., Verma, A. K., Barba, F. J., Singh, V. P., Kumar, P., & Lorenzo, J. M. (2020). Edible films/coating with tailored properties for active packaging of meat, fish and derived products. *Trends in Food Science and Technology, 98*, 10–24. https://doi.org/10.1016/j.tifs.2020.01.032

Vélez-Erazo, E. M., Pasquel-Reátegui, J. L., Dorronsoro-Guerrero, O. H., & Martínez-Correa, H. A. (2021). Phenolics and carotenoids recovery from agroindustrial mango waste using microwave-assisted extraction: Extraction and modeling. *Journal of Food Process Engineering, 44*(9), e13774. https://doi.org/10.1111/jfpe.13774

Vinitha, U. G., Sathasivam, R., Muthuraman, M. S., & Park, S. U. (2022). Intensification of supercritical fluid in the extraction of flavonoids: A comprehensive review. *Physiological and Molecular Plant Pathology, 118*, Article 101815. https://doi.org/10.1016/j.pmpp.2022.101815

Wang, R., Hu, X., Agyekumwaa, A. K., Li, X., Xiao, X., & Yu, Y. (2021). Synergistic effect of kojic acid and tea polyphenols on bacterial inhibition and quality maintenance of refrigerated sea bass (*Lateolabrax japonicus*) fillets. *Lebensmittel-Wissenschaft & Technologie, 137*, Article 110452. https://doi.org/10.1016/j.lwt.2020.110452

Wang, M., Zhou, J., Castagnini, J. M., Berrada, H., & Barba, F. J. (2023). Pulsed electric field (PEF) recovery of biomolecules from *Chlorella*: Extract efficiency, nutrient relative value, and algae morphology analysis. *Food Chemistry, 404*, Article 134615. https://doi.org/10.1016/j.foodchem.2022.134615

Weng, M., Li, Y., Wu, L., Zheng, H., Lai, P., Tang, B., & Luo, X. (2021). Effects of passion fruit peel flour as a dietary fibre resource on biscuit quality. *Food Science and Technology, 41*(1), 65–73. https://doi.org/10.1590/fst.33419

Wijeratne, S. S. K., Abou-Zaid, M. M., & Shahidi, F. (2006). Antioxidant polyphenols in almond and its coproducts. *Journal of Agricultural and Food Chemistry, 54*(2), 312–318. https://doi.org/10.1021/jf051692j

Wu, Y., Bai, J., Zhong, K., Huang, Y., Qi, H., Jiang, Y., & Gao, H. (2016). Antibacterial activity and membrane-disruptive mechanism of 3-p-trans-coumaroyl-2-hydroxyquinic acid, a novel phenolic compound from pine needles of *Cedrus deodara*, against *Staphylococcus aureus*. *Molecules, 21*(8), 1084. https://doi.org/10.3390/molecules21081084

Wu, Q., & Zhou, J. (2021). The application of polyphenols in food preservation. In C. Gallegos-montes, & M.-V. Ruiz-Méndez (Eds.), *Advances in food and nutrition research* (pp. 35–99). Elsevier Inc.

Xia, B., Yan, D., Bai, Y., Xie, J., Cao, Y., Liao, D., & Lin, L. (2015). Determination of phenolic acids in *Prunella vulgaris* L.: A safe and green extraction method using alcohol-based deep eutectic solvents. *Analytical Methods, 7*(21), 9354–9364. https://doi.org/10.1039/C5AY02035B

Xu, B., Feng, M., Tiliwa, E. S., Yan, W., Wei, B., Zhou, C., Ma, H., Wang, B., & Chang, L. (2022). Multi-frequency power ultrasound green extraction of polyphenols from *Pingyin rose*: Optimization using the response surface methodology and exploration of the underlying mechanism. *Lebensmittel-Wissenschaft & Technologie, 156*, 113037. https://doi.org/10.1016/j.lwt.2021.113037

Yaghoubi, A., Ghojazadeh, M., Abolhasani, S., Alikhah, H., & Khaki-Khatibi, F. (2019). Anti-aging and sunscreens: Paradigm shift in cosmetics. *Avanced Pharmaceutical Bulletin, 9*(3), 348–359. https://doi.org/10.15171/jcvtr.2015.24

Yahfoufi, N., Alsadi, N., Jambi, M., & Matar, C. (2018). The immunomodulatory and anti-inflammatory role of polyphenols. *Nutrients, 10*(11), 1–23. https://doi.org/10.3390/nu10111618

Yi, J., Li, S., Wang, C., Cao, N., Qu, H., Cheng, C., Wang, Z., Wang, L., & Zhou, L. (2019). Potential applications of polyphenols on main ncRNAs regulations as novel therapeutic strategy for cancer. *Biomedicine & Pharmacotherapy, 113*, Article 108703. https://doi.org/10.1016/j.biopha.2019.108703

Yildiz, E., Sumnu, G., & Kahyaoglu, L. N. (2021). Monitoring freshness of chicken breast by using natural halochromic curcumin loaded chitosan/PEO nanofibers as an intelligent package. *International Journal of Biological Macromolecules, 170*, 437–446. https://doi.org/10.1016/j.ijbiomac.2020.12.160

Zeb, A. (2020). Concept, mechanism, and applications of phenolic antioxidants in foods. *Journal of Food Biochemistry, 44*(9), e13394. https://doi.org/10.1111/jfbc.13394

Zhang, Q., De Oliveira Vigier, K., Royer, S., & Jérôme, F. (2012). Deep eutectic solvents: Syntheses, properties and applications. *Chemical Society Reviews, 41*(21), 7108–7146. https://doi.org/10.1039/c2cs35178a

Zhang, X., Zeng, Q., Liu, Y., & Cai, Z. (2021). Enhancing the resistance of anthocyanins to environmental stress by constructing ovalbumin-propylene glycol alginate nanocarriers with novel configurations. *Food Hydrocolloids, 118*, Article 106668. https://doi.org/10.1016/j.foodhyd.2021.106668

Zheng, Y., Liu, S., Xie, J., Chen, Y., Dong, R., Zhang, X., Liu, S., Xie, J., Hu, X., & Yu, Q. (2020). Antioxidant, α-amylase and α-glucosidase inhibitory activities of bound polyphenols extracted from mung bean skin dietary fiber. *Lebensmittel-Wissenschaft & Technologie, 132*, Article 109943. https://doi.org/10.1016/j.lwt.2020.109943

Ziółkiewicz, A., Kasprzak-Drozd, K., Rusinek, R., Markut-Miotła, E., & Oniszczuk, A. (2023). The influence of polyphenols on atherosclerosis development. *International Journal of Molecular Sciences, 24*(8), 7146. https://doi.org/10.3390/ijms24087146

Zin, M. M., Anucha, C. B., & Bánvölgyi, S. (2020). Recovery of phytochemicals via electromagnetic irradiation (microwave-assisted-extraction): Betalain and phenolic compounds in perspective. *Foods, 9*, Article 918. https://doi.org/10.3390/FOODS9070918

SECTION 1

Extraction and purification processes

CHAPTER

Natural deep eutectic solvent for the extraction of plant polyphenols 2

Alena Stupar[1], Nemanja Teslić[1], Aleksandra Mišan[1], Milica Pojić[1], Aleksandra Cvetanović Kljakić[2] and Branimir Pavlić[2]

[1]*University of Novi Sad, Institute of Food Technology, Novi Sad, Serbia;* [2]*University of Novi Sad, Faculty of Technology, Novi Sad, Serbia*

1. Introduction

1.1 Phenolic compounds: importance of phenolic compounds in nutrition, biological activity

Polyphenols are considered one of the largest groups of secondary metabolites in the plant kingdom, as more than 8000 polyphenolic compounds have been identified (Rocchetti et al., 2022). Based on their chemical structure, they are divided into several categories: phenolic acids, stilbenes, lignans, and flavonoids (Figueroa-Robles et al., 2021). Despite the large variety of structures, the common characteristic of all compounds of this class is the presence of at least one aromatic ring with one or more hydroxyl groups.

Phenolic acids, divided into hydroxybenzoic and hydroxycinnamic acids, belong to the major class of phenolic compounds in plants and are present in free and bound forms. For example, is widely distributed in plants and occurs caffeic acid in the form of amides, glycosides, esters, and sugar esters presents approximately 75–100% of total hydroxycinnamic acids; thus, it is considered a major phenolic acid found in plants (Aijaz et al., 2023).

Stilbenes and their hydroxylated derivatives, stilbenoids, are naturally found in a wide variety of dietary sources, but the main dietary origins are wines, peanuts, grapes, and peanut products. There are several well-known stilbenes that have attracted scientific attention in recent years, including resveratrol, pterostilbene, 3′-hydroxypterostilbene and ε-viniferin (Beaumont et al., 2022).

Lignans are derived from the oxidative dimerization of 2 phenylpropane units. Probably the richest source of lignans is linseed, while they are present in vegetables, fruits, nuts, oilseeds, garlic, olive oil, wine, tea, beer, and coffee (Aijaz et al., 2023).

Flavonoids are currently the most diverse group of phenolic compounds, responsible, among other biological functions, for the flower pigmentation in plants. They are divided into several groups: anthocyanins, flavones, flavonols, flavanones, isoflavones, flavanonols, and other subclasses (Rocchetti et al., 2022).

Plant Polyphenols in the Food and Cosmetic Industries. https://doi.org/10.1016/B978-0-443-19137-4.01002-4

A number of studies have shown that stress (i.e., attack by pathogens or herbivores, UV light) triggers the additional biosynthesis of phenolic compounds, suggesting that these compounds play an important protective role in plants. Accordingly, some phenolic compounds may be classified as (Rocchetti et al., 2022).

At the same time, a number of published papers suggest that people can benefit from plant phenolics obtained either by the diet or through topical application. In addition to their potential health-promoting effects following dietary intake, some phenolic compounds may alleviate symptoms associated with skin ageing, acne, inflammation, and other dermatological conditions. Due to their natural origin and generally favourable safety profile, phenolic compounds are considered a promising tool in dealing with both minor skin problems (e.g., wrinkles and acne) and serious, potentially life-threatening diseases such as cancer (Zhang et al., 2022).

As an example, the biological activity of ε-viniferin, which serves as the basis for its use in the development of added-value food, is listed in Table 2.1. Referring to Rocchetti et al. (2022), ε-viniferin presents the most significant representative of resveratrol dimers as it attracts great scientific interest due to its biological activity, which has recently been described in more than 100 scientific publications.

Not only ε-viniferin, but also the other polyphenolic compounds possess antioxidant, antiinflammatory, antitumor, antimicrobial properties, etc., which makes them perfect candidates for the formulation of nutraceuticals and cosmeceuticals (Lavoie et al., 2022). Another example is caffeic acid, which was recently demonstrated by in vivo experiments to possess cardioprotective and antioxidant activities as it influences systolic blood pressure, heart rates, angiotensin-1-converting enzyme (ACE), acetylcholinesterase (AChE), butrylcholinesterase (BChE), and arginase in cyclosporine (Agunloye et al., 2019).

In general, the biological activity of phenolic compounds is tested by in vitro and in vivo experiments. In vitro studies are more frequently used and often clearly demonstrate that phenolic compounds and plant extracts rich in phenolic compounds possess various health-promoting effects. However, when in vivo experiments are used to assess the actual potential of these compounds and extracts, the results are not always equally convincing, because they are greatly dependent on the bioaccessibility and bioavailability of the tested compounds in the digestive tract and circulatory system. Unfortunately, the mechanisms of action and pharmacokinetic parameters underlying phenolic compound uptake remain insufficiently understood (Shahidi & Peng, 2018).

Apart from the problems connected with their low absorption and bioavailability in the gastrointestinal tract, there is a problem of their quick oxidation, which leads to food browning and formation of disagreeable odors that is accompanied by a loss in activity. In addition, phenolic compounds show a limited solubility in water and have an unpleasant taste, which must be masked before their incorporation into foodstuff or oral medicines (Cosme et al., 2020).

Nowadays, new forms of administration, such as nano encapsulation, prodrugs, and phytosomes, have been developed in an attempt to control and maintain the release of phenolic compounds in the gastrointestinal tract, improve their solubility, and increase their bioavailability (Figueroa-Robles et al., 2021). In addition to that, the application

Table 2.1 Biological activity of ε-viniferin as the basis for the development of nutraceuticals and cosmeceuticals.

Phenolic compound	Group	Activity	Comment	References
ε-Viniferin	Stilbenoid	Antiinflammatory activity	In vivo administration of viniferin decreased inflammatory mediators TNF-α, MIP-2, IL-6, IL-1 β, and HMGB1, neutrophil influx, and severity of endotoxin-mediated acute lung injury.	Kurundkar et al. (2019)
		Antioxidant	In vitro testing effectively protects human cells against stress induced by H_2O_2	Medrano-Padial et al. (2021)
		Antiatherogenic	In vitro testing reduces the rate of atherosclerosis by inducing nitric oxide (NO) production and antioxidant enzyme expression in vascular endothelial cells.	Wu et al. (2020)
		Antiobesity	In vivo, a decrease in the concentrations of fasting blood glucose, total cholesterol, triglycerides, and low density lipoprotein cholesterol, and improved the glucose tolerance of diabetic rats.	Liu et al. (2020)
		Anticancer	In vitro, cytotoxic activity on HepG2 and Hcp3B human hepatoma cell lines.	Aja et al. (2020)
		Neuroprotective	In vitro, protects neuronal dopaminergic cells from 6hydroxydopamine -induced cytotoxicity and apoptosis.	Sergi et al. (2021)

of NADES (natural deep eutectic solvents) can improve solubility, bioavailability (Mišan et al., 2019), and stability of polyphenolic compounds (Mišan & Pojić, 2021).

1.2 Necessity of natural extracts in food/pharma/cosmetic industry

The growing need for natural extracts nowadays arises from the desire for healthier, sustainable, and environmentally friendly ingredients, whose incorporation in everyday products fosters innovation and product development.

Ongoing climate change has created a need for innovation in the use and obtaining high-value-added compounds from byproducts of the food industry, as many of the abundantly available by-products end up as waste. This is not only an environmental demand, but it can also help the food industry gain benefits from previously discarded materials. Furthermore, traditionally used natural resources of plant phenolics such as medicinal and aromatic plants are not fully valorized by the modern food, cosmetic, and pharmaceutical industry, although the production of herbal medicaments, herbal cosmetics, and dietetic products has increased in recent years (Pavlić et al., 2023).

In this context, our group has created a database—PhInd that collects and displays the data on the content of polyphenols in agri-food waste and by-products (http://phind.uns.ac.rs/). Data are collected from 705 scientific publications published from 2015 to date, containing a total of 6454 composition data. The majority of the entries originate from fruit and wine processing (39.51 and 18.26%, respectively), followed by agricultural waste (11.55%), herb (6.42%) and nut processing (5.85%), while other types of industry processing are represented by less than 5% of entries (Teslić et al., 2024a, 2024b, 2025).

Apart from total polyphenol content, information about extraction methods, type of agri-food waste and by-products and its botanical origin, source of agri-food waste and by-products (e.g., branch of industry, industrial process and scale of process), pretreatments applied to assist the extraction (drying, defeating, hydrolysis, application of assisting technologies), extraction solvents and technique used and individual polyphenol compounds are provided, which can help the reader to discover trends, identify correlations, and monitor the research gaps. These data can help users identify trends, correlations, and research gaps. The database may support industry in new product development, process optimization, the exploration of new opportunities, and the establishment of new value chains. It may also help consumers understand the potential benefits of polyphenol-rich foods and the importance of reducing household food waste (Teslić et al., 2024b).

2. Potential of NADES solvents in the food, pharmaceutical, and cosmetic industry

2.1 Relevance to green and sustainable chemistry, comparison to the traditional solvent

In response to consumer demand, the modern cosmetic industry increasingly seeks to use natural components as active ingredients in cosmetic products. These components are used not only in products for basic care but also in cosmeceutical formulations containing highly potent active ingredients and advanced delivery systems. Consumer preferences in the food and pharmaceutical sectors are also shifting towards natural components that may complement or, in some applications, replace synthetic ingredients. Research indicates that the use of natural medicines and foods without artificial colours and additives is increasing, including in highly industrialized countries (Cvetanović et al., 2015). Such natural components are most often obtained by extracting from natural sources, among which plant sources are the most common. Until recently, the process of extracting ingredients from plant sources was carried out using organic solvents, the use of which can have a whole series of unwanted consequences. The majority of used solvents are easily volatile, the most common of which are methanol, chloroform, benzene, toluene, acetone, and dimethylformamide, which, with long-term use, can have an adverse effect on human health and the environment. Although solvent recovery and reuse can reduce solvent consumption and emissions, these operations require additional energy, equipment, and processing steps and may generate waste streams requiring further treatment. Conventional organic solvents may also raise occupational health and safety concerns, as many of them require strict handling procedures and additional protective measures. These requirements increase process complexity and operating costs and may consequently affect the price of the final product (Cvetanović et al., 2018). In addition to their potential environmental impact, conventional organic solvents may raise concerns related to human health and occupational safety (Cvetanović & Yan, 2021). Many solvents used in industrial processes are flammable, explosive, or toxic and therefore require strict handling procedures and additional precautionary measures. These requirements increase the complexity and cost of the technological process and may consequently affect the price of the final product. Human health, occupational safety, energy consumption, and environmental protection should therefore be considered throughout the solvent life cycle, including production, purification, transport, use, recovery, recycling, and final waste management. Nevertheless, conventional organic solvents continue to be widely used because their physicochemical properties are well characterised, extraction procedures are established, and their generally low viscosity facilitates mixing, mass transfer, filtration, and solvent removal. Their volatility, although associated with environmental and safety concerns, can also facilitate solvent recovery and extract concentration by evaporation or distillation.

The designation of NADES as green, safe, biodegradable, or non-toxic should therefore be approached critically. The epithet of safe and green solvent implies biodegradability and absence of toxicity, and in the case of NADES solvents, this is closely related to the type and proportion of constituents. In general, NADES solvents are composed of natural components whose biodegradability is high. Previous investigations into the cytotoxicity of NADES solvents involved testing the cytotoxicity of the components from which these solvents are composed. One of the most common ingredients of NADES solvent is choline chloride (ChCl), which is a natural component of the cell membrane, but for the production of NADES, it is obtained chemically in a reaction that does not create residues, which makes the E-factor in the process of obtaining it to be 0, and is considered a sustainable and clean process. Although the components that make up NADES are mostly nontoxic, research has been conducted to determine the potential increased toxicity due to the synergistic action of the components (Paiva et al., 2014). The results showed the absence of toxicity and cytotoxicity (Popović et al., 2023; Radošević et al., 2015), which opens the door to the future application of these solvents, especially in high-demand industries.

Meaningful comparisons between NADES and conventional solvents should therefore extend beyond extraction yield alone. Relevant criteria include raw-material sourcing, solvent production and purification, energy requirements, solvent-to-solid ratio, extraction efficiency and selectivity, mass-transfer behaviour, solvent recovery or direct use, downstream processing, toxicity, biodegradability, recyclability, scale-up feasibility, and end-of-life management (Benvenutti et al., 2019). By leveraging the versatility of NADES, the industry can embrace greener and more sustainable approaches while achieving desired outcomes. Their successful industrial implementation, however, depends on demonstrating technological performance, safety, regulatory compliance, and environmental benefits at the level of the complete process.

Beyond the food industry, NADES exhibits tremendous potential in various sectors. From pharmaceuticals (Albertini et al., 2023) to cosmetics (Alishlah et al., 2019), from agriculture (Mišan et al., 2019) to renewable energy (Florindo et al., 2017), the ability to design tailor-made solvents opens doors to novel applications. Researchers and industries can explore the unique properties of NADES, finding innovative ways to improve processes, enhance product quality, and reduce environmental impact. While, their potential advantages depend on the selected components, extraction process, downstream requirements, and intended final application, their sustainability should be assessed throughout the entire process and solvent life cycle rather than inferred solely from their natural origin, low volatility, or extraction performance.

2.2 Designing NADES solvents for food/pharma/cosmetic applications

Natural deep eutectic solvents (NADES) have emerged as potential alternatives to conventional extraction solvents. Compared with many conventional organic solvents NADES may offer advantages such as low volatility, reduced flammability, and the possibility of being prepared from naturally occurring and renewable components. However, their safety, biodegradability, and suitability for food, pharmaceutical, or cosmetic applications depend on the identity, purity, proportion, and concentration of their constituents (Vrgović et al., 2022; Bosanac et al., 2026). Compared with many volatile organic solvents, NADES generally exhibit low volatility and reduced flammability. Their compositional versatility also allows their physicochemical properties and extraction selectivity to be adjusted according to the target compounds and intended application. An important potential advantage of NADES is that they may function not only as extraction media but also as solubilizing, stabilizing, or carrier systems for bioactive compounds. When their composition is compatible with the intended product, NADES-based extracts may potentially be incorporated directly into food, pharmaceutical, or cosmetic formulations. This approach may reduce or eliminate solvent-removal steps, decrease energy consumption, and limit the degradation or loss of sensitive compounds during downstream processing (Ilić et al., 2025). However, NADES are not universally superior to conventional solvents. Their relatively high viscosity may restrict mixing, mass transfer, pumping, filtration, and solid–liquid separation, particularly during scale-up. Recovery of extracted compounds and regeneration of the solvent may also be more difficult than the removal of conventional volatile solvents. When NADES remain in the final product, their concentration, sensory characteristics, formulation compatibility, stability, and biological safety must be carefully evaluated. A direct comparison from our previous research illustrates these trade-offs. In the extraction of polyphenols and ellagic acid from defatted raspberry seeds, an optimized citric acid–betaine–water NADES provided higher recoveries than 80% ethanol. However, acidified methanol achieved the highest ellagic acid recovery under more intensive acidic and thermal conditions (Teslić et al., 2022). This example demonstrates that the solvent providing the highest analytical recovery is not necessarily the most suitable when safety, energy consumption, downstream processing, and the intended application of the extract are also considered.

Traditional, generally safe solvents, such as water, have important limitations when it comes to dissolving certain bioactive compounds used in the food/pharmaceutical or cosmetic industry. Many bioactive molecules, including pharmaceuticals, plant extracts, and nutraceuticals, often exhibit poor solubility in water due to their complex chemical structures. This low solubility poses significant challenges in terms of effective delivery, bioavailability, and overall efficacy of bioactive compounds (Albertini et al., 2023; Mišan et al., 2019; Stupar et al., 2021).

Natural deep eutectic solvents are formed by combining two or more components, a hydrogen bond donor and acceptor hydrogen bond at a specific molar ratio (Mišan et al., 2019). The most prominent functional groups of NADES constituents are carboxylic acids, hydroxyl groups, and carbonyl groups, the formation of NADES involves hydrogen bonding between the receptor and donor compounds, as well as other intermolecular interactions like van der Waals and electrostatic forces (Liu et al., 2018). These intermolecular interactions play a crucial role in enhancing the solubility and dispersibility of bioactive compounds or other substances, typically insoluble or poorly soluble in conventional solvents like water. Moreover, the intermolecular interactions in NADES contribute to the modulation of viscosity, conductivity, surface tension, and other NADES properties (Santana et al., 2019; Teslić et al., 2022).

By selecting and combining appropriate natural components, NADES with tailored properties can be prepared. In this context, quaternary ammonium salts (such as choline chloride) and amino acids (such as betaine, proline, glycine, and alanine) are commonly used as hydrogen bond acceptors (HBAs) in natural deep eutectic solvents. On the contrary, sugars, sugar alcohols (glucose, fructose, sucrose, maltose, trehalose, glycerol, sorbitol, and xylitol), and organic acids (primarily lactic acid, citric acid, oxalic acid, malic acid, levulinic acid, and ascorbic acid) serve as hydrogen bond donors (HBD) in NADES. In addition, acids, amino acids, and alcohols can also exhibit HBA behavior. However, the HBA or HBD role of a component is not always exclusive and may depend on its chemical environment and the other components of the mixture. The wide range of possible combinations provides considerable flexibility for tailoring NADES to different technological applications, The composition of a NADES should be selected according to the chemical characteristics of the target compounds, the properties of the plant matrix, the intended extraction conditions, and the final application of the extract. The identity and molar ratio of the components, together with the water content, can be adjusted to influence solvent performance and extraction selectivity. Therefore, NADES prepared from the same components may exhibit different extraction behaviour when their molar ratio or water content is changed (Mišan et al., 2019; Mohd Fuad et al., 2021; Pojić et al., 2024).

For food applications, NADES containing sugars, polyalcohols, amino acids, and organic acids commonly present in foods are of particular interest. Sugar-based NADES have been investigated for extraction, enzyme stabilization, and the preparation of food-grade microemulsions (Vasyliev et al., 2022). They may also provide a suitable medium for incorporating extracted bioactive compounds, including polyphenols, into food formulations. Organic acids such as lactic, citric, malic, and tartaric acids are commonly used in food products as acidity regulators, and organic-acid-based NADES have shown potential in the extraction and preservation of natural compounds (Teslić et al., 2022). It's important to note that the specific composition of NADES for food industry applications may depend on the desired functionality and the targeted food component. In addition, regulatory requirements and safety considerations should be thoroughly assessed when utilizing NADES in food-related applications to ensure compliance with applicable food safety standards and regulations (Vrgović et al., 2022).

Appropriate safety and regulatory requirements must likewise be considered when selecting NADES components applications. In addition to NADES that may be suitable for several industrial sectors, therapeutic deep eutectic solvents THEDES, been developed for pharmaceutical applications. In THEDES, an active pharmaceutical ingredient represents one of the components used to form the eutectic system (Gutiérrez et al., 2019). Therapeutic deep eutectic solvents based on ibuprofen, a nonsteroidal antiinflammatory drug, and menthol or thymol, for application as THESES itself or for controlled drug delivery (da Silva et al., 2021). For the cosmetic industry, menthol-based, glycerol- or fatty acid-based NADES. These NADES can serve as solvents for natural compounds, including essential oils, botanical extracts, or other bioactive ingredients used in cosmetics formulations. Their selection should be based not only on extraction performance but also on compatibility with the complete formulation and the intended route of application (Dwamena & Dwamena, 2019; van Osch, 2018).

NADES are typically prepared using various physical methods, including heating and stirring, freeze-drying, evaporation, grinding, but also with the assistance of ultrasound and microwave. Among these methods, stirring with heating and freeze-drying are most common in scientific literature. Interesting and very important studies have shown that NADES prepared using different methods, such as heating and stirring, ultrasound-assisted, and microwave-assisted preparation, did not exhibit significant differences in terms of density and viscosity, and NMR profile. However, the microwave-assisted method was found to be the fastest, requiring less than 30 s for NADES preparation (Popovic et al., 2022; Santana et al., 2019). Additionly, NADES preparation has advantage over others modern alternative solvents. The process of obtaining DES solvent involves the use of a temperature of 50–120°C and requires a mixing time of 30 min to 6 h. On the other hand, obtaining IL requires a temperature of 25–100°C and a time of up to 48 hours (Benvenutti et al., 2019). Nevertheless, the effects of the preparation method should not be generalized to all NADES compositions. Depending on the selected components, NADES formation may occur at room temperature or require moderate heating to facilitate mass transfer and homogenization. Microwave- and ultrasound-assisted methods may reduce preparation time, but their effectiveness depends on the chemical and physicochemical characteristics of the components and preparation conditions should be optimized for each individual system.

NADES have demonstrated their versatility as predominantly polar liquids, being utilized for the extraction of various compounds such as alkaloids, anthocyanins, glycosides of phenyletanes and phenylpropanoids, and polysaccharides. Moreover, NADES can be tailored for the extraction of less polar compounds including aglycones of flavonoids, anthraquinones, astaxanthin, curcumin, glycyrrhizic acid, iridoids, and steroidal saponins (Benvenutti et al., 2020; Jones et al., 2022; Panić et al., 2019; Popovic et al., 2022; Punzo et al., 2021) have enabled the isolation of carotenoids, free fatty acids, and pheophytin (Florindo et al., 2018; Stupar et al., 2021). Recent advancements include the proposal of a biphasic system using deep eutectic solvents for the simultaneous extraction of high-polarity chlorogenic acid, quercetin, anthocyanidins, and the low-polarity compound artemisinin from (Ilić et al., 2025; Rebocho et al., 2022).

NADES are mostly composed of two components; however, water is often added as a third compound. Water plays an important role in enhancing the interactions between the NADES components, promoting the formation of stable hydrogen bonds that contribute to the solubilization, homogeneity of the solvent system, and lower viscosity. On the contrary, excess water may reduce the extraction efficiency by weakening or breaking the intermolecular hydrogen bond structure of NADES components. For example, the extraction yield of antioxidant compounds increased significantly with water increase; however, with further water increase, decreased of antioxidant compounds was detected, pointing to a compromise on the extractability of more and less polar antioxidant compounds (Bener et al., 2022; Cui et al., 2015). The same was detected for rutin, quercitin, and cinnamic acid; their solubility significantly increased in choline chloride and glucose-based NADES; however, when water content increased in glucose-choline chloride-water significantly decreased rutin solubility (Dai et al., 2013).

The use of multicomponent NADES has shown promising results in various applications, particularly in the extraction of specific compounds. In the case of steviol glycosides, multicomponent NADES have demonstrated higher extraction yields compared with binary NADES and conventional extraction methods. This improvement in extraction efficiency can be attributed to the ability of multicomponent NADESs to mimic the intracellular environment, providing a favorable microenvironment for the solubility and stability of the targeted metabolites (Suresh et al., 2023). In addition, ternary NADES have shown superior performance in the extraction of polyphenolic compounds compared with binary NADES, indicating their enhanced capacity for extracting specific target compounds (Alves et al., 2022).

The selection and combination of hydrogen bond acceptors (HBA) and hydrogen bond donors (HBD) in natural deep eutectic solvents (NADES) have traditionally relied on a trial-and-error approach, observing the formation of clear solutions and their stability over time (Albertini et al., 2023). However, recent studies have started to employ computational methods to gain a better understanding of the properties of hydrogen-bonded networks in NADES. Computational modeling, such as the COSMOtherm software, has been used to predict the solubility of specific molecules in a wide range of solvents, including NADES. These predictions have shown close agreement with experimental data, providing a more efficient and time-saving approach for solvent screening (Gómez-Urios et al., 2022). Among the computational models, the COSMO-RS model has been extensively utilized for solubility predictions, solvent screening, and multiphase chemical equilibria. COSMO-RS was shown to be a more reliable tool to theoretically approximate the solubility of curcumin in mixed, complex systems than the Hansen Solubility Parameters were (Huber et al., 2022). However, the accuracy of this approach is often not sufficient for reproducing experimental data or screening purposes. To address this, machine learning techniques, particularly artificial neural networks (ANNs), have emerged as powerful tools for solubility predictions. Machine learning protocols combined with molecular descriptors characterizing the properties of solutes in various solvents have shown significant efficiency in solubility computations, including for active pharmaceutical ingredients (Cysewski, Jeliński, Cymerman, & Przybyłek, 2021; Cysewski, Jeliński, Procek, & Dratwa, 2021). In addition, Giner et al. (2020) highlight the potential of mixing rules in exploring the ecotoxicity of NADES. The results highlight the importance of computational methods, such as QSAR models, in improving and optimizing NADES, enabling the identification of environmentally friendly and efficient solvents for various applications. This presents a promising approach to streamline solvent selection, minimize time and energy usage, and identify suitable solvents for specific compounds (Giner et al., 2020).

2.3 Physicochemical properties of NADES

The successful implementation of NADES in industrial settings is highly dependent on its distinctive properties, rheological behavior, viscosity, density, polarity and pH.

These properties, which are influenced by the chemical composition, molar ratio, water content, temperature, and interactions of NADES components, and are pivotal in determining their effectiveness in various applications. A comprehensive understanding of these physicochemical properties is therefore essential for the appropriate selection, processing, and application of NADES across different industries. (Pojić et al., 2024; Mišan et al., 2019).

2.3.1 Rheological properties and Viscosity

In the food industry, rheological properties dictate how food flows, mixe, and behaves during manufacturing processes like pumping, mixing, extrusion, and filling. Appropriate rheological properties contribute to uniform ingredient distribution and help prevent undesirable effects such as phase separation, creaming, or sedimentation, while also maintaining the desired appearance and texture of the final product. Rheological properties also influence sensory attributes such as consistency, creaminess, chewiness, smoothness, and mouthfeel. Similarly, in the cosmetics and pharmaceutical industry, rheological characteristics guide the selection and combination of ingredients, including thickeners, emulsifiers, and stabilizers, to achieve the desired viscosity, consistency, and stability. By manipulating rheology, formulations can be designed to promote controlled release, enhanced absorption, and increased effectiveness of specific ingredients on the skin or hair.

There have been several studies investigating the rheological properties of the NADES solvents (Altamash et al., 2017; Elhamarnah et al., 2022; Gupta et al., 2021). However, when considering their application in food and cosmetic products, the focus shifts toward the concentration of the NADES extract in the final product and the rheological properties of the product itself, rather than solely on the NADES-based extract. The rheological properties of pure NADES or NADES-based extract are also important in the process of adding it to the final product, such as pumping dosing, mixing, filtration, and incorporation into the final product.

For instance, the viscosity of NADES influences its flow characteristics and handling during industrial processes and often has to be adjusted to meet specific requirements, enabling efficient extraction, separation, or formulation processes. NADES are normally characterized by a high viscosity, with kinematic viscosity values (up to 720 mm^2/s at 40°C), and this represents an important limit considering future industrial applications (Wu et al., 2022). High viscosity can be reduced by increasing the temperature or the water content. An increase in temperature leads to a decrease in viscosity. For instance, when the temperature was increased from 40°C to 70°C, the viscosity of NADES consisting of L-isoleucine and tetrabutyl ammonium chloride was decreased from 1.58 Pa· s to 0.2 Pa· s (Mjalli & Al-Azzawi, 2021). Aroso et al. observed a significant viscosity reduction in NADES containing choline chloride and xylose (1:1) from 100 Pa· s to 0.5 Pa· s as the temperature increased from 27°C to 87°C (Aroso et al., 2017). However, although increasing the temperature improves solvent fluidity and mass transfer, it may also promote the degradation of thermolabile target compounds. Temperature should therefore be optimized by balancing viscosity reduction against the stability and recovery of the compounds being extracted (Stupar et al., 2021). The addition of water represents another effective approach for reducing NADES viscosity. In particular, the viscosity of choline-based NADES was reduced by 32 – 36% compared with the original value after addition of 1%w/w water and decreased by 80 – 81% when diluted with 5% water. After the addition of 10% water, the viscosity decreased by about 93% of the original values for choline-based (Albertini et al., 2023). However, excessive water addition may progressively weaken the interactions between the original NADES constituents. At sufficiently high water contents, the components may behave predominantly as individually hydrated solutes rather than as an integrated eutectic system (Mehariya et al., 2021).

2.3.2 Density

The density of NADES, along with its viscosity, influences their handling and behavior during extraction, phase separation, pumping, mixing, and formulation. Density differences are particularly relevant in liquid–liquid and solid–liquid systems because they influence phase behaviour, sedimentation, and separation efficiency. In cosmetic and pharmaceutical formulations, density may also affect the physical stability of emulsions and suspensions.

Similar to viscosity, NADES density generally decreases with increasing temperature because thermal expansion increases the average intermolecular distance within the liquid. In the systems investigated by Gómez et al. (2019), all measured NADES densities were higher than that of water. The higher densities were found for NADES prepared with a combination of sugars and organic acids (citric acid:sugar 1.442 g cm^{-3}) and lower when combined with choline chloride and organic acids (malic acid:choline chloride 1.229 g cm^{-3}). Nevertheless, densities of all tested NADES decreased linearly with the increasing water content (Gómez et al., 2019). Lower densities were measured in combination with choline chloride: urea $\left(1.1692\ g \cdot cm^{-3}\right)$. The addition of L-arginine as a third component increased the density of the investigated mixtures, possibly because of changes in molecular packing and intermolecular interactions (Chemat et al., 2016).

2.3.3 Polarity

The polarity of NADES is a crucial physicochemical property influencing their solubilization capacity, extraction selectivity, and miscibility with other solvents. It depends on the chemical structure and molar ratio of the constituent components, their intermolecular interactions, and the water content of the system. NADES polarity is commonly evaluated using solvatochromic probes. When Nile red is used as the probe, polarity is expressed as the Nile red molar transition energy ENR reported in kcal/mol, with lower values indicating higher solvent polarity. Reported ENR values for hydrophilic NADES range from approximately 44.81 to 51.89 kcal/mol. Organic acid-based NADES are generally among the most polar systems, whereas sugar- and polyalcohol-based systems tend to exhibit lower polarity. However, polarity cannot be predicted solely from the functional groups of the individual components, as it is also strongly governed by their intermolecular interactions and the resulting supramolecular structure of the solvent (Leng & Suyin, 2019; Liu et al., 2018).The polarity and hydrophobicity of these eutectic systems can be tailored through the selection of appropriate components and molar ratios. Hydrophobic deep eutectic solvents (HDES) are commonly formed from constituents containing long hydrocarbon chains or other non-polar structural moieties, such as fatty acids. In fatty acid-based systems, increasing the hydrocarbon-chain length generally enhances hydrophobicity and reduces miscibility with water. Nevertheless, the overall polarity and hydrophobicity depend on the properties of all components, their molar ratio, and the intermolecular interactions established within the eutectic mixture. These characteristics should therefore be carefully considered when designing HDES for the selective extraction of compounds with different polarities (Florindo et al., 2018, 2019).

2.3.4 pH

The acid–base behavior of NADES depends on their composition and water content. Because conventional pH is defined for aqueous media, pH values reported for neat or nearly anhydrous NADES should be interpreted with caution. Some organic acids or amines present in NADES compositions can contribute acidity or basicity. Consequently, the acidity or basicity of the individual components can influence the overall pH behavior of NADES. This should be taken into account when considering the application of NADES, whether in their pure form or as diluted solutions in specific conditions. Understanding the pH range in which these components are stable is crucial for maintaining the integrity of the NADES mixture. The acid–base character of NADES may also influence reaction kinetics, selectivity, catalytic activity, and the ionization state of extracted compounds.

In biomedical and pharmaceutical applications, the acid–base behavior of NADES is particularly important because it may affect enzyme activity, drug ionization and solubility, and interactions with biological systems. Paiva et al. (2014) evaluated the cytotoxic profile of products based on choline chloride and citric or tartaric acid. The results indicated that the presence of tartaric acid had a negative effect on cellular metabolic activity, which was closely related to pH. Although mild changes in extracellular pH may not immediately compromise cell survival, they can affect cellular metabolism and proliferation. Hence, understanding the pH range within which NADES components remain stable is crucial to maintain the integrity of the NADES mixture (Mbous et al., 2016; Paiva et al., 2014).

2.3.5 Toxicity and biodegradability

NADES are frequently described as green solvents, however, this designation should be supported by evidence of their biodegradability, toxicity, renewability, and overall environmental performance. Biodegradability and renewability are important sustainability considerations, while both biodegradation and toxicity depend strongly on the chemical composition and concentration of the NADES system. Under the reported test conditions, biodegradation values ranging from 69.3 to 97.1% were obtained for selected quaternary-ammonium-based eutectic systems containing different HBDs, with the urea-based system showing the highest value and the triethylene-glycol-based system the lowest. Choline chloride, one of the most commonly used components in eutectic systems, was reported to undergo 93% degradation within 14 days. In addition, the biodegradability of NADES, based on choline chloride as the hydrogen bond acceptor (HBA), varies depending on the hydrogen bond donor (HBD). For example, when the hydrogen bond donor (HBD) was glycerol, the degradation level reached 96%, and for oxalic acid as an HBD, the degradation level was 68% (Radošević et al., 2018; Wu et al., 2022; Zhao et al., 2015).

Water content may also influence NADES biodegradation. Undiluted systems may degrade more slowly, whereas dilution with water can reduce viscosity and improve the accessibility of the components to microorganisms, thereby facilitating biodegradation. However, this effect depends on the composition of the system and the conditions of the biodegradation test (Mišan et al., 2019).

In addition to biodegradability, the composition of NADES also affects its toxicity. Research discovered that a NADES composed of betaine, glycerol, sucrose, and water exhibited higher toxicity compared with a NADES composed of betaine, glycerol, trehalose, and water, even when both had the same proportions (2 : 1 : 3 : 5). In addition, studies have indicated that NADES containing organic acids, such as malonic acid and tartaric acid, display higher in vitro cytotoxicity compared with sugar-based NADES. The chemical composition and concentration of the NADES in the exposure medium were identified as key factors affecting cytotoxicity, together with the cell model and duration of exposure (Popović et al., 2023).

Because experimental (eco)toxicological information remains limited for many NADES, quantitative structure–activity relationship (QSAR) models may support preliminary toxicity screening. These models use molecular descriptors for the individual components together with appropriate mixture rules to estimate the behavior of the complete system. However, such predictions may not fully capture interactions arising within the eutectic mixture and should not replace experimental toxicological and ecotoxicological assessment (Giner et al., 2020).

2.4 NADES-based approaches for sustainable plant extraction

2.4.1 NADES in the extraction of plant polyphenols

As summarized in Table 2.2, the majority of research focused on the utilization of NADES extraction for the isolation of various polyphenolic compounds was published in recent years. It could be observed that polyphenols could be isolated from common plant resources such as herbs and fruits (green tea leaves, mate tea leaves, yarrow, bilberry, etc.) by NADES extraction.

Several studies were focused on the extraction of polyphenols from green tea. Cui et al. (2021) applied the conductor-like screening model for segment activity coefficient (COSMO-SAC) to evaluate the affinity of green tea polyphenols in various NADES systems. Epigallocatechin gallate (EGCG) was employed as the model solute. The evaluated NADES contained choline chloride (ChCl) as the hydrogen-

Table 2.2 Recent applications of NADES extraction on the recovery of polyphenols and comparison with conventional extraction in terms of yield and bioactive potential.

Matrix	NADES extraction		Conventional extraction	Target compounds		Bioactivity		References
	Solvents	Conditions		NADES Extraction	Covnetional Extraction	NADES Extraction	Covnetional Extraction	
Coffee husk	ChCl-Gly ChCl-Glu ChCl-CA-H2O ChCl-Pro-H_2O	SS = 120rpm WC = 20 – 60% t = 25 – 150 min T = 40 – 80°C S/S ratio = 0.25 – 0.1	–	EY = 1.44 – 5.88% TP = 71.79 – 294.02 mg/L	–	Antioxidant IC_{50}(DPPH) = 12 3.86 – 472.93 mg/L	–	Maimulyanti et al. (2023)
Tobacco leaves	Suc-LA Suc-LA-H_2O CA-PG Glu-Fru-Suc H_2O ChCl-U-H_2O	SS = 100 rpm WC = ~7.4 – 22% t = 30 min T = 25°C S/L ratio = 0.1 g/mL	S = H_2O, 70% EtOH, 33% Ace SS = 100 rpm t = 30 min T = 25°C S/L ratio = 0.1 g/mL	TP = 963–2198 μg GAE /mL TF = 46.4–134 μgQE/mL	TP = 824–1300 μg GAE/mL TF = 14.9–151.3 μg QE /mL	Antioxidant IC_{50}(ABTS) = 1.6 0 – 41.9μ g GAE/mL $IC_{50}(H_2O_2)$ = 3.8 – 40.0μ g QE /mL	Antioxidant IC_{50}(ABTS) = 5.0–11.5μ g GAE/mL $IC_{50}(H_2O_2)$ = 21.0 – 44.0μ g QE/mL	Leal et al. (2023)
Saffron floral byproducts	Bet-LA Glu-LA Bet-Gly-H_2O Pro-CA-H_2O Pro-Gly	UAE WC = 10–70% t = 5 – 35 min S/L = 0.05g/mL	–	TP = 18.90 – 88.96 mg GAE/g TF = 1.16 – 4.93 mg CE/g	–	Antioxidant IC_{50}(DPPH) = 2.0 6 – 6.64mg/mL	–	Cerdá-Bernad et al. (2023)
Sandy everlasting inflorescences	ChCl-LA ChCl-Prop ChCl-Fru-H_2O ChCl-U	VA MSPDE WC = 4.46% t = 1 – 4 min S/L = 0.025 – 0.1 g/mL	UAE S = 80% MeOH t = 50°C T = 1 h S/L = 0.02 g/mL	TP = 26.88 –38.34mgGA/g ChA = 37.87 –48.99mg/10 g Iso = 10.44 36.23 mg/10 g	TP = 32.73mg GA/g ChA = 43.10 mg/10 g Iso = 30.68 mg/10 g	Antioxidant ABTS ≈ 8 – 35 mg TE/g FRAP ≈ 70 – 165mg Fe^{2+}/g	–	Ivanović, Krajnc, et al. (2022)
Bilberry fruit				TP = 21.1530.94 mg GAE/g TF ≈ 5 – 8mg RE/g ChA = 1.40 – 1.51 mg/g	TP ≈ 16 – 19mg GAE/g TF ≈ 6 – 16mg RE/g ChA = 0.83 – 0.86 mg/g	Antioxidant IC_{50}(DPPH) = 1.0 5 – $\frac{2.43mg}{mL}$ IC_{50}(ABTS) = 43.27 – 92.72 μg/mL FRAP = 0.14–0.47 mmol Fe^{2+}/g	Antioxidant IC_{50}(DPPH) = 3.2 4 – **4.**$\frac{47mg}{mL}$ IC_{50}(ABTS) = 78.55 – 128.17 μg/mL FRAP = 0.25–0.3 mmol Fe^{2+}/g	

Continued

Table 2.2 Recent applications of NADES extraction on the recovery of polyphenols and comparison with conventional extraction in terms of yield and bioactive potential.—*cont'd*

Matrix	NADES extraction		Conventional extraction	Target compounds		Bioactivity		References
	Solvents	Conditions		NADES Extraction	Covnetional Extraction	NADES Extraction	Covnetional Extraction	
Bilberry leaves	Bet-U MA-Gly TA-Sor CA-Sor	UAE WC = 30% t = 30 min T = 50°C S/L = 0.05g/mL	UAE S = H_2O, 50% EtOH t = 30 min T = 50°C S/L = 0.05g/mL	TP = 63.51 – 77.64 mg GAE/g TF ≈ 20 – 30 mg RE/g ChA = 17.13 22.48 mg/g	TP ≈ 57 – 89 mg GAE/g TF ≈ 3 – 25 mg RE/g ChA = 8.37 – 22.51 mg/g	Antioxidant IC_{50}(DPPH) = 0.4 1 – $\frac{1.2mg}{mL}$ IC_{50}(ABTS) = 23 75 – 39.38 μg/mL FRAP = 0.510.94 mmol Fe^{2+}/g	Antioxidant IC_{50}(DPPH) = 0.4 8 – $\frac{0.78mg}{mL}$ IC_{50}(ABTS) = 15 56 – 49.49 μg/mL FRAP = 0.410.66 mmol Fe^{2+}/g	Martinović et al. (2022)
Green tea leaves				TP = 71.66 – 133.55 mg GAE/g TF ≈ 8 – 10 mg RE/g EGC = 32.08 –60.88 mg/g	TP ≈ 60 – 85 mg GAE/g TF ≈ 2 – 21 mg RE/g EGC = 34.65–36.85 mg/g	Antioxidant IC_{50}(DPPH) = 0.09 – 0.44mg/mLIC_{50} (ABTS) = 7.03 – 20.14μ g/mL FRAP = 0.96 – 1.91 mmol Fe^{2+}/g	Antioxidant IC_{50}(DPPH) = 0.24 – 0.42mg/mLIC_{50} (ABTS) = 8.1 7 – 16.01μ g/mL FRAP = 0.87 – 1.49 mmol Fe^{2+}/g	
Habanero pepper leaves	ChCl-Glu	UAE WC = 40% t = 30 min S/L = 0.1g/mL	UAE S = MeOH t = 30 min S/L = 0.1 g/mL	TP = 11.00 – 22.38 mg GAE/100 g C = 9.64 – 39.72mg/100 g ChA = 7.57 –17.35mg/100 g	TP = 37.75 – 63.80 mg GAE/100 g C = 12.83 – 16.62 $\frac{mg}{100}$ g ChA = 0.98 – 1.77 mg /100 g	Antioxidant DPPH = 28.48 – 36.99 %Inh ABTS = 7.44–8.63 mg TE/g	Antioxidant DPPH = 77.84 – 78.32 %Inh ABTS = 1.58 –2.62 mg TE/g	Avilés-Betanzos et al. (2022)
Habanero pepper steams				TP = 7.60 – 9.18 mg GAE/100 g C = 50.20–59.70 mg/100 g ChA = 15.74 – 27.84 mg/100 g	TP = 19.37–33.90 mg GAE/100 g C = 0.12 – 0.14 mg /100 g ChA = 0.00 mg/100 g	Antioxidant DPPH = 53.88 – 54.21 %Inh ABTS = 8.139.55 mg TE/g	Antioxidant DPPH = 81.36 – 82.83 %Inh ABTS = 3.58–3.94 mg TE/g	

Thyme herbal dust	CA-Glu CA-Suc CA-Bet-H_2O ChCl-Glu Gly-Bet Bet-Gly-H_2O Bet-Glu Gly-Fru ChCl-Gly ChCl-Gly-H_2O LA-Glu-H_2O ChCl-LA Glu-TA LA-Fru Pro-LA Pro-Gly-H_2O MA-Bet-H_2O TA-Bet-H_2O ChCl-CA Prop-ChCl H_2O	SS = 700 rpm WC = 5.5C25% t = 60 – 180 min T = 40 – 70°C S/L = 0.03 – 0.1 g/g	–	TP = 35.33 – 71.43 mg GAE/g TF = 4.82 – 19.43 mg CE/g RA = 524.18 mg/100 g	–	Antioxidant DPPH = 37.79 – 188.01 mg TE/g ABTS = 70.81– 147.87 mg TE/g FRAP = 19.85– 44.70 mg Fe^{2+}/g	–	Pavlić et al. (2022)
Raspberry seeds	LA-Fru-H_2O LA-Glu-H_2O MA-Bet-H_2O TA-Bet-H_2O CA-Bet-H_2O LA-Bet-H_2O	WC = 2025% t = 30 – 240 min T = 60 – 95°C S/L = 0.03 –0.1 g/g	S = 80% EtOH, Acidic MeOH t = 30 min T = 60 S/L = 0.05 g/g	TP = 26.91 – 44.77 mg GAE/ g EA = 60.13– 147.02 mg/100 g	TP = 20.67 – 26.54 GAE/g EA = 45.23 –703.28 mg/100 g	Antioxidant DPPH = 995.18 –348.14 μ mol TE/g Antiproliferative IC_{50} (Caco2) = 3.80 – 4.63 $\frac{mg}{mL}$ IC_{50}(HT29) = 2.16 – 3.84mg/mL	Antioxidant DPPH = 197.78 –597.10 μ mol TE/g Antiproliferative IC_{50} (Caco-2) = 103.10 –120.60 mg/mL IC_{50}(HT29) = 22. 93–48.65 mg/mL	Teslić et al. (2022)
Mate tea leaves	CA-Bet-H_2O CA-Pro-H_2O ChCl – CA – H_2O ChCl-Suc LA-Glu-H_2O LA-Glyc-H_2O MA-Pro Me-LaA OxA-Bet-Pro	UAE WC = ~9.9 – 31.7 t = 30 – 60 min T = 40 – 60°C S/L 0.05 – 0.1 g/g	SLE/SE S = 70% MeOH, MeOH t = 20 – 180 min S/L = 0.04 – 0.05 g/mL	TP = 0.00 – 13.5 mg GAE/100 mg ChA = 2.25 – 4.28 $\frac{g}{100}$ g FA = 2.48 –4.52 g/100 g	TP = 9.4 – 12mg GAE/100 mg ChA = 1.50 – 1.91 g/100 g FA = 0.95 –1.5 g/100 g	Antioxidant IC_{50}(DPPH) = 1.06 – 4.02 μ g GAE/mL	Antioxidant IC_{50}(DPPH) = 6.42 – 9.12 μ g GAE/mL	Rebocho et al. (2022)
Sour cherry pomace	ChCl-MA ChCl-U ChCl-Fru	UAE/MAE/SLE SS = 650 rpm WC = 20% t = 0.25 – 30 min T = 40°C S/L = 0.075 g/mL	UAE/MAE/SLE S = 50% EtOH, Acidic MeOH SS = 200 rpm t = 8 – 25 min T = 45 – 75°C S/L = 0.029 – 0.083 g/mL	TP ≈ 1750 – 3250 μg TP/g TA ≈ 1450 – 2450 μg TA/g CGR = 773.72 – 1483.20μ g/g	TP ≈ 1900 – 2200 μg TP/g TA ≈ 1350 – 1600 μg TA/g	Antioxidant IC_{50}(DPPH) = 0.9 4 – 1.40mg/mL Antimicrobial MIC(Pseudomonas aeruginosa) = 0 781 – 12.5 %NADES MIC(*Escherichia coli*) = 0.781 –12.5 %NADES	–	Popovic et al. (2022)

Continued

Table 2.2 Recent applications of NADES extraction on the recovery of polyphenols and comparison with conventional extraction in terms of yield and bioactive potential.—*cont'd*

Matrix	NADES extraction		Conventional extraction	Target compounds		Bioactivity		References
	Solvents	Conditions		NADES Extraction	Covnetional Extraction	NADES Extraction	Covnetional Extraction	
Yarrow aerial parts	ChCl-LA ChCl-1,4-But ChCl-Prop ChCl-Fru ChCl-U	UAE WC = 25% t = 30 min T = 50°C $\frac{S}{L} = 0.025g/mL$	UAE S = 80%EtOH, 30%MeOH, H_2O t = 30 min T = 50°C S/L = 0.025 g/mL	TP = 12.66–35.44 mg GAE/g DCQA isomer = 4.98 –5.79 mg/g ChA = 2.65–2.97 mg/g	TP = 21.67–28.92 mg GAE/g DCQA isomer = 3.83 – 6.45mg/g ChA = 2.88–3.29 mg/g	Antioxidant ABTS = 12.66 – 35.44 mg TE/g FRAP = 21.98 –58.52 mg Fe^{2+}/g Antibacterial MIC(*E.* coli) = 0.1–3.2 mg/mL	Antioxidant ABTS = 21.67–28.92 mg TE/g FRAP = 18.84 –22.50 mg Fe^{2+}/g Antibacterial MIC(*E.* coli) = 2.4 – 3.2 mg/mLmg/mL	Ivanović, Grujić, et al. (2022)
Sea knotgrass leaves and stems	ChCl-Fru ChCl-Glu ChCl-Xyl ChCl-Suc	UAE WC = 30–40% t = 30 min S/L = 0.025 g/mL	UAE S=Ace, EtOH S/L = 0.025g/mL	MyRh = 18.71 –23.85 mg/g QGal = 3.23–3.66 mg/g EGC = 0.37–0.84 mg/g	MyRh = 96.19–62.71 mg/g QGal = 12.26–24.27 mg/g EGC = 2.30–5.92 mg/g	Antioxidant IC_{50} (DPPH) = 0.73–3.37 mg/mL IC_{50} (ORAC) = 148.12–153.51 mg TE/gIC_{50}(CCA) = 1.01 – 3.87mg/mL	Antioxidant IC_{50} (DPPH) = 0.42–1.73 mg/mL IC_{50} (CCA) = 49.34 – 328mg TE/g IC_{50} (CCA) = 1.17 –6.78mg/mL	Rukavina et al. (2021)
Kratom leaves	CA-Glu	MAE t = 10 – 20 min S/L = 0.05 0.1 g/mL	–	TP = 106.63 –358.59 mg GAE/g	–	–	–	Herman et al. (2021)
Green tea	ChCl-EG ChCl-Gly ChCl-Glu ChCl-OxaA ChCl-CA	WC = 20– 30% t = 180 min T = 70 – 95°C S/L = 0.016 –0.05 g/g	S = EtOH, H_2O t = 180 min T = 70°C S/L = 0.025 g/g	EY ≈ 13 – 20.4%	EY ≈ 11.8 – 15%	–	–	Cui et al. (2021)
Orange peel	ChCl-EG ChCl-Glu ChCl-Gly Glu-Gly Glc-EG Sor-EG Glu-Fru-Suc	WC = 30 –80% S/L = 0.25 g/mL	S = EtOH S/L = 0.25 g/mL	TP ≈ 26 – 57 mg/g	TP ≈ 41mg/g	–	–	Panić et al. (2021)
Tears of the virgin leaves	CA-Glu	MAE WC = 16.6750% t = 5 – 15 min S/L = 0.083–0.125 g/g	–	TP = 27.09 –85.25 mg GAE/g	–	–	–	Yusuf et al. (2021)

Citrus sinensis peel	ChChl-Glu ChChl-Fru ChChl-Xyl ChChl-Gly ChChl-MA ChChl-TA ChChl-LA ChChl-CA	SS = magnetic stirrer WC = 25% t = 20 min T = 45°C S/L ratio = 1:10 g/mL	S = 50% EtOH SS = magnetic stirrer t = 20 min T = 45°C S/L ratio = 1:10 g/mL	TP = 330.57 –1053.97 mg/100 mL TF = 34.23 –168.55 mg/100 mL AA = 159.08 –430.71 mg/100 mL	–	Antioxidant FRAP = 474–664 μM TE ORAC = 7.60 –11.37μM TE DPPH = 909–1065 μM TE TEAC = 439–978 μM TE	Antioxidant FRAP = 981μM TE ORAC = 9.59 μM TE DPPH = 1037 μM TE TEAC = 718μM TE	Gómez-Urios et al. (2023)
Spent coffee ground	Bet-TEG ChCl-Prop	SS = magnetic stirrer WC = 30–50% t = 150 min T = 65°C S/L ratio = 1:15 g/mL	S = 60%EtOH SS = magnetic stirrer t = 2 h T = 60°C S/L ratio = 1:8 g/mL	TP = 1.4% GA = 78.16– 154.73 mg/kg CGA = 71.7– 131.34 mg/kg CA = 40.28 –63.17 mg/kg	TP = 1.3% GA = 83.45 mg/ kg CGA = 89.33 mg/ kg CA = 37.14 mg/ kg	Antimicrobial Slight inhibition toward *Bacillus subtilis* and *Bacillus cereus*	Antimicrobial Clear inhibition toward Bacillus subtilis and *Bacillus cereus and* *slight toward* *Escherichia coli*	García-Roldán et al. (2023)
Rhodiola rosea	ChChl-MA ChChl-TA ChChl-MOA ChChl-CA	UAE WC = 10–50% t = 20 – 60 min T = 30 – 60°C S/L ratio = 1:10 g/mL	UAE S = 90%EtOH WC = 10–50% t = 20 – 60 min T = 30 – 60°C S/L ratio = 1:10 g/mL	Optimal TP = 248.2 mg GAE/g	TP = 122 mg GAE/g	Antioxidant TAC = 29.6 mg AAE/G DPPH = 35.6%	Antioxidant TAC = 43 mg AAE/G DPPH = 61%	Tsvetov et al. (2023)
Date palm *(Phoenix dactylifera)* seeds	ChChl-Glu ChChl-Suc ChChl-Xyl ChChl-Xylit ChChl-LA ChChl-MA ChCl-1,4-But	UAE WC = 30–70% t = 10 – 30 min T = 35 – 40°C S/L ratio = 3.3:1 00 g/mL	–	TP = 89.92 –152.69 mg GAE/g		Antioxidant DPPH = 285–747 mM TE/g	–	Osamede Airouyuwa et al. (2022)
Green tea leaves	ChChl-LA ChChl-Gly ChChl-LA-Gly	SS = magnetic stirrer WC = 0.41.5% t = 15 min T = 80°C S/L ratio = 1:20 g/g	S = H_2O SS = magnetic stirrer t = 15 minT = 80°C S/L ratio = 1:20 g/g	TP = 0.089 0.165 g/L	TP = 0.257 g/L	Antioxidant DPPH = 10.6–34 μM TE FRAP = 212–397 μM TE	Antioxidant DPPH = 25.7μM TE FRAP = 71μM TE	Alves et al. (2022)
Melia azedarach fruit	ChChl-Gly	UAE WC = 1050% t = 20 min T = 30 – 50°C S/L ratio = 1:50 g/L	–	TP = 7.45 –10.95 mg GAE/g TF = 9.25 –23.66 mg/g	–	Antioxidant DPPH = 26.71% ICA = 11.9946.24%	–	Jamshaid and Ahmed (2022)

Continued

Table 2.2 Recent applications of NADES extraction on the recovery of polyphenols and comparison with conventional extraction in terms of yield and bioactive potential.—*cont'd*

Matrix	NADES extraction		Conventional extraction	Target compounds		Bioactivity		References
	Solvents	Conditions		NADES Extraction	Covnetional Extraction	NADES Extraction	Covnetional Extraction	
Orange (*Citrus sinensis*) peel	LA-Glu MA-Pro	SS = magnetic stirrer WC = 1570% t = 5 – 30 min T = 45°C S/L ratio = 5 25 g/mL	–	TP = 1107 –4108 mg GAE/ 100 g TF = 20.6–167.3 mg/100 g	–	–	–	Gómez-Urios et al. (2022)
Mango peels	ChChl-Gly ChChl-Glu ChChl-Suc ChChl-LA ChChl-MA ChChl-EG LA-Glu LA-EG	UAE WC = 0–40% t = 30 min T = 30°C S/L ratio = 1:40 g/mL	UAE S = 80% EtOH t = 60 min T = 50°C S/L ratio = 1:40 g/mL	TP = 20 –69.84 mg GAE/g TF = 5 – 15.5 mg QE/g	TP = 49.89 mg GAE/g TF = 9.64 mgQE/ g	Antioxidant IC_{50}(DPPH) = 35. 37 mg/L	Antioxidant IC_{50}(DPPH) = 67. 43 mg/L	Lanjekar et al. (2022)
Carya cathayensi s peels	ChChl-MA ChCl-1,3-But ChCl-1,4-But ChChl-PG ChChl-LA ChChl-EG ChChl-Gly TA-EG TA-Gly MA-Gly MA-Glu LA-SA LA-AA LA-Glu LA-Gly	UAE t = 10 min T = 65°C S/L ratio = 1:40 g/mL	UAE S=Acidic EtOH t = 10 min T = 65°C S/L ratio = 1:40 g/mL	TP = 46.47 –76.26 mg GAE/g TF = 457.16 –793.17 mg QE/g	TP = 43.91 mg GAE/g TF = 613.53 mg QE/g	Antioxidant DPPH = 50.41 –75.13 mg GAE/g ABTS = 57.50 –102.45 mg GAE/g Enzyme inhibition α -glucosidase = 3. 30 – 12.44 mg AE/g α -amylase = 5.90– 22.64 mg AE/g	Antioxidant DPPH = 57.22 mg GAE/g ABTS = 80.36 mg GAE/g Enzyme inhibition α -glucosidase = 6. 61 mg AE/g α-amylase = 15.24 mg AE/g	Fu et al. (2022)
Apple pomace	ChChl-Gly ChChl-LA ChChl-Glu ChChl-U ChChl-CA U-Gly TA-Glu	UAE WC-10%–50% t = 10 – 50 min T = 20 – 70°C S/L ratio = 1: 10–1 : 50 g/mL	UAE S = 70%EtOH t = 30 min T = 40°C S/L ratio = 1:30 g/mL	TP = 2.5 – 5.6 mg GAE/g TF = 2.5 – 5.4 mg QE/g	TP = 4.5 mg GAE/g TF = 4.1 mg QE/g	Antioxidant DPPH = 30–80% FRAP = 5–10.5 mM AAE/mg LP = 20–45%	Antioxidant DPPH = 70% FRAP = 6 mM AAE/ mg LP = 35%	Rashid et al. (2023)
Mitragyna speciosa leaves	LA-Suc ChChl-Sor CA-Glu MA-Glu	MAE t = 10 – 20 min S/L ratio = 1:10 1 : 30 g/mL	S = 96%EtOHt room temperature S/L ratio = 1:10 g/mL	TP = 42.12 –495.12 μg GAE/g	TP = 23.17μg GAE/g	–	–	Prabowo et al. (2022)

Olive leaves	Glu-Fru-Suc H_2O Glu-Fru-H_2O Glu-Suc-H_2O Fru-Suc-H_2O ChChl-Glu H_2O ChChl-Fru H_2O ChChl-Suc H_2O ChChl-LA ChChl-MOA ChChl-EG ChChl-Gly	UAE WC = 2580% t = 60 min T = 55 – 75°C S/L ratio = 1: 20–1 : 70 g/mL	UAE S = 50% MeOH t = 60 min T = 55 – 75°C S/L ratio = 1:20– 1 : 70 g/mL	TP = 99.45 –187.31 mg GAE/g TF = 8.12 – 12.75 mg/g Oleuropein = 29 0.07–1630.80 mg/kg	Oleuropein = 12 21.17 mg/kg	Antioxidant DPPH = 435–480 μM DPPH/g	–	Ünlü (2021)
Hazelnut skin	ChChl-U ChChl-LA ChChl-Gly ChChl-Xylit ChChl-1,6- Hex ChChl-TEG ChChl-EG ChChl-PG ChChl-MA Bet-Gly Bet-EG Bet-TEG Bet-LA Bet-MA	UAE WC = 15– 35% T = 50 – 80°C S/L ratio = 1:3.33–1:25 g/ mL	UAE S = 50% EtOH t = 30 min T = room S/L ratio = 1:10 g/mL	TP = 6.14–16.95 g GAE/100 g	TP = 10.22 g GAE/100 g	–	–	Fanali et al. (2021)
Buddleja Globosa leaves	ChChl-LA LA-SA LA-AA LA-Gly ChChl-Prop ChChl-Gly Pro-CA	SS = 340 rpm WC = 20% t = 50 min T = 60°C S/L ratio = 1:100 g/mL	SS = 340 rpm S = MeOH t = 50 min T = 60°C S/L ratio = 1:100 g/mL	TP = 125–165 mg GAE/g L7G = 3.7–4.3 mg/g Verbascoside = 15 – 44 mg/g	TP = 153 mg GAE/g L7G = 3.1 mg/g Verbascoside = 25 mg/g	–	–	Torres-Vega et al. (2021)

NADES compounds: ChCl, *choline chloride;* Gly, *glycerol;* Glu, *glucose;* CA, *citric acid;* Pro, *proline;* LA, *lactic acid;* CA, *citric acid;* Suc, *sucrose;* PG, *propylene glycol;* Fru, *fructose;* U, *urea;* Bet, *betaine;* Prop, *1,2-Propanediol;* MA, *malic acid;* TA, *tartaric acid;* Sor, *sorbitol;* Glyc, *glycine;* Me, *menthol;* LaA, *lauric acid;* OxA, *oxalic acid;* Xyl, *xylose;* EG, *ethylene glycol;* TEG, *triethylene glycol;* MOA, *malonic acid;* Xylit, *xylitol;* 1,4-But, *1,4-butanediol;* 1,3-But, *1,3-butanediol;* SA, *sodium acetate;* AA, *ammonium acetate;* 1,6-Hex, *1,6-hexanediol;* SS, *stirring speed;* WC, *water content in NADES;* t, *extraction time;* T, *temperature;* S/L ratio, *sample to solvent ratio;* S, *solvents;* UAE, *ultrasound-assisted;* VA MSPDE, *vortex assisted matrix solid phase dispersion extraction;* EtOH, *ethanol;* Ace, *acetone; MeOH –methanol;* SLE, *solid liquid extraction;* SE, *Soxhlet extraction;* MAE, *microwave-assisted extraction;* EY, *extraction yield;* TP, *total phenols content;* TF, *total flavonoids;* ChA, *chlorogenic acid;* Iso, *isosalipurposide;* EGC, *epigallocatechin;* C, *catechin;* RA, *rosmarinic acid;* EA, *ellagic acid;* FA, *ferulic acid;* CGR, *cyaniding-3-glucosylrutinoside;* MyRh, *Myricetin 3-O-rhamnoside;* QGal, *quercetin-3-O-galactoside;* L7G, *luteolin-7-O-glucoside;* DCQA, *1,5-dicaffeoylquinic acid;* IC_{50}, *half maximal inhibitory concentration;* GAE, *gallic acid equivalent;* QE, *quercetin equivalent;* CE, *catechin equivalent;* TE, *trolox equivalent;* RE, *rutin equivalent;* %Inh, *% of inhibition;* MIC, *minimum inhibitory concentration;* AA, *ascorbic acid;* TAC, *total antioxidant capacity;* AAE, *ascorbic acid equivalents;* ICA, *iron chelating activity;* AE, *acarbose equivalents;* LP, *lipid peroxidation.*

bond acceptor, while EG, Gly, Glu, OA, and CA were used as hydrogen-bond donors at different molar ratios. The highest extraction yield was achieved by ChCl-EG and ChCl-Gly NADES systems. ChCl-EG was selected for further research, and other extraction parameters (time, temperature, liquid–solid ratio, and water content) were optimized by response surface methodology (RSM). Yield of polyphenols at optimized conditions was 20.12%, which highlighted the potential of this technique in the isolation of green tea polyphenols. For the systems evaluated, COSMO-SAC calculations suggested that the combination of a strong HBA, such as ChCl, and a relatively weak HBD, such as EG, provided favorable affinity for green tea polyphenols. Alves et al. (2022) evaluated three NADES solvents (ChChl-LA, ChChl-Gly, and ChChl-LA-Gly) for the isolation of green tea polyphenols. The ternary ChCl–LA–Gly system provided the highest total polyphenol content. The resulting NADES extract was subsequently used both as a plasticizing agent and as a source of polyphenols in chitosan-based packaging films. In this specific application, direct incorporation of the extract enabled the use of naturally derived antioxidants and avoided a separate solvent-removal step (Martinović et al., 2022). The highest content of both TP and TF was achieved with Bet-U solvent, while the predominant polyphenols determined in these extracts were epigallocatechin (54.00 mg/g) and EGCG (29.20 mg/g). Rebocho et al. (2022) applied a broad range of NADES systems for the isolation of bioactive compounds from mate tea leaves (Table 2.3). A menthol-based NADES system was used to selectively extract chlorophyll in the first step, while the residue was further extracted with other solvents for the isolation of polyphenols and purine alkaloids. LA-Glu-H_2O and LA-Gly H_2O provided extracts rich in polyphenols (up to 13.5 mg GAE /100mg). Chlorogenic acid, rutin, and ferulic acid were the major polyphenols in these extracts, while NADES solvents contributed to the coextraction of caffeine and theobromine in particularly high yields. Besides improved biocompatibility of NADES extracts and stability of extracted polyphenols in them, the authors highlighted that the main improvement in the application of these extracts is that they could be directly used, thus avoiding purification or separation steps.

Common aromatic and medicinal plants represent important matrices for the extraction of structurally diverse polyphenols. Choline-chloride-based NADES were evaluated for the ultrasound-assisted extraction of polyphenols from yarrow aerial parts (Ivanović, Grujić, et al., 2022). Authors exhibited significant improvement in TP yield (12 – 66 – 35.44 mg GAE/g) compared with the UAE process with conventional solvents, while 1,5-dicaffeoylquinic acid was the predominant compound (4.95 – 5.79mg/g). Special emphasis is being put on alternative raw materials based on medicinal plants since their utilization as raw material for polyphenolic BACs could provide new value chains. For example, saffron floral by-products were used for recovery of polyphenols using five NADES solvents (Bet-LA, Glu-LA, Bet-Gly H_2O, Pro-CA-H_2O and Pro-Gly) (Cerdá-Bernad et al., 2023). The highest TP was achieved by using Bet-Gly-H_2O and Pro-Gly, while Pro-Gly was selected as a promising green solvent to implement the extraction process and its optimization through the experimental design. According to the same study,

Table 2.3 Different NADES-based inventions in the food and cosmetic sector owned by academic institutions.

Application	Patent applicant	Product/Invention	Key benefits
Preservative	Sericulture and Agri-Food Research Institute GAAS	NADES compounded with the folium mori extract	Bacteria inhibiting and anticorrosive functions
Plant extract	Chungbuk National University Industry-Academic Cooperation Foundation	Method for preparing a green tea extract	Ingredients for cosmetic, food, and pharmaceutical products for skin moisturizing, skin barrier strengthening, skin whitening, skin pigmentation suppression, skin elasticity enhancement, or skin wrinkle improvement.
Removal of free fatty acids from grease	Sericulture and Agri-Food Research Institute GAAS	Method for removing free fatty acids in grease by enzyme catalysis and NADES	Obtaining the crude vegetable oil without the free fatty acids. Method for removing free fatty acids in grease by enzyme catalysis
Extraction and preservation	Sericulture and Agri-Food Research Institute GAAS	Extraction and preservation method of rice bran lipase.	Facilitation of the release of the rice bran lipase, improvement of the extraction efficiency, and stability of the rice bran lipase.
By-product extract	Consejo Superior de Investigaciones Científicas (CSIC), Universidad de Sevilla	Method for obtaining an extract rich in phenolic compounds from fresh alpeorujo (olive oil byproduct)	The phenolic extract is used as a nutraceutical, food additive, and phytosanitary product, particularly for the treatment of microbial diseases; for the fertilizer and/or organic soil conditioner.
By-product extract	Guangxi Academy of Agricultural Sciences	Method for simultaneous extraction of natural pigment and pectin from passion fruit peel.	Highly stable natural pigment

Continued

Table 2.3 Different NADES-based inventions in the food and cosmetic sector owned by academic institutions.—*cont'd*

Application	Patent applicant	Product/Invention	Key benefits
Utilization of agricultural products	Sericulture and Agri-Food Research Institute GAAS	Method for separating insect proteins, oil, and chitin in one step.	Under simple processes at mild conditions, the solvent can be recycled, cost and energy consumption are reduced, and the processes have better economy and environmental protection.
Utilization of agricultural byproducts	Universidad de Sevilla	Method for extracting oleuropein from olive leaf.	Functional ingredient with an oleuropein content of 8085% for the food, cosmetic, or pharmaceutical industry.
Natural preservative	Zhejiang Gongshang University	Method for formulation of natural preservative based on octyl gallate prepared by an enzymatic method in a DES	Preservative is green, safe, tasteless, nontoxic, and edible; the water solubility and antibacterial activity of the octyl gallate are improved, preventing spoilage and deterioration in the process of storage and transportation of the caviar.
Extraction of biocomponents	Universidade do Minho	Natural deep eutectic mixture for extraction of biocomponents from natural sources	The obtained extract contains a topical active compound for cosmetic formulations.

optimized extracts obtained from saffron floral by-products were used for the production of novel chitosan/alginate hydrogels, providing added value to the currently unexploited biomass. According to Pavlić et al. (2023), special emphasis is being put on the utilization of alternative natural resources (agricultural waste, food industry by-products, and under-utilized medicinal plants) as sustainable raw materials for the isolation of bioactive compounds for high-value applications. NADES extraction was utilized for the isolation of polyphenols from wild thyme (*Thymus serpyllum* L.) herbal dust, which is being generated as a by-product during filter tea production

(Pavlić et al., 2022). The process was developed in three stages. First, 20 NADES systems were screened, and the two systems providing the highest total polyphenol contents and antioxidant activities were selected. A fractional factorial design was then used to identify the most influential extraction variables and define their experimental ranges. Finally, temperature, extraction time, and liquid-to-solid ratio were optimized using a face-centred central composite design, with total polyphenol and flavonoid contents and antioxidant activity measured by DPPH, ABTS, and FRAP assays as the responses. The optimized NADES process provided high polyphenol recovery and antioxidant activity. Comparisons with extraction results reported in other studies should nevertheless be interpreted cautiously because of differences in the raw materials, analytical methods, and processing conditions (Pavlić et al., 2022). Optimized extract was obtained by Pro-Gly-H_2O as solvent, and it was particularly rich in rosmarinic acid. A similar study was performed with defatted raspberry seeds as raw material (Teslić et al., 2022). Screening experiments with different NADES systems ($n = 10$) was performed in the first step and CA BET-H_2O provided the best results in terms of ellagic acid content. A fractional factorial design followed by response surface methodology was used to optimize the extraction conditions. Two sets of optimal conditions were established, one maximizing ellagic acid yield and the other maximizing its concentration in the extract. The optimized NADES provided higher total polyphenol and ellagic acid contents than 80% ethanol. However, acidified methanol produced the highest ellagic acid recovery, although under more intensive acidic and thermal conditions. Sour cherry pomace was investigated as a source of polyphenols, which were isolated by ChCl -based solvents (ChCl-MA, ChCl-U, and ChCl-Fru) (Popovic et al., 2022). Authors investigated several approaches (heating and stirring, ultrasound and microwave methods) for the NADES preparation and extraction, and it was found that the microwave-assisted approach was the fastest, requiring less than 30 s. It was reported that the particularly high yields of TP (≈ 1750; 3250μ g/g) and TA ($\approx 1450 - 2450\mu$ g/g) were achieved, calculated as a sum of compounds identified by HPLC-PDA analysis. Microwave-assisted extraction also reduced the extraction time to less than 5 min while providing high polyphenol recovery. Among the evaluated NADES, ChCl–MA showed the best overall performance for polyphenol extraction from sour cherry pomace. Neochlorogenic acid and cyanidin 3-glucosylrutinoside were the predominant phenolic acid and anthocyanin, respectively.

Orange (*Citrus sinensis*) peel was studied as a potential source of polyphenols in several studies. Panić et al. (2021) applied 7 NADES systems for the isolation of polyphenols, and TP ranged from 26–57 mg/g. ChCl-EG solvent provided excellent ability to extract polyphenols (45.7 mg/g), while D-limonene was determined as another valuable compound coextracted with polyphenols. Furthermore, a broad range of ChCl -based solvents ($n = 8$) A combination of sugars, polyalcohols, or acids was used for the isolation of polyphenols from sweet orange peel by Gómez-Urios et al. (2023). It was observed that TP and TF in extracts were obtained in particularly high content ($330.57 - 1053.97$mg/100 mL and

34.23 – 168.55mg/100 mL, respectively). The best extraction yields of bioactive compounds were achieved by ChCl-MA and ChCl-Gly with low water content (75% of NADES).

Recent studies suggested that the approach with NADES solvents could be an excellent alternative for the isolation of polyphenols from coffee processing by-products compared with the aforementioned conventional and emerging extraction techniques. Maimulyanti et al. (2023) used ChCl-based solvents ($ChCl - Gly, ChCl - Glu, ChCl - CA - H_2O$ and $ChCl - Pro - H_2O$) for isolation of polyphenolic compounds from coffee husk waste. TP varied from 71.79 to 294.02 mg/L, and a decrease in TP was observed with utilization of the following compounds with the composition of ChCl:Pro > Gly > Glu > CA. Authors used a one-factor-at-a-time approach to optimize water content, time, S/L ratio and temperature, and suggested following optimized conditions for TP recovery, while HPLC analysis suggested that chlorogenic acid is the main polyphenolic compound that could be obtained in 0.13% yield calculated on coffee husk waste. On the contrary, García-Roldán et al. (2023) applied BetTEG and ChCl-Prop solvents for the isolation of polyphenols from spent coffee grounds. Results suggested that TP can be achieved in 1.4% yield, while the predominant compounds were observed in the following yields: gallic acid (78.19 – 154.73 mg/kg), chlorogenic acid (71.7 – 131.34 mg/g) and caffeic acid (40.28 –; 63.17 mg/kg) calculated on raw material.

According to Panić et al. (2021), the main limitations and challenges of proposed innovative biorefinery concepts based on the use of NADES might be economic, environmental, social, or ethical constraints. On the contrary, same authors highlighted the main strengths of NADES extraction approach on the case study of orange peel waste valorization: (1) NADES properties: low price, stability, nonvolatility, nonflammability, low ecological footprint; (2) NADES extraction: high dissolution capacity and/or extraction selectivity which could be tuned for a broad range of polyphenols; (3) NADES extract: stabilization of valuable bioactive compounds such as polyphenols and enzymes; (4) possibility to fine-tune NADES properties to fit the process requirements.

2.4.2 Biological activity, stability, and bioavailability of polyphenolic compounds in NADES extracts

Polyphenols are widely known compounds, as more than 10,000 different compositions have been identified so far (Ofosu et al., 2020). Due to their variability in chemical composition, polyphenols possess a broad spectrum of bioactivities such are antioxidant, antimicrobial (bacteria, viruses, molds and yeasts) anticancer, antiinflammatory, immunomodulatory, antiaging, hepatoprotective, enzyme inhibitory, antidiabetic and many other biological activities (Amanat et al., 2021; Fu et al., 2022; Luo et al., 2021; Marić et al., 2023; Pavlić et al., 2017, 2021; Sun et al., 2020; Yahfoufi et al., 2018). Exactly for this reason, plant-based extracts enriched with polyphenols have been extensively used by the food and cosmetic industry. Depending on their composition and intended application, these extracts may contribute antioxidant or antimicrobial properties, support product stability, or provide functional activity in food and cosmetic formulations. The type and magnitude of the biological activity of an extract depend on several factors, including the plant species and tissue, polyphenol profile, co-extracted constituents, extraction conditions, solvent composition, and concentration used in the biological assay. The efficacy of the extraction process and selection of optimal conditions (e.g., appropriate NADES) also greatly impact the bioactivity (Teslić et al., 2022). In certain cases, even pure NADES could contribute to the bioactive potential of extracts. This was reported in several recent studies where pure NADES exhibited antiproliferative and antibacterial activity (Ivanović, Grujić, et al., 2022; Popovic et al., 2022; Teslić et al., 2022). The bioactivity of plant-based extract is not only related to polyphenols but also to the synergetic effect of polyphenols with other coextracted compounds. For example, DPPH free radical scavenging activity was enhanced when polyphenols were co-extracted with terpenes by microwave or ultrasound-assisted extraction compared with microwave hydrodistillation or hydrodistillation (Pavlić et al., 2021). The biological activity of the obtained extract tends to be highest when samples are freshly prepared, while activity often deteriorates with time. The results from a recent study clearly indicated that the stability of antioxidant activity was more pronounced after 90 days of storage for NADES-based mate extract than in mate extract obtained with 70% methanol (Rebocho et al., 2022). Gómez-Urios et al. (2022) also reported that phenol stability was enhanced after 30 days in NADES composed of lactic acid and glucose in comparison with 50% ethanol. However, stabilization of phenolics was temperature dependent, as the degradation effect was higher and the difference between NADES and 50% ethanol was lower at room temperature (25°C), while the stability of phenols in NADES medium truly excels at lower temperatures (4°C) (Gómez-Urios et al., 2022). This stabilizing effect may be associated with hydrogen-bonding interactions between the NADES components and polyphenols, together with reduced molecular mobility and limited exposure to oxygen. However, the underlying mechanism is likely to depend on the solvent composition, water content, target compounds, and storage conditions. These findings indicate that appropriately selected NADES may support the storage stability of polyphenols and their associated biological activity. Nevertheless, the protective effect should be confirmed for each extract, formulation, and set of storage conditions. Another important aspect of phenols' bioactivity is their bioavailability in the human body. One of the good examples is resveratrol, a phenolic compound that possesses antioxidant, antitumor, antihypertensive, antiinflammatory, antiplatelet aggregation, cardioprotective, and immunomodulatory activities per se; however, it also has low bioavailability due to low poor solubility in water (Santos et al., 2019). In a recent study, the pharmacokinetic profile of blueberry polyphenolic compounds isolated with conventional solvents and a NADES mixture composed of choline chloride, glycerol, citric acid, and water was tested on rats (da Silva et al., 2021). Authors reported that total anthocyanins bioavailability was still low even in NADES medium (0.82%); however, it was significantly higher compared with the extract initially recovered with a methanol-water-formic acid mixture, then evaporated and redissolved in water (0.34%). Da Silva et al. (2021) suggested that the reasons for such an outcome are enhanced solubility of anthocyanins and higher intestinal stability of anthocyanins in NADES medium compared with anthocyanins extracted/dissolved in conventional solvents. Another group also reported that orally administered rutin dissolved in NADES (proline-glutamic acid mixture) caused an increase in rutin content and extended retention of this phenolic compound in the blood plasma of tested mice when compared to an aqueous solution of rutin (Faggian et al., 2016). These animal studies indicate that appropriately selected NADES may enhance the oral bioavailability of certain poorly water-soluble polyphenols. However, the available evidence remains limited and is largely based on animal studies; therefore, the effects of NADES composition, dose, digestion, intestinal transport, metabolism, and long-term safety require further investigation before these findings can be extrapolated to humans.

2.4.2.1 Evaluation of extraction performance of NADES in comparison with conventional solvents

According to the literature review, there are several dozen articles published in the last 3 years related to the extraction of polyphenols using NADES and the bioactivity (Table 2.2). The most commonly used analytical method for the determination of polyphenols quantity was the well-known Folin-Ciocalteu method, while in vitro antioxidant tests (e.g., DPPH, ABTS, and FRAP) were the most often used assays to examine the bioactivity of obtained extracts. Furthermore, a wide range of different plant species (green tea, raspberry, sour cherry, tobacco, orange, olives, etc.) and plant parts (e.g., husks, leaves, fruit, seeds, and stems) were used as raw material for polyphenols recovery, whereas some of the plant materials were by-products (sour cherry pomace and saffron floral by-products). There were also many different extraction techniques, solvents, and extraction parameter setups used to isolate bioactive compounds (Table 2.2). Particularly interesting are articles with a direct comparison between extracts obtained with NADES and conventional solvents in terms of polyphenols quantity and bioactivity. In a recently published study, several NADES combinations, water, 70% ethanol, and 33% acetone were used to recover polyphenols from tobacco leaves and compared in terms of total polyphenolic (TP), total flavonoids concentration (TF), and 2 antioxidant assays (ABTS and H_2O_2) (Leal et al., 2023). Generally, almost all selected NADES extracts had higher TP than conventional solvents, while 70% ethanol and 33% acetone had higher TF compared with all tested NADES. Solvent made from choline chloride (ChCl), urea (U), and water, which had the highest TF and relatively high TP, exhibited the highest ABTS free radical scavenging ability but also relatively low H_2O_2 scavenging activity in comparison with other solvents (Leal et al., 2023). Furthermore, 2 of 5 tested NADES extracts exhibited higher antioxidant activity (ABTS test) in comparison with 3 conventional solvent extracts (Leal et al., 2023). All mentioned results suggest that indeed NADES can be superior compared with conventional solvents in terms of polyphenols content and bioactivity, but it will greatly depend on the selected NADES combination. A somewhat similar conclusion could be made from Martinović et al. (2022) study in which 4 different NADES extracts were compared with water and 50% ethanol extracts in terms of TP, TF, various individual polyphenols content, and antioxidant activity (DPPH, ABTS, and FRAP tests). For most of the tested responses, one of the examined NADES was superior in comparison with organic solvents. However, TF and the concentration of quercetin-3-O-glucoside, rutin, hyperoside, quercitrin, and chlorogenic acid were higher in the 50% ethanol extract than in all 4 tested NADES extracts. Thus, other NADES combinations should be used to maximize the flavonoids recovery. NADES extracts were also compared with organic solvent extracts in terms of antiproliferative, antimicrobial activity, and enzyme inhibition activity. In particular, Teslić et al. (2022) tested raspberry seed extract obtained with NADES (citric acid, betaine, and water) and 80% ethanol, and reported that selected NADES had higher antiproliferative activity toward HT29 cancer cells but also higher cytotoxicity toward healthy human Caco-2 cells. The authors also reported that the selectivity index (ratio between EC_{50} from the cytotoxicity and antiproliferative activity were lower for NADES extract in comparison with ethanolic extract, suggesting certain therapeutic properties of the former extract. In a recently published study, 5 different NADES combinations, 80% methanol, 80% ethanol, and water, were used to isolate polyphenols with antifungal and antibacterial activity from A. millefolium (Ivanović, Grujić, et al., 2022). Authors reported that a combination of ChCl and lactic acid exhibited the highest antimicrobial activity in comparison with all other extracts, while the aqueous extract exhibited the lowest antimicrobial activity. The ethanolic extract exhibited the highest antimicrobial activity in comparison with other organic solvent extracts, and it was superior to all NADES extracts apart from NADES made from ChCl and lactic acid. It is noteworthy that pure solvents were also tested, and NADES composed of lactic acid had significantly higher antibacterial and antifungal activity than other examined solvents, suggesting that NADES with organic acids tends to be more toxic toward various types of cells. In a comprehensive study, 15 NADES extracts and 5 organic solvent extracts of *C. cathayensis* were compared in terms α-glucosidase and α-amylase activity among many other responses. The extract with the highest enzymatic activity was obtained with NADES composed of ChCl and malic acid, while the water extract exhibited the lowest activity. Furthermore, other tested NADES combinations had generally equal or superior enzymatic inhibitory activity in comparison with extracts obtained with conventional solvents, suggesting that *C. cathayensis* extract obtained with NADES could be an efficient antidiabetic agent.

3. Industrial NADES extract application

The food and animal-feed processing industries still rely heavily on established products and conventional processes, which may limit the rate of innovation. The implementation of novel processing technologies may enhance their competitiveness and support the production of sustainable, safe, and nutritious food. NADES are increasingly investigated as potentially more sustainable extraction and formulation media for the food and cosmetic sectors, not only as alternative solvents, but also as stabilizing and protective media for enzymes, natural colorants, aromas, and bioactive metabolites. Certain NADES formulations have also demonstrated antimicrobial or cryoprotective effects. However, these properties depend strongly on the chemical composition, component ratio, water content, and applied concentration and should therefore not be generalized to all NADES systems. Their potential contribution to reducing the use of volatile organic solvents, improving resource efficiency, and facilitating the development of more stable products may support broader sustainability objectives. Nevertheless, the actual environmental performance of each NADES-based process should be confirmed through appropriate life-cycle, safety, and techno-economic assessments.

The potential of NADES to reduce undesirable chemical, microstructural, and textural changes in foods, cosmetics, and their ingredients may be relevant to both small- and large-scale processing. This potential has been associated with properties such as tunable composition, relatively simple preparation, low volatility, and, in some systems, enhanced stability of extracted compounds. However, thermal stability, biodegradability, affordability, and safety are formulation-specific characteristics rather than universal properties of all NADES. To support the progression of NADES-based technologies from laboratory and pilot-scale studies towards higher technology readiness levels, greater attention should be paid to process scale-up, reproducibility, energy requirements, downstream processing, solvent recovery, regulatory compliance, and product safety. The commercialization of NADES is still in its early stages, and further research is needed to fully understand its potential in various industrial applications. Nevertheless, several start-ups, spin-offs, and established companies have explored the commercial application of NADES in natural-product extraction, pharmaceutical screening, animal feed, and cosmetic formulations. One such example is the spin-off of the Nova University, School of Science and Technology, Lisbon, Portugal—“Des Solutio,” whose business idea has already attracted the interest of several large industrial groups, including a global beauty products giant, with over €26 billion a year in sales, and a large pharmaceutical company.[1] A collaborative start-up company “NADES Solutions”[2] emerged from research conducted in LEI related to tailoring high-performance natural solvents for in vitro testing of new drugs for the pharmaceutical industry. Another LEI spin-off, “In Ovo,” in collaboration with the company Pas Reform, the Netherlands, launched a new chicken feed product—SmartStart that utilizes NADES to retain water inside a semimoist feed and prevent drying under prolonged heating conditions

[1] https://www.fct.unl.pt/en/news/2019/04/new-fct-nova-spin-will-make-cosmetics-greener.
[2] http://www.smbpc.nl/nades-solutions/.

for a practical, simple posthatch feeding solution.[3] The company Gattefossé, France, launched a novel antiaging cosmetics product.

Gatuline Link n Lift[4] for which active ingredients were obtained by extraction of bioactives from the horse chestnut flower by NADES technology. The French company Naturex launched the product Eutectys based on a patented extraction process—*eutectigenesis*—through the formation of NADES for cosmetic applications. The number of NADEs is described in the literature, but only a limited number of them can actually be used for cosmetic applications due to very strict safety and regulatory issues (Benoit et al., 2021). A patent filed by Guangzhou Inspection and Testing and Certification Group Co., Ltd. describes the use of a deep eutectic solvent for the removal of heavy metals and rare-earth elements from plant-derived foods (CN113736583). This invention demonstrated the average simultaneous removal rate of multiple heavy metals and multiple rare earth elements of 90% and 80%, respectively, without remaining on a plant food matrix and generating secondary pollution to food. Patents filed by NESTEC and LUCTA describe the use of deep eutectic or NADES-based reaction media for flavor generation and the production of savoury and mouthfulness-enhancing compounds and LUCTA (Hartl et al., 2023). SYMRISE AG patented a high-purity phyllodulcin extract obtained from bigleaf hydrangea (*Hydrangea macrophylla*) for potential use as a taste modulator in sugar-reduced products (WO2023078551). CELLACURE LLC protected the solution related to the making of biopolymers and nanoparticle biopolymers from biowaste using DES/NADES used to stimulate plant growth and development, stimulate root growth, increase adsorption of nutrients, increase crop yield, improve the quality and taste of crops, increase specific amino acid or other nutrient content, and protect against various pathogens.

The implementation of NADES-based technologies in the food and cosmetic sectors may support the development of processes that use fewer volatile organic solvents and enable the recovery and direct formulation of natural compounds. However, environmental, social, and health benefits cannot be assumed solely from the use of NADES and should be demonstrated for each process and product. Many NADES are composed of primary or secondary metabolites, including sugars, sugar alcohols, organic acids, amino acids, amines, terpenes, and phenolic compounds. The use of renewable and low-volatility constituents may support certain green-chemistry objectives. Nevertheless, natural origin alone does not guarantee low toxicity, biodegradability, or favourable life-cycle performance, and direct generalized comparisons with ionic liquids or conventional organic solvents should be avoided (Tobiszewski et al., 2017). In addition, the concept of NADES for enhancing the solubility and stability of plant-derived metabolites as candidates for sustainable agriculture will provide an additional opportunity for more efficient

[3] https://www.pasreform.com/us/knowledge/120/post-hatch-feeding-made-easy-practical-and-flexible-1?query=NADES (accessed 19th April 2023).

[4] https://www.gattefosse.com/personal-care-actives/gatuline-link-n-lift (accessed 19th April 2023).

addressing of environmental hazards caused by pesticides. On the contrary, NADES have revealed promising therapeutic properties—antibacterial and antibiofilm, antifungal, anticancer, antioxidant, and wound healing (Oliveira & Duarte, 2021). Patent activity provides an indication of technological and commercial interest in NADES-based applications but does not by itself demonstrate industrial maturity or process viability. Considering that fact, the viability on a laboratory scale has been demonstrated, unlike the viability of the process at a pilot or industrial scale (Rente et al., 2022). Patents owned by academic institutions that are based on the use of NADES in the food and cosmetics sector are presented in Table 2.3.

To increase the industrial applicability of NADES and determine the viability of a process, key factors must be considered—economic, technological, social, and regulatory. Consideration of economic factors should include the cost of raw materials, investments, energy consumption of all required upstream and downstream processing operations, and the cost of the treatment of generated byproducts and wastes (Rente et al., 2022). Consideration of technological factors should include the treatment of the raw material to be extracted (drying, grinding, sieving, stabilization, pretreatment), the recovery of the liquid phase after extraction, the purification of extracted molecules, the extraction kinetics, and the feasibility of the solid/liquid separation since NADES is often of high viscosity. An important process-design decision is whether the target compounds should be separated from the NADES or whether the complete NADES-based extract can be incorporated directly into the final formulation. Direct use may reduce downstream processing, but its suitability depends on the identity and concentration of the NADES components, the intended application, sensory and functional compatibility, exposure level, safety, and regulatory requirements (Rente et al., 2022). Equipment design should account for the energy required for agitation and pumping, heat and mass transfer, and the compatibility of construction materials with the selected NADES. The potential corrosivity of individual formulations should be evaluated experimentally rather than inferred solely from the presence of choline chloride (Rublova et al., 2019). To support the utilization of NADES on a larger scale, regulatory factors must be considered as well. Regulatory requirements depend on the jurisdiction, composition, intended use, exposure route, and whether the NADES remains in the final product. Individual components may already be covered by existing food, pharmaceutical, or cosmetic legislation, but this does not automatically establish the safety or regulatory acceptability of the complete eutectic mixture (Altamash et al., 2017). To support the wider utilization of NADES at a larger scale, an open online database was developed that is available at https://database.dessolutio.com/. To boost scientific legitimacy among the citizens, especially in cases that involve complexity and technological uncertainty, as is the case with NADES, democratic participatory processes are welcome. Therefore, by pursuing a people-centric approach and giving the consumers a more central role in the innovation process, and by reconnecting them with other partners within the food/cosmetic sector through a quadruple helix network of communication, the consumer awareness and social acceptance of innovative products will be achieved. This concept will generate not only new product ideas addressing the specific requirements of consumers that could be introduced to the market, but also engage consumers in the collective innovation venture. This interactive approach and utilization of design thinking will stimulate their creativity and foster the acceptance of new NADES-based products. When consumers are involved in the creation process, the generation of new and unexpected ideas is possible, as well as better financial performance and time savings. However, the problems that may hinder these activities are the accusations that co-creation exploits consumers and other stakeholders who engage their time and creativity for the benefit of industrial partners (Vrgović et al., 2022). That this makes sense is also confirmed by the fact that over the past decade, consumers demand natural products, especially cosmetics and food, more and more (Saulais et al., 2023; Suphasomboon & Vassanadumrongdee, 2022).

4. Conclusion and further perspectives

Plant polyphenols are of considerable interest to the food, pharmaceutical, and cosmetic industries because of their chemical diversity, technological functionality, and biological activity.Research on polyphenol recovery is increasingly focused on reducing the use of hazardous and volatile organic solvents and on valorizing agri-food by-products as alternative sources of bioactive compounds. Nevertheless, to effectively harness the potential of food by-products as a viable and economically sustainable source of bioactive compounds, it is crucial to carefully select and optimize environmentally friendly extraction technologies that enable the recovery and sustainability of the desired analytes. This selection process becomes indispensable to make these by-products appealing and economically feasible for commercial utilization.

The utilization of NADES in the food, pharmaceutical, and cosmetic industries offers numerous opportunities for innovation and sustainable extraction processes of phenolic compounds. NADES can serve as a platform for the formulation and development of new products. Depending on their composition, NADES may improve the solubility and storage stability of selected polyphenols and may also serve as components of delivery systems. Evidence of improved oral bioavailability is promising but remains limited and is predominantly based on animal studies. This can lead to the production of high-quality ingredients for food, pharmaceutical, and cosmetic formulations.

The studies reviewed in this chapter show that appropriately selected NADES can provide extraction yields or polyphenol recoveries comparable to, and in some cases higher than, those obtained with conventional solvents. However, their relative performance depends strongly on the plant matrix, target compounds, solvent composition, and extraction conditions. Multicomponent NADES provide additional possibilities for adjusting polarity, viscosity, and interactions with target compounds. Nevertheless, increasing the number of components does not necessarily improve extraction performance and may complicate solvent characterization, reproducibility, safety assessment, and regulatory evaluation. In addition, a methodology that proposes the use of different NADES for fractionated extraction of different classes of bioactive compounds from plant sources contributes to complete utilization with maximal plant exhaustion (Ilić et al., 2025; Rebocho et al., 2022). Furthermore, the combination of ultrasound with NADES, which can be a feasible means to enhance the extraction kinetics, increases the global extraction yield and reduces the extraction time and thus operational costs, without compromising the quality of the final extracts (Stupar et al., 2021). Sequential combinations of supercritical fluid or pressurized liquid extraction with subsequent NADES extraction may provide a route for recovering compounds with complementary polarities from the same plant material. The technical and environmental advantages of such integrated processes require confirmation through direct comparison with simpler extraction schemes.

The low vapor pressure of eutectic mixtures reduces solvent emissions but may complicate solvent removal and recovery after extraction, as conventional evaporation is generally inefficient. One of the key challenges in developing a cost-effective extraction process using NADES solvents lies in the recycling and reuse of these solvents after the extraction. Recycling NADES solvents can be a crucial aspect in reducing operational costs and minimizing waste. Anti-solvent-based approaches, including the addition of acetone, have been reported for separating extracted compounds or assisting NADES recovery (Mišan et al., 2019). However, the introduction of an additional organic solvent creates further separation, recovery, safety, and waste-management requirements. This suggests a potential solution for the recovery and reutilization of NADES solvents, enhancing their sustainability and economic viability. Direct incorporation of a complete NADES-based extract into a final formulation may avoid solvent-removal and compound-isolation steps. Nevertheless, this approach is suitable only when the composition, concentration, safety, sensory properties, functionality, and regulatory status of the complete extract are compatible with its intended use.

When direct incorporation is not appropriate or purified target compounds are required, switchable eutectic systems may provide an alternative strategy for product–solvent separation (Stupar et al., 2021; Zhang et al., 2022).Stupar et al. (2021) demonstrated the extraction of pumpkin carotenoids using a hydrophobic eutectic solvent followed by polarity switching to facilitate carotenoid separation from the solvent phase. This approach illustrates the potential of switchable systems to simplify product recovery, although recovery efficiency, solvent losses, energy requirements, and environmental performance should be confirmed at larger scales (Stupar et al., 2021). Water uptake by hydrophilic NADES is another important consideration because it may alter solvent composition, viscosity, intermolecular interactions, and extraction performance (Cysewski, Jeliński, Procek, & Dratwa, 2021). An additional solution, suitable for oral application, presents encapsulation of liquid NADES formulations. Albertini et al. showed that some NADES formulations containing active components can be incorporated into gelatine shell capsules without any interaction of NADES to act as a plasticizer/solvent toward the gelatine shell (Albertini et al., 2023).

The main advantages of applying the NADES concept are increased product quality and safety, reduced waste and impact on the environment, contributing that way to the green shift in the processing sector. Commercial development appears to be more advanced in cosmetic applications, where several NADES-derived ingredients and extraction technologies have been introduced. The rapid adoption of NADES-based extracts in the cosmetic industry is a testament to their effectiveness and market potential.

Although great progress has been made in the application of NADES in the food field, there are still some limitations and challenges. The toxicological assessment of many NADES remains incomplete. This lack of comprehensive exposure, metabolism, long-term safety, and regulatory data is an important limitation for their direct incorporation into food products (Popović et al., 2023).

Sensory properties of NADES and NADES-based extracts have received comparatively limited attention, although they are critical when the solvent remains in a food, oral pharmaceutical, or cosmetic formulation. One possible reason for this lack of sensorial analysis is that the primary focus of NADES research has been on their physicochemical properties, such as solubility, stability, and extraction capabilities. These aspects are crucial for establishing their suitability for different applications. Future studies should assess taste, odour, colour, mouthfeel, texture, and consumer acceptability at concentrations relevant to the final product.

Industrial implementation requires detailed characterization of NADES physicochemical properties, but solvent selection cannot be based on viscosity, density, and polarity alone. Extraction performance, stability, safety, sensory compatibility, solvent recovery, scale-up behaviour, and process economics must be evaluated together. However, assessing the suitability of a NADES based solely on physicochemical properties like viscosity, density, and polarity is challenging, as their effects on extraction can be complex and difficult to predict. High viscosity may be reduced by adjusting temperature or water content, but such modifications may also change the structure and selectivity of the solvent and affect the stability of thermolabile compounds. Consequently, these variables should be optimized rather than treated as universally applicable solutions. Lastly, it is crucial to address the recycling potential of NADES. Solvent recovery and reuse should be evaluated through recovery yield, solvent purity, component losses, energy consumption, product contamination, and performance over repeated cycles. Switchable eutectic systems offer a potentially useful approach, but their practical advantages require validation at pilot and industrial scales.

Future research should prioritise standardized NADES characterization, application-relevant toxicological and sensory assessment, direct comparison with conventional processes, solvent recovery and reuse, life-cycle and techno-economic analysis, and validation at pilot scale. Collaboration among academia, industry, regulators, and end users will be essential for identifying applications in which NADES provide measurable technical, environmental, and economic advantages without compromising product safety or functionality.

Funding

This research was funded by the Science Fund of the Republic of Serbia for the funding of the IDEAS project 'Novel extracts and bioactive compounds from under-utilized resources for high-value applications-BioUtilize', number '77501680', and the Ministry of Science, Technological Development and Innovation of the Republic of Serbia under the Agreements on the Implementation and Financing of Research (nos. 451-03-33/2026-03/200222).

References

Agunloye, O. M., Oboh, G., Ademiluyi, A. O., Ademosun, A. O., Akindahunsi, A. A., Oyagbemi, A. A., Omobowale, T. O., Ajibade, T. O., & Adedapo, A. A. (2019). Cardio-protective and antioxidant properties of caffeic acid and chlorogenic acid: Mechanistic role of angiotensin converting enzyme, cholinesterase and arginase activities in cyclosporine induced hypertensive rats. *Biomedicine and Pharmacotherapy, 109*, 450–458. https://doi.org/10.1016/J.BIOPHA.2018.10.044

Aijaz, M., Keserwani, N., Yusuf, M., Ansari, N. H., Ushal, R., & Kalia, P. (2023). Chemical, biological, and pharmacological prospects of caffeic acid. *Biointerface Research in Applied Chemistry, 13*(4). https://doi.org/10.33263/BRIAC134.324

Aja, I., Begoña Ruiz-Larrea, M., Courtois, A., Krisa, S., Richard, T., & Ruiz-Sanz, J. I. (2020). Screening of natural stilbene oligomers from Vitis vinifera for anticancer activity on human hepatocellular carcinoma cells. *Antioxidants, 9*, 469. https://doi.org/10.3390/ANTIOX9060469

Albertini, B., Bertoni, S., Sangiorgi, S., Nucci, G., Passerini, N., & Mezzina, E. (2023). NaDES as a green technological approach for the solubility improvement of BCS class II APIs: An insight into the molecular interactions. *International Journal of Pharmaceutics, 634*, Article 122696. https://doi.org/10.1016/J.IJPHARM.2023.122696

Alishlah, T., Mun'im, A., & Jufri, M. (2019). Optimization of urea-glycerin based NADES-UAE for oxyresveratrol extraction from Morus alba roots for preparation of skin whitening lotion. *Journal of Young Pharmacists, 11*(2), 155–160. https://doi.org/10.5530/jyp.2019.11.33

Altamash, T., Nasser, M. S., Elhamarnah, Y., Magzoub, M., Ullah, R., Anaya, B., Aparicio, S., & Atilhan, M. (2017). Gas solubility and rheological behavior of natural deep eutectic solvents (NADES) via combined experimental and molecular simulation techniques. *ChemistrySelect, 2*(24), 7278–7295. https://doi.org/10.1002/SLCT.201701223

Alves, T. F. P., Teixeira, N., Vieira, J., Vicente, A. A., Mateus, N., de Freitas, V., & Souza, H. K. S. (2022). Sustainable chitosan packaging films: Green tea polyphenolic extraction strategies using deep eutectic solvents. *Journal of Cleaner Production, 372*, Article 133589. https://doi.org/10.1016/J.JCLEPRO.2022.133589

Amanat, M., Reza, M. S., Shuvo, M. S. R., Ahmed, K. S., Hossain, H., Tawhid, M., Saifuzzaman, M., Islam, M. S., Mazumder, T., Islam, M. A., & Daula, A. F. M. S. U. (2021). Zingiber roseum Rosc. rhizome: A rich source of hepatoprotective polyphenols. *Biomedicine & Pharmacotherapy, 139*, Article 111673. https://doi.org/10.1016/j.biopha.2021.111673

Aroso, I. M., Paiva, A., Reis, R. L., Rita, A., Duarte, C., & Duarte, A. R. C. (2017). Natural deep eutectic solvents from choline chloride and betaine– physicochemical properties.

Journal of Molecular Liquids, 241, 654–661. https://www.sciencedirect.com/science/article/pii/S0167732217304427.

Avilés-Betanzos, K. A., Oney-Montalvo, J. E., Cauich-Rodríguez, J. V., González-Ávila, M., Scampicchio, M., Morozova, K., Ramírez-Sucre, M. O., & Rodríguez-Buenfil, I. M. (2022). Antioxidant capacity, vitamin C and polyphenol profile evaluation of a capsicum chinense by-product extract obtained by ultrasound using eutectic solvent. *Plants, 11*(15), 2060. https://doi.org/10.3390/plants11152060

Beaumont, P., Courtois, A., Atgié, C., Richard, T., & Krisa, S. (2022). In the shadow of resveratrol: Biological activities of epsilon-viniferin. *Journal of Physiology and Biochemistry 2022, 78*(2), 465–484. https://doi.org/10.1007/S13105-022-00880-X

Bener, M., Sen, F. B., Önem, A. N., Bekdeser, B., Çelik, S. E., Lalikoglu, M., Capanoglu, E., & Apak, R. (2022). Microwave-assisted extraction of antioxidant compounds from by-products of Turkish hazelnut (Corylus avellana L.) using natural deep eutectic solvents: Modeling, optimization and phenolic characterization. *Food Chemistry, 385*, Article 132633. https://doi.org/10.1016/J.FOODCHEM.2022.132633

Benoit, C., Virginie, C., & Boris, V. (2021). The use of NADES to support innovation in the cosmetic industry. *Advances in Botanical Research, 97*, 309–332. https://doi.org/10.1016/BS.ABR.2020.09.009

Benvenutti, L., Antonio, A., Zielinski, F., Regina, S., Ferreira, S. R. S., Zielinski, A. A. F., & Ferreira, S. R. S. (2019). Which is the best food emerging solvent: IL, DES or NADES? *Trends in Food Science and Technology, 90*, 133–146. https://doi.org/10.1016/J.TIFS.2019.06.003

Benvenutti, L., Sanchez-Camargo, A., del, P., Zielinski, A. A. F., & Ferreira, S. R. S. (2020). NADES as potential solvents for anthocyanin and pectin extraction from Myrciaria cauliflora fruit by-product: In silico and experimental approaches for solvent selection. *Journal of Molecular Liquids, 315*, Article 113761. https://doi.org/10.1016/J.MOLLIQ.2020.113761

Bosanac, M., Andrejić Višnjić, B., Popović, A., Ljubković, M., Martić, N., Miljković, D., … Cvetković, B. (2026). Hydrophobic NADES-Derived Pumpkin Carotenoid Extract Attenuates Oxidative Stress and Mitochondrial Dysfunction in a Rat Model of Doxorubicin-Induced Cardiotoxicity. *Pharmaceutics, 18*(6), 662. https://www.mdpi.com/1999-4923/18/6/662.

Cerdá-Bernad, D., Pitterou, I., Tzani, A., Detsi, A., & Frutos, M. J. (2023). Novel chitosan/alginate hydrogels as carriers of phenolic-enriched extracts from saffron floral by-products using natural deep eutectic solvents as green extraction media. *Current Research in Food Science, 6*(December 2022), Article 100469. https://doi.org/10.1016/j.crfs.2023.100469

Chemat, F., Anjum, H., Shariff, A. M., Kumar, P., & Murugesan, T. (2016). Thermal and physical properties of (Choline chloride + urea + l-arginine) deep eutectic solvents. *Journal of Molecular Liquids, 218*, 301–308. https://doi.org/10.1016/J.MOLLIQ.2016.02.062

Cosme, P., Rodríguez, A. B., Espino, J., & Garrido, M. (2020). Plant phenolics: Bioavailability as a key determinant of their potential health-promoting applications. *Antioxidants, 9*, 1263. https://doi.org/10.3390/ANTIOX9121263

Cui, Z., Enjome Djocki, A. V., Yao, J., Wu, Q., Zhang, D., Nan, S., Gao, J., & Li, C. (2021). COSMO-SACsupported evaluation of natural deep eutectic solvents for the extraction of tea polyphenols and process optimization. *Journal of Molecular Liquids, 328*, Article 115406. https://doi.org/10.1016/j.molliq.2021.115406

Cui, Q., Peng, X., Yao, X.-H., Wei, Z.-F., Luo, M., Wang, W., Zhao, C.-J., Fu, Y.-J., & Zu, Y.-G. (2015). Deep eutectic solvent-based microwave-assisted extraction of genistin, genistein and apigenin from pigeon pea roots. *Separation and Purification Technology, 150*, 63–72. https://doi.org/10.1016/j.seppur.2015.06.026

Cvetanović, A., Švarc-Gajić, J., Mašković, P., Savić, S., & Nikolić, L. (2015). Antioxidant and biological activity of chamomile extracts obtained by different techniques: Perspective of using superheated water for isolation of biologically active compounds. *Industrial Crops and Products, 65*, 582–591. https://doi.org/10.1016/J.INDCROP.2014.09.044

Cvetanović, A., Švarc-Gajić, J., Zeković, Z., Gašić, U., Tešić, Ž., Zengin, G., Mašković, P., Mahomoodally, M. F., & Đurović, S. (2018). Subcritical water extraction as a cutting edge technology for the extraction of bioactive compounds from chamomile: Influence of pressure on chemical composition and bioactivity of extracts. *Food Chemistry, 266*, 389–396. https://doi.org/10.1016/J.FOODCHEM.2018.06.037

Cvetanović, A., & Yan, L. (2021). A closer look on polyphenols isolation by subcritical water extraction as first choice green technique for their extraction from medicinal plants. In P. Bertollini (Ed.), *A closer look at polyphenolics*. USA: Nova Science Publisher Inc. https://doi.org/10.52305/QQNR6474

Cysewski, P., Jeliński, T., Cymerman, P., & Przybyłek, M. (2021). Solvent screening for solubility enhancement of theophylline in neat, binary and ternary NADES solvents: New measurements and ensemble machine learning. *International Journal of Molecular Sciences, 22*(14), 7347. https://doi.org/10.3390/IJMS22147347/S1

Cysewski, P., Jeliński, T., Procek, D., & Dratwa, A. (2021). Solubility of sulfanilamide and sulfacetamide in neat solvents: Measurements and interpretation using theoretical predictive models, first principle approach and artificial neural networks. *Fluid Phase Equilibria, 529*, Article 112883. https://doi.org/10.1016/J.FLUID.2020.112883

da Silva, D. T., Smaniotto, F. A., Costa, I. F., Baranzelli, J., Muller, A., Somacal, S., Monteiro, C. S. A., Vizzotto, M., Rodrigues, E., Barcia, M. T., & Emanuelli, T. (2021). Natural deep eutectic solvent (NADES): A strategy to improve the bioavailability of blueberry phenolic compounds in a ready-to-use extract. *Food Chemistry, 364*, Article 130370. https://doi.org/10.1016/J.FOODCHEM.2021.130370

Dai, Y., Witkamp, G.-J., Verpoorte, R., & Choi, Y. H. (2013). Natural deep eutectic solvents as a new extraction media for phenolic metabolites in Carthamus tinctorius L. *Analytical Chemistry, 85*, 6272–6278. https://doi.org/10.1021/ac400432p

Dwamena, A. K., & Dwamena, K. A. (2019). Recent advances in hydrophobic deep eutectic solvents for extraction. *Separations, 6*(1), 9. https://doi.org/10.3390/separations6010009

Elhamarnah, Y., Alrasheedi, M., Almarri, W., Albadr, A., Almalki, A., Mohamed, N., Fatima, I., Nasser, M., & Qiblawey, H. (2022). An experimental investigation on the thermo-rheological behaviors of lactic acid-based natural deep eutectic solvents. *Materials, 15*(11), 4027. https://doi.org/10.3390/MA15114027/S1

Faggian, M., Sut, S., Perissutti, B., Baldan, V., Grabnar, I., Dall'Acqua, S., Dall Acqua, S., Faggian, M., Sut, S., Perissutti, B., Baldan, V., Grabnar, I., & Dall'Acqua, S. (2016). Natural deep eutectic solvents (NADES) as a tool for bioavailability improvement: Pharmacokinetics of rutin dissolved in proline/glycine after oral administration in rats: Possible application in nutraceuticals. *Molecules, 21*(11), 1531. https://doi.org/10.3390/molecules21111531

Fanali, C., Gallo, V., Posta, S. D., Dugo, L., Mazzeo, L., Cocchi, M., Piemonte, V., & De Gara, L. (2021). Choline chloride-lactic acid-based NADES as an extraction medium in a response surface methodology-optimized method for the extraction of phenolic compounds from hazelnut skin. *Molecules, 26*, 2652. https://doi.org/10.3390/MOLECULES26092652

Figueroa-Robles, A., Antunes-Ricardo, M., & Guajardo-Flores, D. (2021). Encapsulation of phenolic compounds with liposomal improvement in the cosmetic industry. *International Journal of Pharmaceutics, 593*, Article 120125. https://doi.org/10.1016/J.IJPHARM.2020.120125

Florindo, C., Branco, L. C. C., & Marrucho, I. M. M. (2017). Development of hydrophobic deep eutectic solvents for extraction of pesticides from aqueous environments. *Fluid Phase Equilibria, 448*, Article 135142. https://doi.org/10.1016/j.fluid.2017.04.002

Florindo, C., Branco, L. C., & Marrucho, I. M. (2019). Quest for green-solvent design: From hydrophilic to hydrophobic (deep) eutectic solvents. *ChemSusChem, 12*(8), 1549–1559. https://doi.org/10.1002/cssc.201900147

Florindo, C., Romero, L., Rintoul, I., Branco, L. C., & Marrucho, I. M. (2018). From phase change materials to green solvents: Hydrophobic low viscous fatty acid-based deep eutectic solvents. *Sustainable Chemistry and Engineering, 6*, 3888–3895. https://doi.org/10.1021/acssuschemeng.7b04235

Fu, X., Belwal, T., He, Y., Xu, Y., Li, L., & Luo, Z. (2022). UPLC-Triple-TOF/MS characterization of phenolic constituents and the influence of natural deep eutectic solvents on extraction of Carya cathayensis Sarg. peels: Composition, extraction mechanism and in vitro biological activities. *Food Chemistry, 370*, Article 131042. https://doi.org/10.1016/J.FOODCHEM.2021.131042

Gómez, A. V., Tadini, C. C., Biswas, A., Buttrum, M., Kim, S., Boddu, V. M., & Cheng, H. N. (2019). Microwave-assisted extraction of soluble sugars from banana puree with natural deep eutectic solvents (NADES). *Lebensmittel-Wissenschaft & Technologie, 107*, 79–88. https://doi.org/10.1016/J.LWT.2019.02.052

Gómez-Urios, C., Viñas-Ospino, A., Puchades-Colera, P., Blesa, J., López-Malo, D., Frígola, A., & Esteve, M. J. (2023). Choline chloride-based natural deep eutectic solvents for the extraction and stability of phenolic compounds, ascorbic acid, and antioxidant capacity from Citrus sinensis peel. *Lebensmittel-Wissenschaft & Technologie, 177*, Article 114595. https://doi.org/10.1016/J.LWT.2023.114595

Gómez-Urios, C., Viñas-Ospino, A., Puchades-Colera, P., López-Malo, D., Frígola, A., Esteve, M. J., & Blesa, J. (2022). Sustainable development and storage stability of Orange By-Products extract using natural deep eutectic solvents. *Foods, 11*(16), 2457. https://doi.org/10.3390/FOODS11162457/S1

García-Roldán, A., Piriou, L., & Jauregi, P. (2023). Natural deep eutectic solvents as a green extraction of polyphenols from spent coffee ground with enhanced bioactivities. *Frontiers in Plant Science, 13*, 5481. https://doi.org/10.3389/FPLS.2022.1072592/BIBTEX

Giner, B., Lafuente, C., Lapeña, D., Errazquin, D., & Lomba, L. (2020). QSAR study for predicting the ecotoxicity of NADES towards Aliivibrio fischeri. Exploring the use of mixing rules. *Ecotoxicology and Environmental Safety, 191*, Article 110004. https://doi.org/10.1016/J.ECOENV.2019.110004

Gupta, V., Thakur, R., & Das, A. B. (2021). Effect of natural deep eutectic solvents on thermal stability, syneresis, and viscoelastic properties of high amylose starch. *International Journal of Biological Macromolecules, 187*, 575–583. https://doi.org/10.1016/J.IJBIOMAC.2021.07.099

Gutiérrez, A., Aparicio, S., & Atilhan, M. (2019). Design of arginine-based therapeutic deep eutectic solvents as drug solubilization vehicles for active pharmaceutical ingredients.

Physical Chemistry Chemical Physics, 21(20), 10621–10634. https://doi.org/10.1039/C9CP01408J

Hartl, D. M., Frank, O., Dawid, C., & Hofmann, T. F. (2023). A new inert natural deep eutectic solvent (NADES) as a reaction medium for food-grade maillard-type model reactions. *Foods, 12*(9), 1877.

Herman, I.,A., Rahayu, B. P., Arifuddin, M., Nur, Y., Prabowo, W. C., Ambarwati, N. S. S., Rijai, L., & Ahmad, I. (2021). Single factor effect of natural deep eutectic solvent citric acid-glucose based microwave-assisted extraction on total polyphenols content from mitragyna speciosa Korth. Havil leaves. *Pharmacognosy Journal, 13*(5), 1109–1115. https://doi.org/10.5530/pj.2021.13.143

Huber, V., Hioe, J., Touraud, D., & Kunz, W. (2022). Uncovering the curcumin solubilization ability of selected natural deep eutectic solvents based on quaternary ammonium compounds. *Journal of Molecular Liquids, 361*, Article 119661. https://doi.org/10.1016/J.MOLLIQ.2022.119661

Ilić, N.Đ., Stupar, A., Matić, M., Đerić, M., Mišan, A., Pojić, M., … Mandić, A. (2025). Enhancing bioactive compound extraction from hemp sprouts: Synergistic effects of NADES and ultrasound. *Industrial Crops and Products, 233*, 121420. https://doi.org/10.1016/j.indcrop.2025.121420

Ivanović, M., Grujić, D., Cerar, J., Razboršek, M. I., Topalić-Trivunović, L., Savić, A., Kočar, D., & Kolar, M. (2022). Extraction of bioactive metabolites from *Achillea millefolium* L. with choline chloride based natural deep eutectic solvents: A study of the antioxidant and antimicrobial activity. *Antioxidants, 11*, 724. https://doi.org/10.3390/ANTIOX11040724

Ivanović, M., Krajnc, P., Mlinarič, A., & Razboršek, M. I. (2022). Natural deep eutectic solvent-based matrix solid phase dispersion (MSPD) extraction for determination of bioactive compounds from sandy everlasting (Helichrysum arenarium L.): A case of stability study. *Plants, 11*(24), 3468. https://doi.org/10.3390/plants11243468

Jamshaid, S., & Ahmed, D. (2022). Optimization of ultrasound-assisted extraction of valuable compounds from fruit of Melia azedarach with glycerol-choline chloride deep eutectic solvent. *Sustainable Chemistry and Pharmacy, 29*, Article 100827. https://doi.org/10.1016/J.SCP.2022.100827

Jones, M., Tzani, A., Pitterou, I., Divani, F., Tsiaka, T., Sotiroudis, G., Zoumpoulakis, P., & Detsi, A. (2022). Green extraction of Greek propolis using natural deep eutectic solvents (NADES) and incorporation of the NADES-extracts in cosmetic formulation. *Sustainable Chemistry, 4*(1), 8–25. https://doi.org/10.3390/SUSCHEM4010002

Kurundkar, D., Kurundkar, A. R., Bone, N. B., Becker, E. J., Liu, W., Chacko, B., Darley-Usmar, V., Zmijewski, J. W., & Thannickal, V. J. (2019). SIRT3 diminishes inflammation and mitigates endotoxin-induced acute lung injury. *JCI Insight, 4*(1). https://doi.org/10.1172/JCI.INSIGHT.120722

Lanjekar, K. J., Gokhale, S., & Rathod, V. K. (2022). Utilization of waste mango peels for extraction of polyphenolic antioxidants by ultrasound-assisted natural deep eutectic solvent. *Bioresource Technology Reports, 18*, Article 101074. https://doi.org/10.1016/J.BITEB.2022.101074

Lavoie, S., Sweeney-Jones, A. M., Mojib, N., Dale, B., Gagaring, K., Mcnamara, C. W., Quave, C. L., Soapi, K., Kubanek, J., Kalasariya, H. S., & Pereira, L. (2022). Dermo-cosmetic benefits of marine macroalgae-derived phenolic compounds. *Applied Sciences, 12*, Article 11954. https://doi.org/10.3390/APP122311954

Leal, M., Moreno, M. A., Albornoz, P. L., Mercado, M. I., Zampini, I. C., & Isla, M. I. (2023). Nicotiana tabacum leaf waste: Morphological characterization and chemical-functional analysis of extracts obtained from powder leaves by using green solvents. *Molecules, 28*(3), 1396. https://doi.org/10.3390/molecules28031396

Leng, K. Y., & Suyin, G. (2019). Natural deep eutectic solvent (NADES) as a greener alternative for the extraction of hydrophilic (polar) and lipophilic (Non-Polar) phytonutrients. *Key Engineering Materials, 797*, 20–28. https://doi.org/10.4028/WWW.SCIENTIFIC.NET/KEM.797.20

Liu, Y., Zhang, Y., Chen, S. N., Friesen, J. B., Nikolić, D., Choules, M. P., McAlpine, J. B., Lankin, D. C., Gemeinhart, R. A., & Pauli, G. F. (2018). The influence of natural deep eutectic solvents on bioactive natural products: Studying interactions between a hydrogel model and Schisandra chinensis metabolites. *Fitoterapia, 127*, 212–219. https://doi.org/10.1016/J.FITOTE.2018.02.024

Liu, R., Zhang, Y., Yao, X., Wu, Q., Wei, M., & Yan, Z. (2020). Cite this. *Food & Function, 11*, Article 10084. https://doi.org/10.1039/dOfo01932a

Luo, J., Si, H., Jia, Z., & Liu, D. (2021). Dietary anti-aging polyphenols and potential mechanisms. *Antioxidants, 10*(2), 283. https://doi.org/10.3390/antiox10020283

Maimulyanti, A., Nurhidayati, I., Mellisani, B., Amelia Rachmawati Putri, F., Puspita, F., & Restu Prihadi, A. (2023). Development of natural deep eutectic solvent (NADES) based on choline chloride as a green solvent to extract phenolic compound from coffee husk waste. *Arabian Journal of Chemistry, 16*(4), Article 104634. https://doi.org/10.1016/J.ARABJC.2023.104634

Marić, B., Abramović, B., Ilić, N., Bodroža-Solarov, M., Pavlić, B., Oczkowski, M., Wilczak, J., ČetojevićSimin, D., Šarić, L., & Teslić, N. (2023). UHPLC-Triple-TOF-MS characterization, antioxidant, antimicrobial and antiproliferative activity of raspberry (Rubus idaeus L.) seed extracts. *Foods, 12*(1), 161. https://doi.org/10.3390/foods12010161

Martinović, M., Krgović, N., Nešić, I., Žugić, A., & Tadić, V. M. (2022). Conventional vs. green extraction using natural deep eutectic solvents—Differences in the composition of soluble unbound phenolic compounds and antioxidant activity. *Antioxidants, 11*(11), 2295. https://doi.org/10.3390/antiox11112295

Mbous, Y. P., Hayyan, M., Hayyan, A., Wong, W. F., Hashim, A., & Looi, C. Y. (2016). Applications of deep eutectic solvents in biotechnology and bioengineering-promises and challenges. *Biotechnology Advances, 35*, 105–134. https://doi.org/10.1016/j.biotechadv.2016.11.006

Medrano-Padial, C., Puerto, M., Richard, T., Cantos-Villar, E., & Pichardo, S. (2021). Protection and reversion role of a pure stilbene extract from grapevine shoot and its major compounds against an induced oxidative stress. *Journal of Functional Foods, 79*, Article 104393. https://doi.org/10.1016/J.JFF.2021.104393

Mehariya, S., Fratini, F., Lavecchia, R., & Zuorro, A. (2021). Green extraction of value-added compounds form microalgae: A short review on natural deep eutectic solvents (NaDES) and related pretreatments. *Journal of Environmental Chemical Engineering, 9*(5), Article 105989. https://doi.org/10.1016/J.JECE.2021.105989

Mišan, A., Nađpal, J., Stupar, A., Pojić, M., Mandić, A., Verpoorte, R., & Choi, Y. H. (2019). The perspectives of natural deep eutectic solvents in agri-food sector. *Critical Reviews in Food Science and Nutrition*, 1–29. https://doi.org/10.1080/10408398.2019.1650717

Mišan, A., & Pojić, M. (2021). Applications of NADES in stabilizing food and protecting food compounds against oxidation. *Advances in Botanical Research, 97*, 333–359. https://doi.org/10.1016/BS.ABR.2020.09.010

Mjalli, F. S., & Al-Azzawi, M. (2021). Aliphatic amino acids as possible hydrogen bond donors for preparing eutectic solvents. *Journal of Molecular Liquids, 330*, Article 115637. https://doi.org/10.1016/J.MOLLIQ.2021.115637

Mohd Fuad, F., Mohd Nadzir, M., & Harun@Kamaruddin, A. (2021). Hydrophilic natural deep eutectic solvent : A review on physicochemical properties and extractability of bioactive compounds. *Journal of Molecular Liquids, 339*, Article 116923. https://doi.org/10.1016/J.MOLLIQ.2021.116923

Ofosu, F. K., Daliri, E. B. M., Elahi, F., Chelliah, R., Lee, B. H., & Oh, D. H. (2020). New insights on the use of polyphenols as natural preservatives and their emerging safety concerns. *Frontiers in Sustainable Food Systems, 4*, Article 525810. https://doi.org/10.3389/fsufs.2020.525810

Oliveira, F. S. N. de, & Duarte, A. R. C. (2021). A look on target-specificity of eutectic systems based on natural bioactive compounds. *Advances in Botanical Research, 97*, 271–307. https://doi.org/10.1016/BS.ABR.2020.09.008

Osamede Airouyuwa, J., Mostafa, H., Riaz, A., & Maqsood, S. (2022). Utilization of natural deep eutectic solvents and ultrasound-assisted extraction as green extraction technique for the recovery of bioactive compounds from date palm (Phoenix dactylifera L.) seeds: An investigation into optimization of process parameters. *Ultrasonics Sonochemistry, 91*, Article 106233. https://doi.org/10.1016/J.ULTSONCH.2022.106233

Paiva, A., Craveiro, R., Aroso, I., Martins, M., Reis, R. L., Duarte, A. R. C., Rita, A., & Duarte, C. (2014). Natural deep eutectic solvents—Solvents for the 21st century. *ACS Sustainable Chemistry & Engineering, 4*(5), 1063–1071. https://doi.org/10.1021/sc500096j

Panić, M., Andlar, M., Tišma, M., Rezić, T., Šibalić, D., Cvjetko Bubalo, M., & Radojčić Redovniković, I. (2021). Natural deep eutectic solvent as a unique solvent for valorisation of orange peel waste by the integrated biorefinery approach. *Waste Management, 120*, 340–350. https://doi.org/10.1016/j.wasman.2020.11.052

Panić, M., Gunjević, V., Cravotto, G., & Radojčić Redovniković, I. (2019). Enabling technologies for the extraction of grape-pomace anthocyanins using natural deep eutectic solvents in up-to-half-litre batches extraction of grape-pomace anthocyanins using NADES. *Food Chemistry, 300*, Article 125185. https://doi.org/10.1016/J.FOODCHEM.2019.125185

Pavlić, B., Aćimović, M., Sknepnek, A., Miletić, D., Mrkonjić, Ž., Kljakić, A. C., Jerković, J., Mišan, A., Pojić, M., Stupar, A., Zeković, Z., & Teslić, N. (2023). Sustainable raw materials for efficient valorization and recovery of bioactive compounds. *Industrial Crops and Products, 193*, Article 116167. https://doi.org/10.1016/J.INDCROP.2022.116167

Pavlić, B., Mrkonjić, Ž., Teslić, N., Cvetanović Kljakić, A., Pojić, M., Mandić, A., Stupar, A., Santos, F., Rita, A., Duarte, C., & Mišan, A. (2022). Natural deep eutectic solvent (NADES) extraction improves polyphenol yield and antioxidant activity of wild thyme (Thymus serpyllum L.) extracts. *Molecules, 27*, 1508. https://doi.org/10.3390/MOLECULES27051508

Pavlić, B., Teslić, N., Vidaković, A., Vidović, S., Velićanski, A., Versari, A., Radosavljević, R., & Zeković, Z. (2017). Sage processing from by-product to high quality powder: I. Bioactive potential. *Industrial Crops and Products, 107*, 81–89. https://doi.org/10.1016/j.indcrop.2017.05.031

Pavlić, B., Teslić, N., Zengin, G., Đurović, S., Rakić, D., Cvetanović, A., Gunes, A. K., & Zeković, Z. (2021). Antioxidant and enzyme-inhibitory activity of peppermint extracts and essential oils obtained by conventional and emerging extraction techniques. *Food Chemistry, 338*, Article 127724. https://doi.org/10.1016/J.FOODCHEM.2020.127724

Pojić, M., Teslić, N., Banjac, V., Mrkonjić, Ž., Stupar, A., Mandić, A., … Pavlić, B. (2024). Natural deep eutectic solvents for efficient recovery of bioactive compounds from by-product of industrial hemp processing: Pretreatment, modeling and optimization. *Industrial Crops and Products, 222*, 119617. https://doi.org/10.1016/j.indcrop.2024.119617

Popović, B. M., Gligorijević, N., Aranđelović, S., Macedo, A. C., Jurić, T., Uka, D., Mocko-Blažek, K., & Serra, A. T. (2023). Cytotoxicity profiling of choline chloride-based natural deep eutectic solvents. *RSC Advances, 13*(6), 3520–3527. https://doi.org/10.1039/D2RA07488E

Popovic, B. M., Micic, N., Potkonjak, A., Blagojevic, B., Pavlovic, K., Milanov, D., & Juric, T. (2022). Novel extraction of polyphenols from sour cherry pomace using natural deep eutectic solvents—Ultrafast microwave-assisted NADES preparation and extraction. *Food Chemistry, 366*, Article 130562. https://doi.org/10.1016/J.FOODCHEM.2021.130562

Prabowo, W. C., Agustina, R., Nur, Y., Hidayati, R., Rahmawati, D., Arifuddin, M., Ambarwati, N. S. S., Purwoko, R. Y., Mun'im, A., & Ahmad, I. (2022). Green and optimum extraction of total polyphenols content from Mitragyna speciosa Korth. Havil leaves using microwave-assisted natural deep eutectic solvent extraction. *Pharmacognosy Journal, 14*(1), 29–38. https://doi.org/10.5530/pj.2022.14.5

Punzo, A., Porru, E., Silla, A., Simoni, P., Galletti, P., Roda, A., Tagliavini, E., Samorì, C., & Caliceti, C. (2021). Grape pomace for topical application: Green NaDES sustainable extraction, skin permeation studies, antioxidant and anti-inflammatory activities characterization in 3D human keratinocytes. *Biomolecules, 11*, 1181. https://doi.org/10.3390/BIOM11081181

Radošević, K., Bubalo, M. C., Srček, V. G., Grgas, D., Dragičević, T. L., & Redovniković, I. R. (2015). Evaluation of toxicity and biodegradability of choline chloride based deep eutectic solvents. *Ecotoxicology and Enviromental Safety, 112*, 46–53. https://doi.org/10.1016/j.ecoenv.2014.09.034

Radošević, K., Čanak, I., Panić, M., Markov, K., Bubalo, M. C., Frece, J., Srček, V. G., & Redovniković, I. R. (2018). Antimicrobial, cytotoxic and antioxidative evaluation of natural deep eutectic solvents. *Environmental Science and Pollution Research, 25*(14), 14188–14196. https://doi.org/10.1007/S11356-018-1669-Z/FIGURES/2

Rashid, R., Mohd Wani, S., Manzoor, S., Masoodi, F. A., & Masarat Dar, M. (2023). Green extraction of bioactive compounds from apple pomace by ultrasound assisted natural deep eutectic solvent extraction: Optimisation, comparison and bioactivity. *Food Chemistry, 398*, Article 133871. https://doi.org/10.1016/J.FOODCHEM.2022.133871

Rebocho, S., Mano, F., Cassel, E., Anacleto, B., Bronze, M. do R., Paiva, A., & Duarte, A. R. C. (2022). Fractionated extraction of polyphenols from mate tea leaves using a combination of hydrophobic/hydrophilic NADES. *Current Research in Food Science, 5*, 571–580. https://doi.org/10.1016/J.CRFS.2022.03.004

Rente, D., Cvjetko Bubalo, M., Panić, M., Paiva, A., Caprin, B., Radojčić Redovniković, I., & Duarte, A. R. C. (2022). Review of deep eutectic systems from laboratory to industry, taking the application in the cosmetics industry as an example. *Journal of Cleaner Production, 380*, Article 135147. https://doi.org/10.1016/J.JCLEPRO.2022.135147

Rocchetti, G., Gregorio, R. P., Lorenzo, J. M., Barba, F. J., Oliveira, P. G., Prieto, M. A., Simal-Gandara, J., Mosele, J. I., Motilva, M. J., Tomas, M., Patrone, V., Capanoglu, E., & Lucini, L. (2022). Functional implications of bound phenolic compounds and phenolics-food interaction: A review. *Comprehensive Reviews in Food Science and Food Safety, 21*(2), 811–842. https://doi.org/10.1111/1541-4337.12921

Rublova, Y. D., Kityk, A. A., Bannyk, N. G., Protsenko, V. S., & Danilov, F. I. (2019). The influence of various factors on corrosion of mild steel in deep eutectic solvents. *Materials Today: Proceedings, 6*, 232–236. https://doi.org/10.1016/J.MATPR.2018.10.099

Rukavina, I., Rodrigues, M. J., Pereira, C. G., Mansinhos, I., Romano, A., Slusarczyk, S., Matkowski, A., & Custódio, L. (2021). Greener is better: First approach for the use of natural deep eutectic solvents (NADES) to extract antioxidants from the. *Molecules, 26*, 6136.

Santana, A. P. R., Mora-Vargas, J. A., Guimarães, T. G. S., Amaral, C. D. B., Oliveira, A., & Gonzalez, M. H. (2019). Sustainable synthesis of natural deep eutectic solvents (NADES) by different methods. *Journal of Molecular Liquids, 293*, Article 111452. https://doi.org/10.1016/J.MOLLIQ.2019.111452

Santos, A. C., Pereira, I., Pereira-Silva, M., Ferreira, L., Caldas, M., Collado-González, M., Magalhães, M., Figueiras, A., Ribeiro, A. J., & Veiga, F. (2019). Nanotechnology-based formulations for resveratrol delivery: Effects on resveratrol in vivo bioavailability and bioactivity. *Colloids and Surfaces B: Biointerfaces, 180*, 127–140. https://doi.org/10.1016/j.colsurfb.2019.04.030

Saulais, L., Corcuff, R., & Boonefaes, E. (2023). Natural and healthy? Consumers knowledge, understanding and preferences regarding naturalness and healthiness of processed foods. *International Journal of Gastronomy and Food Science, 31*, Article 100662. https://doi.org/10.1016/J.IJGFS.2023.100662

Shahidi, F., & Peng, H. (2018). Bioaccessibility and bioavailability of phenolic compounds. *Journal of Food Bioactives, 4*, 11–68. https://doi.org/10.31665/JFB.2018.4162

Sergi, D., Gélinas, A., Beaulieu, J., Renaud, J., Tardif-Pellerin, E., Guillard, J., Martinoli, M.-G., & AntiApoptotic, M. (2021). Anti-apoptotic and anti-inflammatory role of trans -viniferin in a neuronGlia co-culture cellucar Modelmof Parkinson's disease. https://doi.org/10.3390/foods10030586.

Stupar, A., Šeregelj, V., Ribeiro, B. D., Pezo, L., Cvetanović, A., Mišan, A., & Marrucho, I. (2021). Recovery of β-carotene from pumpkin using switchable natural deep eutectic solvents. *Ultrasonics Sonochemistry, 76*, Article 105638. https://doi.org/10.1016/J.ULTSONCH.2021.105638

Sun, C., Zhao, C., Guven, E. C., Paoli, P., Simal-Gandara, J., Ramkumar, K. M., Wang, S., Buleu, F., Pah, A., Turi, V., Damian, G., Dragan, S., Tomas, M., Khan, W., Wang, M., Delmas, D., Portillo, M. P., Dar, P., Chen, L., & Xiao, J. (2020). Dietary polyphenols as antidiabetic agents: Advances and opportunities. *Food Frontiers, 1*(1), 18–44. https://doi.org/10.1002/fft2.15

Suphasomboon, T., & Vassanadumrongdee, S. (2022). Toward sustainable consumption of green cosmetics and personal care products: The role of perceived value and ethical concern. *Sustainable Production and Consumption, 33*, 230–243. https://doi.org/10.1016/J.SPC.2022.07.004

Suresh, P. S., Singh, P. P., Sharma, M., & Sharma, U. (2023). Multicomponent natural deep eutectic solvents: Super solvents for the efficient extraction of steviol glycosides (rebaudioside A) from Stevia rebaudiana. *Journal of Cleaner Production, 385*, Article 135639. https://doi.org/10.1016/J.JCLEPRO.2022.135639

Teslić, N., Pojić, M., Stupar, A., Mandić, A., Mišan, A., & Pavlić, B. (2024a). PhInd database–Polyphenol content in agri-food by-products and trends in extraction technologies: A critical review. *Food Chemistry, 458*, 140474.

Teslić, N., Pojić, M., Stupar, A., Mandić, A., Pavlić, B., & Mišan, A. (2024b). PhInd—database on polyphenol content in agri-food by-products and waste: Features of the database. *Antioxidants, 13*(1), 97.

Teslić, N., Santos, F., Oliveira, F., Stupar, A., Pojić, M., Mandić, A., Pavlić, B., Kljakić, A. C., Duarte, A. R. C., Paiva, A., & Mišan, A. (2022). Simultaneous hydrolysis of ellagitannins and extraction of ellagic acid from defatted raspberry seeds using natural deep eutectic solvents (NADES). *Antioxidants, 11*, 254. https://doi.org/10.3390/ANTIOX11020254

Tobiszewski, M., Namieśnik, J., & Pena-Pereira, F. (2017). Environmental risk-based ranking of solvents using the combination of a multimedia model and multi-criteria decision analysis. *Green Chemistry, 19*(4), 1034–1042. https://doi.org/10.1039/C6GC03424A

Torres-Vega, J., Gómez-Alonso, S., Pérez-Navarro, J., Alarcón-Enos, J., & Pastene-Navarrete, E. (2021). Polyphenolic compounds extracted and purified from Buddleja globosa hope (Buddlejaceae) leaves using natural deep eutectic solvents and centrifugal partition chromatography. *Molecules, 26*, 2192. https://doi.org/10.3390/MOLECULES26082192

Tsvetov, N., Paukshta, O., Fokina, N., Volodina, N., & Samarov, A. (2023). Application of natural deep eutectic solvents for extraction of bioactive components from Rhodiola rosea (L.). *Molecules, 28*, 912. https://doi.org/10.3390/MOLECULES28020912

Ünlü, A. E. (2021). Green and non-conventional extraction of bioactive compounds from olive leaves: Screening of novel natural deep eutectic solvents and investigation of process parameters. *Waste and Biomass Valorization, 12*(10), 5329–5346. https://doi.org/10.1007/S12649-021-014113/TABLES/8

van Osch, D. J. G. P. (2018). *Design and applications of hydrophobic deep eutectic solvents*. Issue 2018) https://pure.tue.nl/ws/portalfiles/portal/103474989/20180918_Osch.pdf.

Vasyliev, G., Lyudmyla, K., Hladun, K., Skiba, M., & Vorobyova, V. (2022). Valorization of tomato pomace: Extraction of value-added components by deep eutectic solvents and their application in the formulation of cosmetic emulsions. *Biomass Conversion and Biorefinery*, 1–17. https://doi.org/10.1007/S13399-022-02337-Z

Vrgović, P., Pojić, M., Teslić, N., Mandícmandíc, A., Cvetanović Kljakić, A., Pavlić, B., Stupar, A., Pestorić, M., Škrobot, D., & Mišan, A. (2022). Communicating function and co-creating healthy food: Designing a functional food product together with consumers. *Foods, 11*, 961. https://doi.org/10.3390/FOODS11070961

Wu, C. W., Nakamoto, Y., Hisatome, T., Yoshida, S., & Miyazaki, H. (2020). Resveratrol and its dimers εviniferin and δ-viniferin in red wine protect vascular endothelial cells by a similar mechanism with different potency and efficacy. *The Kaohsiung Journal of Medical Sciences, 36*(7), 535–542. https://doi.org/10.1002/KJM2.12199

Wu, K., Ren, J., Wang, Q., Nuerjiang, M., Xia, X., & Bian, C. (2022). Research progress on the preparation and action mechanism of natural deep eutectic solvents and their application in food. *Foods, 11*(21), 3528. https://doi.org/10.3390/FOODS11213528

Yahfoufi, N., Alsadi, N., Jambi, M., & Matar, C. (2018). The immunomodulatory and anti-inflammatory role of polyphenols. *Nutrients, 10*(11), 1618. https://doi.org/10.3390/nu10111618

Yusuf, B., Astati, S. J., Ardana, M., Herman, I.,A., Rijai, L., Nainu, F., & Ahmad, I. (2021). Optimizing natural deep eutectic solvent citric acid-glucose based microwave-assisted extraction of total polyphenols content from eleutherine bulbosa (Mill.) bulb. *Indonesian Journal of Chemistry, 21*(4), 797–805. https://doi.org/10.22146/ijc.58467

Zhang, Z., Li, X., Sang, S., McClements, D. J., Chen, L., Long, J., Jiao, A., Jin, Z., & Qiu, C. (2022). Polyphenols as plant-based nutraceuticals: Health effects, encapsulation, nano-delivery, and application. *Foods, 11*(15), 2189. https://doi.org/10.3390/FOODS11152189

Zhao, B.-Y., Xu, P., Yang, F.-X., Wu, H., Zong, M.-H., & Lou, W.-Y. (2015). Biocompatible deep eutectic solvents based on choline chloride: Characterization and application to the extraction of rutin from Sophora japonica. *Sustainable Chemistry and Engineering, 3*, 2746–2755. https://doi.org/10.1021/acssuschemeng.5b00619

CHAPTER

By-products as rich sources of plant polyphenols

3

Paulo Eduardo Sichetti Munekata

Centro Tecnológico de la Carne de Galicia, Ourense, Spain

1. Introduction

Polyphenols are naturally occurring compounds widely distributed across vegetables around the globe. Their main role is to support secondary metabolism in plant tissue, which includes defense against oxidative damage, herbivory, and pathogens; and protection from environmental stressors (UV radiation, heavy metals, and temperature fluctuation, for instance) (Singh et al., 2021). The natural occurrence of polyphenols involves the induction of the phenylpropanoid pathway (Sharma et al., 2019). The general pathway involves enzymatic reactions that initiate with the deamination of phenylalanine by the action of phenylalanine ammonia lyase. Sequential reactions catalyzed by several enzymes (such as 4-hydroxycinnamoyl-CoA ligase and caffeic acid/5-hydroxyferulic acid O-methyltransferase for the general synthesis of polyphenols and cinnamoyl-CoA reductase, exclusively for the synthesis of lignans) lead to the formation of the myriad of compounds reported in scientific literature (Yadav et al., 2020).

Among the main sources of polyphenols are the fruits and vegetables composed the diet of global population. Numerous papers support the biological and technological importance of these compounds with potential application in food, pharmaceutical, and cosmetic sectors (Rana et al., 2022). Polyphenols can delay oxidative reactions, slow microbial spoilage, and have also been suggested to assist in the preservation of health and management of diseases (Dias et al., 2021).

Although the consumption of fruits and vegetables is widely advised, their harvesting and processing are also regarded as an important environmental concern and a waste of high-added value compounds, especially polyphenols (Fierascu et al., 2020; Majerska et al., 2019). The presence of phenolic compounds in fruits and vegetable by-products is a known fact that could be explored and improved the sustainability of food sector, especially by means of circular economy and strengthening connections within food industry and other industrial sectors, especially pharmaceutical and cosmetic (Campos et al., 2020; Fidelis et al., 2019). Thus, this chapter aims to overview the residues generated from harvesting and processing of fruits and vegetables in terms of their phenolic profile, total phenolic content, and factors affecting their diversity and content.

Plant Polyphenols in the Food and Cosmetic Industries. https://doi.org/10.1016/B978-0-443-19137-4.00006-5

2. Polyphenol classification

Polyphenols is a diverse group of compounds characterized by the presence of phenol group and organized according to structural similarity (common skeleton). Due to the several categories, some of the most relevant categories for this chapter are listed and include *Phenolic acids* (Hydroxycinnamic acids and Hydroxybenzoic acids), *Flavonoids* (Anthocyanins, Flavanols, Flavanones, Flavones, Flavonols, Isoflavonoids, Chalcones, and Dihydrochalcones), *Stilbenes*, and *Other polyphenols* (Alkylphenols, Hydroxybenzaldehydes, Hydroxycoumarins, Tyrosols, and Xanthones) (Crozier et al., 2009; Phenol-Explorer, 2015; Singla et al., 2019). Some other classes of polyphenols have also been listed: Lignans, Ellagitannins, Curcuminoids, Furanocoumarins, Methoxyphenols, and Naphtoquinones (Phenol-Explorer, 2015). Examples of compounds belonging to each category/subcategory are presented on Table 3.1. Following section will disclose the main groups and/or main phenolic compounds found across different sources of fruits and vegetables by-products.

3. Phenolic profile of tropical and subtropical fruits by-products

The production of tropical fruits accounts for a major share of its market (about 60%) (Md Nor & Ding, 2020). Many factors seem to explain the importance of this sector in the fruit market such as reduced fluctuation in productivity due to climate consistency throughout the year and global demand due to its popularity and taste (Cheok et al., 2018).

3.1 Orange and citrus fruits

The characterization of harvesting and processing residues comprises an importance source of phenolic acids (especially hydroxycinnamic acids) and flavonoids (Table 3.2). The quantification of polyphenols in orange peel indicated hesperetin 7-O-rutinoside and narirutin as main compounds (Barrales et al., 2018). In another study, the identification of compounds also indicated the presence of other flavonoids (naringenin 7-O-neohesperidoside and quercetin 3-O-rutinoside), and phenolic acids (*p*-coumaric acid, ferulic acid, and gallic acid) (Ozturk et al., 2018). Regarding Total Phenolic Content (TPC), different research groups indicate values between 1 and 10 mg GAE/g (Barrales et al., 2018; Ozturk et al., 2018). Pomace consists in the whole residue generated from orange pressing and is also considered as an important source of polyphenols especially hesperetin 7-O-rutinoside and narirutin (Barrales et al., 2018; Multari et al., 2020). Other polyphenols found in orange pomace include flavonoids (especially quercetin 3-O-rhamnoside, followed by narirutin, luteolin, and isorhamnetin 3-O-rutinoside) and vanillic acid (Multari et al., 2020).

Table 3.1 Categories of polyphenols and examples related to fruits and vegetables by-products.

Category	Subcategory	Examples
Phenolic acids	Hydroxycinnamic acids	Caffeic acid, Caftaric acid/Caffeoyl tartaric acid, Chicoric acid/Dicaffeoyl tartaric acid, 1-Caffeoylquinic acid, 3-Caffeoylquinic acid/Neochlorogenic acid, 4-Caffeoylquinic acid/Cryptochlorogenic acid, 5-Caffeoylquinic acid/Chlorogenic acid, Isochlorogenic acid, Caffeoyltartaric acid, *p*-Coumaric acid/*p*-Hydroxycinnamic acid, Ferulic acid, and Sinapic acid/Sinapinic acid
	Hydroxybenzoic acids	Benzoic acid, 4-Hydroxybenzoic acid/p-Hydroxybenzoic acid, Ellagic acid/Benzoaric acid, Gallic acid, Protocatechuic acid, Salicylic acid/2-Hydroxybenzoic acid, Syringic acid, and Vanillic acid
Flavonoids	Anthocyanins	Cyanidin, Cyanidin 3-*O*-glucoside, Cyanidin 3-*O*-galactoside, Cyanidin 3-*O*-xyloside, Cyanidin 3-*O*-arabinoside, Cyanidin 3-*O*-rutinoside, Delphinidin 3-*O*-glucoside, Delphinidin, Delphinidin 3-*O*-rutinoside, Malvidin 3-*O*-glucoside, Peonidin, Peonidin 3-*O*-glucoside, and Petunidin 3-*O*-glucoside
	Flavanols	Catechin, Epicatechin, (Epi)Catechin gallate, Epigallocatechin, Procyanidin dimers, Procyanidin trimers, Procyanidin tetramers, Procyanidin pentamers, Proanthocyanidins dimers, Proanthocyanidins trimers, Proanthocyanidins tetramers, Proanthocyanidins pentamers, Proanthocyanidins hexamers, Proanthocyanidins heptamers, Proanthocyanidins octamers, and Prodelphinidin dimer
	Flavanones	Eriodictyol 7-*O*-rutinoside/Eriocitrin, Hesperetin 7-*O*-rutinoside/Hesperidin, Naringenin, Naringenin 7-*O*-neohesperidoside/Naringin, Naringenin 7-*O*-rutinoside/Narirutin, and Narirutin
	Flavones	Apigenin, Apigenin glucuronide, Apigenin 7-*O*-rutinoside/Isorhoifolin, Apigenin glucoside, Luteolin, Luteolin 7-*O*-glucuronide, Luteolin 7-*O*-glucoside, Luteolin 7-*O*-rutinoside, and Maysin
	Flavonols	Quercetin, Quercetin 3-*O*-arabinoside/Avicularin, Quercetin 3-*O*-galactoside/Hyperin, Quercetin 3-*O*-glucoside/Isoquercitrin/Isoquercetin, Quercetin 3-*O*-glucuronide, Quercetin 3-*O*-rutinoside/Rutin, Quercetin 3-rhamnoside/Quercitrin, Kaempferol, Kaempferol 3-*O*-rutinoside, Myricetin, Isorhamnetin 3-*O*-glucoside, and Isorhamnetin 3-*O*-rutinoside
	Isoflavonoids	Daidzein/Daidzeol, Daidzin/Daidzein 7-*O*-glucoside, Formononetin/Biochanin B/Neochanin/Pratol, Genistein/Prunetol/Sophoricol/Genisteol, Genistin/Genistein 7-*O*-glucoside, Glycitein, and Stevenin
	Chalcones	Butein and 4′-4-5-Hydroxy-2′-methoxychalcone
	Dihydrochalcones	Phloretin, Phloretin 2′-O-glucoside/Phloridzin, and Phloretin 2′-xyloglucoside
Stilbenes	–	Resveratrol
Other polyphenols	Tyrosols	Tyrosol, Hydroxytyrosol, Hydroxytyrosol glucoside, Hydroxytyrosol 4-β-glucoside, and Oleuropein
	Xanthones	Mangiferin, Mangiferin gallate, and Isomangiferin
	Hydroxycoumarins	Coumarin, 4-Hydroxycoumarin, and Esculetin
	Hydroxybenzaldehydes	4-Hydroxybenzaldehyde
	Alkylphenols	4-Vinylphenol

From Crozier, A., Jaganath, I. B., & Clifford, M. N. (2009). Dietary phenolics: Chemistry, bioavailability and effects on health. Natural Product Reports, *26(8), 1001–1043. http://doi.org/10.1039/b802662a; Phenol-Explorer. (2015). Polyphenol classes. Retrieved November 29, 2023, from http://phenol-explorer.eu/compounds/classification; Singla, R. K., Dubey, A. K., Garg, A., Sharma, R. K., Fiorino, M., Ameen, S. M. et al. (2019). Natural polyphenols: Chemical classification, definition of classes, subcategories, and structures.* Journal of AOAC International, *102(5), 1397–1400. http://doi.org/10.5740/jaoacint.19-0133.*

Table 3.2 Phenolic compounds found in fruit harvesting and processing by-products.

Fruit	By-product	TPC	TFC	Main identified phenolic compounds (concentration)	Ref.
Tropical and subtropical fruits					
Orange	Peel	1–6 mg GAE/g	n.d.	*p*-Coumaric acid (n.q.) Ferulic acid (n.q.) Gallic acid (n.q.) Naringenin 7-*O*-neohesperidoside (n.q.) Quercetin 3-*O*-rutinoside (n.q.)	Ozturk et al. (2018)
Orange	Peel	5–6 mg GAE/g	n.d.	Hesperetin 7-*O*-rutinoside (18 mg/g DW) Narirutin (11 mg/g DW)	Barrales et al. (2018)
Orange	Peel	2–10 mg GAE/g	n.d.	Hesperetin 7-*O*-rutinoside (12–58 mg/g DW) Narirutin (2.4–15 mg/g DW)	Barrales et al. (2018)
Orange	Peel	6 mg GAE/g	n.d.	Hesperetin 7-*O*-rutinoside (7–8 mg/g DW) Narirutin (5–6 mg/g DW)	Barrales et al. (2018)
Orange	Pomace	n.d.	n.d.	Hesperetin 7-*O*-rutinoside (3 mg/g DW) Narirutin (2 mg/g DW)	Barrales et al. (2018)
Orange	Pomace	n.d.	n.d.	Hesperetin 7-*O*-rutinoside (2–19 mg/g DW) Narirutin (1–5 mg/g DW)	Barrales et al. (2018)
Orange	Pomace	n.d.	n.d.	Hesperetin 7-*O*-rutinoside (2 mg/g DW) Narirutin (1 mg/g DW)	Barrales et al. (2018)
Orange	Pomace	≈800 mg/kg	n.d.	Hesperetin 7-*O*-rutinoside (314 mg/kg) Quercetin 3-*O*-rhamnoside (125 mg/kg) Narirutin (43 mg/kg) Vanillic acid (20 mg/kg) Luteolin (19 mg/kg) Isorhamnetin 3-*O*-rutinoside (18 mg/kg)	Multari et al. (2020)
Orange	Wastewater	50 mg/100 g	n.d.	Hesperetin 7-*O*-rutinoside (13 mg/100 g) Quercetagetin (4 mg/100 g) Peonidin (3 mg/100 g) Apigenin 7-*O*-glucoside (3 mg/100 g) Cyanidin (2 mg/100 g) Quercetin + naringenin (2 mg/100 g)	Argun et al. (2022)

Bergamot	Pomace	410 mg/kg	n.d.	Hesperetin 7-*O*-rutinoside (254 mg/kg) Eriocitrin (156 mg/kg) Quercetin 3-*O*-rutinoside (141 mg/kg) Narirutin (130 mg/kg) Quercetin 3-*O*-rhamnoside (82 mg/kg) Isorhamnetin 3-*O*-rutinoside (81 mg/kg)	Multari et al. (2020)
Lemon	Pomace	886 mg/kg	n.d.	Hesperetin 7-*O*-rutinoside (377 mg/kg) Quercetin 3-*O*-rhamnoside (150 mg/kg) Narirutin (141 mg/kg) Isorhamnetin 3-*O*-rutinoside (42 mg/kg) Luteolin (42 mg/kg) Quercetin 3-*O*-rutinoside (33 mg/kg)	Multari et al. (2020)
Yuzu (*Citrus junos*)	Pomace	1712–1894 mg/100 g DW	n.d.	Hesperetin 7-*O*-rutinoside (1048–1267 mg/100 g DW) Naringenin 7-*O*-neohesperidoside (155–220 mg/100 g DW) Kaempferol rutinoside isomer (76–95 mg/100 g DW) Sinapoyl hexoside derivative (26–32 mg/100 g DW) Hesperetin 7-*O*-rutinoside isomer (1–26 mg/100 g DW)	Tao et al. (2014)
Mandarin	Pomace	≈600 mg/kg	n.d.	Quercetin (126 mg/kg) Eriocitrin (109 mg/kg) Narirutin (68 mg/kg) Hesperetin 7-*O*-rutinoside-7-*O*-glucoside (39 mg/kg) Luteolin 7-*O*-glucoside (18 mg/kg) Hesperetin 7-*O*-rutinoside (10 mg/kg)	Multari et al. (2020)
Grapefruit	Peel	75–147 mg GAE/100 g	n.d.	Gallic acid (42 μg/g DW) Naringenin 7-*O*-neohesperidoside (22 μg/g DW) Ferulic acid (18 μg/g DW)	Peng et al. (2021)

Continued

Table 3.2 Phenolic compounds found in fruit harvesting and processing by-products.—*cont'd*

Fruit	By-product	TPC	TFC	Main identified phenolic compounds (concentration)	Ref.
Avocado	Peel	16–25 mg CE/g DW	n.d.	Procyanidin dimer B (149 μg/g DW) Quercetin 3-*O*-galactoside (31–144 μg/g DW) Quercetin 3-*O*-arabinoside (94 μg/g DW) Quercetin derivative (82 μg/g DW) 5-Caffeoylquinic acid (77–82 μg/g DW)	Kosińska et al. (2012)
Avocado	Peel	7 mg/100 g	6 mg/100 g	Procyanidin dimer B (2262 μg/100 g) Epicatechin (1891 μg/100 g) 5-Caffeoylquinic acid (969 μg/100 g) Procyanidin trimer B (383 μg/100 g) Quercetin arabinoside glucoside (374 μg/100 g) Quercetin diglucoside (294 μg/100 g) Kaempferol arabinoside glucoside (161 μg/100 g) Procyanidin tetramer (106 μg/100 g)	Velderrain-Rodríguez et al. (2021)
Avocado	Peel	29–78 mg GAE/g	1–3 mg QE/g	Caffeic acid (n.q.) Catechin (n.q.) 5-Caffeoylquinic acid (n.q.) Gallic acid (n.q.) *p*-Hydroxybenzoic acid (n.q.) Protocatechuic acid (n.q.) Quercetin (n.q.)	Lyu et al. (2023)
Avocado	Seed	9–13 mg CE/g DW	n.d.	Catechin/epicatechin gallate (105–153 μg/g DW) Procyanidin trimer A (II) (73–89 μg/g DW) Procyanidin trimer A (I) (82–99 μg/g DW) 3-Caffeoylquinic acid (53–57 μg/g DW) 3-*p*-Coumaroylquinic acid (8–14 μg/g DW)	Kosińska et al. (2012)
Avocado	Seed	2 mg/100 g	1 mg/100 g	Caffeoylshikimic acid (2210 μg/100 g) Epicatechin (360 μg/100 g) Catechin (281 μg/100 g) Procyanidin trimer A (231 μg/100 g) Procyanidin dimer B (208 μg/100 g) 3-Caffeoylquinic acid (176 μg/100 g) Salidroside (148 μg/100 g)	Velderrain-Rodríguez et al. (2021)

Avocado	Seed	27–45 mg GAE/g	1–3 mg QE/g	Caffeic acid (n.q.) Catechin (n.q.) 5-Caffeoylquinic acid (n.q.) Epicatechin gallate (n.q.) Gallic acid (n.q.) *p*-Hydroxybenzoic acid (n.q.) Quercetin (n.q.)	Lyu et al. (2023)
Avocado	Leaf	27 mg GAE/g DW	n.d.	Dihydroferulic acid 4-*O*-glucuronide (517 µg/g DW) Quercetin 3-*O*-rutinoside isomer c (332 µg/g DW) Quercetin glucuronide (313 µg/g DW) 5-Caffeoylquinic acid (311 µg/g DW) Protocatechuic acid-4-glucoside (302 µg/g DW) Quercetin 3-*O*-arabinosyl-glucoside isomer a (269 µg/g DW) Quercetin 3-*O*-glucoside isomer b (222 µg/g DW) Quercetin-diglucoside isomer a (215 µg/g DW)	De Montijo-Prieto et al. (2023)
Banana	Flower	1075 mg/100 DW	n.d.	Caffeic acid (10 mg/100 g DW) Hydroxybenzoic acid (7 mg/100 g DW)	Ramírez-Bolaños et al. (2021)
Banana	Pseudo-stem	886 mg/100 DW	n.d.	Protocatechuic acid (49 mg/100 g DW) Isoferulic acid (6 mg/100 g DW)	Ramírez-Bolaños et al. (2021)
Banana	Outer bract	8 mg/g DW	n.d.	Ferulic acid (267 mg/kg) Salicylic acid (106 mg/kg) *p*-Coumaric acid (35 mg/kg) Quercetin dihydrate (5 mg/kg) Quinic acid (4 mg/kg)	Begum and Deka (2019)
Banana	Inner bract	9 mg/g DW	n.d.	Ferulic acid (144 mg/kg) Salicylic acid (39 mg/kg) Quinic acid (9 mg/kg) 5-Caffeoylquinic acid (7 mg/kg) *p*-Coumaric acid (6 mg/kg) Catechin hydrate (3 mg/kg)	Begum and Deka (2019)

Continued

Table 3.2 Phenolic compounds found in fruit harvesting and processing by-products.—*cont'd*

Fruit	By-product	TPC	TFC	Main identified phenolic compounds (concentration)	Ref.
Mango	Peel	≈45 g/kg DW	≈3 g/kg DW	Mangiferin (11–13 g/kg DW) Valoneic acid dilactone (6–7 g/kg DW) Ethyl 2,4-dihydroxy-3-((3,4,5-trihydroxybenzoyl)oxy)-benzoate (2–3 g/kg DW) *m*-Digallic acid methyl ester (2–3 g/kg DW) Galloyl glucose (2–3 g/kg DW) Ellagic acid (1–2 g/kg DW)	de Ancos et al. (2018)
Mango	Peel	4 g/kg DW	n.d.	Mangiferin (1690 mg/kg DW) Quercetin 3-*O*-galactoside (651 mg/kg DW) Quercetin 3-*O*-glucoside (558 mg/kg DW) Mangiferin gallate (322 mg/kg DW) Quercetin 3-*O*-xyloside (207 mg/kg DW) Isomangiferin (134 mg/kg DW) Quercetin 3-*O*-arabinofuranoside (104 mg/kg DW) Quercetin 3-*O*-arabinopyranoside (102 mg/kg DW) Rhamnetin 3-*O*-galactoside/glucoside (94 mg/kg DW)	Berardini et al. (2005)
Mango	Paste	≈40 g/kg DW	≈2 g/kg DW	Valoneic acid dilactone (3 g/kg DW) Mangiferin (2 g/kg DW) Galloyl glucose (1 g/kg DW) Ethyl gallate (1 g/kg DW) Ethyl 2,4-dihydroxy-3-((3,4,5-trihydroxybenzoyl)oxy)-benzoate (1 g/kg DW)	de Ancos et al. (2018)
Pineapple	Peel	405 mg GAE/100 g DW	n.d.	Caffeic acid (144–218 mg/100 g DW) Ferulic acid (146–148 mg/100 g DW) *trans*-Cinnamic acid (113–138 mg/100 g DW) Coumaric acid (97–129 mg/100 g DW) Syringic acid (90–97 mg/100 g DW) Gallic acid (57–66 mg/100 g DW)	Zampar et al. (2022)

Pineapple	Crown	1533–1868 mg GAE/100 g	306–551 mg CE/100 g	Caffeic acid (n.q.) 5-Caffeoylquinic acid I (n.q.) *p*-Coumaric acid (n.q.) Daidzein 7-*O*-glucoside (n.q.) Stevenin I (n.q.) Ferulic acid (n.q.) Isorhamnetin 3-*O*-glucoside (n.q.)	Brito et al. (2021)
Cashew apple	Pomace	8–10 mg GAE/g	n.d.	Myricetin (43–44 mg/kg) Vanillic acid (31–32 mg/kg) Gallic acid (11–29 mg/kg) Quercetin 3-*O*-rutinoside (7–14 mg/kg) *p*-Coumaric acid (8–12 mg/kg) Protocatechuic acid (3–8 mg/kg)	Freitas et al. (2023)
Cocoa	Bean shells	22–36 mg GAE/g	n.d.	Epicatechin (1–2 mg/g) Protocatechuic acid (1 mg/g) Caffeic acid (0.7–0.9 mg/g) 5-Caffeoylquinic acid (0.8 mg/g) Ferulic acid (0.3–0.5 mg/g)	Mellinas et al. (2020)
Small fruits					
Grape	Seed	312–492 mg GAE/g	n.d.	Catechin (8–17 mg/g) Epicatechin (9–18 mg/g) Procyanidin A (6–11 mg/g) Procyanidin B (5–7 mg/g)	Cheng et al. (2012)
Grape	Skin	14–57 mg GAE/g	n.d.	Malvidin 3-*O*-glucoside (2–11 mg/g) Delphinidin 3-glucoside (0.1–5 mg/g) Quercetin methyl-glucoside (0.4–2 mg/g) Quercetin (0.1–2 mg/g) Petunidin 3-*O*-glucoside (0.2–1 mg/g) Peonidin 3-*O*-glucoside (0.7–1 mg/g)	Cheng et al. (2012)

Continued

Table 3.2 Phenolic compounds found in fruit harvesting and processing by-products.—*cont'd*

Fruit	By-product	TPC	TFC	Main identified phenolic compounds (concentration)	Ref.
Grape	Pomace	55–148 mg GAE/g	n.d.	Catechin (3–12 mg/g) Epicatechin (2–7 mg/g) Malvidin 3-*O*-glucoside (2–6 mg/g) Procyanidin A (0.7–4 mg/g) Procyanidin B (0.5–4 mg/g)	Cheng et al. (2012)
Grape	Pomace	76–281 mg GAE/g	n.d.	Catechin (3366–8973 μg/g) 5-Caffeoylquinic acid (526–5327 μg/g) Gallic acid (675–1525 μg/g) Quercetin 3-*O*-rutinoside (110–1662 μg/g) Quercetin (105–1144 μg/g) Kaempferol (60–713 μg/g) Epicatechin (22–504 μg/g)	Sagdic et al. (2011)
Grape	Wine lees	380–691 mg/L	242–542 mg/L	Gallic acid (96–121 mg/L) Catechin (39–98 mg/L) Procyanidin dimer (32–64 mg/L) Malvidin 3-*O*-glucoside (27–61 mg/L) Epicatechin (13–43 mg/L) Quercetin (25–37 mg/L) Procyanidin dimer B2 (9–35 mg/L)	López-Fernández-Sobrino et al. (2021)
Grape	Wine lees	4–73 mg GAE/g DW	n.d.	Caftaric acid (0.3–9 mg/g) Quercetin (0.6–7 mg/g) Catechin (0.1–6 mg/g) Procyanidin A (0.1–5 mg/g) Quercetin methyl-glucoside (0.2–4 mg/g) Epicatechin gallate (0.1–4 mg/g) Procyanidin C (0.1–3 mg/g)	Ye et al. (2022)
Black chokeberry	Pomace	12 g/100 g DW	n.d.	Procyanidins (9586 mg/100 g DW) Cyanidin 3-*O*-galactoside (1190 mg/100 g DW) Cyanidin 3-*O*-arabinoside (454 mg/100 g DW) 3-Caffeoylquinic acid (220 mg/100 g DW) 5-Caffeoylquinic acid (219 mg/100 g DW) Cyanidin 3-*O*-glucoside (96 mg/100 g DW) Cyanidin 3-*O*-xyloside (90 mg/100 g DW)	Rodríguez-Werner et al. (2019)

Blackcurrant	Pomace	208–2094 mg GAE/100 g	n.d.	Delphinidin 3-*O*-rutinoside (54–629 mg/100 g DW) Cyanidin 3-*O*-rutinoside (32–373 mg/100 g DW) Delphinidin 3-*O*-glucoside (22–251 mg/100 g DW) *p*-Coumaric acid (29–126 mg/100 g DW) Caffeic acid (26–84 mg/100 g DW) Myricetin (9 mg/100 g DW)	Azman et al. (2022)
Saskatoon berry	Pomace	404 mg/100 g	n.d.	Cyanidin 3-*O*-galactoside (114 mg/100 g) Cyanidin 3-*O*-glucoside (21 mg/100 g) Cyanidin 3-*O*-arabinoside (12 mg/100 g) Cyanidin 3-*O*-xyloside (8 mg/100 g)	de Souza et al. (2019)
Saskatoon berry	Pomace	16–43 mg CAE/g DW	4–10	Quercetin-arabinoglucoside (162–164 mg/100 g DW) Delphinidin 3-*O*-glucoside (45–150 mg/100 g DW) Cyanidin-glucoside (33–87 mg/100 g DW) 5-Caffeoylquinic acid (33–67 mg/100 g DW) Cyanidin-galactoside (25–64 mg/100 g DW) 4-Caffeoylquinic acid (30–34 mg/100 g DW)	Li et al. (2014)
Blackberry	Pomace	4 mg GAE/100 g	n.d.	5-Caffeoylquinic acid (111 μg/g) *p*-Coumaric acid (60 μg/g) Gallic acid (72 μg/g)	Rojas-Ocampo et al. (2021)
Blueberry	Pomace	3 mg GAE/100 g	n.d.	5-Caffeoylquinic acid (1277 μg/g) Epicatechin (1114 μg/g) Caffeic acid (144 μg/g) Gallic acid (103 μg/g) Catechin (90 μg/g) *p*-Coumaric acid (78 μg/g)	Rojas-Ocampo et al. (2021)
Elderberry	Pomace	5 mg GAE/100 g	n.d.	Epicatechin (984 μg/g) 5-Caffeoylquinic acid (510 μg/g) Catechin (153 μg/g) *p*-Coumaric acid (58 μg/g) Caffeic acid (44 μg/g)	Rojas-Ocampo et al. (2021)

Continued

Table 3.2 Phenolic compounds found in fruit harvesting and processing by-products.—*cont'd*

Fruit	By-product	TPC	TFC	Main identified phenolic compounds (concentration)	Ref.
Goldenberry	Pomace	1 mg GAE/100 g	n.d.	Epicatechin (145 μg/g) 5-Caffeoylquinic acid (108 μg/g) *p*-Coumaric acid (60 μg/g)	Rojas-Ocampo et al. (2021)
Kiwi berry	Peel	276–447 mg GAE/100 g	69–156 mg CE/100 g	Caffeic acid (72–109 μg/g) 5-Caffeoylquinic acid (60–99 μg/g) Protocatechuic acid (67–96 μg/g) Quinic acid (34–70 μg/g) Quercetin 3-*O*-galactoside (12–23 μg/g) Quercetin 3-*O*-rutinoside (6–20 μg/g) Quercetin 3-*O*-glucoside (5–17 μg/g)	Zhang et al. (2021)
Kiwi	Pomace	2–86 mg CE/g DW	1–34 mg QE/g DW	5-Caffeoylquinic acid (n.q.) *p*-Coumaric acid (n.q.) Ferulic acid (n.q.) Protocatechuic acid (n.q.) Caffeic acid derivative (n.q.) Catechin (n.q.) Epicatechin (n.q.) Quercetin (n.q.)	Kheirkhah et al. (2019)
Stone fruits					
Peach	Stone	0.5 mg GAE/g	0.2 mg QE/g	Quercetin (20 mg/g) *p*-Hydroxybenzoic acid (19 mg/g) 5-Caffeoylquinic acid (16 mg/g) Catechin (7 mg/g) Kaempferol (6 mg/g) Gallic acid (3 mg/g)	Hong et al. (2021)
Peach	Stone	n.d.	n.d.	Catechin (3–18 mg/kg) *p*-Hydroxybenzoic acid (10–15 mg/kg) Ferulic acid (3–11 mg/kg) *p*-Hydroxyphenylacetic acid (4–9 mg/kg) 5-Caffeoylquinic acid (1–6 mg/kg) *p*-Coumaric acid (1–4 mg/kg) Protocatechuic acid (2–4 mg/kg)	Koprivica et al. (2018)

Peach	Peel	89–277 mg GAE/100 g	39–245 mg CE/100 g	5-Caffeoylquinic acid (7–31 mg/100 g) 3-Caffeoylquinic acid (1–8 mg/100 g) Cyanidin 3-*O*-glucoside (1–5 mg/100 g) Quercetin 3-*O*-galactoside + Quercetin 3-*O*-rutinoside (1–2 mg/100 g)	Saidani et al. (2017)
Plum	Stone	0.9 mg GAE/g	0.6 mg QE/g	Quercetin (15 mg/g) Catechin (15 mg/g) *p*-Hydroxybenzoic acid (14 mg/g) 5-Caffeoylquinic acid (12 mg/g) Kaempferol (8 mg/g) Caffeic acid (3 mg/g)	Hong et al. (2021)
Plum	Pomace	5–50 mg ECE/100 g DW	n.d.	3-Caffeoylquinic acid (6 mg/100 g) Quercetin glucoside and rutinoside (3 mg/100 g) 5-Caffeoylquinic acid (2 mg/100 g)	Sójka et al. (2015)
Apricot	Stone	0.6 mg GAE/g	0.2 mg QE/g	Catechin (8 mg/g) Quercetin (7 mg/g) 5-Caffeoylquinic acid (5 mg/g) *p*-Hydroxybenzoic acid (4 mg/g) Protocatechuic acid (2 mg/g) Kaempferol (2 mg/g)	Hong et al. (2021)
Nectarine	Stone	0.3 mg GAE/g	0.2 mg QE/g	Quercetin (12 mg/g) *p*-Hydroxybenzoic acid (10 mg/g) Catechin (10 mg/g) Caffeic acid (5 mg/g) 5-Caffeoylquinic acid (4 mg/g) Kaempferol (3 mg/g)	Hong et al. (2021)
Pome Fruits					
Apple	Pomace	182–375 mg/100 g DW	n.d.	5-Caffeoylquinic acid (66–114 mg/100 g DW) Epicatechin (12–31 mg/100 g DW) Phloretin 2′-xyloglucoside (11–20 mg/100 g DW) Phloridzin (9–39 mg/100 g DW) Procyanidin dimer B (16–29 mg/100 g DW) Quercetin 3-O-glucoside (13–24 mg/100 g DW)	Arraibi et al. (2021)

Continued

Table 3.2 Phenolic compounds found in fruit harvesting and processing by-products.—*cont'd*

Fruit	By-product	TPC	TFC	Main identified phenolic compounds (concentration)	Ref.
Apple	Pomace	10–132 mg/100 g DW	n.d.	Quercetin-*O*-pentoside (15–19 mg/100 g DW) Quercetin-*O*-hexoside II (15–19 mg/100 g DW) Quercetin 3-*O*-rhamnoside (15–18 mg/100 g DW) Quercetin-*O*-hexoside I (18 mg/100 g DW) Isorhamnetin 3-*O*-rutinoside (18 mg/100 g DW) 4-Caffeoylquinic acid (4–15 mg/100 g DW)	Arraibi et al. (2021)
Apple	Pomace	40–49 mg GAE/g	n.d.	Quercetin (1–10 mg/g) Phloretin (1–9 mg/g) Phloridzin (1–2 mg/g) Coumaric acid (n.q.) Caffeic acid (n.q.) Epicatechin (n.q.)	S. Rana et al. (2014)
Apple	Pomace	1–6 g GAE/kg DW	n.d.	Phloridzin (3–380 mg/kg DW) Quercetin 3-*O*-galactoside (1–254 mg/kg DW) Quercetin 3-*O*-arabinoside (2–185 mg/kg DW) Phloretin 2′-xyloglucoside (1–171 mg/kg DW) 5-Caffeoylquinic acid (0–171 mg/kg DW) Protocatechuic acid (5–134 mg/kg DW) Quercetin 3-rhamnoside (3–132 mg/kg DW) Quercetin 3-*O*-rutinoside + Isoquecitrin (1–122 mg/kg DW)	Suárez et al. (2010)
Apple	Pomace	4–14 mg GAE/kg	n.d.	5-Caffeoylquinic acid (96–1416 mg/kg) Procyanidin trimer C1 + tetramer (50–1316 mg/kg) Phloretin 2′-xyloglucoside (20–996 mg/kg) Procyanidin B2 (92–590 mg/kg) Quercetin 3-*O*-galactoside (113–463 mg/kg) Phloridzin (160–435 mg/kg)	García et al. (2009)

Apple	Pomace	200–500 mg GAE/kg	n.d.	Procyanidin trimer (4–86 mg/kg) Procyanidin B2 (2–66 mg/kg) Quercetin 3-*O*-galactoside (6–44 mg/kg) Quercetin 3-*O*-rhamnoside (7–40 mg/kg) Procyanidin trimer III (1–36 mg/kg) 5-Caffeoylquinic acid (2–35 mg/kg) Quercetin 3-*O*-glucoside (3–34 mg/kg)	Persic et al. (2017)
Melons					
Melon	Peel	45 mg GAE/100 g DW	n.d.	Luteolin 7-*O*-glucoside (753 mg/100 g DW) Luteolin 6-glucoside (206 mg/100 g DW) *p*-Hydroxybenzoic acid (161 mg/100 g DW) Gallic acid (113 mg/100 g DW) *p*-Coumaric acid (62 mg/100 g DW) Caffeic acid (30 mg/100 g DW) Ferulic acid (23 mg/100 g DW)	Gómez-García et al. (2021)
Melon	Peel	111–703 mg GAE/100 g	104–262 μg CE/100 g	Gallic acid (9–100 μg/mL) Catechin (4–27 μg/mL)	Rolim et al. (2018)
Melon	Peel	25 mg GAE/g	15 mg CE/g	Gallic acid (2 mg/g) Ellagic acid (0.6 mg/g) Kaempferol (0.3 mg/g)	Vella et al. (2019)
Melon	Seed	70–112 mg GAE/100 g	93–109 μg CE/100 g	Gallic acid (11 μg/mL) Catechin (1–2 μg/mL) Vanillin (1 μg/mL) Vanillic acid (1 μg/mL)	Rolim et al. (2018)
Melon	Seed	1.5 mg GAE/g	0.7 mg CE/g	Ferulic acid (1.5 mg/g) Kaempferol (0.5 mg/g)	Vella et al. (2019)
Melon	Seed	10–80 mg GAE/100 g	n.d.	Caffeic acid (66 mg/100 g) Syringic acid derivative (9 mg/100 g) Gallic acid (7 mg/100 g) Ellagic acid (6 mg/100 g) Catechin derivative (5 mg/100 g) Catechin (4 mg/100 g)	Zeb (2016)

CE, *Catechin Equivalents;* CEA, *Chlorogenic acid Equivalents;* DW, *Dry weight;* ECE, *Epicatechin Equivalents;* GAE, *Gallic Acid Equivalents;* n.d., *Not determined;* n.q., *Not quantified;* QE, *Quercetin Equivalents;* TFC, *Total Flavonoid Content;* TPC, *Total Phenolic Content.*

A study with wastewater obtained from orange processing is another evidence that polyphenols are also wasted in this residue (Argun et al., 2022). In a similar way to reported to orange peel and pomace, hesperetin 7-*O*-rutinoside is indicated as main polyphenol found in orange processing wastewater, followed by other flavonoids: quercetagetin, peonidin, and apigenin 7-*O*-glucoside, for instance. A recent study indicated the presence of several flavonoids in the pomace obtained from bergamot (Multari et al., 2020). Hesperetin 7-*O*-rutinoside was the main polyphenol, which was followed by eriocitrin, quercetin 3-*O*-rutinoside, and narirutin. A similar profile was reported in the same study for lemon wherein hesperetin 7-*O*-rutinoside was the main polyphenol, followed by quercetin 3-*O*-rhamnoside and narirutin. The predominant content of hesperetin 7-*O*-rutinoside was also observed in another with yuzu (*Citrus junos*) pomace (Tao et al., 2014). Differently than aforementioned citrus, yuzu pomace also contained naringenin 7-*O*-neohesperidoside, a kaempferol rutinoside isomer and a sinapoyl hexoside derivative.

In the case of mandarin pomace, the main phenolic compounds were quercetin and eriocitrin (Multari et al., 2020). Other compounds indicated in this study were narirutin, hesperetin 7-*O*-rutinoside-7-*O*-glucoside, luteolin 7-*O*-glucoside, and hesperetin 7-*O*-rutinoside. Grapefruit peel was analyzed in a recent study, and both flavonoids (Naringenin 7-*O*-neohesperidoside) and phenolic acids (ferulic acid and gallic acid) were identified and quantified (Peng et al., 2021). In addition, comparison of TPC on pomace across different citrus fruits revealed that orange and lemon showed higher contents than mandarin and bergamot (Multari et al., 2020).

3.2 Avocado

Another relevant tropical fruit is avocado, which is widely known for its taste and nutritional value. Harvesting and processing of avocado generate different residues (leaf, peel, and seed) that can be explored as source of polyphenols. Studies carried out on avocado peel revealed that pro procyanidins (dimers, trimers, and tetramers) and epicatechin are major components (Kosińska et al., 2012; Velderrain-Rodríguez et al., 2021). Another major group found in avocado is flavonols due to the elevated content of kaempferol arabinoside glucoside, quercetin 3-*O*-arabinoside, quercetin 3-*O*-galactoside, quercetin arabinoside glucoside, and quercetin diglucoside. In addition, 5-caffeoylquinic acid was also indicated as one main phenolic compound in this residue. Scientific evidence suggests that avocado peel also contains other phenolic acids such as caffeic acid, gallic acid, *p*-hydroxybenzoic acid, and protocatechuic acid (Lyu et al., 2023).

Avocado seed has a relevant phenolic profile that is largely composed of flavanols (catechin, epicatechin, catechin/epicatechin gallate, and procyanidin dimers and trimers) and phenolic acids (3-caffeoylquinic acid, caffeoylshikimic acid, and 3-*p*-coumaroylquinic acid) (Kosińska et al., 2012; Velderrain-Rodríguez et al., 2021). Other phenolic acids reported in avocado seed may include gallic acid and *p*-hydroxybenzoic acid (Lyu et al., 2023).

A recent study characterized the phenolic profile of avocado leaves, a residue produced during harvesting of this fruit. Differently than peel and seeds, leaves seem to contain mainly phenolic acids and flavonols (De Montijo-Prieto et al., 2023). Among phenolic acids, the main compound was dihydroferulic acid 4-*O*-glucuronide, followed by 5-caffeoylquinic acid and protocatechuic acid-4-glucoside. In the group of flavonols, quercetin 3-*O*-rutinoside and quercetin glucuronide were found to have the highest levels in comparison to other compounds of this category (quercetin 3-*O*-arabinosyl-glucoside, quercetin 3-*O*-glucoside, and quercetin-diglucoside). The content of phenolic compounds also varies among different avocado by-products, and peel seems to contain more polyphenols than seed (Kosińska et al., 2012; Lyu et al., 2023; Velderrain-Rodríguez et al., 2021). In addition, some evidence suggests that variety can affect the TPC of both avocado peel and seed (Kosińska et al., 2012; Lyu et al., 2023).

3.3 Banana

Two of the potential sources of phenolic compounds from banana harvesting are the flowers (found on the bract of banana plant inflorescence) and pseudo-stem that are left on the field after fruit harvesting (Ramírez-Bolaños et al., 2021). Characterization of phenolic compounds indicated the presence of phenolic acids and differences among these sources. Caffeic acid and hydroxybenzoic acid were the main phenolic compounds in flower whereas protocatechuic acid and isoferulic acid were the major compounds in pseudo-stem. Another source of phenolic compounds found in banana harvesting residues is the bracts of banana (Begum & Deka, 2019). The phenolic compounds found in these residues are predominantly composed of phenolic acids, especially ferulic, *p*-coumaric, quinic, and salicylic acids. Some flavonoids have also been observed and include catechin and quercetin dihydrates. In terms of TPC, different studies indicate that values between 8 and 11 mg/g dry weight (DW) (Begum & Deka, 2019; Ramírez-Bolaños et al., 2021).

3.4 Mango

Mango processing generates diverse products such as beverages (nectar and juice, for instance), ingredients (concentrate and dehydrated fruit), and processed foods (such as fruit bars and jams). Some of the main residues generated are peels and paste (Berardini et al., 2005; de Ancos et al., 2018). Polyphenol composition in mango peels varies across different studies indicating the predominance of xanthones (mainly mangiferin, flowed by mangiferin gallate and isomangiferin) (Berardini et al., 2005; de Ancos et al., 2018).

Another group of polyphenols found in mango peels is flavonoids. Different quercetin derivatives linked to sugars were reported including quercetin 3-*O*-galactoside, quercetin 3-*O*- glucoside, and quercetin 3-*O*-xyloside (Berardini et al., 2005). Another class of compounds reported in mango peel is phenolic acids and derivatives due to the identification and quantification of ellagic acid, m-digallic acid methyl ester, and valoneic acid dilactone (de Ancos et al., 2018). The evaluation of mango paste (another mango processing by-product) revealed the presence of many compounds concomitantly found in peels such as valoneic acid dilactone, mangiferin, galloyl glucose, and ethyl gallate (de Ancos et al., 2018). In terms of TPC, the values of 4 and 45 g/kg DW have been reported and indicate a large variation between studies (Berardini et al., 2005; de Ancos et al., 2018).

3.5 Other tropical fruits

The evaluation of phenolic profile in pineapple peel revealed the presence of phenolic acids, especially caffeic acid, ferulic acid, and *trans*-cinnamic acid (Zampar et al., 2022). Other phenolic compounds include coumaric acid, syringic acid, and gallic acid. In the case of pineapple crown, some of identified compounds were hydroxycinnamic acids (caffeic acid, ferulic acid, *p*-coumaric acid, and 5-caffeoylquinic acid), isoflavonoids (daidzein 7-*O*-glucoside and stevenin I), and flavonoids (isorhamnetin 3-*O*-glucoside) (Brito et al., 2021).

Cashew is a tropical fruit characterized by its characteristic pear-like shape with juicy flesh, yellow to red peel, and astringent taste when fruits are fully developed (Das & Arora, 2017). The evaluation of cashew pomace revealed the presence of several polyphenols, especially myricetin and vanillic acid (Freitas et al., 2023). This paper also indicated the gallic acid, quercetin 3-*O*-rutinoside, and *p*-coumaric acid as relevant phenolic compounds in this residue. In addition, significant differences were indicated among cultivars for TPC and individual polyphenols. It was found in cocoa bean shells that epicatechin and protocatechuic acid were the main polyphenols followed by phenolic acids such as caffeic acid, 5-caffeoylquinic acid, and ferulic acid (Mellinas et al., 2020). TPC in shells seems to range between 22 and 36 mg gallic acid equivalents (GAE)/g.

4. Phenolic profile of small fruits by-products

Another category with major importance as source of phenolic compounds is comprised of small fruits. In this group, fruits are usually produced in mountain or elevated regions (Kondakova et al., 2009).

4.1 Grapes

Equally relevant is the production of residues from grape processing that include seeds, skin, pomace, and processing by-products such as wine lees (Table 3.2). A study characterized the phenolic profile of grape skin, seed, and pomace (mainly composed of seeds and skin) and indicated differences (Cheng et al., 2012). In grape skin, the main polyphenol identified was malvidin 3-*O*-glucoside, followed by delphinidin 3-glucoside, peonidin 3-*O*-glucoside, petunidin 3-*O*-glucoside, quercetin, and quercetin methyl-glucoside. A different profile was obtained from seeds wherein the highest contents were obtained for catechin and epicatechin followed by procyanidins. Due to the combination of both sources (skin and seeds), the pomace showed the predominant polyphenols of these two sources were observed: catechin, epicatechin, and malvidin 3-*O*-glucoside. The predominance of catechin in the phenolic portion of grape pomace was reported in another study (Sagdic et al., 2011). Other compounds reported included both phenolic acids (5-caffeoylquinic acid and gallic acid) and flavonols (quercetin and quercetin 3-*O*-rutinoside).

Recent studies indicated that lees phenolic composition is composed by a mixture of phenolic acids (caftaric acid and gallic acid), flavanols (catechin, epicatechin, their derivatives and procyanidins with different degrees of polymerization), flavonols (quercetin and quercetin methyl-glucoside) and anthocyanins (malvidin 3-*O*-glucoside) (López-Fernández-Sobrino et al., 2021; Ye et al., 2022). In addition, the grape variety was indicated an important factor associated with significant effect in the concentration of individual polyphenols (López-Fernández-Sobrino et al., 2021). Comparatively, the content of polyphenols in grape seeds is indicated as higher than obtained from skin (Cheng et al., 2012). Pomace is indicated to contain an intermediary content of polyphenols between these two residues. In addition, some evidence across different studies seems to indicate that wine lees may contain less polyphenols (Ye et al., 2022) than other residues obtained from processing of fresh grapes (Cheng et al., 2012). Another factor to be take into account in the effect of grape variety, which can affect TPC and content of individual polyphenols (Sagdic et al., 2011).

4.2 Berries

Another important group in small fruits is composed of berries such as blackcurrant, blackberry, and blueberry. Their processing generally produce a pomace with high polyphenol content but with variable polyphenol composition among species (Table 3.2). A recent experiment determined the phenolic profile of black chokeberry and revealed the presence of flavanols, anthocyanins, and phenolic acids (Rodríguez-Werner et al., 2019). The main compounds were procyanidins, followed by cyanidin 3-*O*-galactoside, cyanidin 3-*O*-arabinoside, 3-caffeoylquinic acid, and 5-caffeoylquinic acid. In the case of blackcurrant pomace, anthocyanins were the predominant group due to the high contents of delphinidin 3-*O*-rutinoside, delphinidin 3-*O*-glucoside, and cyanidin 3-*O*-rutinoside (Azman et al., 2022). Moreover, phenolic acids (*p*-coumaric acid and caffeic acid) were also identified in this residue.

Scientific evidence indicated that saskatoon berry pomace is mainly composed of anthocyanins: cyanidin 3-*O*-arabinoside, cyanidin 3-*O*-galactoside, cyanidin 3-*O*-glucoside, cyanidin 3-*O*-xyloside, and delphinidin 3-*O*-glucoside (de Souza et al., 2019; Li et al., 2014). In addition, other major compounds found in saskatoon berry pomace were 5-caffeoylquinic acid, 4-caffeoylquinic acid, and quercetin-arabinoglucoside (Li et al., 2014). A recent publication compared the pomace of 4 berries (goldenberry, elderberry, blueberry, and blackberry) and indicated that some differences exist, especially in terms of content of individual polyphenols (Rojas-Ocampo et al., 2021). 5-Caffeoylquinic acid was indicated as one of the predominant main polyphenols for all four berries. Another noticeable polyphenol was epicatechin observed at high content in blueberry, elderberry, and goldenberry. Other relevant polyphenols were caffeic acid, *p*-coumaric acid, and gallic acid in across the four berries.

Another potential source of polyphenols is kiwi berry, an emerging fruit consumed in Asia (Zhang et al., 2021). The kiwi berry peel was found to contain caffeic acid, 5-caffeoylquinic acid, protocatechuic acid, and quinic acid. Other polyphenols reported in this by-product were quercetin derivatives. A similar high content of phenolic acids was reported in a study with kiwi pomace where the presence of 5-caffeoylquinic acid, protocatechuic acid, *p*-coumaric acid, and ferulic acid was observed (Kheirkhah et al., 2019). This study also indicated the presence of flavonoids (catechin, epicatechin, and quercetin). The quantification of TPC in a recent study also indicated that phenolic content varies among berries species wherein elderberry showed higher content than goldenberry (Rojas-Ocampo et al., 2021). In addition, intermediary values for TPC analysis were obtained for blackberry and blueberry in this study.

5. Phenolic profile of stone fruits by-products

The category of stone fruits is characterized by the presence of the characteristic woody endocarp (stone) and fleshy mesocarp or epicarp that encloses the endocarp (Famiani et al., 2020). The stone is a major residue generated from peach processing, and its phenolic composition is mainly composed of flavonoids (especially catechin) and phenolic acids (particularly *p*-hydroxybenzoic acid and 5-caffeoylquinic acid) (Hong et al., 2021; Koprivica et al., 2018). Other polyphenols observed in peach stone were quercetin and kaempferol in the group of flavonoids and ferulic acid and *p*-hydroxyphenylacetic acid in phenolic acids category. Regarding peach peel, phenolic acids (3-caffeoylquinic acid and 5-caffeoylquinic acid) predominate the phenolic profile followed by flavonoids from different subcategories (cyanidin 3-*O*-glucoside, quercetin 3-*O*-galactoside, and quercetin 3-*O*-rutinoside) (Saidani et al., 2017).

Studies with plum processing by-products seem to indicate a unique phenolic profile for residue. In the case of plum stone, the main phenolic compounds were quercetin and catechin followed by *p*-hydroxybenzoic acid and 5-caffeoylquinic acid (Hong et al., 2021). Conversely, the characterization of phenolic composition of plum peel revealed the predominance of 3-caffeoylquinic acid in relation to quercetin glucoside and quercetin rutinoside (Sójka et al., 2015).

Another stone fruit with residues showing a relevant phenolic profile is apricot. A recently published paper indicated catechin and quercetin as main polyphenols in this by-product (Hong et al., 2021). Other polyphenols reported in this study were 5-caffeoylquinic acid, *p*-hydroxybenzoic acid, and protocatechuic acid. A similar study with nectarine stone pointed out that quercetin, *p*-hydroxybenzoic acid, and catechin were the main polyphenols (Hong et al., 2021). In addition, kaempferol, caffeic acid, and 5-caffeoylquinic acid were also observed in this residue. In terms of phenolic content, some evidence indicates that differences may exist among species of fruits of the stone group. A comparative study indicates that plum stone may contain more polyphenol than apricot and peach (Hong et al., 2021). This paper also indicated that nectarine stone seems to have the lowest TPC among the four species.

6. Phenolic profile of pome fruits by-products

The category of pome fruits has the characteristic many small seeds covered by a protective membrane located in the core of fruit, which is surrounded by an edible flesh (Gharaghani et al., 2016). Some examples of this category are hawthorn, pears, and apples. As one of the most produced fruits at global scale, apple processing generates residues in the form of pomace (Table 3.2). The analysis of phenolic composition of apple pomace revealed the presence of some prevalent groups (principally flavonols, phenolic acids, dihydrochalcones, and flavanols) with variable content among studies. One of the key groups found in flavonols that is found as quercetin derivatives: quercetin 3-*O*-arabinoside (Suárez et al., 2010), quercetin 3-*O*-galactoside (García et al., 2009; Persic et al., 2017; Suárez et al., 2010), quercetin 3-*O*-glucoside (Arraibi et al., 2021; Persic et al., 2017), quercetin 3-*O*-rhamnoside (Arraibi et al., 2021; Persic et al., 2017; Suárez et al., 2010), and quercetin 3-*O*-rutinoside + Isoquercitrin (Suárez et al., 2010). In the case of phenolic acids, the main compound found is 5-caffeoylquinic acid (Arraibi et al., 2021; García et al., 2009; Persic et al., 2017; Suárez et al., 2010). Other phenolic acids obtained from apple pomace are 4-caffeoylquinic acid (Arraibi et al., 2021), caffeic acid and coumaric acid (Rana et al., 2014), and protocatechuic acid (Suárez et al., 2010).

The dihydrochalcones is a group of characteristic polyphenols for apple that also found in its pomace. Different studies with apple pomace reported phloretin 2′-xyloglucoside (Arraibi et al., 2021; García et al., 2009; Suárez et al., 2010) and phloridzin (Arraibi et al., 2021; García et al., 2009; Rana et al., 2014; Suárez et al., 2010) as main dihydrochalcones. Apple pomace is also composed of flavanols, mainly epicatechin (Arraibi et al., 2021; Rana et al., 2014) and procyanidins dimers and trimers (Arraibi et al., 2021; García et al., 2009; Persic et al., 2017).

In terms of TPC, apple pomace is characterized by large variations of values across studies (Arraibi et al., 2021; García et al., 2009; Persic et al., 2017). For example, the comparison of TPC of pomaces obtained from apple processing industries located in Spain and Belgium revealed that the combination of varieties can affect the content of polyphenols in pomace (Arraibi et al., 2021). The residue obtained from Belgium company seems to contain more polyphenols than that collected from Spanish industry. The effect of variety in the TPC and phenolic composition of apple by-products is also supported by other authors (García et al., 2009; Persic et al., 2017).

7. Phenolic profile of melons by-products

Melons are fruits characterized by the sweet taste and juicy flesh when ripe. The diversity of species can be observed in terms of their form, size, color of rind and flesh (Nuñez-Palenius et al., 2008). Processing of melons generates by-products rich in polyphenols such as reported in studies with peels and seeds (Table 3.2). Melon peels are indicated as sources of phenolic acids. One of the common phenolic acids reported in this by-product in multiple studies is gallic acid (Gómez-García et al., 2021; Rolim et al., 2018; Vella et al., 2019). Other phenolic acids found in melon peels include caffeic acid, *p*-coumaric acid, ferulic acid, and *p*-hydroxybenzoic acid (Gómez-García et al., 2021) and ellagic acid (Vella et al., 2019). Flavonoids were also observed in this by-product, such as luteolin 6-glucoside and luteolin 7-*O*-glucoside (Gómez-García et al., 2021), kaempferol (Vella et al., 2019), and catechin (Rolim et al., 2018).

A similar comprehensive perspective can be observed about the phenolic composition of melon seeds. Diverse phenolic acids were found in this by-product wherein gallic acid was reported in more than one study (Rolim et al., 2018; Zeb, 2016). Other phenolic acids reported in melon seeds were vanillic acid (Rolim et al., 2018), ferulic acid (Vella et al., 2019), caffeic acid and ellagic acid (Zeb, 2016). Although some similarity exists in terms of phenolic profile, the TPC of peels and seeds seems to differ. Peels seem to contain more polyphenols than seeds, as indicated by different studies, which suggest that melon peels may be preferable polyphenol source of polyphenols than their seeds (Rolim et al., 2018; Vella et al., 2019).

8. Phenolic profile of flower vegetables by-products

8.1 Artichoke

The harvesting and processing of artichoke produce different residues (leaves, bracts, floral stem, and pomace) rich in polyphenols that are wasted (Table 3.3). The phenolic composition of Artichoke leaves was found to contain two major classes: hydroxycinnamic acids and flavones. In the case of artichoke leaves, the evaluation of phenolic composition revealed the presence of two main phenolic groups: hydroxycinnamic acids and flavones. Regarding the hydroxycinnamic acids, the most prevalent compound was 5-caffeoylquinic acid, followed by 1,5-dicaffeoylquinic acid and 3,5-dicaffeoyl quinic acid (Pagano et al., 2016; Pandino et al., 2011, 2013; Sihem et al., 2015). In the group of flavones, Luteolin and many of its derivatives (luteolin 7-*O*-glucoside, luteolin 7-*O*-glucuronide, luteolin 7-*O*-rutinoside, and luteolin 7-*O*-malonyl-glucoside). Similarly, apigenin and their derivatives (apigenin glucoside, apigenin glucuronide, and apigenin 7-*O*-rutinoside) were found in artichoke leaves.

Artichoke bracts are another residue rich in polyphenols, which also has a similar phenolic profile found in artichoke leaves, indicated above. Hydroxycinnamic acids

Table 3.3 Phenolic compounds found in vegetables harvesting and processing by-products.

Source	By-product	TPC	TFC	Main identified phenolic compounds (concentration)	Ref.
Flower vegetables					
Artichoke	Leaves	469–1746 mg/100 g DW	n.d.	1,5-Dicaffeoylquinic acid (108–547 mg/100 g DW) 5-Caffeoylquinic acid (277–491 mg/100 g DW) Luteolin 7-*O*-rutinoside (37–354 mg/100 g DW) Luteolin glucoside (21–216 mg/100 g DW) Luteolin 7-*O*-glucuronide (3–47 mg/100 g DW) 1-Caffeoylquinic acid (9–36 mg/100 g DW) 3,4-Dicaffeoylquinic acid (2–22 mg/100 g DW) 3-Caffeoylquinic acid (8–11 mg/100 g DW) Apigenin glucoside (0.8–3 mg/100 g DW)	Pagano et al. (2016)
Artichoke	Leaves	5–13 g/kg DW	n.d.	Luteolin 7-*O*-rutinoside (0.1–6 g/kg DW) Luteolin glucoside (0.1–4 g/kg DW) Apigenin glucuronide (2–3 g/kg DW) Luteolin 7-*O*-glucuronide (2 g/kg DW) 5-Caffeoylquinic acid (0.2–2 g/kg DW) Luteolin malonylglucoside (0.6–2 g/kg DW) Apigenin (0.2–1 g/kg DW)	Pandino et al. (2011)
Artichoke	Leaves	2–8 g/kg DW	n.d.	Luteolin 7-*O*-rutinoside (891–2956 mg/kg DW) 5-Caffeoylquinic acid (122–2135 mg/kg DW) Luteolin (98–2087 mg/kg DW) Luteolin 7-*O*-glucoside (460–1758 mg/kg DW) Apigenin 7-*O*-rutinoside (405–1059 mg/kg DW) Luteolin 7-*O*-malonyl-glucoside (284–1047 mg/kg DW)	Pandino et al. (2013)
Artichoke	Leaves	8–11 mg/g DW	n.d.	5-Caffeoylquinic acid (2–4 mg/g DW) Luteolin 7-*O*-glucoside (2–3 mg/g DW) Luteolin 7-*O*-glucuronide (1–2 mg/g DW) 3,5-Dicaffeoylquinic acid (0.4–1 mg/g DW) 1,5-Dicaffeoylquinic acid (0.8–0.9 mg/g DW)	Sihem et al. (2015)

Continued

Table 3.3 Phenolic compounds found in vegetables harvesting and processing by-products.—*cont'd*

Source	By-product	TPC	TFC	Main identified phenolic compounds (concentration)	Ref.
Artichoke	Bracts	515–852 mg/100 g DW	n.d.	5-Caffeoylquinic acid (101–444 mg/100 g DW) 1,5-Dicaffeoylquinic acid (273–299 mg/100 g DW) Apigenin glucuronide (27–84 mg/100 g DW) 1-Caffeoylquinic acid (13–22 mg/100 g DW) 3-Caffeoylquinic acid (7–14 mg/100 g DW) Luteolin 7-*O*-glucuronide (12 mg/100 g DW) 3,4-Dicaffeoylquinic acid (7–10 mg/100 g DW)	Pagano et al. (2016)
Artichoke	Bracts	7–15 mg/g DW	n.d.	3,5-Dicaffeoylquinic acid (1–8 mg/g DW) 1,5-Dicaffeoylquinic acid (2–3 mg/g DW) 5-Caffeoylquinic acid (3 mg/g DW) Monosuccinylcaffeoylquinic acid (0.4–0.6 mg/g DW) Apigenin 7-*O*-glucoside (0.3–0.4 mg/g DW)	Sihem et al. (2015)
Artichoke	Floral stem	10–11 mg/g DW	n.d.	1,5-Dicaffeoylquinic acid (4–5 mg/g DW) 5-Caffeoylquinic acid (2–3 mg/g DW) 3,5-Dicaffeoylquinic acid (2–3 mg/g DW) Monosuccinylcaffeoylquinic acid (0.3–0.7 mg/g DW) 3-Caffeoylquinic acid (0.2 mg/g DW)	Sihem et al. (2015)
Artichoke	Pomace	8341 mg/kg	n.d.	3,5-Dicaffeoylquinic acid (5479 mg/kg) 5-Caffeoylquinic acid (1802 mg/kg) Caffeic acid (552 mg/kg) Protocatechuic acid (321 mg/kg) Luteolin 7-*O*-glucoside (105 mg/kg)	Vaz et al. (2022)
Broccoli	Leaves	99–163 mg/g DW	11–40 mg/g DW	5-Caffeoylquinic acid derivatives (42–127 mg/g DW) Sinapic acid derivatives (10–52 mg/g DW)	Domínguez-Perles et al. (2010)
Broccoli	Leaves	25–28 mg GAE/g DW	8–10 mg CE/g DW	Esculetin (6 mg/g) Myricetin (3 mg/g) Quercetin (1 mg/g) Ferulic acid (0.2 mg/g) Gallic acid (0.2 mg/g)	Le et al. (2020)

Broccoli	Stalk	7–12 mg/g DW	1–3 mg/g DW	5-Caffeoylquinic acid derivatives (5–44 mg/g DW) Sinapic acid derivatives (1 mg/g DW)	Domínguez-Perles et al. (2010)
Broccoli	Stalk	74 mg GAE/g DW	n.d.	3-Caffeoyl quinic acid (35 mg CAE/kg) Sinapic acid (21 mg SAE/kg) 5-Caffeoylquinic acid (18 mg CAE/kg)	Thomas et al. (2018)
Broccoli	Stalk	n.d.	n.d.	1,2,2′-Tri-sinapoyl-gentiobioside (13 g/kg DW) Feruloyl-caffeoyl derivative (6 g/kg DW) Feruloyl-caffeoyl derivative (5 g/kg DW) Di-sinapoyl-gentiobioside II (5 g/kg DW) 5-Caffeoylquinic acid (4 g/kg DW) Di-sinapoyl-diglucose (2 g/kg DW)	Costa-Pérez et al. (2022)
Broccoli	Leaves and stems	n.d.	n.d.	5-Caffeoylquinic acid (58–128 μg/g DW) 3-Caffeoylquinic acid (56–125 μg/g DW) Ferulic acid (40–88 μg/g DW) Sinapic acid (37–81 μg/g DW) Quercetin 3-*O*-rutinoside (4–22 μg/g DW) Kaempferol 3-*O*-rutinoside (3–20 μg/g DW)	Aires et al. (2017)
Broccoli	Leaves and stems	n.d.	n.d.	5-Caffeoylquinic acid (39–86 μg/g DW) 3-Caffeoylquinic acid (38–85 μg/g DW) Ferulic acid (29–64 μg/g DW) Sinapic acid (27–60 μg/g DW) Quercetin 3-*O*-rutinoside (7–19 μg/g DW) Kaempferol 3-*O*-rutinoside (3–17 μg/g DW)	Aires et al. (2017)
Leafy vegetables					
Lettuce	Leaves	49–55 μg/mL	n.d.	Chicoric acid (34–38 μg/mL) Caffeoyltartaric acid (6–8 μg/mL) 3-Caffeoylquinic acid (3–5 μg/mL) Luteolin 7-*O*-glucuronide (0.5–4 μg/mL) Isochlorogenic acid (1–2 μg/mL) Quercetin 3-*O*-glucuronide (0.6–0.9 μg/mL)	Plazzotta and Manzocco (2018)

Continued

Table 3.3 Phenolic compounds found in vegetables harvesting and processing by-products.—*cont'd*

Source	By-product	TPC	TFC	Main identified phenolic compounds (concentration)	Ref.
Lettuce	Leaves	52–81 μg/mL	n.d.	Chicoric acid (39–49 μg/mL) Caffeoyltartaric acid (7–12 μg/mL) 3-Caffeoylquinic acid (3–9 μg/mL) Isochlorogenic acid (2–5 μg/mL) Luteolin 7-*O*-glucuronide (0.5–4 μg/mL) Quercetin 3-*O*-glucuronide (0.6–1 μg/mL)	Plazzotta and Manzocco (2018)
Lettuce	Leaves	37–41 μg/mL	n.d.	Chicoric acid (25–29 μg/mL) Caffeoyltartaric acid (6 μg/mL) 3-Caffeoylquinic acid (2–3 μg/mL) Isochlorogenic acid (2 μg/mL) Quercetin 3-*O*-glucuronide (1 μg/mL) Luteolin 7-*O*-glucuronide (0.5–1 μg/mL)	Plazzotta and Manzocco (2018)
Lettuce	Leaves	40–61 μg/mL	n.d.	Chicoric acid (24–33 μg/mL) Luteolin 7-*O*-glucuronide (3–10 μg/mL) Caffeoyltartaric acid (5–9 μg/mL) Isochlorogenic acid (2–4 μg/mL) 3-Caffeoylquinic acid (2–4 μg/mL) Quercetin 3-*O*-glucuronide (0.6–2 μg/mL)	Plazzotta and Manzocco (2018)
Cabbage	Stalk	454–521 mg GAE/100 g	29–50 mg CE/100 g	Quercetin 3-*O*-sophoroside (n.q.) 4-*p*-Coumaroylquinic acid (n.q.) 5-Caffeoylquinic acid (n.q.) 4-Hydroxybenzaldehyde (n.q.) *p*-Coumaric acid (n.q.)	Brito et al. (2021)

Stem vegetables					
Spinach	Leaves out of commercial shape standard	474 mg GAE/100 g DW	n.d.	Quercetin 3-*O*-rutinoside derivative 2 (21 mg/100 g DW) Epigallocatechin gallate derivative (11 mg/100 g DW) Caffeic acid derivative (8 mg/100 g DW) Lutein derivative 1 (4 mg/100 g DW) Lutein derivative 2 (3 mg/100 g DW) Quercetin 3-*O*-rutinoside derivative 1 (2 mg/100 g DW)	Araújo-Rodrigues et al. (2021)
Rocket	Leaves out of commercial shape standard	470 mg GAE/100 g DW	n.d.	Apigenin derivative (137 mg/100 g DW) Quercetin 3-*O*-glucoside (48 mg/100 g DW) Quercetin 3-*O*-rutinoside (36 mg/100 g DW) *trans*-Ferulic acid (16 mg/100 g DW) Quercetin derivative (12 mg/100 g DW) Ferulic acid (6 mg/100 g DW) 5-Caffeoylquinic acid (6 mg/100 g DW) Quercetin (4 mg/100 g DW) *p*-Coumaric acid (1 mg/100 g DW)	Araújo-Rodrigues et al. (2021)
Watercress	Leaves out of commercial shape standard	630 mg GAE/100 g DW	n.d.	Caffeic acid derivative (215 mg/100 g DW) 5-Caffeoylquinic acid derivative (58 mg/100 g DW) *p*-Coumaric acid derivative 1 (54 mg/100 g DW) Quercetin (92 mg/100 g DW) Quercetin 3-*O*-glucoside (38 mg/100 g DW) Lutein derivative (31 mg/100 g DW) Ferulic acid (14 mg/100 g DW) *p*-Coumaric acid derivative 2 (12 mg/100 g DW) *p*-Coumaric acid (5 mg/100 g DW)	Araújo-Rodrigues et al. (2021)
Wild cardoon	Leaves	28–58 mg/g	19–31 mg/g	*p*-Coumaric acid (3–17 mg/g) Kaempferol 3-*O*-rutinoside (4–7 mg/g) Kaempferol-*O*-rhamnoside-*O*-pentoside (3–6 mg/g) Kaempferol-*O*-dirhamnoside (3–6 mg/g) Caffeoyl malate (4 mg/g) Feruloyl malate (1–2 mg/g)	Chihoub et al. (2019)

Continued

Table 3.3 Phenolic compounds found in vegetables harvesting and processing by-products.—*cont'd*

Source	By-product	TPC	TFC	Main identified phenolic compounds (concentration)	Ref.
Underground vegetables					
Carrot	Peel	4.8 mg/g	n.d.	5-Caffeoylquinic acid (2024 μg/g) Phloretin 2′-*O*-glucoside (1768 μg/g) 4-Caffeoylquinic acid (508 μg/g) 3-Caffeoylquinic acid (408 μg/g) 5-*p*-Coumaroylquinic acid (33 μg/g) 4-*p*-Coumaroylquinic acid (26 μg/g)	Delpino-Rius et al. (2015)
Carrot	Peel	5 mg GAE/g	1 mg CE/g	Pyrogallol (601 mg/100 g) Catechin (384 mg/100 g) 5-Caffeoylquinic acid (234 mg/100 g) *p*-Hydroxybenzoic acid (203 mg/100 g) Salicylic acid (196 mg/100 g) Catechol (138 mg/100 g) Ellagic acid (121 mg/100 g)	El-Sawi et al. (2022)
Carrot	Pomace	217 mg/kg	n.d.	5-Caffeoylquinic acid (178 mg/kg) *o*-Salicylic acid (13 mg/kg) Protocatechuic acid (11 mg/kg) Caffeic acid (7 mg/kg) 3,5-Dicaffeoylquinic acid (4 mg/kg) Vanillic acid (2 mg/kg) Ferulic acid (2 mg/kg)	Vaz et al. (2022)
Carrot	Product out of commercial shape standard	≈150 mg GAE/100 g DW	n.d.	5-Caffeoylquinic acid (2 mg/100 g DW) *trans*-Ferulic acid (3 mg/100 g DW)	Araújo-Rodrigues et al. (2021)
Black carrot	Pomace	≈5 mg GAE/g DW	2–3 mg CE/g DW	5-Caffeoylquinic acid (81 mg/100 g DW) Cyanidin 3-*O*-glucoside (70 mg/100 g DW) Malvidin 3-*O*-glucoside (35 mg/100 g DW) Cyanidin 3-*O*-galactoside (18 mg/100 g DW) Ferulic acid (5 mg/100 g DW) Sinapic acid (2 mg/100 g DW)	Gulsunoglu et al. (2019)

Black carrot	Pomace	511 mg/kg DW	n.d.	5-Caffeoylquinic acid (186 mg/kg DW) Di-caffeic acid derivative (66 mg/kg DW) Caffeic acid derivative 1 (65 mg/kg DW) Caffeic acid derivative 2 (51 mg/kg DW) Quercetin 3-*O*-galactoside (23 mg/kg DW) Protocatechuic acid (16 mg/kg DW) Cyanidin 3-xylosyl-feruloy -glucoside-galactoside (15 mg/kg DW)	Polat et al. (2022)
Potato	Peel	3–9 mg GAE/g DW	n.d.	5-Caffeoylquinic acid (16–135 mg/100 g DW) Caffeic acid (29–82 mg/100 g DW) Ferulic acid (6–82 mg/100 g DW) *p*-Coumaric acid (3–6 mg/100 g DW)	Albishi et al. (2013)
Potato	Peel	5 mg GAE/g	n.d.	5-Caffeoylquinic acid (3.9 mg/g) Caffeic acid (0.9 mg/g) 3-Caffeoylquinic acid (0.4 mg/g) Isochlorogenic acid B (0.3 mg/g) Vanillic acid (0.3 mg/g) Isochlorogenic acid A (0.2 mg/g)	Lima et al. (2021)
Potato	Peel	30–46 mg/100 g	n.d.	5-Caffeoylquinic acid (9–29 mg/100 g) Caffeic acid (5–12 mg/100 g) *p*-Coumaric acid (0.8–6 mg/100 g) Ferulic acid (1–5 mg/100 g) Protocatechuic acid (2 mg/100 g) Syringic acid (0.8–2 mg/100 g) Gallic acid (0.5–0.6 mg/100 g)	P. P. Singh and Saldaña (2011)
Potato	Peel	23–82 mg/100 g	n.d.	Gallic acid (7–30 mg/100 g) 5-Caffeoylquinic acid (6–16 mg/100 g) Protocatechuic acid (1–14 mg/100 g) *p*-Hydroxyl benzoic acid (0.7–11 mg/100 g) Caffeic acid (2–9 mg/100 g) Syringic acid (2–4 mg/100 g)	P. P. Singh and Saldaña (2011)

Continued

Table 3.3 Phenolic compounds found in vegetables harvesting and processing by-products.—*cont'd*

Source	By-product	TPC	TFC	Main identified phenolic compounds (concentration)	Ref.
Red and purple sweet potato	Peel	0.3–2.0 mg/g DW	0.05–0.1 mg/g DW	Caffeic acid (0.1–1.0 mg/g DW) 4-Caffeoylquinic acid (0.03–0.3 mg/g DW) Bis(dihydrocaffeoyl)spermidine (0.2–0.3 mg/g DW) Kaempferol-O-hexoside-O-deoxyhexoside-hexoside (0.03–0.1 mg/g DW) Salvianolic acid F (0.02–0.1 mg/g DW) Kaempferol 3-*O*-rutinoside (0.04–0.05 mg/g DW) Kaempferol-*O*-deoxyhexosyl-hexoside (0.02 mg/g DW)	Sampaio et al. (2021)
Radish	Leaves and stem	29–38 mg/g	12–15 mg/g	4-Caffeoylquinic acid (10–14 mg/g) Luteolin-*O*-malonylhexoside (6–7 mg/g) 3,4-Dicaffeoylquinic acid (4–6 mg/g) Luteolin-*O*-hexoside (2–6 mg/g) Acetylapigenin-*O*-hexoside (2–5 mg/g) 5-Hydroxyferuloylglycoside (1 mg/g)	Chihoub et al. (2019)
Black radish	Peel	306 mg GAE/g DW	172 mg CE/g DW	Ferulic acid (28.0 mg/g DW) Epicatechin (20 mg/g DW) Sinapic acid (11 mg/g DW) Gallic acid (7 mg/g DW) 5-Caffeoylquinic acid (3 mg/g DW)	Yücetepe et al. (2021)
Turnip	Leaves and stem	18–25 mg/g	14–19 mg/g	Isorhamnetin-*O*-dihexoside (2–3 mg/g) Synapoylmalic acid (1–3 mg/g) Kaempferol 3-*O*-sinapoyl-sophoroside-7-*O*-glucoside (1–2 mg/g) Kaempferol-*O*-dihexoside (2 mg/g) Isorhamnetin-*O*-pentoside (1 mg/g) 3-Caffeoylquinic acid (1 mg/g) 3-*p*-Coumaroylquinic acid (1 mg/g) Kaempferol 3-*O*-(sinapoyl)-sophoroside (1 mg/g)	Chihoub et al. (2019)
Tiger nuts	Wastes	223 mg GAE/100 g DW	n.d	4-Vinylphenol (20–353 μg/kg) Spinacetin 3-*O*-glucosyl-(1 → 6)-glucoside (23–95 μg/kg) Sinensetin (50–78 μg/kg) *p*-Coumaric acid (0–48 μg/kg) Dihydro-*p*-coumaric acid (10–38 μg/kg) 4-Hydroxybenzaldehyde + Benzoic acid (1–34 μg/kg) Ethyl vanillin (10–26 μg/kg)	Roselló-Soto et al. (2019)

Oilseeds and olive by-products					
Corn	Pericarp	n.d.	n.d.	Hesperetin 7-*O*-rutinoside (102 mg/g) Cyanidin 3-*O*-glucoside (71 mg/g) Quercetin (46 mg/g) Cyanidin chloride (45 mg/g) Delphinidin (30 mg/g) Kaempferol (14 mg/g) Caffeic acid (10 mg/g) Naringenin 7-*O*-neohesperidoside (8 mg/g) 5-Caffeoylquinic acid (6 mg/g) Ferulic acid (6 mg/g)	Boateng et al. (2023)
Corn	Pericarp	121 g GAE/kg	n.d.	Hesperetin 7-*O*-rutinoside (104 mg/g) Cyanidin 3-*O*-glucoside (81 mg/g) Cyanidin chloride (56 mg/g) Delphinidin (30 mg/g) Quercetin (42 mg/g) Kaempferol (15 mg/g) Caffeic acid (11 mg/g) 5-Caffeoylquinic acid (9 mg/g) Naringenin 7-*O*-neohesperidoside (8 mg/g) Ferulic acid (7 mg/g)	Boateng et al. (2023)
Corn	Bracts	441 mg GAE/100 g	192 mg CE/100 g	Ferulic acid (943 mmol/100 g) *p*-Coumaric acid (490 mmol/100 g) Gallic acid (128 mmol/100 g) Quercetin 3-rhamnoside (33 mmol/100 g) 4-Hydroxycoumarin (23 mmol/100 g) Coumarin (16 mmol/100 g)	Yang et al. (2019)
Corn	Cob	915 mg GAE/100 g	198 mg CE/100 g	Ferulic acid (1164 mmol/100 g) *p*-Coumaric acid (933 mmol/100 g) Kaempferol (124 mmol/100 g) Gallic acid (109 mmol/100 g) Quercetin 3-rhamnoside (91 mmol/100 g) 4-Hydroxycoumarin (33 mmol/100 g)	Yang et al. (2019)

Continued

Table 3.3 Phenolic compounds found in vegetables harvesting and processing by-products.—*cont'd*

Source	By-product	TPC	TFC	Main identified phenolic compounds (concentration)	Ref.
Corn	Silk	354 mg GAE/ 100 g	264 mg CE/ 100 g	Ferulic acid (591 mmol/100 g) Quercetin (142 mmol/100 g) Quercetin 3-rhamnoside (120 mmol/100 g) Gallic acid (114 mmol/100 g) 4-Hydroxycoumarin (83 mmol/100 g) *p*-Coumaric acid (41 mmol/100 g) Coumarin (22 mmol/100 g) Kaempferol (8 mmol/100 g)	Yang et al. (2019)
Corn	Silk	21–43 mg CGAE/g	18–36 mg CE/ g	5-Caffeoylquinic acid (27–53 mg/100 g DW) 3-Caffeoylquinic acid (23–50 mg/100 g DW) *p*-Coumaroylquinic acid (31–39 mg/100 g DW) 4-Caffeoylquinic acid (21–29 mg/100 g DW) Maysin (5–17 mg/100 g DW) Methoxymaysin derivative (1–5 mg/100 g DW) Maysin derivative (1–4 mg/100 g DW)	Žilić et al. (2016)
Corn	Tassel	n.d.	n.d.	Vanillic acid (1046 μg/L) Caffeic acid (670 μg/L) Butein (60 μg/L) Myricetin (29 μg/L) Resveratrol (20 μg/L) Catechin hydrate (15 μg/L) *p*-Hydroxybenzoic acid (11 μg/L)	Al-Khayri et al. (2022)
Corn	Tassel	8–11 mg/g	n.d.	Gallic acid (315–1694 μg/g) 5-Caffeoylquinic acid (98–1172 μg/g) Caffeic acid (10–662 μg/g) Naringenin (119–270 μg/g) Quercetin 3-O-rutinoside (160–211 μg/g) Catechin (115–198 μg/g) Ferulic acid (20–180 μg/g) Syringic acid (66–167 μg/g)	Elsayed et al. (2022)

Canola	Hulls	11 mg SAE/g	n.d.	Sinapoyl glucose (1 mg/g) Sinapine (1 mg/g) Sinapic acid (0.04 mg/g)	Mayengbam et al. (2014)
Canola	Oil press residue	1169 mg/100 g DW	n.d.	Sinapine (509 mg/100 g DW) Sinapoyl malate (270 mg/100 g DW) Kaempferol 3-*O*-(sinapoyl)-sophoroside-7-*O*-glucoside (97 mg/100 g DW) Sinapoyl glucose (64 mg/100 g DW) Kaempferol 3-*O*-sophoroside-7-*O*-glucoside (42 mg/100 g DW) Kaempferol-sinapoylhexoside-hexoside-hexoside (42 mg/100 g DW)	Tian et al. (2022)
Canola	Oil press residue	68 mg/g	n.d.	Sinapine derivatives (0.2 mmol/g)	Vidal et al. (2022)
Soy	Oil press residue	5–12 mg GAE/g	7–14 mg GE/g	Genistin-β-glucoside (0.09–0.5 mg/g DW) Genistein (0.08–0.4 mg/g DW) Daidzein (0.09–0.4 mg/g DW) Glycitein (0.09–0.4 mg/g DW) Daidzin-β-glucoside (0.06–0.3 mg/g DW)	Silva and Perrone (2015)
Sunflower	Oil press residue	59 mg/g	n.d.	5-Caffeoylquinic acid derivatives (0.2 mmol/g)	Vidal et al. (2022)
Peanut	Skin	92 mg/g DW	n.d.	Proanthocyanidins trimers (21 mg/g DW) Proanthocyanidins tetramers (20 mg/g DW) Proanthocyanidins hexamers (15 mg/g DW) Proanthocyanidins heptamers (14 mg/g DW) Proanthocyanidins pentamers (8 mg/g DW) Proanthocyanidins octamers (7 mg/g DW) Proanthocyanidins A-type dimer (6 mg/g DW)	Ma et al. (2014)
Peanut	Skin	48–68 mg CE/g DW	n.d.	Procyanidin trimer A (3629–6494 μg/g DW) Procyanidin dimer B (1862–3059 μg/g DW) Procyanidin dimer A (3749–2372 μg/g DW) (+)-Catechin (1247–1257 μg/g DW) Procyanidin tetramer A (828–1050 μg/g DW) (−)-Epicatechin (527–574 μg/g DW)	de Camargo et al. (2015)

Continued

Table 3.3 Phenolic compounds found in vegetables harvesting and processing by-products.—*cont'd*

Source	By-product	TPC	TFC	Main identified phenolic compounds (concentration)	Ref.
Pistachio	Skin	116 mg GAE/g	70 mg CE/g	Cyanidin 3-O-galactoside (5865 μg/g) Gallic acid (1453 μg/g) Catechin (377 μg/g) Eriodictyol 7-*O*-glucoside (366 μg/g) Naringenin 7-*O*-neohesperidoside (119 μg/g) Epicathechin (105 μg/g)	Tomaino et al. (2010)
Olive	Mill wastewater	6–10 g TYE/L	3–7 g CE/L	Hydroxytyrosol (2127–3766 mg TYE/L) Tyrosol (246–2491 mg TYE/L) *p*-Coumaric acid (549–785 mg TYE/L) Gallic acid (331–583 mg TYE/L) Hydroxytyrosol 4-β-glucoside (168–226 mg TYE/L)	El-Abbassi et al. (2012)
Olive	Mill wastewater	17–33 mg GAE/g DW	n.d.	Verbascoside (0.9–4 mg/g) Hydroxytyrosol glucoside (0.4–3 mg/g) Hydroxytyrosol (0.4–0.8 mg/g) Comselogoside (0.06–0.3 mg/g) Luteolin (0.07–0.2 mg/g) Oleuropein (0.06–0.2 mg/g)	Obied et al. (2008)
Legumes					
Lentil	Hulls	17–54 mg/g	5–16 mg/g	Catechin (4–9 mg/g) Catechin glucoside (0–7 mg/g) Prodelphinidin dimer (1–6 mg/g) Procyanidin dimer (2–4 mg/g) Quercetin (0.1–4 mg/g) Quercetin-*O*-pentoside (0–3 mg/g) Luteolin 7-*O*-glucoside (0.2–3 mg/g) Procyanidin trimer (2–3 mg/g)	Yeo and Shahidi (2020)

Lentil	Hulls	96–167 mg/g	46–71 mg/g	Syringic acid (8–32 mg/g) Protocatechuic acid (16–19 mg/g) *cis*-Ferulic acid (11–16 mg/g) Quercetin (11–15 mg/g) Caffeic acid (7–13 mg/g) Catechin 3-*O*-gallate dimer (8–12 mg/g) Catechin (10–12 mg/g) *trans*-Ferulic acid (4–11 mg/g)	Yeo et al. (2021)
Lentil	Hulls	81–85 mg GAE/g	3–4 mg CE/g	Formononetin (724–921 μg/g) Quercetin 3-*O*-glucoside (39–884 μg/g) Quercetin (167–807 μg/g) Myricetin (188–704 μg/g) Gallic acid (108–586 μg/g) Catechin (486–546 μg/g) Protocatechuic acid (289–452 μg/g) Caffeic acid (245–412 μg/g) *trans-p*-Coumaric acid derivative (202–253 μg/g) Luteolin glucoside (74–239 μg/g)	Paranavitana et al. (2021)
Lentil	Hulls	1–20 mg GAE/g DW	n.d.	Procyanidin dimer 1 (762–1137 μg/g DW) Dimethoxybenzoic acid derivative (631–954 μg/g DW) Procyanidin dimer 2 (449–593 μg/g DW) *p*-Coumaric acid (122–416 μg/g DW) Hydroxymandelic acid derivative 1 (186–309 μg/g DW) Catechin (119–292 μg/g DW)	Sun et al. (2020)
Vine-grown vegetables					
Tomato	Peel	34–155 mg/100 g DW	n.d.	Naringenin chalcone (7–88 mg/100 g DW) Quercetin 3-rutinoside (6–51 mg/100 g DW) 3,4-Dicaffeoylquinic acid (6–16 mg/100 g DW) Naringenin (5–13 mg/100 g DW) 3,4,5-Tri-Caffeoylquinic acid (5–9 mg/100 g DW) Quercetin-triglucoside (4–5 mg/100 g DW)	Szabo et al. (2019)

Continued

Table 3.3 Phenolic compounds found in vegetables harvesting and processing by-products.—*cont'd*

Source	By-product	TPC	TFC	Main identified phenolic compounds (concentration)	Ref.
Tomato	Peel	n.d.	n.d.	4′-4-5-Hydroxy-2′-methoxychalcone (n.q.) Quercetin 3-*O*-rutinoside (n.q.) Naringenin (n.q.)	Benakmoum et al. (2008)
Tomato	Peel and seeds	1230 mg GAE/kg	415 mg QE/kg	Ellagic acid (143 mg/kg) 5-Caffeoylquinic acid (76 mg/kg) Salicylic acid (67 mg/kg) Myricetin (64 mg/kg) Quercetin 3-*O*-rutinoside (29 mg/kg) Vanillic acid (27 mg/kg)	Nour et al. (2018)
Tomato	Peel and seeds	104 mg GAE/g DW	n.d.	Naringenin (28 mg/g DW) Quercetin (14 mg/g DW) Caffeic acid (12 mg/g DW) *p*-Coumaric acid (10 mg/g DW) Catechin (5 mg/g DW) Ferulic acid (3 mg/g DW) Gallic acid (1 mg/g DW)	Perea-Domínguez et al. (2018)
Tomato	Pomace	1 mg GAE/g DW	n.d.	Gallic acid (86 mg/100 g DW) 5-Caffeoylquinic acid (64 mg/100 g DW) Naringenin (42 mg/100 g DW) Quercetin 3-*O*-rutinoside (18 mg/100 g DW) *trans*-Ferulic acid (7 mg/100 g DW) Caffeic acid (3 mg/100 g DW) Quercetin (3 mg/100 g DW) Quercetin 3-*O*-glucoside (3 mg/100 g DW)	Bao et al. (2020)
Tomato	Pomace	1 mg GAE/g DW	n.d.	Gallic acid (79–89 mg/100 g DW) 5-Caffeoylquinic acid (72–76 mg/100 g DW) Naringenin (42–50 mg/100 g DW) Quercetin 3-*O*-rutinoside (17–19 mg/100 g DW) *trans*-Ferulic acid (7 mg/100 g DW) Caffeic acid (3 mg/100 g DW) Quercetin (3 mg/100 g DW) Quercetin 3-*O*-glucoside (3 mg/100 g DW)	Bao et al. (2020)

Tomato	Leaves and stems	n.d.	n.d.	Kaempferol 3-*O*-rutinoside (160–260 μg/g DW) Caffeic acid (52–91 μg/g DW) Quercetin 3-*O*-rutinoside (26–45 μg/g DW) 5-Caffeoylquinic acid (25–44 μg/g DW) *p*-Coumaric acid (5–9 μg/g DW)	Aires et al. (2017)
Tomato	Leaves and stems	n.d.	n.d.	Kaempferol 3-*O*-rutinoside (58–88 μg/g DW) Caffeic acid (17–35 μg/g DW) Quercetin 3-*O*-rutinoside (15–23 μg/g DW) 5-Caffeoylquinic acid (8–17 μg/g DW) *p*-Coumaric acid (2–4 μg/g DW)	Aires et al. (2017)
Tomato	Product out of commercial shape standard	≈400 mg GAE/100 g DW	n.d.	5-Caffeoylquinic acid (27 mg/100 g DW) Quercetin 3-*O*-rutinoside (11 mg/100 g DW) Quercetin derivative (17 mg/100 g DW) Naringenin (500 mg/100 g DW)	Araújo-Rodrigues et al. (2021)
Pumpkin	Peel	2–9 mg/g	1–5 mg/g	Epicatechin (0.3–5 mg/g) Isorhamnetin-*O*-dideoxyhexosyl-hexoside (0.4–2 mg/g) Isorhamnetin-*O*-deoxyhexosyl-hexoside (0.9 mg/g) Kaempferol-*O*-deoxyhexosyl-hexoside (0.5–0.6 mg/g) Quercetin-*O*-dideoxyhexosyl-hexoside (0.5 mg/g) 7 4-*O*-(6′-*O*-Glucosyl-4″-hydroxybenzoyl)-4-hydroxybenzyl alcohol (0.2 mg/g)	Leichtweis et al. (2022)
Pumpkin	Peel	3 mg/g	2 mg/g	Isorhamnetin-*O*-dideoxyhexosyl-hexoside (0.4–0.5 mg/g) Kaempferol-*O*-dideoxyhexosyl-hexoside (0.4–0.5 mg/g) Isorhamnetin-*O*-deoxyhexosyl-hexoside (0.4–0.5 mg/g) Kaempferol-*O*-deoxyhexosyl-hexoside (0.4–0.5 mg/g) *cis*-4-*O*-*p*-Coumaroylquinic acid (0.4–0.5 mg/g) *trans*-4-*O*-*p*-Coumaroylquinic acid (0.4 mg/g) Epicatechin (0.2 mg/g)	Leichtweis et al. (2023)
Pumpkin	Peel	1 mg/g	1 mg/g	Isorhamnetin-*O*-dideoxyhexosyl-hexoside (0.5 mg/g) Kaempferol-*O*-dideoxyhexosyl-hexoside (0.4 mg/g) Isorhamnetin-*O*-deoxyhexosyl-hexoside (0.4 mg/g) Epicatechin (0.1 mg/g) *cis*-4-*O*-*p*-Coumaroylquinic acid (0.1 mg/g)	Leichtweis et al. (2023)

Continued

Table 3.3 Phenolic compounds found in vegetables harvesting and processing by-products.—*cont'd*

Source	By-product	TPC	TFC	Main identified phenolic compounds (concentration)	Ref.
Pumpkin	Fibrous strands	2–5 mg/g	0.5–2 mg/g	Epicatechin (1–3 mg/g) 7 4-*O*-(6′-*O*-Glucosyl-4″-hydroxybenzoyl)-4-hydroxybenzyl alcohol (2 mg/g) Isorhamnetin-*O*-dideoxyhexosyl-hexoside (0.5 mg/g) Kaempferol-*O*-deoxyhexosyl-hexoside (0.5 mg/g)	Leichtweis et al. (2022)
Pumpkin	Seeds	0.6–4 mg/g	0.5–1 mg/g	Epicatechin (3 mg/g) Isorhamnetin-*O*-dideoxyhexosyl-hexoside (0.5 mg/g) Kaempferol-*O*-deoxyhexosyl-hexoside (0.5 mg/g) 7 4-*O*-(6′-*O*-Glucosyl-4″-hydroxybenzoyl)-4-hydroxybenzyl alcohol (0.1 mg/g)	Leichtweis et al. (2022)
Squash	Peel	18 mg GAE/g	n.d.	Epicatechin (4 mg/g) Epigallocatechin (3 mg/g) Hydroxytyrosol (3 mg/g) Catechin gallate (2 mg/g) Resveratrol (1 mg/g)	Mansour et al. (2023)
Squash	Seed	14 mg GAE/g	n.d.	Epicatechin (2 mg/g) Catechin gallate (2 mg/g) Epigallocatechin (1 mg/g) 5-Caffeoylquinic acid (1 mg/g) Resveratrol (1 mg/g)	Mansour et al. (2023)
Squash	Endocarp	16 mg GAE/g	n.d.	Epigallocatechin (22 mg/g) Epicatechin (10 mg/g) Catechin gallate (2 mg/g) Hydroxytyrosol (2 mg/g) Vanillic acid (1 mg/g) 5-Caffeoylquinic acid (1 mg/g) Quercetin 3-O-rutinoside (1 mg/g)	Mansour et al. (2023)
Red pepper	Placenta	60 mg GAE/g DW	57 mg NE/g DW	Quercetin (10 mg/g DW) Myricetin (1 mg/g DW) Vanillic acid (1 mg/g DW) *p*-Coumaric acid (1 mg/g DW)	Chen and Kang (2013)
Red pepper	Stem	70 mg GAE/g DW	68 mg NE/g DW	Quercetin (7 mg/g DW) 5-Caffeoylquinic acid (4 mg/g DW) *p*-Coumaric acid (3 mg/g DW)	Chen and Kang (2013)

Red pepper	Pomace	304 mg/kg	n.d.	Hesperetin 7-*O*-rutinoside (105 mg/kg) Quercetin 3-rhamnoside (95 mg/kg) Eriodictyol 7-*O*-rutinoside (41 mg/kg) Ferulic acid (11 mg/kg) Luteolin 8-C-glucoside (10 mg/kg) Apigenin 6,8-di-C-glucoside (9 mg/kg)	Vaz et al. (2022)
Yellow bell pepper	Product out of commercial shape standard	11 mg GAE/g DW	271–1111 μg/g	Kaempferol 3-(3″-acetyl-alpha-L-arabinopyranosyl)-glucoside (88–215 μg/g DW) Quercetin dihexoside isomer b (8–175 μg/g DW) Quercetin 3-rhamnopyranoside or Luteolin 8-glucoside isomer c (0–173 μg/g DW) Luteolin 7-*O*-(2-apiofuranosyl-4-glucopyranosyl-6-malonyl)-glucopyranoside (35–88 μg/g DW) Luteolin 7-(2-*O*-apiosyl-6-O-malonyl)-glucoside (32–71 Quercetin 3-*O*-rutinosyl-pentoside (20–66 μg/g DW) Quercetin-3-rhamnopyranoside or Luteolin-8-glucoside isomer b (13–62 μg/g DW) Luteolin 6-C-β-D-glucopyranoside-8-C-α-L-arabinopyranoside or Luteolin 7-*O*-[2-(β-D-apiofuranosyl)-β-D-glucopyranoside isomer a (10–58 μg/g DW) Erigeroflavanone (29–55 μg/g DW)	Razola-Díaz et al. (2022)
Cucumber	Pomace	196 mg/kg	n.d.	Hesperetin 7-*O*-rutinoside (136 mg/kg) *p*-Salicylic acid (11 mg/kg) Phloretin 2′-*O*-glucoside (10 mg/kg) Naringenin 7-*O*-rutinoside (9 mg/kg) Quercetin 3-*O*-galactoside (6 mg/kg) Quercetin 3-rhamnoside (5 mg/kg)	Vaz et al. (2022)
Other by-products					
Brewery	Waste stream	449 mg GAE/g DW	n.d.	Gallocatechin (162 mg/g) Epigallocatechin (51 mg/g) Catechin (41 mg/g) Ferulic acid (40 mg/g) Epicatechin (17 mg/g) Protocatechuic acid (17 mg/g)	Barbosa-Pereira et al. (2013)

CAE, *Caffeic Acid Equivalents;* CE, *Catechin Equivalents;* GAE, *Gallic Acid Equivalents;* GE, *Genistein equivalents;* n.d., *Not determined;* n.q., *Not quantified;* NE, *Narigin Equivalents;* QE, *Quercetin Equivalents;* SAE, *Sinapic Acid Equivalents;* TFC, *Total Flavonoid Content;* TPC, *Total Phenolic Content.*

predominate the phenolic composition (5-caffeoylquinic acid, 1,5-dicaffeoylquinic acid, and 3,4-dicaffeoylquinic acid), and some flavones (apigenin 7-*O*-glucoside, apigenin glucuronide, and luteolin 7-*O*-glucuronide) have been found as well (Pagano et al., 2016; Sihem et al., 2015). A study with artichoke floral stem revealed that phenolic profile is largely composed of hydroxycinnamic acids (3-caffeoylquinic acid, 5-caffeoylquinic acid, 1,5-dicaffeoylquinic acid, 3,5-dicaffeoyl quinic acid) similar to leaves and bracts (Sihem et al., 2015).

A recent study characterized the pomace generated from artichoke processing and indicated that 3,5-dicaffeoylquinic acid and 5-caffeoylquinic acid as main components, followed by caffeic acid, protocatechuic acid, and luteolin 7-*O*-glucoside (Vaz et al., 2022). Another relevant information regarding the TPC of artichoke processing by-products, there is no clear trend toward a major source of polyphenols when leaves and bracts are compared (Pagano et al., 2016; Sihem et al., 2015). This condition indicates that both residues can be seen as equally relevant sources of polyphenols from artichoke processing. Conversely, there is scientific evidence indicating that variety and plant section can affect the TPC and polyphenol profile of artichoke by-products (Pagano et al., 2016; Pandino et al., 2013).

Broccoli is another flower vegetable associated with rich phenolic profile, especially its leaves and stalk that are discarded. In the case of leaves, the identification of polyphenols seems to differ among studies. While a paper indicated the presence of derivates from two phenolic acids (5-caffeoylquinic acid and sinapic acid) (Domínguez-Perles et al., 2010), another more recent study indicated esculetin (a hydroxycoumarin) as the main polyphenol, followed by flavonoids myricetin and quercetin (Le et al., 2020).

A different outcome can be observed for studies reporting the phenolic composition of broccoli stalk (Costa-Pérez et al., 2022; Domínguez-Perles et al., 2010; Thomas et al., 2018). In this case, phenolic acids are found to be the main polyphenols, particularly for the presence of 5-caffeoylquinic acid. Moreover, another polyphenol found in broccoli stalk 3-caffeoylquinic acid and sinapic acid (Thomas et al., 2018), 1,2,2′-tri-sinapoyl-gentiobioside, feruloyl-caffeoyl derivative, di-sinapoyl-gentiobioside, and di-sinapoyl-diglucose (Costa-Pérez et al., 2022).

Additional evidence regarding the combination of leaves and stems, as a single residue, corroborates with the predominance of phenolic acids in the composition of broccoli by-products (Aires et al., 2017). In this study, 5-caffeoylquinic acid and 3 -caffeoylquinic acid were the main polyphenols, followed by ferulic acid and sinapic acid. In addition, some flavonoids were also isolated from broccoli leaves and stems: quercetin 3-*O*-rutinoside and kaempferol 3-*O*-rutinoside. Results from TPC analysis indicate contrasting results about which by-product from broccoli would contain more polyphenols. One study indicates that leaves may contain 10 times more polyphenols than stalks (Domínguez-Perles et al., 2010) whereas other studies suggest that stalk may have a higher content TPC than leaves (Le et al., 2020; Thomas et al., 2018).

9. Phenolic profile of leafy vegetables by-products

Due to the processing, external leaves and core of lettuce are discarded due to commercialize in the minimally processed form for modern consumers. In this sense, the residues generated are discarded, and polyphenols are wasted as well. The characterization of phenolic profile of discarded lettuce leaves indicated chicoric acid as the main compound (Plazzotta & Manzocco, 2018). Other relevant polyphenols found in spent lettuce leaves were caffeoyl tartaric acid, 3-caffeoylquinic acid, and isochlorogenic acid among phenolic acids and luteolin 7-*O*-glucuronide and quercetin 3-*O*-glucuronide in the class of flavonoids. Another potential source of polyphenols among leafy vegetable is the stalk of cabbage. A recent study indicates the presence of phenolic acids (4-hydroxybenzaldehyde, *p*-coumaric acid, 4-*p*-coumaroylquinic acid, and 5-caffeoylquinic acid) and flavonoids (quercetin 3-*O*-sophoroside) in this residue (Brito et al., 2021).

10. Phenolic profile of stem vegetables by-products

A recent experiment indicated that the phenolic profile of discarded spinach leaves (out of commercial size) is mainly composed of flavonoids, especially derivates of quercetin 3-*O*-rutinoside, epigallocatechin gallate, and lutein (Araújo-Rodrigues et al., 2021). A similar profile was observed for rocket in the same study regarding the predominance of flavonoids wherein an apigenin derivative was indicated as major polyphenol, followed by quercetin 3-*O*-glucoside, quercetin 3-*O*-rutinoside, and quercetin. Moreover, phenolic acids (ferulic acid, 5-caffeoylquinic acid and *p*-coumaric acid) were also found among the main polyphenols. Conversely, discarded leaves of watercress were found to contain phenolic acids, especially derivatives from caffeic acid, 5-caffeoylquinic acid, and *p*-coumaric acid (Araújo-Rodrigues et al., 2021). Another group of compounds found in this by-product was flavonoids, particularly quercetin and quercetin 3-*O*-glucoside.

A related experiment with wild cardoon leaves indicated *p*-coumaric acid as main phenolic compound (Chihoub et al., 2019). Kaempferol linked to different sugar substituents was also obtained from wild cardoon leaves: kaempferol 3-*O*-rutinoside, kaempferol-*O*-rhamnoside-*O*-pentoside, and kaempferol-*O*-dirhamnoside. In addition, this study also identified caffeoyl malate and feruloyl malate among the main polyphenols. Comparing the TPC of different leafy vegetables can highlight sources with the highest potential for exploitation of polyphenols. Some scientific evidence indicates that watercress discarded leaves may contain more polyphenols than rocket and spinach (Araújo-Rodrigues et al., 2021). Although differences exist among these leafy vegetables, all seem to have high content of polyphenols and should received attention on further experiments.

11. Phenolic profile of underground vegetables by-products

11.1 Carrot

Different studies indicate that the main group in carrot processing by-products is phenolic acids. One example is their presence in peels of carrots, which includes 3-caffeoylquinic acid, 4-caffeoylquinic acid, 5-caffeoylquinic acid, 4-*p*-coumaroylquinic acid, and 5-*p*-coumaroylquinic acid (Delpino-Rius et al., 2015). Another recent experiment also indicates that phenolic acid comprises one of the main groups of polyphenols in carrot peel and reported the presence of 5-caffeoylquinic acid, *p*-hydroxybenzoic acid, salicylic acid, and ellagic acid (El-Sawi et al., 2022). Other polyphenols reported in carrot peels were phloretin 2′-*O*- glucoside (Delpino-Rius et al., 2015), catechol and catechin (El-Sawi et al., 2022). Carrot pomace, obtained from juice pressing, is also mainly composed of phenolic acids, especially 5-caffeoylquinic acid, *o*-salicylic acid, protocatechuic acid, and caffeic acid (Vaz et al., 2022). A similar result was reported for the phenolic composition of carrots out of commercial shape wherein 5-caffeoylquinic acid and *trans*-ferulic acid were indicated as key polyphenols (Araújo-Rodrigues et al., 2021).

In the case of black carrot, characterized by its dark color ranging from purple to black, the predominance of phenolic acids (Gulsunoglu et al., 2019; Polat et al., 2022). Recent evidence indicates that phenolic acids are the major class of polyphenols, especially 5-caffeoylquinic acid. Other phenolic acids found in black carrot pomace were ferulic acid, sinapic acid, and derivatives of caffeic acid. Anthocyanins are also found in black carrot pomace and the compounds responsible for the intense color of this vegetable. Some of the main anthocyanins found in black carrot pomace are cyanidin 3-*O*-galactoside, cyanidin 3-*O*-glucoside, cyanidin 3-xylosyl-feruloyl-glucoside-galactoside, and malvidin 3-*O*-glucoside (Gulsunoglu et al., 2019; Polat et al., 2022).

11.2 Potato

Potato is a major crop that generates residues in the form of peels that are rich in phenolic acids, especially 5-caffeoylquinic acid and caffeic acid (Albishi et al., 2013; Lima et al., 2021; Singh & Saldaña, 2011). Other polyphenols found in potato peel include ferulic acid and *p*-coumaric acid (Albishi et al., 2013; Singh & Saldaña, 2011), isochlorogenic acid (Lima et al., 2021), protocatechuic acid, syringic acid, and gallic acid (Singh & Saldaña, 2011).

Another study related to potato peel, indicated that this residue is source of phenolic acids and flavonoids in different varies of red and purple sweet potato (Sampaio et al., 2021). In this case, caffeic acid was the main polyphenols, followed by 4-caffeoylquinic acid. Among the flavonoids, some kaempferol was found (kaempferol 3-*O*-rutinoside, kaempferol-*O*-deoxyhexosyl-hexoside, and kaempferol-*O*-hexoside-*O*-deoxyhexoside-hexoside). Moreover, some studies also support the importance of potato variety in the composition and content of polyphenols in the by-products generated from this food (Albishi et al., 2013; Sampaio et al., 2021).

11.3 Other underground vegetables

Radish leaves and stem were also studied as potential source of polyphenols, wherein phenolic acids and flavones were the main groups (Chihoub et al., 2019). In these residues, remarkable phenolic acids were 4-caffeoylquinic acid and 3,4-dicaffeoylquinic acid whereas among flavones the main compounds were luteolin-*O*-malonylhexoside, luteolin-*O*-hexoside, and acetylapigenin-*O*-hexoside. An experiment with black radish peel revealed a different phenolic profile than observed for common radish leaves and stems (Yücetepe et al., 2021). In this case, black radish peels phenolic composition was largely composed of ferulic acid, gallic acid, sinapic acid, and epicatechin.

Another potential source of polyphenols is the residue generated by turnip harvesting and processing, which is mainly composed of leaves and stems (Chihoub et al., 2019). The determination of phenolic profile revealed the presence of isorhamnetin-*O*-dihexoside, synapoylmalic acid, kaempferol 3-*O*-sinapoyl-sophoroside-7-*O*-glucoside, and kaempferol-*O*-dihexoside as main compounds. Tiger nuts are tubers used in the production of Spanish traditional beverage known as orchata de chufa that generates wastes rich in polyphenols (Roselló-Soto et al., 2019). In this residue, the main polyphenols include 4-vinylphenol, spinacetin 3-O-glucosyl-(1 → 6)-glucoside, sinensetin, and *p*-coumaric acid.

12. Phenolic profile of oilseeds and olive by-products

12.1 Corn

Corn is a major crop produced to attend the global demand for its grain and oil. Its harvesting and processing generates many residues rich in polyphenols as indicated by studies with corn pericarp, tassel, bracts, cobs, and silk (Table 3.3). Regarding corn pericarp, a recent experiment indicated hesperetin 7-*O*-rutinoside as main polyphenol, followed by other flavonoids: delphinidin, cyanidin 3-*O*-glucoside, cyanidin chloride, kaempferol, and quercetin (Boateng et al., 2023). Some phenolic acids were also indicated as major components of corn pericarp phenolic composition, which include caffeic acid, ferulic acid, and 5-caffeoylquinic acid.

A study characterized the phenolic composition of corn bracts, cob, and silk and found that ferulic acid was a major phenolic compound in all these by-products of corn harvesting and processing (Yang et al., 2019). Gallic acid and 4-hydroxycoumarin were other phenolic compounds found all three by-products whereas *p*-coumaric acid and quercetin 3-rhamnoside were major polyphenols only on bracts and cob. Another relevant phenolic compound in more than one residue (bract and silk) was coumarin. A related experiment with corn silk indicated a different phenolic profile than observed for the aforementioned study (Žilić et al., 2016). In this case, the phenolic acids (3-caffeoylquinic acid, 4-caffeoylquinic acid, 5-caffeoylquinic acid, and *p*-coumaroylquinic acid) were the major polyphenols. This outcome indicates that variations, probably associated with corn variety, may affect the phenolic composition of corn silk.

Corn tassel is another residue generated from corn that could be explored as potential source of polyphenols. Although scientific evidence indicates differences among the phenolic profiles reported from papers recently published, the main polyphenol group is phenolic acids. In one case, corn tassel was mainly composed of vanillic acid and caffeic acid, followed by flavonoids (especially butein, myricetin, and resveratrol) (Al-Khayri et al., 2022). In another study, the major polyphenols were gallic acid, 5-caffeoylquinic acid, and caffeic acid, followed by catechin, naringenin, and quercetin 3-*O*-rutinoside (Elsayed et al., 2022). Corn variety and extraction method may explain the differences observed between these two studies. In addition, the analysis of TPC of different corn sections suggests that cob may contain more polyphenols than bracts and silk (Yang et al., 2019). The variety of corn was also found to affect the individual polyphenols and TPC in corn silk (Žilić et al., 2016).

12.2 Canola

Canola processing can generate important residues with relevant phenolic composition and content. Studies indicate sinapic acids derivatives (especially sinapine and sinapoyl glucose) and sinapic acid as main polyphenols in canola hulls (Mayengbam et al., 2014). A similar outcome was observed of the evaluation of phenolic profile of canola oil pressing in a related study (Tian et al., 2022). Sinapine, sinapoyl malate, and sinapoyl glucose were found among the main compounds, which were followed by some kaempferol derivatives. In addition, a related experiment also suggested the presence of sinapine derivatives in canola oil pressing residue (Vidal et al., 2022).

12.3 Other oilseeds

The evaluation of soy residue from pressing reveals da unique phenolic profile in relation to other residues from oil seeds (Silva & Perrone, 2015). In this case, phenolic profile was largely composed of isoflavonoids wherein the main compounds were genistein derivatives (daidzein, daidzin-β-glucoside, genistein, genistin-β-glucoside, and glycitein). It also worth mentioning that derivates of 5-caffeoylquinic acid were found in sunflower oil press residue (Vidal et al., 2022).

Peanuts are also important crops that can be used in the production of vegetable oils. One of its main processing by-products consists of skin that is rich in polyphenols. The analysis of polyphenol composition indicates that polymeric forms of (epi) catechin are the main compounds found in peanut skin (de Camargo et al., 2015; Ma et al., 2014). The degree of polymerization of procyanidins varies from dimers to octamers. Their monomer forms (catechin and epicatechin) were also reported in peanut skin along with protocatechuic acid, gallic acid, and isorhamnetin-glucoside (de Camargo et al., 2015). A related residue is the skin of pistachio seeds that are discarded during its processing and contain polyphenols (Tomaino et al., 2010). In this case, a more diverse profile has been reported due to the presence of cyanidin 3-*O*-galactoside and gallic acid as main polyphenols, followed by catechin and eriodictyol 7-*O*-glucoside.

12.4 Olives

Olives are a traditional crop used for the production of vegetable oil with known potential health benefits. Its processing generates residues that are rich in polyphenols, especially the characteristic group known as tyrosols. The wastewater from olive pressing contains many tyrosols (tyrosol, hydroxytyrosol, and hydroxytyrosol glucoside) as major group, as well as phenolic acids (*p*-Coumaric acid, gallic acid, and verbascoside) (El-Abbassi et al., 2012; Obied et al., 2008) and flavonoids (luteolin) (Obied et al., 2008). In addition, a study also indicated that both TPC and phenolic profile of olive mill waste waster are affected by olive variety (Obied et al., 2008).

13. Phenolic profile of legumes by-products

During processing of lentil, hulls are discarded, and polyphenols are wasted. Scientific evidence indicates differences in phenolic profile across studies (Table 3.3). In one study, the main polyphenol found was syringic acid, followed by protocatechuic acid and *cis*-ferulic acid along with flavonoids (quercetin and catechin) (Yeo & Shahidi, 2020). Conversely, another experiment supports the indication that flavonoids are the main group of polyphenols such as formononetin, quercetin, quercetin 3-*O*-glucoside, and myricetin (Paranavitana et al., 2021). A related study with lentil hulls quantified procyanidin dimers as major polyphenols, followed by derivatives of dimethoxybenzoic acid and *p*-coumaric acid (Sun et al., 2020). Regarding the factors affecting the polyphenols in lentils hulls, variety was indicated as source of variation for TPC and individual polyphenols (Paranavitana et al., 2021; Sun et al., 2020).

14. Phenolic profile of vine-grown vegetables by-products

Another commercially relevant class of vegetables generating by-products rich in polyphenols comprises vine-grown vegetables such as tomatoes, pumpkins, squashes, peppers, and cucumber (Table 3.3).

14.1 Tomato

The by-products of tomato processing are found to contain polyphenols that could be extracted from peels, seeds, stems, and pomace. Regarding the phenolic profile of tomato peels, the main compounds found in a recent study belong to flavonoid class (naringenin, naringenin chalcone, quercetin 3-rutinoside, and quercetin-triglucoside) (Szabo et al., 2019). Other polyphenols found in tomato peels were 3,4-dicaffeoylquinic acid and,4,5-tri-caffeoylquinic acid. An early published study corroborates with the presence of flavonoids, especially for the presence of quercetin 3-*O*-rutinoside and naringenin (Benakmoum et al., 2008).

The evaluation of by-products that combine more than one section of tomato such as reported in a study with peels and seeds or the pomace from pressing revealed phenolic acids predominate along with flavonoids in the phenolic profile. Main flavonoids found in tomato peel (quercetin, quercetin 3-O-rutinoside, and naringenin) (Bao et al., 2020; Nour et al., 2018; Perea-Domínguez et al., 2018) remains as major constituents while phenolic acids become a major share of phenolic composition, particularly due to the presence of 5-caffeoylquinic acid (Bao et al., 2020; Nour et al., 2018), caffeic acid and gallic acid (Bao et al., 2020; Perea-Domínguez et al., 2018) and ellagic acid (Nour et al., 2018).

Removal of leaves and stem is a common practice and composes a major source of wasted polyphenols from tomato processing chain. Characterization of polyphenol composition of this residue revealed the presence of flavonoids (kaempferol 3-*O*-rutinoside and quercetin 3-*O*-rutinoside) and phenolic acids (caffeic acid, *p*-coumaric acid, and 5-caffeoylquinic acid) among the main compounds (Aires et al., 2017). Another potential source of polyphenols from tomatoes is the products out of commercial shape that may be discarded. In this case, 5-caffeoylquinic acid was found to be the main polyphenol, followed by quercetin 3-*O*-rutinoside and naringenin (Araújo-Rodrigues et al., 2021).

14.2 Pumpkin

Another interesting source of polyphenols among vegetable by-products is those derived from pumpkin processing, which include peel, fibrous strands, and seeds (Table 3.3). In the case of pumpkin peels, flavanols compose most of polyphenols found in this residue, especially isorhamnetin-*O*-deoxyhexosyl-hexoside and isorhamnetin-*O*-dideoxyhexosyl-hexoside, followed by epicatechin (Leichtweis et al., 2022, 2023). Another major group of polyphenols found in pumpkin peels is comprised of flavonols (kaempferol-*O*-deoxyhexosyl-hexoside, kaempferol-*O*-dideoxyhexosyl-hexoside, and quercetin-*O*-dideoxyhexosyl-hexoside). Particularly for pumpkin fibrous strands and seeds, epicatechin stands out as main polyphenol (Leichtweis et al., 2022).

14.3 Squash

Squash is another member of the group of vine-grown vegetables that generates residues. A recent experiment three of the main residues from squash (peel, seed, and endocarp) (Mansour et al., 2023). Similarities between the phenolic profiles among these residues were found due to the presence of epicatechin, epigallocatechin, and catechin gallate. Other major polyphenols were hydroxytyrosol, resveratrol, and 5-caffeoylquinic acid found in more than one residue. Moreover, this experiment also indicated significant differences were obtained for TPC and phenolic profile among seeds, peel, and endocarp of squash.

14.4 Red pepper, yellow pepper and cucumber

Red pepper by-products can also be considered as potential sources of polyphenols. A comparative study indicated that quercetin is the major phenolic compound found in red pepper placenta and stem (Chen & Kang, 2013). Other polyphenols that can also be found in these residues are 5-caffeoylquinic acid and *p*-coumaric acid. A related study with red pepper pomace indicated a different phenolic profile (Vaz et al., 2022). In this by-product, the main polyphenols were hesperetin 7-*O*-rutinoside and quercetin 3-rhamnoside, followed by eriodictyol 7-*O*-rutinoside. A similar experiment characterized the phenolic found in yellow bell peppers out of commercial shape (Razola-Díaz et al., 2022). The main compounds belonged to flavonoids group and included several derivatives from kaempferol, quercetin, and luteolin.

Scientific evidence also indicates that cucumber pomace contains many polyphenols (Vaz et al., 2022). The main polyphenol found was hesperetin 7-*O*-rutinoside whereas other polyphenols (particularly *p*-salicylic acid, phloretin 2′-*O*-glucoside, and naringenin 7-*O*-rutinoside) seem to have a minor contribution to this fraction.

15. Phenolic profile of by-products from other vegetable sources

In addition to the by-products generated from fresh produce harvesting and processing, the residues generated from brewery is a relevant example of the diversity of sources that can be explored to recovery polyphenols and prevent their loss. In this case, a study with the waste stream from brewery revealed that flavanols are the main polyphenols in this residue (Barbosa-Pereira et al., 2013). Gallocatechin was found to compose most of the polyphenol fraction of waste stream, which also contained epigallocatechin, catechin, and ferulic acid among the main polyphenols.

16. Final considerations and conclusion

The by-products produced from harvesting and processing of fruits and vegetables can be considered as rich sources of polyphenols. The variability of compounds comprises a major advantage in terms of exploring specific groups and individual compounds for application in downstream industries, such as food, pharmaceutical, and cosmetic.

Although the comparison among sources is particularly complex due to the variations associated with many factors related to sample themselves (particularly variety, and plant section), some sources seem to stand out among all compiled papers in this chapter. Among fruits, some noteworthy residues include apple pomace, avocado peel and seed, banana flower and pseudo-stem, citrus pomace, cocoa bean shell, grape pomace, kiwi berry peel, melon peel, and pineapple crown. Current knowledge also suggests that some residues from vegetable can also be seen as potential sources of polyphenols: artichoke leaves, broccoli leaves, canola oil pressing residue, corn silk, lentil hulls, pistachio skin, radish leaves and stem, soy oil press residue, and wild cardoon leaves.

As information grows for fruits and vegetable products at large scale and commercialized around the globe, more information on less-known and locally produced fruits and vegetables is also of great scientific value. Unveiling new sources and characterizing their by-products can assist in the generation of knowledge to support further studies, especially for commercial applications.

References

Aires, A., Carvalho, R., & Saavedra, M. J. (2017). Reuse potential of vegetable wastes (broccoli, green bean and tomato) for the recovery of antioxidant phenolic acids and flavonoids. *International Journal of Food Science and Technology, 52*(1), 98–107. https://doi.org/10.1111/ijfs.13256

Al-Khayri, J. M., Yüksel, A. K., Yüksel, M., Isık, M., & Dikici, E. (2022). Phenolic profile and antioxidant, anticholinergic, and antibacterial properties of corn tassel. *Plants, 11*(15). https://doi.org/10.3390/plants11151899

Albishi, T., John, J. A., Al-Khalifa, A. S., & Shahidi, F. (2013). Phenolic content and antioxidant activities of selected potato varieties and their processing by-products. *Journal of Functional Foods, 5*(2), 590–600. https://doi.org/10.1016/j.jff.2012.11.019

Araújo-Rodrigues, H., Santos, D., Campos, D. A., Guerreiro, S., Ratinho, M., Rodrigues, I. M., & Pintado, M. E. (2021). Impact of processing approach and storage time on bioactive and biological properties of rocket, spinach and watercress byproducts. *Foods, 10*(10). https://doi.org/10.3390/foods10102301

Argun, M. E., Argun, M.S., Arslan, F. N., Nas, B., Ates, H., Tongur, S., & Cakmakcı, O. (2022). Recovery of valuable compounds from orange processing wastes using supercritical carbon dioxide extraction. *Journal of Cleaner Production, 375*. https://doi.org/10.1016/j.jclepro.2022.134169

Arraibi, A. A., Liberal, Â., Dias, M. I., Alves, M. J., Ferreira, I. C. F. R., Barros, L., & Barreira, J. C. M. (2021). Chemical and bioactive characterization of Spanish and Belgian apple pomace for its potential use as a novel dermocosmetic formulation. *Foods, 10*(8). https://doi.org/10.3390/foods10081949

Azman, E. M., Nor, N. D. M., Charalampopoulos, D., & Chatzifragkou, A. (2022). Effect of acidified water on phenolic profile and antioxidant activity of dried blackcurrant (Ribes nigrum L.) pomace extracts. *Lebensmittel-Wissenschaft & Technologie, 154*. https://doi.org/10.1016/j.lwt.2021.112733

Bao, Y., Reddivari, L., & Huang, J.-Y. (2020). Development of cold plasma pretreatment for improving phenolics extractability from tomato pomace. *Innovative Food Science & Emerging Technologies, 65*, Article 102445. https://doi.org/10.1016/j.ifset.2020.102445

Barbosa-Pereira, L., Angulo, I., Paseiro-Losada, P., & Cruz, J. M. (2013). Phenolic profile and antioxidant properties of a crude extract obtained from a brewery waste stream. *Food Research International, 51*(2), 663–669. https://doi.org/10.1016/j.foodres.2013.01.042

Barrales, F. M., Silveira, P., Barbosa, P.d. P. M., Ruviaro, A. R., Paulino, B. N., Pastore, G. M., Macedo, G. A., & Martinez, J. (2018). Recovery of phenolic compounds from citrus by-products using pressurized liquids — An application to orange peel. *Food and Bio-products Processing, 112*, 9–21. https://doi.org/10.1016/j.fbp.2018.08.006

Begum, Y. A., & Deka, S. C. (2019). Chemical profiling and functional properties of dietary fibre rich inner and outer bracts of culinary banana flower. *Journal of Food Science and Technology. 09758402, 56*(12), 5298–5308. https://doi.org/10.1007/s13197-019-04000-4

Benakmoum, A., Abbeddou, S., Ammouche, A., Kefalas, P., & Gerasopoulos, D. (2008). Valorisation of low quality edible oil with tomato peel waste. *Food Chemistry, 110*(3), 684–690. https://doi.org/10.1016/j.foodchem.2008.02.063

Berardini, N., Knödler, M., Schieber, A., & Carle, R. (2005). Utilization of mango peels as a source of pectin and polyphenolics. *Innovative Food Science and Emerging Technologies, 6*(4), 442–452. https://doi.org/10.1016/j.ifset.2005.06.004

Boateng, I. D., Kumar, R., Daubert, C. R., Flint-Garcia, S., Mustapha, A., Kuehnel, L., Agliata, J., Li, Q., Wan, C., & Somavat, P. (2023). Sonoprocessing improves phenolics profile, antioxidant capacity, structure, and product qualities of purple corn pericarp extract. *Ultrasonics Sonochemistry, 95*. https://doi.org/10.1016/j.ultsonch.2023.106418

Brito, T. B. N., Lima, L. R. S., Santos, M. C. B., Moreira, R. F. A., Cameron, L. C., Fai, A. E. C., & Ferreira, M. S. L. (2021). Antimicrobial, antioxidant, volatile and phenolic profiles of cabbage-stalk and pineapple-crown flour revealed by GC-MS and UPLC-MSE. *Food Chemistry, 339*, Article 127882. https://doi.org/10.1016/j.foodchem.2020.127882

Campos, D. A., Gómez-García, R., Vilas-Boas, A. A., Madureira, A. R., & Pintado, M. M. (2020). Management of fruit industrial by-products—A case study on circular economy approach. *Molecules, 25*(2). https://doi.org/10.3390/molecules25020320

Chen, L., & Kang, Y. H. (2013). Anti-inflammatory and antioxidant activities of red pepper (Capsicum annuum L.) stalk extracts: Comparison of pericarp and placenta extracts. *Journal of Functional Foods, 5*(4), 1724–1731. https://doi.org/10.1016/j.jff.2013.07.018

Cheng, V. J., Bekhit, A. E. D. A., McConnell, M., Mros, S., & Zhao, J. (2012). Effect of extraction solvent, waste fraction and grape variety on the antimicrobial and antioxidant activities of extracts from wine residue from cool climate. *Food Chemistry, 134*(1), 474–482. https://doi.org/10.1016/j.foodchem.2012.02.103

Cheok, C. Y., Mohd Adzahan, N., Abdul Rahman, R., Zainal Abedin, N. H., Hussain, N., Sulaiman, R., & Chong, G. H. (2018). Current trends of tropical fruit waste utilization. *Critical Reviews in Food Science and Nutrition, 58*(3), 335–361. https://doi.org/10.1080/10408398.2016.1176009

Chihoub, W., Dias, M. I., Barros, L., Calhelha, R. C., Alves, M. J., Harzallah-Skhiri, F., & Ferreira, I. C. F. R. (2019). Valorisation of the green waste parts from turnip, radish and wild cardoon: Nutritional value, phenolic profile and bioactivity evaluation. *Food Research International, 126*. https://doi.org/10.1016/j.foodres.2019.108651

Costa-Pérez, A., Moreno, D. A., Periago, P. M., García-Viguera, C., & Domínguez-Perles, R. (2022). A new food ingredient rich in bioaccessible (poly)phenols (and glucosinolates) obtained from stabilized broccoli stalks. *Foods, 11*(12). https://doi.org/10.3390/foods11121734

Crozier, A., Jaganath, I. B., & Clifford, M. N. (2009). Dietary phenolics: Chemistry, bioavailability and effects on health. *Natural Product Reports, 26*(8), 1001–1043. https://doi.org/10.1039/b802662a

Das, I., & Arora, A. (2017). Post-harvest processing technology for cashew apple – A review. *Journal of Food Engineering, 194*, 87–98. https://doi.org/10.1016/j.jfoodeng.2016.09.011

de Ancos, B., Sánchez-Moreno, C., Zacarías, L., Rodrigo, M. J., Sáyago Ayerdí, S., Blancas Benítez, F. J., Domínguez Avila, J. A., & González-Aguilar, G. A. (2018). Effects of two different drying methods (freeze-drying and hot air-drying) on the phenolic and carotenoid profile of 'Ataulfo' mango by-products. *Journal of Food Measurement and Characterization, 12*(3), 2145–2157. https://doi.org/10.1007/s11694-018-9830-4

de Camargo, A. C., Regitano-d'Arce, M. A. B., Gallo, C. R., & Shahidi, F. (2015). Gamma-irradiation induced changes in microbiological status, phenolic profile and antioxidant activity of peanut skin. *Journal of Functional Foods, 12*, 129–143. https://doi.org/10.1016/j.jff.2014.10.034

De Montijo-Prieto, S., Razola-Díaz, M.d. C., Barbieri, F., Tabanelli, G., Gardini, F., Jiménez-Valera, M., Ruiz-Bravo, A., Verardo, V., & Gómez-Caravaca, A. M. (2023). Impact of lactic acid bacteria fermentation on phenolic compounds and antioxidant activity of avocado leaf extracts. *Antioxidants, 12*(2). https://doi.org/10.3390/antiox12020298

de Souza, D. R., Willems, J. L., & Low, N. H. (2019). Phenolic composition and antioxidant activities of Saskatoon berry fruit and pomace. *Food Chemistry, 290*, 168–177. https://doi.org/10.1016/j.foodchem.2019.03.077

Delpino-Rius, A., Eras, J., Vilaró, F., Cubero, M.Á., Balcells, M., & Canela-Garayoa, R. (2015). Characterisation of phenolic compounds in processed fibres from the juice industry. *Food Chemistry, 172*, 575–584. https://doi.org/10.1016/j.foodchem.2014.09.071

Dias, R., Oliveira, H., Fernandes, I., Simal-Gandara, J., & Perez-Gregorio, R. (2021). Recent advances in extracting phenolic compounds from food and their use in disease prevention and as cosmetics. *Critical Reviews in Food Science and Nutrition, 61*(7), 1130–1151. https://doi.org/10.1080/10408398.2020.1754162

Domínguez-Perles, R., Martínez-Ballesta, M. C., Carvajal, M., García-Viguera, C., & Moreno, D. A. (2010). Broccoli-derived by-products - A promising source of bioactive ingredients. *Journal of Food Science, 75*(4), Article C383. https://doi.org/10.1111/j.1750-3841.2010.01606.x

El-Abbassi, A., Kiai, H., & Hafidi, A. (2012). Phenolic profile and antioxidant activities of olive mill wastewater. *Food Chemistry, 132*(1), 406–412. https://doi.org/10.1016/j.foodchem.2011.11.013

El-Sawi, S. A., Ibrahim, M. E., Sleem, A. A., Farghaly, A. A., Awad, G. E. A., & Merghany, R. M. (2022). Development of alternative medicinal sources from golden berry, bananas and carrot wastes as antioxidant, cytotoxic and antimicrobial agents. *Acta Ecologica Sinica, 42*(3), 224–232. https://doi.org/10.1016/j.chnaes.2021.04.006

Elsayed, N., Marrez, D. A., Ali, M. A., El-maksoud, A. A. A., Cheng, W., & Abedelmaksoud, T. G. (2022). Phenolic profiling and in-vitro bioactivities of corn (Zea mays L.) tassel extracts by combining enzyme-assisted extraction. *Foods, 11*(14), 2145.

Famiani, F., Bonghi, C., Chen, Z. H., Drincovich, M. F., Farinelli, D., Lara, M. V., Proietti, S., Rosati, A., Vizzotto, G., & Walker, R. P. (2020). Stone fruits: Growth and nitrogen and organic acid metabolism in the fruits and seeds—A review. *Frontiers in Plant Science, 11*. https://doi.org/10.3389/fpls.2020.572601

Fidelis, M., De Moura, C., Kabbas, T., Pap, N., Mattila, P., Mäkinen, S., Putnik, P., Kovačević, D. B., Tian, Y., Yang, B., & Granato, D. (2019). Fruit seeds as sources of bioactive compounds: Sustainable production of high value-added ingredients from by-products within circular economy. *Molecules, 24*(21). https://doi.org/10.3390/molecules24213854

Fierascu, R. C., Sieniawska, E., Ortan, A., Fierascu, I., & Xiao, J. (2020). Fruits by-products – A source of valuable active principles. A short review. *Frontiers in Bioengineering and Biotechnology, 8*. https://doi.org/10.3389/fbioe.2020.00319

Freitas, B. S. M de, Haas, I. C.da S., Pereira-Coelho, M., Almeida, A. B.de, de Almeida, J.dos S. O., Egea, M. B., Silva, F. G.da, Petkowicz, C. L.de O., de Francisco, A., Madureira, L. A.dos S., Freire, C. B. F., & Amboni, R. D. (2023). Cerrado cashew (Anacardium othonianum Rizz) apple pomace: Chemical characterization and optimization of enzyme-assisted extraction of phenolic compounds. *Food Science and Technology, 43*. https://doi.org/10.5327/fst.90222

García, Y. D., Valles, B. S., & Picinelli Lobo, A. (2009). Phenolic and antioxidant composition of by-products from the cider industry: Apple pomace. *Food Chemistry, 117*(4), 731–738. https://doi.org/10.1016/j.foodchem.2009.04.049

Gharaghani, A., Solhjoo, S., & Oraguzie, N. (2016). A review of genetic resources of pome fruits in Iran. *Genetic Resources and Crop Evolution, 63*(1), 151–172. https://doi.org/10.1007/s10722-015-0334-3

Gómez-García, R., Campos, D. A., Oliveira, A., Aguilar, C. N., Madureira, A. R., & Pintado, M. (2021). A chemical valorisation of melon peels towards functional food ingredients: Bioactives profile and antioxidant properties. *Food Chemistry, 335*. https://doi.org/10.1016/j.foodchem.2020.127579

Gulsunoglu, Z., Karbancioglu-Guler, F., Raes, K., & Kilic-Akyilmaz, M. (2019). Soluble and insoluble-bound phenolics and antioxidant activity of various industrial plant wastes. *International Journal of Food Properties, 22*(1), 1501–1510. https://doi.org/10.1080/10942912.2019.1656233

Hong, Y., Wang, Z., Barrow, C. J., Dunshea, F. R., & Suleria, H. A. R. (2021). High-throughput screening and characterization of phenolic compounds in stone fruits waste by lc-esi-qtof-ms/ms and their potential antioxidant activities. *Antioxidants, 10*(2), 1–22. https://doi.org/10.3390/antiox10020234

Kheirkhah, H., Baroutian, S., & Quek, S. Y. (2019). Evaluation of bioactive compounds extracted from Hayward kiwifruit pomace by subcritical water extraction. *Food and Bioproducts Processing, 115*, 143–153. https://doi.org/10.1016/j.fbp.2019.03.007

Kondakova, V., Tsvetkov, I., Batchvarova, R., Badjakov, I., Dzhambazova, T., & Slavov, S. (2009). Phenol compounds - Qualitative index in small fruits. *Biotechnology and Biotechnological Equipment, 23*(4), 1444–1448. https://doi.org/10.2478/V10133-009-0024-4

Koprivica, M. R., Trifković, J., Dramićanin, A. M., Gašić, U. M., Akšić, M. M. F., & Milojković-Opsenica, D. M. (2018). Determination of the phenolic profile of peach (Prunus persica L.) kernels using UHPLC–LTQ OrbiTrap MS/MS technique. *European Food Research and Technology, 244*(11), 2051–2064. https://doi.org/10.1007/s00217-018-3116-2

Kosińska, A., Karamać, M., Estrella, I., Hernández, T., Bartolomé, B., & Dykes, G. A. (2012). Poland Phenolic compound profiles and antioxidant capacity of persea Americana mill. Peels and seeds of two varieties. *Journal of Agricultural and Food Chemistry, 60*(18), 4613–4619. https://doi.org/10.1021/jf300090p

Le, T. N., Sakulsataporn, N., Chiu, C. H., & Hsieh, P. C. (2020). Polyphenolic profile and varied bioactivities of processed Taiwanese grown broccoli: A comparative study of edible and non-edible parts. *Pharmaceuticals, 13*(5). https://doi.org/10.3390/ph13050082

Leichtweis, M. G., Molina, A. K., Petropoulos, S. A., Carocho, M., Pires, T. C. S. P., Dias, M. I., Calhelha, R., Oliveira, M. B. P. P., Pereira, C., & Barros, L. (2023). Valorization of pumpkin peel as a source of bioactive compounds: Optimization of heat- and ultrasound-assisted extraction. *Molecules, 28*(7). https://doi.org/10.3390/molecules28073168

Leichtweis, M. G., Molina, A. K., Pires, T. C. S., Dias, M. I., Calhelha, R., Bachari, K., Ziani, B. E. C., Oliveira, M. B. P. P., Pereira, C., & Barros, L. (2022). Biological activity of pumpkin byproducts: Antimicrobial and antioxidant properties. *Molecules, 27*(23). https://doi.org/10.3390/molecules27238366

Li, R., Hettiarachchy, N., Rayaprolu, S., Eswaranandam, S., Howe, B., Davis, M., & Jha, A. (2014). Phenolics and antioxidant activity of Saskatoon berry (Amelanchier alnifolia) pomace extract. *Journal of Medicinal Food, 17*(3), 384–392. https://doi.org/10.1089/jmf.2012.0278

Lima, M.d. A., Andreou, R., Charalampopoulos, D., & Chatzifragkou, A. (2021). Supercritical carbon dioxide extraction of phenolic compounds from potato (Solanum tuberosum) peels. *Applied Sciences, 11*(8). https://doi.org/10.3390/app11083410

López-Fernández-Sobrino, R., Soliz-Rueda, J. R., Margalef, M., Arola-Arnal, A., Suárez, M., Bravo, F. I., & Muguerza, B. (2021). Ace inhibitory and antihypertensive activities of wine lees and relationship among bioactivity and phenolic profile. *Nutrients, 13*(2), 1–17. https://doi.org/10.3390/nu13020679

Lyu, X., Agar, O. T., Barrow, C. J., Dunshea, F. R., & Suleria, H. A. R. (2023). Phenolic compounds profiling and their antioxidant capacity in the peel, pulp, and seed of Australian grown avocado. *Antioxidants, 12*(1). https://doi.org/10.3390/antiox12010185

Ma, Y., Kosińska-Cagnazzo, A., Kerr, W. L., Amarowicz, R., Swanson, R. B., & Pegg, R. B. (2014). Separation and characterization of phenolic compounds from dry-blanched peanut skins by liquid chromatography-electrospray ionization mass spectrometry. *Journal of Chromatography A, 1356*, 64–81. https://doi.org/10.1016/j.chroma.2014.06.027

Majerska, J., Michalska, A., & Figiel, A. (2019). A review of new directions in managing fruit and vegetable processing by-products. *Trends in Food Science and Technology, 88*, 207–219. https://doi.org/10.1016/j.tifs.2019.03.021

Mansour, R. B., Falleh, H., Hammami, M., Barros, L., Petropoulos, S. A., Tarchoun, N., & Ksouri, R. (2023). The use of response surface methodology to optimize assisted extraction of bioactive compounds from Cucurbita maxima fruit by-products. *Processes, 11*(6). https://doi.org/10.3390/pr11061726

Mayengbam, S., Aachary, A., & Thiyam-Holländer, U. (2014). Endogenous phenolics in hulls and cotyledons of mustard and canola: A comparative study on its sinapates and antioxidant capacity. *Antioxidants, 3*(3), 544–558. https://doi.org/10.3390/antiox3030544

Md Nor, S., & Ding, P. (2020). Trends and advances in edible biopolymer coating for tropical fruit: A review. *Food Research International, 134*, Article 109208. https://doi.org/10.1016/j.foodres.2020.109208

Mellinas, A. C., Jiménez, A., & Garrigós, M. C. (2020). Optimization of microwave-assisted extraction of cocoa bean shell waste and evaluation of its antioxidant, physicochemical and functional properties. *Lebensmittel-Wissenschaft & Technologie, 127*, Article 109361. https://doi.org/10.1016/j.lwt.2020.109361

Multari, S., Carlin, S., Sicari, V., & Martens, S. (2020). Differences in the composition of phenolic compounds, carotenoids, and volatiles between juice and pomace of four citrus fruits from Southern Italy. *European Food Research and Technology, 246*(10), 1991–2005. https://doi.org/10.1007/s00217-020-03550-8

Nour, V., Panaite, T. D., Ropota, M., Turcu, R., Trandafir, I., & Corbu, A. R. (2018). Nutritional and bioactive compounds in dried tomato processing waste. *CyTA – Journal of Food, 16*(1), 222–229. https://doi.org/10.1080/19476337.2017.1383514

Nuñez-Palenius, H. G., Gomez-Lim, M., Ochoa-Alejo, N., Grumet, R., Lester, G., & Cantliffe, D. J. (2008). Melon fruits: Genetic diversity, physiology, and biotechnology features. *Critical Reviews in Biotechnology, 28*(1), 13–55. https://doi.org/10.1080/07388550801891111

Obied, H. K., Bedgood, D., Mailer, R., Prenzler, P. D., & Robards, K. (2008). Impact of cultivar, harvesting time, and seasonal variation on the content of biophenols in olive mill waste. *Journal of Agricultural and Food Chemistry, 56*(19), 8851–8858. https://doi.org/10.1021/jf801802k

Ozturk, B., Parkinson, C., & Gonzalez-Miquel, M. (2018). Extraction of polyphenolic antioxidants from orange peel waste using deep eutectic solvents. *Separation and Purification Technology, 206*, 1–13. https://doi.org/10.1016/j.seppur.2018.05.052

Pagano, I., Piccinelli, A. L., Celano, R., Campone, L., Gazzerro, P., De Falco, E., & Rastrelli, L. (2016). Chemical profile and cellular antioxidant activity of artichoke by-products. *Food & Function, 7*(12), 4841–4850. https://doi.org/10.1039/c6fo01443g

Pandino, G., Lombardo, S., & Mauromicale, G. (2013). Globe artichoke leaves and floral stems as a source of bioactive compounds. *Industrial Crops and Products, 44*, 44–49. https://doi.org/10.1016/j.indcrop.2012.10.022

Pandino, G., Lombardo, S., Mauromicale, G., & Williamson, G. (2011). Phenolic acids and flavonoids in leaf and floral stem of cultivated and wild Cynara cardunculus L. genotypes. *Food Chemistry, 126*(2), 417–422. https://doi.org/10.1016/j.foodchem.2010.11.001

Paranavitana, L., Oh, W. Y., Yeo, J. D., & Shahidi, F. (2021). Determination of soluble and insoluble-bound phenolic compounds in dehulled, whole, and hulls of green and black lentils using electrospray ionization (ESI)-MS/MS and their inhibition in DNA strand scission. *Food Chemistry, 361*. https://doi.org/10.1016/j.foodchem.2021.130083

Peng, G., Gan, J., Dong, R., Chen, Y., Xie, J., Huang, Z., Gu, Y., Huang, D., & Yu, Q. (2021). Combined microwave and enzymatic treatment improve the release of insoluble bound phenolic compounds from the grapefruit peel insoluble dietary fiber. *Lebensmittel-Wissenschaft & Technologie, 149*, Article 111905. https://doi.org/10.1016/j.lwt.2021.111905

Perea-Domínguez, X. P., Hernández-Gastelum, L. Z., Olivas-Olguin, H. R., Espinosa-Alonso, L. G., Valdez-Morales, M., & Medina-Godoy, S. (2018). Phenolic composition of tomato varieties and an industrial tomato by-product: Free, conjugated and bound phenolics and antioxidant activity. *Journal of Food Science and Technology, 55*(9), 3453–3461. https://doi.org/10.1007/s13197-018-3269-9

Persic, M., Mikulic-Petkovsek, M., Slatnar, A., & Veberic, R. (2017). Chemical composition of apple fruit, juice and pomace and the correlation between phenolic content, enzymatic activity and browning. *Lebensmittel-Wissenschaft & Technologie, 82*, 23–31. https://doi.org/10.1016/j.lwt.2017.04.017

Phenol-Explorer. (2015). In *Polyphenol classes*.

Plazzotta, S., & Manzocco, L. (2018). Effect of ultrasounds and high pressure homogenization on the extraction of antioxidant polyphenols from lettuce waste. *Innovative Food Science and Emerging Technologies, 50*, 11–19. https://doi.org/10.1016/j.ifset.2018.10.004

Polat, S., Guclu, G., Kelebek, H., Keskin, M., & Selli, S. (2022). Comparative elucidation of colour, volatile and phenolic profiles of black carrot (Daucus carota L.) pomace and powders prepared by five different drying methods. *Food Chemistry, 369*, Article 130941. https://doi.org/10.1016/j.foodchem.2021.130941

Ramírez-Bolaños, S., Pérez-Jiménez, J., Díaz, S., & Robaina, L. (2021). A potential of banana flower and pseudo-stem as novel ingredients rich in phenolic compounds. *International*

Journal of Food Science and Technology, 56(11), 5601–5608. https://doi.org/10.1111/ijfs.15072

Rana, S., Rana, A., Gulati, A., & Bhushan, S. (2014). RP-HPLC-DAD determination of phenolics in industrial apple pomace. *Food Analytical Methods, 7*(7), 1424–1432. https://doi.org/10.1007/s12161-013-9765-7

Rana, A., Samtiya, M., Dhewa, T., Mishra, V., & Aluko, R. E. (2022). Health benefits of polyphenols: A concise review. *Journal of Food Biochemistry, 46*(10). https://doi.org/10.1111/jfbc.14264

Razola-Díaz, M. D. C., Gómez-Caravaca, A. M., de Andrés, J. L., Voltes-Martínez, A., Zamora, A., Pérez-Molina, G. M., Castro, D. J., Marchal, J. A., & Verardo, V. (2022). Evaluation of phenolic compounds and pigments content in yellow bell pepper wastes. *Antioxidants, 11*(3). https://doi.org/10.3390/antiox11030557

Rodríguez-Werner, M., Winterhalter, P., & Esatbeyoglu, T. (2019). Phenolic composition, radical scavenging activity and an approach for authentication of Aronia melanocarpa berries, juice, and pomace. *Journal of Food Science, 84*(7), 1791–1798. https://doi.org/10.1111/1750-3841.14660

Rojas-Ocampo, E., Torrejón-Valqui, L., Muñóz-Astecker, L. D., Medina-Mendoza, M., Mori-Mestanza, D., & Castro-Alayo, E. M. (2021). Antioxidant capacity, total phenolic content and phenolic compounds of pulp and bagasse of four Peruvian berries. *Heliyon, 7*(8). https://doi.org/10.1016/j.heliyon.2021.e07787

Rolim, P. M., Fidelis, G. P., Padilha, C. E. A., Santos, E. S., Rocha, H. A. O., & Macedo, G. R. (2018). Phenolic profile and antioxidant activity from peels and seeds of melon (Cucumis melo L. var. reticulatus) and their antiproliferative effect in cancer cells. *Journal of Medical and Biological Research, 51*(4). https://doi.org/10.1590/1414-431x20176069

Roselló-Soto, E., Martí-Quijal, F. J., Cilla, A., Munekata, P. E. S., Lorenzo, J. M., Remize, F., & Barba, F. J. (2019). Influence of temperature, solvent and pH on the selective extraction of phenolic compounds from tiger nuts by-products: Triple-TOF-LC-MS-MS characterization. *Molecules, 24*(4). https://doi.org/10.3390/molecules24040797

Sagdic, O., Ozturk, I., Ozkan, G., Yetim, H., Ekici, L., & Yilmaz, M. T. (2011). RP-HPLC-DAD analysis of phenolic compounds in pomace extracts from five grape cultivars: Evaluation of their antioxidant, antiradical and antifungal activities in orange and apple juices. *Food Chemistry. 03088146, 126*(4), 1749–1758. https://doi.org/10.1016/j.foodchem.2010.12.075

Saidani, F., Giménez, R., Aubert, C., Chalot, G., Betrán, J. A., & Gogorcena, Y. (2017). Phenolic, sugar and acid profiles and the antioxidant composition in the peel and pulp of peach fruits. *Journal of Food Composition and Analysis, 62*, 126–133. https://doi.org/10.1016/j.jfca.2017.04.015

Sampaio, S. L., Petropoulos, S. A., Dias, M. I., Pereira, C., Calhelha, R. C., Fernandes, Â., Leme, C. M. M., Alexopoulos, A., Santos-Buelga, C., Ferreira, I. C. F. R., & Barros, L. (2021). Phenolic composition and cell-based biological activities of ten coloured potato peels (Solanum tuberosum L.). *Food Chemistry, 363*. https://doi.org/10.1016/j.foodchem.2021.130360

Sharma, A., Shahzad, B., Rehman, A., Bhardwaj, R., Landi, M., & Zheng, B. (2019). Response of phenylpropanoid pathway and the role of polyphenols in plants under abiotic stress. *Molecules, 24*(13). https://doi.org/10.3390/molecules24132452

Sihem, D., Samia, D., Gaetano, P., Sara, L., Giovanni, M., Hassiba, C., Laura, G., & Noureddine, H. A. (2015). In vitro antioxidant activities and phenolic content in crop residues of Tunisian globe artichoke. *Scientia Horticulturae, 190*, 128–136. https://doi.org/10.1016/j.scienta.2015.04.014

Silva, F.de O., & Perrone, D. (2015). Characterization and stability of bioactive compounds from soybean meal. *LWT - Food Science and Technology, 63*(2), 992–1000. https://doi.org/10.1016/j.lwt.2015.04.032

Singh, S., Kaur, I., & Kariyat, R. (2021). The multifunctional roles of polyphenols in plant-herbivore interactions. *International Journal of Molecular Sciences, 22*(3), 1–19. https://doi.org/10.3390/ijms22031442

Singh, P. P., & Saldaña, M. D. A. (2011). Subcritical water extraction of phenolic compounds from potato peel. *Food Research International, 44*(8), 2452–2458. https://doi.org/10.1016/j.foodres.2011.02.006

Singla, R. K., Dubey, A. K., Garg, A., Sharma, R. K., Fiorino, M., Ameen, S. M., Haddad, M. A., & Al-Hiary, M. (2019). Natural polyphenols: Chemical classification, definition of classes, subcategories, and structures. *Journal of AOAC International, 102* (5), 1397–1400. https://doi.org/10.5740/jaoacint.19-0133

Sójka, M., Kołodziejczyk, K., Milala, J., Abadias, M., Viñas, I., Guyot, S., & Baron, A. (2015). Composition and properties of the polyphenolic extracts obtained from industrial plum pomaces. *Journal of Functional Foods, 12*, 168–178. https://doi.org/10.1016/j.jff.2014.11.015

Suárez, B., Álvarez, A. L., García, Y. D., Barrio, G.d., Lobo, A. P., & Parra, F. (2010). Phenolic profiles, antioxidant activity and in vitro antiviral properties of apple pomace. *Food Chemistry. 03088146, 120*(1), 339–342. https://doi.org/10.1016/j.foodchem.2009.09.073

Sun, Y., Deng, Z., Liu, R., Zhang, H., Zhu, H., Jiang, L., & Tsao, R. (2020). A comprehensive profiling of free, conjugated and bound phenolics and lipophilic antioxidants in red and green lentil processing by-products. *Food Chemistry, 325*, Article 126925. https://doi.org/10.1016/j.foodchem.2020.126925

Szabo, K., Diaconeasa, Z., Cătoi, A. F., & Vodnar, D. C. (2019). Screening of ten tomato varieties processing waste for bioactive components and their related antioxidant and antimicrobial activities. *Antioxidants, 8*(8). https://doi.org/10.3390/antiox8080292

Tao, B., Ye, F., Li, H., Hu, Q., Xue, S., & Zhao, G. (2014). Phenolic profile and in vitro antioxidant capacity of insoluble dietary fiber powders from citrus (Citrus junos sieb. ex tanaka) pomace as affected by ultrafine grinding. *Journal of Agricultural and Food Chemistry, 62*(29), 7166–7173. https://doi.org/10.1021/jf501646b

Thomas, M., Badr, A., Desjardins, Y., Gosselin, A., & Angers, P. (2018). Characterization of industrial broccoli discards (Brassica oleracea var. italica) for their glucosinolate, polyphenol and flavonoid contents using UPLC MS/MS and spectrophotometric methods. *Food Chemistry, 245*, 1204–1211. https://doi.org/10.1016/j.foodchem.2017.11.021

Tian, Y., Kriisa, M., Foste, M., Kütt, M. L., Zhou, Y., Laaksonen, O., & Yang, B. (2022). Impact of enzymatic pre-treatment on composition of nutrients and phytochemicals of canola (Brassica napus) oil press residues. *Food Chemistry, 387*. https://doi.org/10.1016/j.foodchem.2022.132911

Tomaino, A., Martorana, M., Arcoraci, T., Monteleone, D., Giovinazzo, C., & Saija, A. (2010). Antioxidant activity and phenolic profile of pistachio (Pistacia vera L., variety Bronte) seeds and skins. *Biochimie, 92*(9), 1115–1122. https://doi.org/10.1016/j.biochi.2010.03.027

Vaz, A. A., Odriozola-Serrano, I., Oms-Oliu, G., & Martín-Belloso, O. (2022). Physicochemical properties and bioaccessibility of phenolic compounds of dietary fibre concentrates from vegetable by-products. *Foods, 11*(17). https://doi.org/10.3390/foods11172578

Velderrain-Rodríguez, G. R., Quero, J., Osada, J., Martín-Belloso, O., & Rodríguez-Yoldi, M. J. (2021). Phenolic-rich extracts from avocado fruit residues as functional food ingredients with antioxidant and antiproliferative properties. *Biomolecules, 11*(7). https://doi.org/10.3390/biom11070977

Vella, F. M., Cautela, D., & Laratta, B. (2019). Characterization of polyphenolic compounds in cantaloupe melon by-products. *Foods, 8*(6). https://doi.org/10.3390/foods8060196

Vidal, N. P., Roman, L., Swaraj, V. J. S., Ragavan, K. V., Simsek, S., Rahimi, J., Kroetsch, B., & Martinez, M. M. (2022). Enhancing the nutritional value of cold-pressed oilseed cakes through extrusion cooking. *Innovative Food Science and Emerging Technologies, 77*. https://doi.org/10.1016/j.ifset.2022.102956

Yadav, V., Wang, Z., Wei, C., Amo, A., Ahmed, B., Yang, X., & Zhang, X. (2020). Phenylpropanoid pathway engineering: An emerging approach towards plant defense. *Pathogens, 9*(4). https://doi.org/10.3390/pathogens9040312

Yang, T., Guang Hu, J., Yu, Y., Li, G., Guo, X., Li, T., & Liu, R. H. (2019). Comparison of phenolics, flavonoids, and cellular antioxidant activities in ear sections of sweet corn (Zea mays L. saccharata Sturt). *Journal of Food Processing and Preservation, 43*(1). https://doi.org/10.1111/jfpp.13855

Ye, Z., Qin, Y., Harrison, R., Hider, R., & Bekhit, A. E. D. A. (2022). Characterization of bioactive compounds in lees from New Zealand wines with different vinification backgrounds. *Antioxidants, 11*(12). https://doi.org/10.3390/antiox11122335

Yeo, J. D., & Shahidi, F. (2020). Identification and quantification of soluble and insoluble-bound phenolics in lentil hulls using HPLC-ESI-MS/MS and their antioxidant potential. *Food Chemistry, 315*, Article 126202. https://doi.org/10.1016/j.foodchem.2020.126202

Yeo, J. D., Tsao, R., Sun, Y., & Shahidi, F. (2021). Liberation of insoluble-bound phenolics from lentil hull matrices as affected by Rhizopus oryzae fermentation: Alteration in phenolic profiles and their inhibitory capacities against low-density lipoprotein (LDL) and DNA oxidation. *Food Chemistry, 363*, Article 130275. https://doi.org/10.1016/j.foodchem.2021.130275

Yücetepe, A., Altin, G., & Özçelik, B. (2021). A novel antioxidant source: Evaluation of in vitro bioaccessibility, antioxidant activity and polyphenol profile of phenolic extract from black radish peel wastes (Raphanus sativus L. var. niger) during simulated gastrointestinal digestion. *International Journal of Food Science and Technology, 56*(3), 1376–1384. https://doi.org/10.1111/ijfs.14494

Zampar, G. G., Zampar, I. C., Beserra da Silva de Souza, S., da Silva, C., & Bolanho Barros, B. C. (2022). Effect of solvent mixtures on the ultrasound-assisted extraction of compounds from pineapple by-product. *Food Bioscience, 50*. https://doi.org/10.1016/j.fbio.2022.102098

Zeb, A. (2016). Phenolic profile and antioxidant activity of melon (Cucumis melo l.) seeds from Pakistan. *Foods, 5*(4), 1–7. https://doi.org/10.3390/foods5040067

Zhang, J., Gao, N., Shu, C., Cheng, S., Sun, X., Liu, C., Xin, G., Li, B., & Tian, J. (2021). Phenolics profile and antioxidant activity analysis of kiwi berry (actinidia arguta) flesh and peel extracts from four regions in China. *Frontiers in Plant Science, 12*. https://doi.org/10.3389/fpls.2021.689038

Žilić, S., Janković, M., Basić, Z., Vančetović, J., & Maksimović, V. (2016). Antioxidant activity, phenolic profile, chlorophyll and mineral matter content of corn silk (Zea mays L): Comparison with medicinal herbs. *Journal of Cereal Science, 69*, 363–370. https://doi.org/10.1016/j.jcs.2016.05.003

CHAPTER

4 Advances in extraction techniques for plant polyphenols: From conventional to green and emerging technologies

Shi-Hui Cheng

School of Biological and Environmental Sciences, Unievrsity of Nottingham Malaysia, Semenyih, Selangor, Malaysia

1. Introduction

Polyphenols are naturally occurring compounds found mainly in various plant-based foods. Foods that are rich in polyphenols include berries, beans, oats, garlic, spinach, and other plant-based foods. Polyphenols possess strong antioxidant properties, and the potential health benefits of dietary plant polyphenols as antioxidants have sparked significant interest (Hano & Tungmunnithum, 2020). Extensive epidemiological studies and associated meta-analyses provided convincing evidence pointing to the potential of reducing the risk of various chronic diseases such as cancer, cardiovascular diseases, diabetes, and neurodegenerative diseases by routine intake of plant-based diets high in polyphenols (Arts & Hollman, 2005). They can also be used as dietary supplements, pharmaceuticals, and cosmetic additives.

Polyphenols are compounds that generally have large molecular weights and at least two phenolic groups organized in complex structures (Zhang et al., 2018). However, this category also includes simple phenolic compounds (sometimes referred to as phenolic acids), which might act as precursors to polyphenols. The definition of polyphenol is commonly used to define compounds derived solely from the shikimate/phenylpropanoid and/or the polyketide pathway (Hano & Tungmunnithum, 2020). These compounds should contain multiple phenolic units of any nitrogen-based functional groups and do not have nitrogen-based functional groups (Quideau et al., 2011). Therefore, polyphenols can be simply classified as flavonoids or nonflavonoids. Phenolic acids, coumarins, stilbenes, and lignans are all subgroups of nonflavonoids (Hano & Tungmunnithum, 2020).

Phenolic acids, also known as phenolcarboxylic acids, are aromatic compounds with an organic carboxylic acid and a phenolic ring. Two basic types of phenolic acids found in nature are hydroxybenzoic acids (C6-C1 backbone) and hydroxycinnamic acids (C6-C3 backbone) (Hano & Tungmunnithum, 2020). This group

Plant Polyphenols in the Food and Cosmetic Industries. https://doi.org/10.1016/B978-0-443-19137-4.00009-0

includes compounds such as benzoic acid and p-coumaric acid, as well as more complex phenolic acids like curcumin derived from turmeric. Coumarins are benzopyrone (1,2-benzopyrones or 2H-1-benzopyran-2-ones) derivatives that are widely found in nature. Chinese cinnamon is well known for its high concentration of coumarin derivatives. Stilbenes, also known as stilbenoids, possess a carbon backbone of C6-C2-C6, specifically referring to the *trans*-((*E*)-stilbenes) and *cis*-((*Z*)-stilbenes) 1,2-diphenylethylene structures. One well-known example of this group is resveratrol, which can be found in wine and grapes (Zhang et al., 2018).

Lignans are biphenolic compounds that result from the oxidative coupling of two monolignol units, which are hydroxycinnamic alcohol components. Plants use the basic monolignol units to create lignin, which is a substance found in the cell walls of the conducting vessels. Lignans include a wide range of compounds that differ in terms of the type of bond that connects the two units and the changes that take place after dimerization. Flaxseeds are high in secoisolariciresinol, a dietary lignan that is commonly found (Hano & Tungmunnithum, 2020).

Flavonoids are phenylpropanoid compounds with a C6-C3-C6 structure, consisting of two phenyl rings (rings A and B) and one heterocyclic ring (ring C). Numerous antioxidants that could be found in fruits and vegetables, such as quercetin, kaempferol, isorhamnetin, fisetin, and genistein, are included in this category. Flavonoids are classified into a number of subgroups based on the number of phenol units in their molecular structure, substituent groups, or the type of linkage between phenol units. The multiple subgroups of flavonoids are chalcones, flavones, flavonols, flavonones, isoflavonoids, and anthocyanins (Hano & Tungmunnithum, 2020).

Plant species samples are considered one of the most complex matrices and are widely studied in the scientific community, with numerous studies employing different ways to extract, separate, and obtain their main constituents, the polyphenols (Zhang et al., 2018). The methodological development phase necessitates time and expertise, particularly when dealing with complex samples such as polyphenol-rich herbal matrices. This chapter provides an overview of conventional and advanced extraction techniques for plant polyphenols, highlighting their advantages, limitations, and applications. Recent developments in efficient, sustainable, and innovative extraction approaches are also discussed.

2. Overview of phenolic extraction

The extraction of bioactive compounds from plant materials is the first step in harnessing the potential of phytochemicals for the creation of dietary supplements, nutraceuticals, food ingredients, pharmaceuticals, and cosmetic items. Fresh, frozen, or dried plant samples can all provide phenolic compounds by extraction. Before extraction, plant samples often go through a number of processing stages, such as milling, grinding, and homogenization. The plant material could be air-dried or freeze-dried before these operations. In comparison with air-drying, freeze-drying generally aids in the preservation of a higher amount of phenolic chemicals in plant samples (Abascal et al., 2005). However, it is crucial to remember that drying techniques such as freeze-drying might affect the make-up of plant samples, potentially changing the profiles of their components. As a result, caution should be taken while planning and evaluating research projects that explore the therapeutic potential of plants.

Solvent extraction methods are frequently used to prepare extracts from plant materials due to their practicality, effectiveness, and broad applicability. The selection of solvents with varying polarity, extraction duration, temperature, sample-to-solvent ratio, as well as the chemical composition and physical characteristics of the samples, all have a significant role in the effectiveness and succession of chemical extraction. The solubility of phenolic compounds is affected by the chemical composition of the plant sample and the polarity of the solvents utilized. Various levels of phenolics can be found in plant materials, ranging from simple compounds like phenolic acids and anthocyanins to more complex and highly polymerized compounds like tannins. In addition, other plant compounds, including proteins and carbohydrates, may also be linked to phenolics. As a result, there is not a single extraction method that works for all types of plant phenolics. Depending on the chosen solvent system, a mixture of phenolic compounds soluble in the solvent will be extracted from the plant materials. However, nonphenolic substances like sugars, organic acids, and lipids might also be included in the mixture. As a result, more processes could be required to remove the undesirable components from the extract.

For the extraction of phenolic compounds from plant materials, a number of solvents such as methanol, ethanol, acetone, ethyl acetate, and their combinations have been used. The quantity and effectiveness of extracting polyphenols from plant materials are substantially impacted by the solvent chosen (Xu & Chang, 2007). It has been observed that methanol typically works better in extracting polyphenols with smaller molecular weights when it comes to polyphenol extraction. On the contrary, aqueous acetone is discovered to be better suited for extracting greater molecular weight flavanols (Prior et al., 2001). Another effective solvent for extracting polyphenols is ethanol, which can also be consumed safely by people (Shi et al., 2007). One method that is frequently used to extract anthocyanin-rich phenolic extracts from plant materials is to utilize an acidified organic solvent, usually methanol or ethanol. Anthocyanins are released more easily because of the disruption of cell membranes caused by this solvent system, which also aids in stabilizing the compounds by dissolving and preserving them. To prevent the hydrolysis of labile acyl and sugar residues, care must be taken not to introduce too much acid during the concentration phases. For the optimal yield in anthocyanin extraction, mild organic acids such as formic acid, acetic acid, citric acid, tartaric acid, and phosphoric acid, as well as low concentrations of strong acids like trifluoroacetic acid (0.5%–3.0%) and hydrochloric acid (1.0%), would be recommended (Nicoué et al., 2007). In addition, sulfured water has been utilized as an extraction solvent in an effort to cut down on the usage of organic solvents and lower extraction costs (Cacace & Mazza, 2002).

The recovery of phenolic compounds from plant sources is affected by factors such as temperature and extraction time. These factors reflect the delicate equilibrium between the solubilization of the analytes and the potential degradation of phenolic compounds through oxidation (Robards, 2003). The solubility and mass transfer rate can both be improved by raising the extraction temperature, which consequently increases the analyte solubility. Higher temperatures also cause the viscosity and surface tension of solvents to decrease, which makes them easier to penetrate sample matrices and speeds up extraction. However, it is crucial to remember that many phenolic compounds are vulnerable to oxidation and hydrolysis. The yield of phenolic compounds in the extracts decreases as a result of phenolic oxidation, which is more likely to occur during prolonged extraction duration and high temperatures. As a result, it is essential to select an extraction technique and procedure that can guarantee the stability of phenolic compounds. In recent years, a number of extraction techniques have been developed as alternatives to traditional techniques for the extraction of phenolic compounds from plant sources. These include techniques that utilize compressed fluids as extracting agents, microwave-assisted extraction, and ultrasound-assisted extraction. Examples of these techniques are subcritical water extraction (SWE), supercritical fluid extraction (SFE), pressurized fluid extraction (PFE), and accelerated solvent extraction (ASE). These novel techniques have been successfully used to extract phenolic compounds from plant sources.

Ultrasound-assisted extraction (UAE) is a promising technology that has several benefits, including ease of use, affordability, and adaptability. UAE is suitable for a variety of applications; unlike other extraction techniques, it does not call for the use of sophisticated equipment. It is suitable for use in the phytopharmaceutical extraction sector since it can be used on both small and large scales (Vinatoru, 2001). The enhancement mechanism of ultrasound involves the generation of shear force resulting from the implosion of cavitation bubbles when acoustic waves propagate in the kilohertz (kHz) range (Laborde et al., 1998). Ultrasound-induced bubble collapse can have a variety of outcomes, including physical, chemical, and mechanical repercussions (Mason et al., 1996). These effects play a significant role in the disruption of biological membranes, facilitating the release of extractable compounds and enhancing the penetration of solvents into cellular materials. This process greatly improves mass transfer, aiding in the extraction of desired compounds (Vinatoru, 2001). UAE has grown significantly in popularity in recent years for its ability to extract a variety of phenolic compounds from different plant components such as leaves, stalks, fruits, and seeds. When extracting phenolic compounds from strawberries using other methods, such as solid-liquid extraction, subcritical water extraction, and microwave-assisted extraction, a comparative study found that UAE exhibited minimal phenolic compound degradation and offered a quicker extraction process (Herrera & Luque De Castro, 2005).

Pressurized liquid extraction (PLE), also known as accelerated solvent extraction (ASE), is a relatively new technique for extracting phytochemicals at high pressure and temperature. Pressurized liquid extraction (PLE) involves the application of pressure, enabling the use of solvents at temperatures higher than their typical boiling point. The combination of high pressures (3.3–20.3 MPa) and temperatures (40–200°C) enables quicker extraction processes with small amounts of solvents used (Mendiola et al., 2007). High pressure and temperature enhance analyte solubility and the desorption kinetics from the matrices (Richter et al., 1996). Hence, in pressurized liquid extraction (PLE), extraction solvents like water show decreased effectiveness in extracting phytochemicals at lower temperatures and can become much more efficient at elevated temperatures. Specifically, the process of using water as an extraction solvent in PLE is known as subcritical water extraction (SWE). Water is heated in SWE to temperatures up to 200°C, which can alter the dielectric constant of the liquid. As a result, water behaves like an organic solvent, improving its ability for extraction. However, phenolic compounds are rapidly oxidized at high temperatures, so it is crucial to show that they will not degrade under the PLE conditions that have been proposed (Palma et al., 2001). PLE has been effectively used in recent years to extract phenolic compounds from a variety of plant sources. Another technology that uses carbon dioxide as a compressed fluid solvent for extraction purposes is supercritical and subcritical fluid extraction (SFE). In this procedure, organic modifiers are added to the fluid to improve its polarity and allow the extraction of phenolic compounds (Klejdus et al., 2005). SFE is done in a controlled environment without either air or light. This controlled environment reduces degradation and oxidation processes, making SFE a better technique than other extraction techniques in terms of preserving the integrity of extracted compounds. Overall, these extraction techniques based on compressed fluid provide notable environmental advantages by decreasing the reliance on organic solvents (e.g., PLE), allowing extractions to be carried out using nontoxic and nonpolluting solvents like water (e.g., SWE) or supercritical CO2 fluid (e.g., SFE), which are more environmentally friendly. Nevertheless, the use of high pressure in these techniques has high equipment requirements, resulting in higher expenses, especially on an industrial scale. The financial factor frequently outweighs the technical advantages associated with these methods.

A technique called microwave-assisted extraction (MAE) uses microwave energy to help in the partitioning of analytes from the sample matrix into the solvent. MAE provides a substantial benefit in terms of faster extraction times and less solvent volume needed compared with traditional techniques (Sparr Eskilsson & Björklund, 2000). Gallic acid, ellagic acid (Li et al., 2004), quercetin (Huang & Zhang, 2004), isoflavone (Rostagno et al., 2007), and trans-resveratrol (Du et al., 2007) are various small-molecule phenolic compounds that have been successfully extracted using MAE. These compounds have been discovered to hold their stability for 20 min at temperatures up to 100°C under MAE heating conditions. Due to the possibility of degradation, phenolic compounds with a higher density of hydroxyl-type substituents, such as tannins, and those that are susceptible to degradation at high temperatures, such as anthocyanins, may not be suitable for extraction using MAE (Liazid et al., 2007).

The extraction of phenolic compounds from plant materials can also be influenced by other factors, such as the solvent-to-solids ratio and sample particle size. According to research, the yield of phenols during extraction has been seen to increase when the solvent-to-solid ratio increases (Pinelo et al., 2005). Nevertheless, to achieve an optimized value, a balance between the use of high and low solvent-to-solid ratios, including a balance between high costs and solvent wastes, and, respectively, the avoidance of saturation effects needs to be achieved (Pinelo et al., 2006). Reducing the particle size of the plant material has been shown to improve the yield of phenolic compounds (Nepote et al., 2005). It has been suggested that several enzymatic techniques can be used to improve the release of bound phenolics during extraction. To aid in the release of phenolic compounds from their bound forms, these techniques make use of various enzyme preparations, such as mixed pectinolytic and cell wall polysaccharide-degrading enzymes (Pinelo et al., 2008). In these enzyme-assisted extraction techniques, the particle size of the mashed samples was found to be a key factor that optimizes the enzyme action and increases the extraction efficiency of phenolic compounds (Meyer, 2005). In addition to enzymatic procedures, acid and alkaline treatments were also efficient in releasing bound phenolics in phenolic extractions (White et al., 2010).

2.1 Phenolic acid extraction

Phenols and phenolic acid are existing as insoluble bound complexes, which are bound to the cell wall polymers via ester and glycosidic links, which make them unable to be extracted by organic solvents (Mattila & Hellström, 2007). Before extraction, bound phenolic acids are usually released through base hydrolysis, acid hydrolysis, or a mix of the two (Mattila & Kumpulainen, 2002). Many extraction techniques include the inclusion of antioxidants as stabilizers to guarantee the stability of the isolated phenolic compounds. For this, various substances have been used, such as butylated hydroxyanisole, tert-butylhydroquinone, and ascorbic acid.

2.1.1 Soxhlet extraction

Soxhlet extraction is a traditional method used to extract the secondary metabolites from plant sources. This technique allows repeated digestion of the plant material and continuous extraction, with a fresh portion of solvent being supplied. In turn, this maximizes the concentration difference between the analyte in the cellular solution and the solvent, enabling complete extraction of the plant material.

Using a Soxhlet system with multiple extraction cycles, phenolic acids were extracted from plant materials (Sambucus nigra L. and Polygonum aviculare). The digesting process took 15 h, and the solvent utilized was methanol. The yielding of p-hydroxybenzoic, vanillic, and ferulic acids from wild common lilac was high and successfully extracted using this modified extraction technique (Waksmundzka-Hajnos et al., 2007). Soxhlet extraction was also utilized for the qualitative and quantitative analysis of phenolic compounds found in Salvia halophila and S. virgata (Akkol et al., 2008). Over an 8-hour period, a variety of solvents, including n-hexane, ethyl acetate, methanol, and a 50% aqueous methanol solution, were used to optimize the extraction conditions. The findings showed that the use of 50% methanol produced the maximum extraction yield of total phenolics, whereas n-hexane produced the lowest yield.

Karasová and Lehotay (2007) also used Soxhlet extraction to isolate benzoic acid derivatives from Melissa officinalis. They investigated several solvent compositions and extraction times (1 h, 4 h, 8 h), especially on a mixture of methanol and water at varied volume ratios (60:40 and 80:20). In the end, they discovered that an extraction time of 1 hour was sufficient and found that the ratio of methanol and water at 80:20 was the most effective for extracting the derivatives. Interestingly, the extraction time did not significantly affect the results, except for gallic acid, where longer extraction times led to the lowest quantities of this compound.

However, Soxhlet extraction is associated with several drawbacks, including how time-consuming it is and how costly and high-quality solvents are needed. Due to these restrictions, alternative techniques that are more efficient in terms of time and solvent usage have become increasingly popular recently.

2.1.2 Ultrasound-assisted extraction (UAE)

UAE is renowned for its simplicity and rapidity, the process usually finishing in a few minutes. The efficiency of the extraction process is influenced by the solvent used as well as other factors like sample size, extract pH, temperature, pressure, particle size, duration of sonification, and ultrasonic amplitude (Khan et al., 2010). Despite these factors, the procedure is believed to be the simplest one possible to be carried out in a laboratory (Casazza et al., 2010). The ability to simultaneously extract several samples in a short period of time is one advantage of this technique. However, it is important to separate the extract by decantation or filtration using suitable paper filters. In addition, in some circumstances, the HPLC column may get clogged due to the presence of an opaque solution after extraction. In practice, the UAE enables the effective extraction of phenolic compounds from plant cells by facilitating the rupture of plant cell walls. Pawar et al. (2011) examined various ginger species. The UAE process was run for 30min with methanol as the solvent.

UAE is not always carried out as a standalone process for extraction. In some circumstances, UAE is followed by another procedure to properly prepare the sample, such as moistening it with a solvent, before UAE is carried out. You et al. (2011) proposed a similar strategy for the extraction of blackberry fruits. According to their procedure, the fruits were extracted for 16 h in darkness using an 80:20 (v/v) mixture of methanol and water with the addition of acetic acid. After applying sonication, the extracted samples underwent two rounds of centrifugation before being analyzed. Benetis et al. (2008) also employed ultrasound in the study of phenolic compounds in Achillea millefolium L. They conducted preliminary investigations using an ethanol/water solution with concentrations ranging from 40% to 96% (v/v) and varying extraction times of 5, 10, 20, 30, and 60 min. The results showed that a solution containing 70% ethanol achieved the best results. It was found that increasing the extraction duration led to a higher yield by observing the extraction profile in relation to ultrasonication time. However, a decline in yield was observed after 60 min. As a result, for the following experiments, a 70% ethanoic solution was chosen as the solvent, and a sample of 0.25g and 25 mL of the solvent was used for the 30-minute ultrasonication.

The efficiency of UAE and traditional liquid-liquid extraction has been compared in the literature (Proestos & Komaitis, 2006). Both methods utilized the following solvents: a 60% methanolic solution, a 60% acetone solution, water, and a 60:30 (v/v) mixture of water and ethyl acetate. The study emphasized aromatic plants, particularly Origanum majorana and Rosmarinus officinalis. The classical digestion took place in a water bath at 90°C for 2 h, while ultrasonication was carried out for 1 h at 25°C and 60°C. Both processes were carried out using each of the solvents mentioned above in triplicate. According to the findings, the methanolic solvent produced the best extraction efficiency at a temperature of 60°C.

2.1.3 Supercritical fluid extraction (SFE)

SFE often precedes analysis by HPLC (Žiaková et al., 2003). SFE is an advanced technique that offers numerous advantages compared with traditional extraction methods (Garcia-Salas et al., 2010). One of these features is the ability to function at low temperatures, which is very useful for analyzing thermally labile substances. Additional benefits of SFE include good selectivity, a reduction in the amount of solvent used for extraction, the ability to extract small sample sizes (usually a few milligrams), and a quick extraction time. SFE may also be easily automated and combined with a number of chromatographic methods (GC and HPLC) both offline and online. Due to its accessibility and low toxicity, carbon dioxide is frequently utilized as the supercritical fluid for SFE. In the context of phenolic compounds, carbon dioxide is not considered the most optimal solvent due to its relatively low polarity compared with phenolic compounds (Pinelo et al., 2007). However, despite this limitation, Castro-Vargas et al. (2010) still managed to use carbon dioxide to extract phenolic fractions from the seeds of *Psidium guajava L.*, a plant in the Myrtaceae family, successfully.

The efficacy of SFE is influenced by a number of factors that can be divided into two groups (Kazazi & Rezaei, 2010). Pressure, temperature, extraction time, sample weight, and flow rate are some of the parameters in the first group that are relevant to the actual extraction process. The second group includes parameters related to the sample matrix, such as its form, homogeneity, solubility, and analyte desorption properties. The pH of the system also affects the SFE process since pH changes can have an impact on the extraction yield and the rate at which the analyte is extracted from an aqueous phase.

SFE finds widespread application in the pharmaceutical industry, as well as the food and cosmetic sectors (Garcia-Salas et al., 2010). It is a useful approach for separating biologically active compounds from plant materials, especially those that are difficult to separate using traditional solvent extraction methods.

2.1.4 Accelerated solvent extraction (ASE)

ASE is a relatively new technique. The terms pressurized fluid extraction (PFE) and pressurized liquid extraction (PLE) are also used in the literature to refer to this technique.

The ASE technique makes use of the same solvents as traditional methods, but at higher pressures and temperatures (about 3.3–20.3 MPa and 40–200°C, respectively) (Garcia-Salas et al., 2010). In this method, the sample is placed into an extraction vessel made of stainless steel. The average analysis time is between 5 and 15 min (Dai & Mumper, 2010).

ASE has several advantages, including the ability to extract samples with high humidity, shorter extraction times, better solvent penetration into the samples, favorable extraction kinetics, process automation, and user-friendly instrumentation (Garcia-Salas et al., 2010). In addition, ASE utilizes lesser solvent volumes than traditional techniques do. In contrast to traditional extraction techniques like Soxhlet, which often call for extraction times of 10–48 h and solvent volume surpassing 400 mL, an ASE extraction takes around 20 min and only needs 10–50 mL of solvent (Dai & Mumper, 2010).

2.2 Flavonoid extraction

Flavonoids are extremely beneficial to our health and crucial to the food-nutraceutical industry. However, due to degradation factors, the amount of flavonoids tends to decrease over storage and different processing methods. For instance, enzymatic reactions happen during storage in various food systems, which causes the quantity of flavonoids to decrease. A considerable majority of the flavonoid content in fruits and vegetables is concentrated in the outer layer, especially the skin, leaving it vulnerable to loss through simple peeling. In addition, processing activities such as heat treatment, maceration, and juice extraction further contribute to the lowering of flavonoid content due to enzymatic and thermal breakdown. Therefore, the extraction of flavonoids from natural sources is important for their application in various fields.

2.2.1 Soxhlet extraction

For the extraction of flavonoids, a number of extraction techniques have been proposed, such as maceration, percolation, hydro-distillation, boiling, reflux, soaking, and Soxhlet (Alara, Abdurahman, & Ukaegbu, 2018). Soxhlet extraction is one of these techniques that has gained popularity due to its convenience of use, low cost, and decreased solvent demand when compared to methods like soaking, boiling, or maceration (Sati et al., 2019). To determine their effect on extraction yields, many solvents, including ethanol, methanol, benzene, chloroform, ethyl acetate, and others, have been studied in Soxhlet extraction. Due to the higher yields obtained in the recovery of flavonoids, ethanol and methanol are the solvents that are most frequently utilized for flavonoid extraction (Martens & Mithöfer, 2005).

The Soxhlet method remains the standard for yield comparison in advanced extraction methods (Ighodaro et al., 2010). Soxhlet extraction has been described briefly in the nutraceutical industry (Wang & Weller, 2006). However, the prolonged extraction time and high temperature associated with Soxhlet extraction pose challenges for the recovery of thermolabile components, particularly relevant for flavonoids (Wang & Weller, 2006). In addition, using a lot of solvents can cause sustainability and environmental issues.

2.2.2 Microwave-assisted extraction

Microwaves are an electromagnetic spectrum of radiation ranging from 300 MHz (radio radiation) to 300 GHz (infrared radiation). This heating method utilizes microwave energy and operates by directly influencing dipole polarization and ion conduction molecules through microwaves (Flórez et al., 2015). Time, temperature, plant material-to-solvent ratio, solvent concentration, solvent polarity, irradiation, frequency intensity, and microwave power are important parameters that might affect the extraction process of flavonoids (Yedhu Krishnan et al., 2016).

In comparison with more traditional techniques like maceration, Soxhlet, or heat reflux, studies have shown that microwave-assisted extraction (MAE) offers several advantages, including a significant reduction in the extraction times for different compounds, decreased solvent volumes, and higher extraction yields of bioactive substances (Jha et al., 2017). The reduction in extraction times and solvent usage aims to enhance the cost-effectiveness of the extraction process (Farzaneh & Carvalho, 2017).

The selection of a solvent in this methodology is not only important but also requires consideration of its dielectric properties (Yedhu Krishnan et al., 2016). The most popular MAEs are ethanol and methanol because they have greater extraction yields, while water has also produced beneficial results (Jha et al., 2017). The effectiveness of MAE is influenced by the specific flavonoid being targeted for recovery. The flavonoid's polarity turns into a key factor, and the extraction's desired polarity is determined by the solvent used. The bioactivity of the recovered flavonoids can also be affected by the extraction solvent (Jha et al., 2017). Due to their nonpolar nature, nonpolar solvents like dichloromethane, ethyl acetate, diethyl ether, and chloroform are frequently used for the extraction of isoflavones, flavones, and methylated flavones. On the contrary, polar flavonoids like flavonoid glycosides and aglycones are extracted using solvents like ethanol or methanol.

2.2.3 Ultrasound-assisted extraction

Ultrasound-assisted extraction (UAE) is a method employed in the food and pharmaceutical industries to disrupt plant material and extract bioactive compounds (Akbari et al., 2019). This method makes use of the acoustic cavitation phenomenon, in which bubbles form and then burst, releasing bioactive compounds. The extent of rupture is influenced by the extraction conditions (Um et al., 2018).

This methodology creates a cavitation effect that not only makes it possible to break down the plant material's cell walls but also promotes the reduction in particle size, which is advantageous for solvent-substrate interaction (Um et al., 2018). The extraction of flavonoids can be affected by a wide range of factors, leading to a considerable number of studies. Experimental matrices are frequently used to improve the extraction procedure and identify the conditions that maximize flavonoid recovery. In a published study, the surface response approach is frequently used to accomplish this goal (Alara, Abdurahman, & Olalere, 2018).

Ultrasound-assisted extraction (UAE) has been found to enhance the yield of bioactive compounds, including flavonoids. This technique has the advantage of shortening extraction periods and is particularly successful at extracting flavonoids and phenolic compounds. UAE has also been demonstrated to enhance the biological properties of the extracted compounds (Singanusong et al., 2015). The extraction of bioactive compounds using UAE is influenced by several factors, such as the frequency of ultrasound, typically ranging from 20 kHz to 100 MHz (Piana et al., 2018), solvent selection, solvent concentration, solid-to-solvent ratio, temperature, and extraction duration.

The following flavonoid compounds that have been extracted and recovered using this technique are as follows: Rutin, narcissin, nicotiflorin, epicatechin, epicatechin gallate, catechin, procyanidin B2, apiofuranosyl($1''' \rightarrow 2''$)-β-D-glucopyranosyl] rhamnocitrin, quercetin-3-O-rhamnoglucoside, quercetin-3-O-β-D-glucopyranoside, myricetin-5′-O-β-D-glucopyranoside, 4′-O-(3‴-O-dihydrophaseoyl-β-D-glucopyranosyl) rhamnocitrin, formononetin-7-O-glucoside, myricomplanoside, kaempferol-3-O-glucosylrutinoside, complantoside A, quercetin-3-O-acylglycoside, etc. (Prasad et al., 2009).

2.2.4 Supercritical fluid extraction

When a substance reaches a temperature and pressure over its thermodynamic critical point, it transforms into a supercritical fluid. High diffusivity and low viscosity are two characteristics of the fluid in this state that significantly improve the transfer of matter during the extraction process (Dassoff & Li, 2019). SFE technique with these features has shown noticeably greater flavonoid extraction yields compared with other techniques.

Carbon dioxide (CO2) is the preferred solvent in this extraction method, primarily because of its numerous advantages. CO2 is nonflammable, nontoxic, cost-effective, and highly volatile, making it easy to remove from the extracted compounds (Belbaki et al., 2017). Because of these qualities, CO2 becomes a superior solvent for supercritical extraction.

This method of extraction has a number of benefits, such as the facile separation of volatile and nonvolatile compounds and the preservation of product integrity due to low temperatures and high solvent volatility. However, there are several restrictions attached to this approach. It can be difficult to achieve equilibrium between the solute and solvent, which calls for additional separation processes. High pressures can make it difficult to keep adding particles to the extract continuously, and they can raise operational and equipment maintenance expenses (Song et al., 2019). The use of lower temperatures is a key component of this process, which aims to retain product stability. A study on the valorization of agro-industrial residues (expellers) derived from pressing soybean oil serves as an illustration of this. At either 35 or 40 °C, the extractions were carried out using CO2 as the solvent at a pressure of 40 MPa. The expellers were also ethanol-impregnated. The optimal results demonstrated a flavonoid content of 65.0 and 31.3 QE/100 gdm (Alvarez et al., 2019). Furthermore, supercritical carbon dioxide and ethanol were used to extract flavonoids from the leaves of Odontonema strictum. The impact of temperature (ranging from 55 to 65 °C) on the overall flavonoid content was carefully optimized. The total flavonoid content ranged from 99.33 to 247.78 mg/g of dried extract, while the flavonoid recovery varied between 10.68 and 18.92 mg/g of dried leaf powder (Ouédraogo et al., 2018). In a different study, supercritical carbon dioxide (CO2) was used to extract Citrus unshiu peels with and without the use of ethanol. 30 MPa of pressure and 59.85 °C of temperature were used for the extraction procedure. The results showed that increasing the extraction period and adding ethanol to supercritical CO2 resulted in better extract yields, including nobiletin. In addition, the importance of pressure in these processes was observed, with the solvent used to determine how it would affect the process. In a different study, a successful approach was developed for extracting Medicago sativa using enzyme-assisted supercritical fluid. The extraction procedure was optimized using the Box-Behnken design, with an emphasis on parameters like pressure (between 100 and 300 bar). 68 °C, 205 bar of pressure, and a co-solvent content of 15.5% were found to be the ideal extraction conditions for maximizing the total polyphenol content. With the help of enzymes, this technique provides a powerful approach for supercritical fluid extraction, enhancing the release of polyphenolic compounds (Krakowska et al., 2018).

For the research study of bioactive compounds, supercritical fluids offer a more effective, safer, and environmentally friendly technique for flavonoid extraction and recovery.

2.2.5 Enzyme and microorganism-assisted extraction

Enzyme-assisted extraction represents a promising alternative to traditional extraction techniques that rely on large quantities of solvents. This technique has several benefits, including the ability to catalyze reactions in mild aqueous conditions and the regioselectivity and specificity of enzyme activity. Moreover, the application of enzymes in extraction processes aligns with environmentally friendly practices by reducing the need for excessive solvent usage. As a result, extraction times are decreased, and the yield and quality of the extracted flavonoids are improved. In addition, enzymes and microorganisms can be employed for pretreatment purposes in various agro-industrial and food industries, further enhancing their utility.

Enzymes obtained from fungi, bacteria, plant extracts, and animal organs, such as pectinases, various glucanases, hemicellulases, cellulases, and more, can be used individually or in mixtures to effectively degrade the cell walls. These enzymes hydrolyze the biopolymeric components, resulting in the breakdown of the components and an increase in the permeability of the cell walls. This enzymatic activity not only enhances the extraction efficiency of flavonoids but also improves the yield of other bioactive compounds with physiological benefits (Zha et al., 2019).

Polysaccharide cell walls can be effectively broken down by hydrolytic enzymes, facilitating the release of intracellular components. Cellulose, for instance, can be hydrolyzed by four types of enzymes: endo- and exoglucanases, cellobiohydrolases, and β-glucosidases. Hemicellulose, on the contrary, can be degraded by enzymes such as endoxylanases, endomannanases, β-xylosidases, and β-mannanases. The degradation of pectin's backbone requires a variety of enzymes, including pectin lyases, pectate lyases, endo- and exo-polygalacturonases, endo- and exo-rhamnogalacturonases, and rhamnogalacturon lyases. Moreover, to hydrolyze the side chains of polysaccharides, additional enzymes like arabinases, galactosidases, and feruloyl esterases are necessary (Lagaert et al., 2009).

When extracting flavonoids with enzymes, careful consideration should be paid to the presence of undesirable enzyme activity. These processes may be brought on by endogenous enzymes found in plant matter or by unintended effects of commercial enzyme preparations. For instance, polyphenol oxidases can trigger oxidative browning reactions, while β-glucosidases, β-galactosidases, and α-l-arabinosidases have the potential to hydrolyze native glycosylated anthocyanins and lead to the formation of unstable aglycones (Landbo & Meyer, 2001).

Considerations for optimizing the extraction yield of flavonoids include treatment time, pH, temperature, and the enzyme-to-substrate concentration ratio. Numerous studies have focused on maximizing the flavonoid yield while enhancing the extraction process by optimizing these parameters (Tomaz et al., 2015).

The limitations of enzymatic techniques have also been observed in various reports. One notable limitation is the incomplete hydrolysis of plant cell walls achieved by the tested enzymes. Scaling up enzyme-assisted extraction to an industrial level poses significant challenges, including the high cost associated with using biocatalysts for processing large volumes of raw materials (Zha et al., 2019). Microorganisms have been investigated as potential enzyme producers for flavonoid extraction in an effort to overcome these restrictions. However, environmental factors, including temperature, dissolved oxygen levels, agitation rate, reactor design, and nutrient availability, have an impact on the development and activity of microorganisms (Huynh et al., 2014).

2.3 Anthocyanin extraction

Anthocyanins are a type of water-soluble flavonoid pigment that is usually found in the vacuoles of plant flowers, fruits, leaves, and stems (Alvarez-Suarez et al., 2021). These substances are highly intriguing due to their potential health benefits because they have a distinctive chemical structure and pleasant scent. Anthocyanins have been shown to have a wide range of biological functions, including the ability to neutralize free radicals, possess antibacterial capabilities, stop the growth of cancer cells, and protect eyesight. Furthermore, in vitro and in vivo studies have confirmed the low toxicity and high safety profile of anthocyanins (Zielinska & Michalska,

2016). Because of this, these compounds have many uses in the industries of food, medicine, cosmetics, and more. Therefore, the application of anthocyanin requires proper extraction and purification.

2.3.1 Soxhlet extraction

Traditional techniques like Soxhlet are based on the utilization of different types of solvents and heat. Solvents which are usually used to extract anthocyanins are as follows: methanol, ethanol, acetone, water, or mixtures thereof. Acid solutions are frequently introduced into these solvents to stabilize the flavylium cation, which is stable under extremely acidic conditions (pH 3). Since the use of strong concentrated acids might lead to the destabilization of the anthocyanin molecule, it is advised to use weak acids instead, such as formic acid, citric acid, or acetic acid. The addition of water to the solvent combination can also increase the yield of the extraction because anthocyanins are polar compounds.

Optimizing the maximum yields requires the careful consideration of additional analytical parameters, including the powder size of the solid, the solvent-to-solid ratio, as well as the extraction time and temperature. These factors play a crucial role in achieving optimal extraction efficiency. To maximize the yield of total anthocyanin content (TAC) from various natural matrices, numerous researchers have carried out optimization studies on particular analytical parameters.

These extraction techniques are currently used extensively across many industries, particularly in the natural dye industry, due to the affordability of the necessary equipment. Despite their widespread use, these techniques have some drawbacks, such as high energy consumption, the use of organic solvents that are harmful to the environment, the requirement for expensive, highly pure solvents, the consumption of large volumes of solvents, the requirement for prolonged extraction times to achieve lower yields, and the requirement for moderate to high temperatures, which may cause the degradation of antioxidants (Zhou et al., 2015).

2.3.2 Ultrasound-assisted extraction

The ultrasound-assisted extraction (UAE) method uses ultrasound waves to create cavitation within the sample tissue, which causes the cell walls to be disrupted (Morata et al., 2021). A large amount of energy is produced in the form of pressure and temperature as a result of the development and subsequent collapse of bubbles caused by the cavitation phenomena. As a result, the cell walls are effectively destroyed, facilitating the transfer of anthocyanins from the plant cells to the solvent (Xu et al., 2017).

The UAE is extensively employed for extracting bioactive compounds from various natural sources (Carrera et al., 2021). However, to achieve the highest yield of anthocyanins using UAE, it is crucial to optimize the extraction conditions. Several parameters need to be optimized, including temperature, extraction time, solvent composition, liquid-to-solid (L/S) ratio, moisture content in the sample, particle size, and ultrasound power (Carrera et al., 2021). To improve the effectiveness of anthocyanin extraction from red cabbage using UAE, Demirdöven et al. (2015)

carried out optimization of extraction conditions, including duration, temperature, and solvent composition (ethanol %).

Temperature and solvent concentrations were shown by Espada-Bellido et al. (2017) to be the main extraction factors impacting the extraction of anthocyanins from black mulberry pulps using UAE. The significance of ultrasonic strength or force and pulse cycle in influencing the yields of anthocyanin extraction has also been highlighted by various studies. In a study involving red cabbage, Ravanfar et al. (2015) discovered that the factors that most significantly affected the extraction yield of anthocyanins were time, temperature, and ultrasonic power. The importance of ultrasonic power in improving the extraction of anthocyanins from black carrot pomace was emphasized by Agcam et al. (2017). Their research also demonstrated how combining temperature and ultrasonication can maximize the yield of anthocyanin extraction in black carrot pomace. Several studies have indicated that the efficiency and yield of anthocyanin extraction are directly influenced by ultrasound power, which encompasses frequency and intensity, as well as time and temperature (Xu et al., 2017).

According to research, the UAE outperforms traditional methods in terms of yield, increasing the extraction of antioxidant compounds by at least 20%. In addition, UAE increases the antioxidant activities of the extracted chemicals, increasing the extraction process's overall effectiveness (Xu et al., 2017). Anthocyanin extraction using UAE has gained widespread acceptance as an effective method. UAE improves the mass transfer between the matrix and the solvent by facilitating solvent-efficient penetration into the solid matrix by utilizing the cavitation process. This feature makes it possible to extract materials with larger yields in less time and at lower temperatures. Due to its increased extraction rate, decreased solvent consumption, environmental friendliness, and cost-effectiveness, UAE stands out as a desirable technique and produces greater yields than traditional extraction techniques.

2.3.3 Microwave-assisted extraction

The microwave-assisted extraction (MAE) method utilizes the energy of electromagnetic radiation. The polar molecules found in both the solvent and the food sample absorb this energy, causing dipolar rotation and ion migration. Microwaves act selectively on plant cells, causing water in the matrix to vaporize and create high pressure inside the cell wall. As a result, heat is produced, which causes the cell wall's material properties to change and ultimately results in its rupture (Belwal et al., 2018). Consequently, the solvent exhibits enhanced penetration into the plant cell, facilitating the mass transfer of anthocyanins from the cell to the solvent. Unlike in the UAE, the heat generated by this process is transmitted from the interior to the exterior of the plant cell.

To enhance the yield of bioactive compounds through MAE, it is crucial to optimize several extraction conditions. These involve selecting a sufficiently polar solvent capable of absorbing microwave radiation, determining the appropriate liquid-to-solid (L/S) ratio, regulating the temperature and duration of extraction, modifying the microwave power, and taking the particle size of the matrix into account. The extraction efficiency can be greatly enhanced by optimizing these parameters.

In a study conducted by Xue et al., MAE was used for the extraction of anthocyanins from blueberries (Xue et al., 2018). The researchers looked into how increasing microwave power would affect the extraction yield and found that increasing microwave power had a minimal effect on the distribution of microwave energy. They concluded that a critical temperature of 50.75 °C was optimal to produce the best anthocyanin yield and suggested that controlling the microwave power is crucial for enhancing both the yield of anthocyanins and the efficiency of microwave energy. Farzaneh and Carvalho (2017) used MAE to extract anthocyanins from fresh Lavandula pedunculata L. plants for their study. They looked into how the extraction yield was affected by the microwave's power, duration, and L/S ratio. They found that the highest antioxidant activity was achieved with a lower power (300 W), shorter irradiation time (107.3 s), and higher L/S ratio (34.807 mL/g) compared with the conditions that resulted in the highest total anthocyanin content (TAC) (Farzaneh & Carvalho, 2017). These results are explained by the fact that higher irradiation power and longer irradiation time may increase the overall extraction yield. They also raised the temperature of samples, but it can potentially harm thermally sensitive antioxidants and lower their antioxidant activities. Sun et al.'s research (Yu et al., 2016) used MAE to extract anthocyanins from powdered blueberries. They noticed that anthocyanins were both acquired and degraded at the same time, both processes following exponential trends. The scientists discovered that 53.3 °C resulted in the least amount of anthocyanin degradation. Anthocyanins self-aggregated more readily at lower temperatures, while degraded at higher temperatures. They showed that MAE outperformed the traditional hot reflux extraction (HRE) method in terms of acquired anthocyanins and degradation (Yu et al., 2016). Therefore, the extraction process is influenced by the relationship between microwave power and irradiation time. Higher power levels can improve the heating effect, shortening the time needed for irradiation and boosting yield. It is crucial to keep in mind, though, that using a lot of power could potentially cause chemicals that are thermally labile to degrade. To get the desired extraction results, it is essential to find the ideal balance between microwave power and irradiation time (Xu et al., 2017). To increase the anthocyanin extraction yield from cranberries, Xue et al. (2018) conducted optimization tests on the irradiation time and resulting temperature. According to their research, an extraction temperature of 50°C and an irradiation time of 8 s produced the highest anthocyanin yield. The outcomes also showed that increasing the temperature within a range below 323 K (50°C) had a favorable effect on the effectiveness of anthocyanin extraction (Xue et al., 2018).

Various studies have highlighted the advantages of MAE over conventional extraction methods (Yiğit et al., 2022). According to the research, MAE offers quicker extraction times and uses less solvent than traditional methods (Yiğit et al., 2022), leading to higher extraction yields (Xu et al., 2017). Due to its quicker processing, MAE is also regarded as an energy-efficient technique. However, a notable drawback of this technology is the potential negative impact of microwave-induced heating on the properties of anthocyanins.

2.3.4 Supercritical fluid extraction

Supercritical fluid extraction (SFE) commonly uses carbon dioxide (CO2) as a solvent due to its nontoxicity, affordability, ability to protect the extracts from atmospheric oxidation, and moderate critical temperature (CT = 31.3 °C) and critical pressure (CP = 72.9 atm). However, a drawback of CO2 as a solvent is its nonpolar nature. To address this limitation and enhance the extraction of polar compounds like anthocyanins, a co-solvent is often employed (Zhang et al., 2018). Ethanol, methanol, and aqueous solutions of these alcohols are the most commonly used co-solvents, typically added in concentrations ranging from 1% to 15% (da Silva et al., 2016).

Several extraction conditions need to be carefully optimized in supercritical fluid extraction (SFE) to ensure maximum efficiency. These parameters include extraction duration, CO2 flow rate, liquid-to-solid (L/S) ratio, particle size, co-solvent quantity, moisture content of the natural material, and temperature. Maran et al. (2014) found that extracting complete monomeric anthocyanins from Indian blackberries under pressures exceeding 200 bar led to a modest reduction in yield. In addition, they discovered that yields decreased at temperatures higher than 50 °C, proving that 50 °C was the ideal temperature for extraction. In addition, their study showed that 2 g/min was the optimal solvent flow rate, while greater flow rates had no significant impact on the extraction yield. For the extraction of anthocyanins from various natural sources, a number of studies have been carried out to optimize these variables and produce the best feasible yield. These studies concluded that when the pressure was fairly high (greater than 100 bar), the temperature was controlled (not higher than 100 °C), and the addition of a co-solvent was taken into consideration, the recovery of antioxidant compounds improved. The quantity of co-solvent used depends on the extraction time and the amount of powdered sample used. Ethanol is widely utilized as the preferred co-solvent (Woźniak et al., 2017). The extraction of anthocyanins from Haskap berry paste using traditional extraction and SFE utilizing CO2 with water as a co-solvent was compared in a study by Jiao and Kermanshahi pour (2018). They discovered that SFE extraction offered greater anthocyanin extraction efficiency (52.7% vs. 38.3%) and increased antioxidant activity (89.8% vs. 72.2%) compared with conventional extraction methods by optimization of the SFE extraction parameters (Jiao & Kermanshahi pour, 2018).

This technique offers a major benefit over other low-pressure extraction methods in terms of the reduced amount of solvent needed, which is more environmentally friendly (Xu et al., 2017). In addition, it facilitates the extraction of extracts with increased antioxidant activity and has been shown to be more effective, selective, and time-effective (Jiao & Kermanshahi pour, 2018). Pazir et al. (2021) conducted a comparative study between conventional extraction and supercritical fluid extraction (SFE) using CO2 and ethanol to extract anthocyanins from Merlot red grape pomace. According to their findings, SFE with CO2 produced an extraction yield of 36% for anthocyanins in a reasonably quick extraction time of 80 min. However, they were unable to attain recovery percentages equivalent to those obtained by traditional extraction (63%), even after extending the extraction period to 180 min. They discovered a significant relationship between the growth in overall antioxidant activity and longer extraction durations and the quantity of extractable anthocyanins (Pazir et al., 2021).

Despite the initial perception that supercritical fluid extraction may be too expensive for large-scale implementation, studies on its feasibility demonstrate that effective optimization of operational parameters, particularly pressure and temperature, can lead to a competitive extraction method in comparison with conventional extraction processes (Solana et al., 2016).

2.3.5 Enzyme-assisted extraction

Enzyme-assisted aqueous extraction (EAE) has been extensively used and improved over time. This method has a number of benefits, including high enzyme selectivity and efficiency, operation at moderate pressures and temperatures, the ability of some enzymes to disintegrate or dissolve cell walls, and being recognized as a green technique (Meini et al., 2019).

The enzymatic extraction process operates by utilizing enzymes to break down the cell walls, allowing access to the bioactive compounds stored within plant cells. To overcome the various barriers, such as extracellular cell walls, cell walls, and oleosomes, present in the plant material, specific enzymes are required. These enzymes target the constituents of each barrier, which are naturally synthesized and hydrolyzed. The commonly employed enzymes for hydrolyzing plant cell wall constituents can be classified into four different families: cellulases, hemicellulases, pectinases, and proteases. Each of the three aforementioned enzyme families has a unique impact on the cell wall and can be used alone or in combination. As a result, using particular enzymes makes it possible to extract different bioactive compounds selectively depending on the desired target (Meini et al., 2019). In comparison with traditional ethanol extraction procedures, the extraction of pigments from various plant materials has been improved and accelerated by the use of cellulases, hemicellulases, pectinases, and proteases (Lotfi et al., 2015). Since proteases have the ability to hydrolyze enzymatic proteins and so reduce or eliminate the specific activities of the enzyme mixture, it is necessary to use them separately from the other enzyme families. Various naturally occurring sources, including bacteria, fungi, vegetable and fruit extracts, or animal organs, can provide the enzymes used in EAE (Xu et al., 2017). Depending on their source, enzymes have different characteristics and circumstances of action. While microorganism-produced enzymes may resist temperatures as high as 65 °C, animal enzymes normally denature at about 40–45 °C. However, there are also thermo-resistant enzymes that can tolerate temperatures exceeding 100 °C for several minutes.

In the extraction process, the enzymatic mixture's composition is extremely important. A quicker and more effective extraction yield is produced by a larger concentration of certain enzymes. To suit the particular matrix and bioactive compound being targeted, the enzymatic combination must be adjusted (Serea et al., 2021). In some circumstances, a multistep enzymatic extraction method is used to extract several chemicals. In order for each enzyme to successfully hydrolyze the cell walls, the extraction conditions, such as pH and temperature, must be optimized.

Several factors make pH optimization during enzymatic extraction essential. Firstly, maintaining an acidic pH is important for preserving the stability of anthocyanins. Secondly, an acidic pH enhances the plasticity of cell walls and facilitates intermolecular sliding. Thirdly, enzyme activity, including ionization, is pH-dependent, and an improper pH can significantly decrease enzymatic activity. Moreover, pH-induced denaturation can be irreversible for certain enzymes. Therefore, when using an enzymatic mixture, it is essential to establish an average pH that encompasses the optimal pH range for each enzyme in the mixture. This ensures the promotion of optimal enzymatic hydrolysis for each specific enzyme.

Similar to pH, temperature plays a significant role in enzymatic extraction. Higher temperatures cause the matrix's viscosity to decrease, which makes extraction easier. However, it's important to note that excessive temperature can negatively impact the stability of the extract. Furthermore, the temperature affects the activity of enzymes, with higher temperatures typically resulting in higher catalytic activity. However, enzymes are proteins; they can get thermally denatured when exposed to high temperatures. Therefore, in enzymatic extraction operations, the temperature must be carefully considered.

The optimal liquid/solid ratio and enzyme/solid ratio are also factors that are frequently discussed in the literature (Binaschi et al., 2018).

However, several drawbacks partially limit its industrial-scale application. There are some commercial and technical restrictions on the use of enzymes in extraction methods that must be taken into consideration. Enzymes can be expensive, and the process requires considerable energy. Moreover, enzyme activity is highly influenced by factors such as pH, temperature, and nutrient availability. As a result, various combinations of enzyme preparations must be tested and optimized, which complicates the procedure.

3. Quantification of polyphenols

Accurately quantifying various phenolic structural groups is still difficult despite numerous studies having been conducted on the topic (Ignat et al., 2011). As a result, there is significant potential for the development of quantification methods specific to each phenolic group (Liu et al., 2008). The most commonly used techniques for quantifying phenolic compounds are high-performance liquid chromatography (HPLC) and gas chromatography (GC), either alone or in combination with mass spectrometry. In addition, spectrophotometric tests are useful techniques in this context (Khoddami et al., 2013).

3.1 Spectrophotometric assays

Spectrophotometry is a relatively straightforward technique employed for the quantification of plant phenolics. For many years, the Folin-Denis and Folin-Ciocalteu methods have been widely used as spectrophotometric assays to measure total phenolics in plant materials (Lapornik et al., 2005). Both methods rely on a chemical reduction process involving reagents containing tungsten and molybdenum (Stalikas, 2007). In the presence of phenolic compounds, the resulting reduction products have a blue color and a broad light absorption spectrum of around 760 nm. It is important to take note that the reagents used in both methods do not specifically react with phenols alone, but also with other substances such as ascorbic acid, aromatic amines, and sugars (Box, 1983).

Proanthocyanidins (condensed tannins), total flavonoids, total phenolic content, and hydrolyzable tannins can all be estimated using colorimetric methods. Total flavonoids can be measured between 410 and 423 nm by combining methanolic or ethanolic extracts of plant phenols with AlC13 (Fernandes et al., 2012).

The level of proanthocyanidins can be determined using vanillin and dimethylaminocinnamaldehyde (DMCA) assays (Khoddami et al., 2013). These methods offer insights into the degree of polymerization, hydroxylation pattern, and stereochemistry of flavan-3-ol subunits (Hartzfeld et al., 2002). It should be emphasized that the vanillin method's use of catechin as a standard could lead to an overestimation of proanthocyanidins. Concerns have also been raised about the DMCA assay's precision in quantifying proanthocyanidins (Norris, 1987).

The methods of butanol-HCl and bovine serum albumin (BSA) can be used to determine proanthocyanidins. The butanol-HCl method involves the cleavage of interflavonoid bonds in proanthocyanidins using hot acid, followed by an autooxidation reaction that converts flavan-3-ols to anthocyanidins. The resulting red extract exhibits maximum absorbance around 550 nm (Porter et al., 1985). By treating samples with bovine serum albumin, the insoluble tannin-protein complexes are precipitated in the BSA method. After being dissolved in an alkaline sodium dodecyl sulfate-triethanolamine solution, the tannin-protein complex is then reacted with a ferric chloride solution to generate a violet complex with a maximum absorbance at 510 nm (Hagerman & Butler, 1978).

Methyl cellulose precipitation of proanthocyanidins is a reliable method for measuring proanthocyanidins in the grape extract. Methyl cellulose and proanthocyanidins react to generate an insoluble polymer (Sarneckis et al., 2006). By measuring the absorbance before and after the addition of methyl cellulose, it is possible to calculate the concentration of proanthocyanidins using this method (Mercurio et al., 2007). On the contrary, methods like the potassium iodate method, rhodanine method, and sodium nitrite method can be used to evaluate hydrolyzable tannins (Dai & Mumper, 2010). The reaction between methyl gallate and potassium iodate yields a red color with a maximum absorbance at 500–550 nm, which is the outcome of the potassium iodate method that is frequently used for screening samples (Hartzfeld et al., 2002). In addition, the presence of gallic acid and ellagic acid in the sample can be determined using the rhodanine and sodium nitrite techniques, respectively (Inoue & Hagerman, 1988). Flavonones and dihydroflavonols can also be measured using spectrophotometry by interacting with the acidic compound 2,4-dinitrophenylhydrazine. The reference substance used in this experiment is pinocembrin, and the absorbance is measured at 486 nm (Botanicae et al., 2010).

Spectrophotometry can also be used to measure anthocyanins constitute and other main classes of phenolic compounds. Giusti and Wrolstad (2003) conducted a comprehensive review of the main spectrophotometric assays used for anthocyanin determination. Anthocyanins are typically quantified in weakly acidic media within the wavelength range of 490–550 nm (Welch et al., 2008). Colorimetric techniques provide a quick and inexpensive way to estimate phenolic compound concentrations, but they are only applicable to samples that contain more phenolic compounds than a particular threshold and do not provide individual phenolic compound quantification. However, these methods are useful for quick and reasonably priced sample screening (Ignat et al., 2011).

3.2 Gas chromatography (GC)

Gas chromatography (GC) is widely used for the separation, identification, and quantification of various phenolic compounds, including phenolic acids (Guilherme et al., 2012), condensed tannins (Shadkami et al., 2009), and flavonoids (Proestos et al., 2006). In contrast to HPLC methods, GC analysis has some drawbacks, including the need for derivatization and the volatility of phenolic compounds. In GC-based quantification of phenolics from food matrices, additional steps may be required, such as the removal of lipids from the extract, the release of phenolics from glycoside and ester bonds using enzymatic (Liggins et al., 1998), alkaline (Siess et al., 1996), or acidic (Wang et al., 2000) media, and chemical modifications to convert them into more volatile derivatives (Liu et al., 2008).

Various reagents are employed to modify phenolic compounds and generate volatile derivatives. Examples include ethyl and methyl chloroformate, diazomethane, and a combination of dimethyl sulfoxide with methyl iodate, which are used to create methyl or ethyl esters of phenolics. However, it is crucial to remember that in some studies, the presence of naturally occurring methyl esters can cause confusion (Analityczna, 2001). Trifluoroacetamide, N-(tert-butyldimethylsilyl)-N-methyltrifluoroacetamide, and trimethylsilyl derivatives are members of the trimethylsilyl family of substances, which represents a more recent generation of reagents with advantages in the production of volatile compounds (Zadernowski et al., 2009). The silylation reaction is straightforward, does not produce unwanted side products, and results in highly volatile products that do not interfere with the analysis (Robbins, 2003). Silyl derivatization is thus a viable method for identifying phenolic compounds. However, further research is still needed to explore the identification of silyl derivatives (Proestos et al., 2006).

Fused silica capillaries with lengths of 30 m, internal diameters ranging from 25 to 32 m, and stationary phase particle sizes of 0.25 m are the most often used columns for phenolic measurement in GC methods. Shadkami et al. (2009) used a column with a length of 15 m and a film thickness of 10 m, showing that there are exceptions to this standard. Table 4.1 summarizes the gas chromatographic techniques for phenolic compound analysis.

3.3 High-performance liquid chromatography (HPLC)

HPLC is recommended for separating and quantifying the phenolic compounds. Sample purification, mobile phase composition, column types, and detectors are some of the variables that can affect the HPLC analysis of phenolics (Stalikas, 2007). Typically, purified phenolics are injected into an HPLC instrument using a reversed-phase C18 column (RP-C18), a photodiode array detector (PDA), and polar acidified organic solvents (Ignat et al., 2011). The use of HPLC for the measurement of phenolics has been discussed in numerous reviews. To achieve optimal sensitivity and detection, HPLC methods often involve the purification and preconcentration of phenolics from complex matrices of crude plant extracts.

Using partitionable solvents and methods like open-column chromatography or an adsorption-desorption technique, the crude extract is purified by removing interference-causing compounds. To separate phenolics from unpurified crude extracts, a variety of materials, including Sephadex LH-20, polyamide, Amberlite, solid phase extraction (SPE) cartridges, styrene-divinylbenzene (XAD 4, XAD16, EXA-90, EXA 118, SP70), and acrylic resins (XAD-7, EXA-31), are frequently used (Antolovich et al., 2000). Among these, SPE is frequently used in studies for partial concentration and purification before HPLC separation (Rostagno et al., 2005).

In HPLC measurement of phenolics, acetonitrile and methanol, and their aqueous forms, are frequently utilized as the primary mobile phases (Zarena & Sankar, 2012). In some circumstances, ethanol, tetrahydrofuran (THF), and 2-propanol have also been used (Reichelt et al., 2010). It is important to maintain the pH of the mobile phase within the range of pH 2-4 to prevent phenolics from undergoing ionization during identification. Therefore, aqueous acidified mobile phases commonly contain acetic acid, although the use of formic acid, phosphoric acid, or buffers such as phosphate, citrate, and ammonium acetate at low pH has also been reported (Stalikas, 2007). Gradient elution systems are more commonly used than isocratic elution systems (Robbins, 2003).

Column selection plays a crucial role in the identification of phenolics. Typically, a normal-phase C18 or reversed-phase (RP-C18) column with a length of 10–30 cm, an inner diameter (ID) of 3.9–4.6 mm, and a particle size of 3–10 μm is used to identify various classes of phenolics based on polarity (Lopes-Lutz et al., 2010). However, advanced HPLC techniques such as ultrahigh pressure chromatography

Table 4.1 Summary of GC conditions to detect molecules belonging to phenolic classes.

Sample	Derivatization	Detected phenolics	Detection	Chromatographic assay details	References
Guarana	Dried phenolic extract derivatized with a mixture of hexamethyldisiloxane and dimethylchlorosilane in pyridine	3-Hydroxybenzoic acid, benzoic acid, gallic acid, syringic acid, isovanillic acid, protocatechuic acid, catechin, caffeine, epicatechin, quercetin	GC–MS	Zebron ZB-5 ms fused silica capillary column (30 m × 0.25 mm I.D. × 0.25 μm film thickness); Oven temperature: 150 °C held for 5 min, to 295 °C at 3 °C/min, held for 18 min; Injector temperature: 300 °C; Carrier gas: helium flow at 1 mL/min; ion source temperature: 200 °C; transfer line temperature: 290 °C	Kuskoski et al. (2012)
Mirabelle plums	Dried phenolic extract derivatized with *N*,*O*-Bis (trimethylsilyl)trifluoro-acetamide	Benzoic acid, *p*-hydroxybenzaldehyde, *p*-hydroxybenzoic acid, vanillin, 3,4-dihydroxybenzoic acid, vanillic acid, gallic acid, syringaldehyde, syringic acid, coniferyl aldehyde, 3,5-dimethoxycinnamaldehyde, dehydrodiconiferyl aldehyde, guajacyl-glycerin-coniferyl aldehyde, guajacyl-glycerin-coniferyl aldehyde	GC–MS	HP 5MS capillary column (30 m × 0.25 mm I.D × 0.25 μm film thickness). Oven temperature: 100–270 °C at 4 °C/min, held for 20 min; Injector temperature: 250 °C; Helium flow at 0.9 mL/s; ion source temperature: 230 °C; transfer line temperature: 280 °C	Khallouki et al. (2012)

Continued

Table 4.1 Summary of GC conditions to detect molecules belonging to phenolic classes.—*cont'd*

Sample	Derivatization	Detected phenolics	Detection	Chromatographic assay details	References
Cranberry	Dried phenolic extract derivatized with a mixture of *N*,*O*-Bis(trimethylsilyl)-trifluoroacetamide and 1% trimethylchlorosilane in pyridine	Benzoic acid, *o*-hydroxybenzoic acid, trans-cinnamic acid, *m*-hydroxybenzoic acid, *p*-hydroxybenzoic acid, *p*-hydroxyphenyl acetic acid	GC-MS	Temperature: 80 °C for 1 min, to 250°C, at 20 °C/min, held 1 min, to 300°C at 6 °C/min, held 5 min, to 310°C at 15 °C/min held 10 min, to 320 °C at 20 °C/min, held 10 min; Injector temperature: 280 °C; transfer line temperature: 280 °C. DB-5 fused-silica capillary column (30 m × 0.32 mm ID × 0.25 μm film thickness)	Wang and Zuo (2011)
Saffron corms	Dried phenolic extract derivatized with a mixture of *N*-methyl-*N*-(trimethylsilyl) trifluoroacetamide and iodotrimethylsilane	Acetic acid, *o*-phthalic acid, 2,3-dihydroxy-benzoic acid, vanillic acid, *o*-hydroxy-cinnamic acid, 2,4-dihydroxy-benzoic acid, *p*-coumaric acid, ferulic acid, caffeic acid, sinapic acid, epicatechin, catechin. Quercetin, myricetin, *m*-methylbenzoic acid. catechol, vanillin, salicylic acid, cinnamic acid, *p*-hydroxybenzoic acid, syringic acid, *p*-coumaric acid, gallic acid, *t*-ferulic acid, caffeic acid, gentisic acid	GC-MS	Oven temperature: 80 °C for 1 min, to 220 °C, at 10 °C/min, to 310 °C, at 20 °C/min, held 6 min; Injector temperature: 280 °C; detector temperature: 305 °C; transfer line temperature: 280 °C. DB-5 capillary column (30 m × 0.25 mm ID × 0.25 μm film thickness); Oven temperature: 140 °C for 2 min, to 270 °C at 5 °C/min	Esmaeili et al. (2011)

Mangosteen fruit	Dried phenolic extract derivatized with *N,O*-bis (trimethylsilyl)acetamide	Hydroxybenzoic acid, protocatechuic acid, vanillic acid, caffeic acid, *p*-coumaric acid, ferulic acid, *p*-hydroxyphenylacetic acid, 3,4-dihydroxymandelic acid, cinnamic acid	GC-MS	Held 20 min; Injector temperature: 270 °C; transfer line temperature: 270 °C. SPB-1 silica-fused capillary column (30 m × 0.25 mm ID × 0.25 μm film thickness); Oven temperature: 120 °C held 2 min, to 260°C at 20 °C/min, held 10 min; Injector temperature: 240 °C; Helium flow at 28 cm^3/min; transfer line temperature: 240 °C.	Zadernowski et al. (2009)
Green tea	Dried phenolic extract derivatized with trimethyl-sulfonium hydroxide and trimethylsilyl diazomethane	Catechin, epicatechin, epigallocatechin, gallocatechin, kaempferol, quercetin	GC-MS	A ZB-5HT Inferno capillary column (15 m × 0.32 mm ID × 0.10 μm film thickness); Oven temperature: 100°C held for 5 min, to 375°C at 20°C/min, held for 5 min; Injector temperature: 350°C; transfer line temperature: 300°C	Shadkami et al. (2009)
Various plant extracts	Dried phenolic extract derivatized with a mixture of trimethylchlorosilane and *N,O*-bis(trimethylsilyl)-trifluoroacetamide with dimethyldichlorosilane in toluene and dimethyldichlorosilane	Gallic acid, *p*-hydroxybenzoic acid, gentisic acid, *p*-coumaric acid, vanillic acid, ferulic acid, syringic acid, catechin	GC-MS	CP-Sil 8 capillary column (30 m × 0.32 mm ID × 0.25 μm film thickness)	Xi and Shouqin (2007)

Continued

Table 4.1 Summary of GC conditions to detect molecules belonging to phenolic classes.—*cont'd*

Sample	Derivatization	Detected phenolics	Detection	Chromatographic assay details	References
Propolis	Dried phenolic extract derivatized with *N*,*O*-Bis (trimethylsilyl)trifluoro-acetamide	Quercetin, apigenin, naringenin, luteolin, caffeic acid, epicatechin, rutin, hydroxytyrosol. Ethyl hydrocinnamate, hydrocinnamic acid, inositol, cinnamic acid, ferulic acid, caffeic acid, pinostrobin	GC-MS	Oven temperature: 70 °C, to 135 °C at 2 °C/min, held for 10 min, to 220 °C at 4°C/min, held for 10 min, to 270 °C at 3.5 °C/min, held for 20 min; Injector temperature: 280 °C; transfer line temperature: 290 °C. Borosilicate capillary column (20 × 0.3 mm ID × 0.1 μm)	Salomão et al. (2004)

(UHPLC), high temperature liquid chromatography (HTLC), and two-dimensional liquid chromatography (LC × LC) employ newer types of columns, including monolithic and superficially porous particle columns (Klejdus et al., 2007). These columns, which offer increased performance for phenolic detection, have lengths ranging from 3 to 25 cm, an ID of 1–4.6 mm, and particle sizes of 1.7–10 μm. Traditionally, HPLC assays for phenolics have been conducted at room temperature. However, there has been a recent recommendation to use higher temperatures in the analysis due to advancements in columns and instrumentation. Another factor that influences the detection of phenolics is the HPLC assays' running times, which can range from 10 to 150 min. Roggero et al. (1997) highlighted the need to maintain a steady temperature for ensuring excellent reproducibility of results, especially when longer analytical periods are used. Table 4.2 lists recent HPLC applications for classifying various phenolic compounds.

3.4 Mass spectrometry (MS)

MS has grown in popularity recently, but the flame ionization detector (FID) is still the most commonly used technique to identify phenolics (Robbins, 2003). Mass spectrometry and gas chromatography work together to increase the sensitivity and selectivity (Ignat et al., 2011). An example of the improved capabilities is seen in the evaluation of flavonoid glycosides, where the challenges faced by conventional GC were overcome by employing high-temperature-high-resolution GC-MS (Dos Santos Pereira et al., 2004). In a separate study, it was discovered that GC-MS analysis of standards for phenolic and flavonoid compounds showed higher efficiency than HPLC. This technique allowed for rapid analysis with improved resolution and complete baseline separation of all standards, minimizing co-elution (Norris, 1987).

Structural characterization and confirmation of various phenolic classes have been accomplished using different types of mass spectrometric (MS) detectors coupled with high performance liquid chromatography (HPLC–MS) (Može et al., 2011), such as electrospray ionization mass spectrometry (ESI-MS) (Pereira Do Amaral et al., 2012), matrix-assisted laser desorption/ionization mass spectrometry (MALDI–MS) (Wei et al., 2010), fast atom bombardment mass spectrometry (FAB-MS) (Edenharder et al., 2001), and electron impact mass spectrometry (Edenharder et al., 2001). These techniques have been employed to provide valuable insights into the identification and analysis of phenolic compounds. The combination of HPLC with MS detectors offers high sensitivity and excellent specificity, owing to the mass selectivity of detection (Bianco et al., 2006). In addition, hydrophilic interaction liquid chromatography (HILIC) represents new trends in the analysis of phenolic compounds, enabling enhanced separation and characterization capabilities. HILIC is expected to gain in popularity due to its compatibility with MS when utilizing the proper mobile phase and its improved accuracy in analyzing polar components with complex matrices (Jandera, 2008).

Table 4.2 Summary of recent HPLC conditions for plant and food phenolic classes.

Sample	Phenolic class	Column/ Detector	Solvent/Flow rate/Injection volume	Temperature (°C)/Detection time (min)	References
Mangosteen pericarp	Gallic acid, gentisic acid, protocatechuic acid, gentisic acid, 4-hydroxybenzoic acid, veratric acid, vanillic acid, caffeic acid, syringic acid, *p*-coumaric acid, sinapic acid, ferulic acid, *t*-cinnamic acid, catechin, epicatechin	Bondapak C18 column (300 × 3.9 mm ID × 5 μm)/PDA, ESI-MS	Water: methanol: acetic acid (85:14:1); Flow rate: 1.0 mL/min; Injection volume: 20 μL	Ambient/45	Zarena and Sankar (2012)
Mulberry fruit	Cyanidin 3-*O*-rutinoside, cyanidin 3-*O*-glucoside, pelargonidin 3-*O*-glucoside, pelargonidin 3-*O*-rutinoside	RP C18 column (250 × 4.6 mm ID, 5 μm)/PDA, ESI-MS	A: water containing 0.1% TFA (trifluoroacetic acid); B: acetonitrile containing 0.1% TFA; Elution profile: 0–2 min, 10% B; 2–35 min, 10%–90% B; 35–40 min, 90%–100% B; 40–60 min, 100% B/Flow rate: 1.0 mL/min; NM	Ambient/60	Qin et al. (2010)
Fruit juice	Cyanidin, peonidin, delphinidin, petunidin, malvidin, pelargonidin	ODS-3 column (250 × 4.6 mm ID × 5 μm)/PDA	A: acetonitrile; B: water containing 10% acetic acid and 1% phosphoric acid; Elution profile: 25 min, 2%–20% A; 5 min, 20%–40%; Flow rate: 1.0 mL/min; Injection volume: 25 μL	NM/50	Lee et al. (2008)
Maytenus aquifolium and *Maytenus ilicifolia* leaves	Quercetin, kaempferol derivatives, rutin	Supelcosil C8 and C18 (250 × 4.6 mm ID × 5 μm) column/PDA	A: water containing 2.0, 2.5, or 3.0% formic acid or 0.3% trifluoroacetic acid; B: acetonitrile or methanol; various elution profiles; Flow rate 1.0 mL/min/Injection volume: 10 μL	35/Different detection times	Diagone et al. (2012)

Apple	Gallic acid, chlorogenic acid, catechin, epicatechin, procyanidin, phloridzin, cyanidin 3-galactoside, quercetin 3-rutinoside, quercetin 3-galactoside, quercetin 3-glucoside, quercetin 3-rhamnoside	RP C18 (250 × 4.6 mm ID × 4 μm) column/PDA	A: water containing 1% TFA, B: ACN containing 1% TFA; Elution profile: 0–10 min, 10% B; 10–45 min, 10%–20% B; 45–50 min, 20%–50% B; 50–55 min, isocratic 50%; 55–60 min, 50%–10% B. Flow rate: 1 mL/min. Injection volume: 10 μL	40/60	Hoang et al. (2011)
Medicinal plants	Cyanidin glucoside, pelargonidin glucoside, gallocatechin-catechin gallate, afzelechin–catechin dimer, gallocatechin catechin gallate, ferulic acid glucoside, rutin, naringenin-7-*O*-rutinoside	RP C18 (250 × 4.6 mm ID × 5 μm) column/PDA, ESI- MS	A: water containing 1% formic acid, B: acetonitrile; Elution profile: 30 min, 90%–75% A; 30–45 min, 75%–40% A; Flow rate: 1 mL/min; Injection volume: 20 μL	25/45	Simirgiotis et al. (2012)
Food samples	Monomeric, dimeric, and trimeric procyanidins, catechin, epicatechin	RP 18 (250 × 2 mm ID × 5 μm) column/ PDA, FLD, ESI-MS/MS	A: water containing 0.1% formic acid; B: acetonitrile containing 0.1% formic acid; Elution profile: 0–10 min, 10% B; 10–30 min, 15% B; 30–65 min, 40% B. Flow rate: 300 μL/min; Injection volume: 20 μL	25/30	Rzeppa et al. (2011)
Oregano	Quercetin, fisetin, kaempferol, luteolin, apigenin, eriodictyol, hesperetin, taxifolin, (+)-catechin, (−)-epicatechin	Hypersil C18 ODS (250 × 4.6 mm ID × 5 μm) column/PDA, ESI-MS-MS	A: water; B: methanol; C: acetonitrile, each containing 0.2% trifluroacetic acid; Elution profile: Initial, 90% A, 6% B, 4% C; 5 min, 85% A, 9% B, 6% C; 5–35 min, 71% A, 17.4% B, 11.6% C; 35–95 min, 0% A, 85% B, 15% C; Flow rate: 1 mL/min; NM	30/NM	Tsimogiannis et al. (2007)

Continued

Table 4.2 Summary of recent HPLC conditions for plant and food phenolic classes.—*cont'd*

Sample	Phenolic class	Column/ Detector	Solvent/Flow rate/Injection volume	Temperature (°C)/Detection time (min)	References
Lotus leaves	Myricetin 3-*O*-glucoside, quercetin 3-*O*-arabinopyranosyl, quercetin 3-*O*-glucuronide, kaempferol 3-*O*-galactoside, astragalin, isorhamnetin 3-*O*-glucoside, kaempferol 3-*O*-glucuronide, quercetin	C18 (150 × 4.6 mm ID × 3.5 μm) column/PDA, ESI-MS	A: water containing 0.5% formic acid; B: acetonitrile containing 0.1% formic acid; Elution profile: 0–10 min, 12% B; 10–32 min, 12%–20% B; 32–40 min, 20%–30% B; 40–48 min, 30%–60% B; 48–49 min, 60%–12% B; 49–53 min 12% B; Flow rate: 0.6 mL/min; NM	30/53	Chen et al. (2012)
Bilberries and blueberries	Delphinidin-3-*O*-glucopyranoside, delphinidin-3-*O*-galactopyranoside, cyanidin-3-*O*-arabinopyranoside, malvidin-3-*O*-arabinopyranoside, petunidin-3-*O*-galactopyranoside	C18 (250 × 4.6 mm ID × 3 μm) column/UV-VISc	A: acetonitrile: water: formic acid (87/3/10); B acetonitrile: water: formic acid (50/40/10); Elution profile: 0–20 min, 2%–14% B; 20–40 min, 14% B; 40–50 min, 15% B; 50–55 min, 19% B; 55–65 min, 20% B/ Flow rate: 0.5 mL/min; Injection volume: 20 μL	Ambient/65	Müller et al. (2012)
Persian walnut	3-caffeoylquinic, 3-*p*-coumaroylquinic, 4-*p*-coumaroylquinic acid, quercetin 3-galactoside, quercetin 3-arabinoside, quercetin 3-xyloside, quercetin 3-rhamnoside, quercetin 3-pentoside, kaempferol 3-pentoside	LiChroCART RP C 18 (250 × 4 mm ID × 5 μm)/PDA, MS–MS	A: water containing 0.1% TFA; B: methanol; Elution profile: 30 min, 30%–50% B; 30–32 min, 70% B; 32–33 min, 80% B; 33–35 min, 80% B; Flow rate: 1 mL/min; Injection volume: 5 μL	NM/35	Jalili (2012)

Rye grain	Sinapic acid, syringic acid, vanillic acid, ferulic acid, caffeic acid, *p*-hydroxybenzoic acid, protocatechuic acid, *p*-coumaric acid ferulic acid dehydrodimers	Inertsil ODS-3 (150 × 4.0 mm ID × 3 μm/PDA	A: 50 mM H_3PO_4 (pH 2.5) B: acetonitrile; Elution profile: 0–5 min, 95% A; 5–17 min, 95%–85% A; 17–40 min, 85%–80% A; 40–60 min, 80%–50% A; 60–65 min 50% A. Flow rate: 0.7 mL/min; Injection volume: 10 μL	35/67	Heiniö et al. (2008)
Pomegranate juices	Delphinidin 3,5-diglucoside, cyanidin 3,5-diglucoside, delphinidin 3-glucoside, pelargonidin 3,5-diglucoside, ellagic acid	RP C18 Nucleosil (125 × 5.0 mm ID × 5.0 μm) column/UV-VIS	A: water containing 2.5% acetic acid; B: methanol containing 2.5% acetic acid; Elution profile: 0–5 min, 100% A; 5–15 min, 90%; 15–45 min, 50% A; 45–55 min, 100% A. Flow rate: 1.0 mL/min; Injection volume: 50 μL	NM/55	Mousavinejad et al. (2009)
Orange juice	Gallic acid, protocatechuic acid, *p*-hydroxybenzoic acid, vanillic acid, caffeic acid, chlorogenic acid, *p*-coumaric acid, ferulic acid, sinapic acid, narirutin, naringin, hesperidin, neohesperidin, didymin	Ultrasphere ODS (250 × 4.6 mm ID × 5 μm) column/UV-VIS	A: water containing 5% formic acid; B: acetonitrile/solvent A (60:40; v/v); Elution profile: 0–10 min, 0% B; 10–40 min, 0%–5% B; 40–58 min, 5%–15% B; 48–62 min, 15%–25%, 62–93 min, 25%–50% B; 93–96 min, 50%–100% B; Flow rate: 1.0 mL/min; NM	25/96	Kelebek and Selli (2011)
Quinoa	Apigenin-7-methyl ether, 1-*O*-galloyl-β-D-glucose, protocatechuic acid 4-*O*-glucoside, vanillic glucoside, penstebioside, ferulic acid 4-*O*-glucoside, ethyl-m-digallate,	Kinetex C18 (100 × 4.6 mm ID × 2.6 μm) column/PDA, ESI-MS	A: water containing 1% acetic acid; B: acetonitrile/solvent A (40:60; v/v); Elution profile: 0–3.5 min, 2% B; 3.5–4.5 min, 2%–6%; 4.5–6 min, 6%–10% B; 6–7.5 min, 10%–17%;	25/30	Gómez-Caravaca et al. (2012)

Continued

Table 4.2 Summary of recent HPLC conditions for plant and food phenolic classes.—*cont'd*

Sample	Phenolic class	Column/ Detector	Solvent/Flow rate/Injection volume	Temperature (°C)/Detection time (min)	References
	gallocatechin, quercetin, kaempferol, rutin		7.5–13 min, 17%–36% B; 13–14 min, 36%–38.5% B; 14–19 min, 38.5%–60% B; 19–24 min, 60%–100% B; 24–30 min, 100% B; 30–32 min, 100%–2% B; Flow rate: 0.8 mL/min; Injection volume: 10 μL		
Pine needle	Catechin, proanthocyanidins	SupelcoSil LC18 (250 × 4.6 mm ID × 5 μm) column/UV	A: acetonitrile; B: water containing 0.3% phosphoric acid; Elution profile: 0–35 min, 10%–20% A; 35–55 min, 20%–90% A; Flow rate: 0.7 mL/min; Injection volume: 10 μL	NM/47	Park et al. (2011)
Apricot fruit	*p*-aminobenzoic acid, chlorogenic acid, caffeic acid, protocatechuic acid, ferulic acid, rutin, resveratrol, quercetin	Gemini C18 (150 × 4.6 mm ID × 3 μm) column/UV-VIS	A: citric acid (75 mM); B: ammonium acetate (25 mM); Elution profile: 0–1 min, 5% B; 1–4 min, 5%–6% B; 4–20 min, 6%–25% B; 20–30 min, 25%–100% B; 30–36 min, 100% B; 36–38 min, 100%–5% B; 38–45 min, 5% B; Flow rate: 1.0 mL/min; Injection volume: 20 μL	35/45	Zitka et al. (2011)
Sage tea	Carnosic acid, epirosmanol, luteolin-rutinoside, salvianolic	RP C18 (150 × 2.1 mm	A: water containing 0.1% formic acid; B: acetonitrile	40/28	Zimmermann et al. (2011)

	acid, apigenin-glucuronide, rosmarinic acid, apigenin-rutinoside, luteolin-rutinoside, luteolin-7-*O*-glucoside, monohydroxy benzoic acid, luteolin-diglucuronide, caffeic acid, caffeoyl-fructosyl-glucose, coumaroyl-hexose, protocatechuic acid	ID × 1.7 μm) column/PDA, MS–MS	containing 0.1% formic acid; Elution profile: 0–14 min, 4%–27% A; 14–28 min, 27%–59.7% A; 28–28.2 min, 59.7%–100% A; 28.2–30.5 min, 100% A; 30.5–31 min, 100%–4% A; 31–34 min, 4% A; Flow rate: 0.4 mL/min; Injection volume: 3 μL		
Almond skin	Quercetin-3-*O*-glucoside, isorhamnetin-3-rutinoside, kampferol-3-rutinoside, naringenin-7-*O*-glucoside, isorhamnetin-3-glucoside, *p*-hydroxybenzoic acid, naringenin, protocatechuic acid, vanillic acid	RP C18 (50 × 2 mm ID, × 2.5 μm) column/ESI-MS	A: water containing 0.1% formic acid. B: acetonitrile containing 0.1% formic acid; Elution profile: 0–9.5 min, 1%–100% B; Flow rate: 0.5 mL/min; Injection volume: 5 μL	35/9.5	Arráez-Román et al. (2010)
Burdock leaves	Quercetin, cynarin, benzoic acid, quercitrin, caffeic acid, luteolin, chlorogenic acid, *p*-coumaric acid, rutin, arctiin	BEH C18 (150 × 2.1 mm ID × 1.7 μm) column/PDA, ESI-MS-MS	A: water containing 0.1% formic acid; B: acetonitrile/methanol (20/80); Elution profile: 0–10 min, 10%–30% B; 10–20 min, 30%–50% B; 20–23 min, 50%–70% B; 23–25 min, 70%–10% B; Flow rate: 0.28 mL/min; NM	NM/25	Lou et al. (2010)
Grape extract	Malvidin glucoside, delphinidin glucoside, cyanidin glucoside, petunidin glucoside, peonidin glucoside, malvidin acetylglucoside delphinidin acetylglucoside, cyanidin acetylglucoside, petunidin acetylglucoside, peonidin acetylglucoside, malvidin coumarylglucoside	Zorbax SB-C18 (50 × 2.1 mm ID × 1.8 μm) column/PDA, MS–MS	A: water containing 10%; B: acetonitrile; Elution profile: 0–1.5 min, 10%–13% B; 1.5–4.5 min, 13%–15% B; 4.5–7.5 min, 15%–22% B; 7.5–15 min, 22% B; Flow rate: 0.2 mL/min; Injection volume: 1 μL	NM/15	Pati et al. (2009)

Continued

Table 4.2 Summary of recent HPLC conditions for plant and food phenolic classes.—*cont'd*

Sample	Phenolic class	Column/Detector	Solvent/Flow rate/Injection volume	Temperature (°C)/Detection time (min)	References
Stem bark of *Acacia confusa*	(+)-catechin, (−)-epicatechin, 4β-(2-aminoethylthio) catechin, 4β-(2-aminoethylthio) epicatechin	Hypersil ODS (250 × 4.6 mm ID × 2.5 μm) column/ESI-MS	A: water containing 0.5% trifluoroacetic acid; B: acetonitrile containing 0.5% trifluoroacetic acid; Elution profile: 0–5 min, 3% B; 5–15 min, 3%–9% B; 15–45 min, 9%–16% B; 45–60 min, 16%–60% B; Flow rate: 1 mL/min; NM	Ambient/NM	Wei et al. (2011)
Lettuce	Caffeoyltartaric acid, *p*-coumaroyltartaric acid, caffeoylquinic acid, chlorogenic acid, *p*-coumaroylquinic acid, caffeoylmalic acid, dicaffeoyltartaric acid, chicoric acid, *p* -coumaroylcaffeoyltartaric acid, di-*p* -coumaroyltartaric acid, quercetin-3-*O*-glucuronide, 3,5-dicaffeoylquinic acid, quercetin malonylglucoside	HSS T3 (100 × 2.1 mm ID × 1.8 μm) column/PDA, ESI-MS	A: water:methanol:formic acid (94.9:5.0:0.1); B: methanol: water:formic acid (60.0:39.9:/0.1); Elution profile: 0–30 min, 100%–50% A; Flow rate: 0.5 mL/min; Injection volume: 10 μL	35/NM	Ribas-Agustí et al. (2011)
Cocoa and chocolate products	Catechin, epicatechin	Diol-based (250 × 4.6 mm ID × 5 μm)/FLD, MS	A: acetonitrile:acetic acid (98: 2); B: methanol:water:acetic acid (95:3:2); 0–35 min, 100%–60% A; 35–39 min, 60% A; 39–41 min, 60%–0% A; 41–47 min, 0.0% A; 47–51 min, 0%–100% A; Flow rate: 1.0 mL/min; Injection volume: 10 μL	30/51	Hurst et al. (2009)
Wild mushroom	Benzoic acids, *p*-hydroxybenzoic,	Spherisorb RP C18	A: water containing 2.5% acetic acid; B: acetic acid	25/NM	Barros et al. (2009)

	protocatechuic, vanillic, cinnamic, *p*-coumaric acids	(150 × 4.6 mm ID × 3 μm) column/PDA, ESI-MS	2.5%: acetonitrile (90:10); C: acetonitrile; Elution profile: 10 min, 100% A; 10–20 min, 50% A and 50% B; 20–35 min, 100% B; 35–45 min, 90% B and 10% C; 45–55 min, 70% B and 30% C; 55–60 min, 50% B and 50% C; 60–65 min, 20% B and 80% C; 65–70 min, 100% A; Flow rate: 0.50 mL/min; NM		
Cocoa, apple	Quercetin, phloridzin, clovamide, *p*-coumaroylquinic acid, caffeoylquinic acid, quercetin-3-*O*-galactoside, quercetin-3-*O*-arabinoside, quercetin-3-*O*-xyloside, dideoxyclovamide, quercetin-3-*O*-rhamnoside	BEH C18 (50 × 2.1 mm ID× 1.7 μm) column/UV, FLD, ESI-MS	A: water containing 0.1 % formic acid; B: acetonitrile; Elution profile: 0–0.25 min, 2% B; 0.25–10.70 min, 2%–18% B; 10.70–18 min, 18%–25% B; 18–20.70 min, 25%–100% B; 20.70–22.5 min, 100% B; Flow rate: 0.80 mL/min; Injection volume: 2 μL	50/22.5	Kalili and de Villiers (2009)
Bean	Ferulic acid, *p*-coumaric acid, sinapic acid, caffeic acid	RP C18 Luna (150 × 4.6 mm ID × 5 μm) column/PDA	A: water containing 0.1% formic acid; B: methanol; Elution profile: 0–50 min, 5%–30% B; 50–65 min, 30% B; 65–75 min, 30%–100% B; Flow rate: 0.7 mL/min; NM	25/65	Luthria and Pastor-Corrales (2006)
Green tea, green coffee, grapefruit	Catechin, epigallocatechin gallate, epicatechin gallate, epicatechine, gallocatechin, catechin gallate, gallic acid, caffeine	RP C18 Atlantis (100 × 4.6 mm ID × 3 μm)/UV	A: water containing 0.1% formic acid; B: methanol containing 0.1% formic acid; Elution profile: 0–5 min, 10% B; 5–14 min, 10%–20% B; 14–20 min, 20%–50% B; 20–22 min, 50%–90% B; 22–26 min, 90% B; 26–30 min, 90%–10% B; Flow rate: 0.5 mL/min; Injection volume: 20 μL	25/NM	Colon and Nerin (2012)

4. Conclusion

Polyphenols have the potential to lower the risk of various chronic diseases, including cancer, cardiovascular diseases, diabetes, and neurodegenerative diseases. The extent of these benefits can vary significantly depending on the phenolic content that is present in the natural sources. The biology and health benefits of phenolics have inspired researchers to explore, develop, enhance, and employ different techniques to extract, separate, and quantify these compounds from natural sources.

When compared to traditional extraction strategies, new extraction methodologies have advantages such as high yield, low energy consumption, short duration, and environmental protection. However, there are several drawbacks, such as high expenditures and advanced equipment needs. The majority of industries use ultrasound-assisted or microwave-assisted extractions. Several new green extraction technologies cannot be applied in large-scale applications because of their increased costs, necessitating an immediate solution.

Accurate analytical technology is critical for the in-depth analysis of plant polyphenols, the promotion of plant polyphenol standardization, and the development of the food industry. Isolation was challenging due to the enormous quantity and structural variability of polymeric polyphenols. Various methods were utilized by the various groups to optimize the pre-, during-, and postpreparation processes. The most difficult aspect of structural elucidation was deciphering the complicated data produced by the analytical tools. The more complex the polyphenols, the more complex the signals generated, needing specific technical-analytical skills to comprehend the data, and its analysis demands careful observation and available time. Furthermore, polyphenol component isolation and purification processes have been enhanced. Aside from standard HPLC methods, HILIC reflects new trends in phenolic chemical analysis, allowing for improved separation and characterization capabilities. Because the organic solvent is substantially more volatile, HILIC can provide a tenfold increase in sensitivity over reversed-phase chromatography.

While conventional methods remain widely used due to their simplicity and accessibility, emerging techniques provide promising alternatives for enhanced recovery of polyphenols with reduced environmental impact. Future research should focus on optimizing these advanced techniques for extracting phenolics from plant-based sources and establishing standardized protocols to ensure consistent quality and functionality of plant polyphenol extracts. These advancements will contribute to the broader utilization of plant-derived polyphenols in the area of food, nutraceutical, cosmetic, and pharmaceutical products.

References

Abascal, K., Ganora, L., & Yarnell, E. (2005). The effect of freeze-drying and its implications for botanical medicine: A review. *Phytotherapy Research, 19*(8), 655–660. https://doi.org/10.1002/PTR.1651

Agcam, E., Akyıldız, A., & Balasubramaniam, V. M. (2017). Optimization of anthocyanins extraction from black carrot pomace with thermosonication. *Food Chemistry, 237*, 461–470. https://doi.org/10.1016/J.FOODCHEM.2017.05.098

Akbari, S., Abdurahman, N. H., Yunus, R. M., & Fayaz, F. (2019). Microwave-assisted extraction of saponin, phenolic and flavonoid compounds from Trigonella foenum-graecum seed based on two level factorial design. *Journal of Applied Research on Medicinal and Aromatic Plants, 14*, Article 100212. https://doi.org/10.1016/J.JARMAP.2019.100212

Akkol, E. K., Göger, F., Kosar, M., & Baser, K. H. C. (2008). Phenolic composition and biological activities of Salvia halophila and Salvia virgata from Turkey. *Food Chemistry, 108* (3), 942–949. https://doi.org/10.1016/J.FOODCHEM.2007.11.071

Alara, O. R., Abdurahman, N. H., & Olalere, O. A. (2018). Optimization of microwave-assisted extraction of flavonoids and antioxidants from Vernonia amygdalina leaf using response surface methodology. *Food and Bioproducts Processing, 107*, 36–48. https://doi.org/10.1016/J.FBP.2017.10.007

Alara, O. R., Abdurahman, N. H., & Ukaegbu, C. I. (2018). Soxhlet extraction of phenolic compounds from Vernonia cinerea leaves and its antioxidant activity. *Journal of Applied Research on Medicinal and Aromatic Plants, 11*, 12–17. https://doi.org/10.1016/J.JARMAP.2018.07.003

Alvarez, M. V., Cabred, S., Ramirez, C. L., & Fanovich, M. A. (2019). Valorization of an agro-industrial soybean residue by supercritical fluid extraction of phytochemical compounds. *The Journal of Supercritical Fluids, 143*, 90–96. https://doi.org/10.1016/J.SUPFLU.2018.07.012

Alvarez-Suarez, J. M., Cuadrado, C., Redondo, I. B., Giampieri, F., González-Paramás, A. M., & Santos-Buelga, C. (2021). Novel approaches in anthocyanin research - Plant fortification and bioavailability issues. *Trends in Food Science and Technology, 117*, 92–105. https://doi.org/10.1016/J.TIFS.2021.01.049

Analityczna, H. S.-C. (2001). In *Application of GC-MS method for analysis of phenolic acids and their esters in chloroformic extracts from some taxons of Polygonum L. genus*. https://www.infona.pl/resource/bwmeta1.element.baztech-article-BPP1-0022-0074.

Antolovich, M., Prenzler, P., Robards, K., & Ryan, D. (2000). Sample preparation in the determination of phenolic compounds in fruits. *Analyst, 125*(5), 989–1009. https://doi.org/10.1039/B000080I

Arráez-Román, D., Fu, S., Sawalha, S. M. S., Segura-Carretero, A., & Fernández-Gutiérrez, A. (2010). HPLC/CE-ESI-TOF-MS methods for the characterization of polyphenols in almond-skin extracts. *Electrophoresis, 31*(13), 2289–2296. https://doi.org/10.1002/ELPS.200900679

Arts, I. C. W., & Hollman, P. C. H. (2005). Polyphenols and disease risk in epidemiologic studies. *The American Journal of Clinical Nutrition, 81*(1 Suppl. 1). https://doi.org/10.1093/AJCN/81.1.317S

Barros, L., Dueñas, M., Ferreira, I. C. F. R., Baptista, P., & Santos-Buelga, C. (2009). Phenolic acids determination by HPLC-DAD-ESI/MS in sixteen different Portuguese wild mushrooms species. *Food and Chemical Toxicology, 47*(6), 1076–1079. https://doi.org/10.1016/J.FCT.2009.01.039

Belbaki, A., Louaer, W., & Meniai, A. H. (2017). Supercritical CO2 extraction of oil from crushed Algerian olives. *The Journal of Supercritical Fluids, 130*, 165–171. https://doi.org/10.1016/J.SUPFLU.2017.08.005

Belwal, T., Ezzat, S. M., Rastrelli, L., Bhatt, I. D., Daglia, M., Baldi, A., Devkota, H. P., Orhan, I. E., Patra, J. K., Das, G., Anandharamakrishnan, C., Gomez-Gomez, L.,

Nabavi, S. F., Nabavi, S. M., & Atanasov, A. G. (2018). A critical analysis of extraction techniques used for botanicals: Trends, priorities, industrial uses and optimization strategies. *TrAC, Trends in Analytical Chemistry, 100*, 82–102. https://doi.org/10.1016/J.TRAC.2017.12.018

Benetis, R., Radušiene, J., Jakštas, V., Janulis, V., & Malinauskas, F. (2008). Development of an RP-HPLC method for the analysis of phenolic compounds in Achillea millefolium L. *Journal of Liquid Chromatography and Related Technologies, 31*(4), 596–610. https://doi.org/10.1080/10826070701815247

Bianco, A., Buiarelli, F., Cartoni, G., Coccioli, F., Muzzalupo, I., Polidori, A., & Uccella, N. (2006). Analysis by HPLC-MS/MS of biophenolic components in olives and oils. *Analytical Letters, 34*(6), 1033–1051. https://doi.org/10.1081/AL-100103612

Binaschi, M., Garrido, G. D., Cirelli, C., & Spigno, G. (2018). Biotechnological strategies to valorise grape pomace for food applications. *Chemical Engineering Transactions, 64* (NA), 367–372. https://doi.org/10.3303/CET1864062

Botanicae, N., Agrobotanici, H., Marghitas, L. A., Mihai, C. M., Chirila, F., Severus Dezmirean, D., & Fit, N. (2010). The study of the antimicrobial activity of transylvanian (Romanian) propolis. *Notulae Botanicae Horti Agrobotanici Cluj-Napoca, 38*(3), 40–44. https://doi.org/10.15835/NBHA3835060

Box, J. D. (1983). Investigation of the Folin-Ciocalteau phenol reagent for the determination of polyphenolic substances in natural waters. *Water Research, 17*(5), 511–525. https://doi.org/10.1016/0043-1354(83)90111-2

Cacace, J. E., & Mazza, G. (2002). Extraction of anthocyanins and other phenolics from black currants with sulfured water. *Journal of Agricultural and Food Chemistry, 50*(21), 5939–5946. https://doi.org/10.1021/JF025614X

Carrera, C., Aliaño-González, M. J., Valaityte, M., Ferreiro-González, M., Barbero, G. F., & Palma, M. (2021). A novel ultrasound-assisted extraction method for the analysis of anthocyanins in potatoes (Solanum tuberosum L.). *Antioxidants, 10*(9). https://doi.org/10.3390/ANTIOX10091375

Casazza, A. A., Aliakbarian, B., Mantegna, S., Cravotto, G., & Perego, P. (2010). Extraction of phenolics from Vitis vinifera wastes using non-conventional techniques. *Journal of Food Engineering, 100*(1), 50–55. https://doi.org/10.1016/J.JFOODENG.2010.03.026

Castro-Vargas, H. I., Rodríguez-Varela, L. I., Ferreira, S. R. S., & Parada-Alfonso, F. (2010). Extraction of phenolic fraction from guava seeds (Psidium guajava L.) using supercritical carbon dioxide and co-solvents. *The Journal of Supercritical Fluids, 51*(3), 319–324. https://doi.org/10.1016/J.SUPFLU.2009.10.012

Chen, S., Wu, B. H., Fang, J. B., Liu, Y. L., Zhang, H. H., Fang, L. C., Guan, L., & Li, S. H. (2012). Analysis of flavonoids from lotus (Nelumbo nucifera) leaves using high performance liquid chromatography/photodiode array detector tandem electrospray ionization mass spectrometry and an extraction method optimized by orthogonal design. *Journal of Chromatography. A, 1227*, 145–153. https://doi.org/10.1016/J.CHROMA.2011.12.098

Colon, M., & Nerin, C. (2012). Role of catechins in the antioxidant capacity of an active film containing green tea, green coffee, and grapefruit extracts. *Journal of Agricultural and Food Chemistry, 60*(39), 9842–9849. https://doi.org/10.1021/JF302477Y

da Silva, R. P. F. F., Rocha-Santos, T. A. P., & Duarte, A. C. (2016). Supercritical fluid extraction of bioactive compounds. *TrAC, Trends in Analytical Chemistry, 76*, 40–51. https://doi.org/10.1016/J.TRAC.2015.11.013

Dai, J., & Mumper, R. J. (2010). Plant phenolics: Extraction, analysis and their antioxidant and anticancer properties. *Molecules, 15*(10), 7313–7352. https://doi.org/10.3390/MOLECULES15107313

Dassoff, E. S., & Li, Y. O. (2019). Mechanisms and effects of ultrasound-assisted supercritical CO2 extraction. *Trends in Food Science and Technology, 86*, 492–501. https://doi.org/10.1016/J.TIFS.2019.03.001

Demirdöven, A., Özdoğan, K., & Erdoğan-Tokatli, K. (2015). Extraction of anthocyanins from red cabbage by ultrasonic and conventional methods: Optimization and evaluation. *Journal of Food Biochemistry, 39*(5), 491–500. https://doi.org/10.1111/JFBC.12153

Diagone, C. A., Colombo, R., Lanças, F. M., & Yariwake, J. H. (2012). CZE/PAD and HPLC-UV/PAD profile of flavonoids from maytenus aquifolium and maytenus ilicifolia "espinheira santa" leaves extracts. *Chromatography Research International, 2012*, 1–7. https://doi.org/10.1155/2012/691509

Dos Santos Pereira, A., Costa Padilha, M., & De Aquino Neto, F. R. (2004). Two decades of high temperature gas chromatography (1983–2003): What's next? *Microchemical Journal, 77*(2), 141–149. https://doi.org/10.1016/J.MICROC.2004.02.008

Du, F. Y., Xiao, X. H., & Li, G. K. (2007). Application of ionic liquids in the microwave-assisted extraction of trans-resveratrol from Rhizma Polygoni Cuspidati. *Journal of Chromatography. A, 1140*(1–2), 56–62. https://doi.org/10.1016/J.CHROMA.2006.11.049

Edenharder, R., Keller, G., Platt, K. L., & Unger, K. K. (2001). Isolation and characterization of structurally novel antimutagenic flavonoids from spinach (Spinacia oleracea). *Journal of Agricultural and Food Chemistry, 49*(6), 2767–2773. https://doi.org/10.1021/JF0013712

Esmaeili, N., Ebrahimzadeh, H., Abdi, K., & Safarian, S. (2011). Determination of some phenolic compounds in Crocus sativus L. corms and its antioxidant activities study. *Pharmacognosy Magazine, 7*(25), 74–80. https://doi.org/10.4103/0973-1296.75906

Espada-Bellido, E., Ferreiro-González, M., Carrera, C., Palma, M., Barroso, C. G., & Barbero, G. F. (2017). Optimization of the ultrasound-assisted extraction of anthocyanins and total phenolic compounds in mulberry (Morus nigra) pulp. *Food Chemistry, 219*, 23–32. https://doi.org/10.1016/J.FOODCHEM.2016.09.122

Farzaneh, V., & Carvalho, I. S. (2017). Modelling of microwave assisted extraction (MAE) of anthocyanins (TMA). *Journal of Applied Research on Medicinal and Aromatic Plants, 6*, 92–100. https://doi.org/10.1016/J.JARMAP.2017.02.005

Fernandes, A. J. D., Ferreira, M. R. A., Randau, K. P., De Souza, T. P., & Soares, L. A. L. (2012). Total flavonoids content in the raw material and aqueous extractives from Bauhinia monandra Kurz (Caesalpiniaceae). *TheScientificWorldJOURNAL, 2012*. https://doi.org/10.1100/2012/923462

Flórez, N., Conde, E., & Domínguez, H. (2015). Microwave assisted water extraction of plant compounds. *Journal of Chemical Technology & Biotechnology, 90*(4), 590–607. https://doi.org/10.1002/JCTB.4519

Gómez-Caravaca, A. M., Iafelice, G., Lavini, A., Pulvento, C., Caboni, M. F., & Marconi, E. (2012). Phenolic compounds and saponins in quinoa samples (Chenopodium quinoa Willd.) grown under different saline and nonsaline irrigation regimens. *Journal of Agricultural and Food Chemistry, 60*(18), 4620–4627. https://doi.org/10.1021/JF3002125

Garcia-Salas, P., Morales-Soto, A., Segura-Carretero, A., & Fernández-Gutiérrez, A. (2010). Phenolic-compound-extraction systems for fruit and vegetable samples. *Molecules, 15*(12), 8813–8826. https://doi.org/10.3390/MOLECULES15128813

Giusti, M. M., & Wrolstad, R. E. (2003). Acylated anthocyanins from edible sources and their applications in food systems. *Biochemical Engineering Journal, 14*(3), 217–225. https://doi.org/10.1016/S1369-703X(02)00221-8

Guilherme, J., Martin, P., Porto, E., Alencar, S. M., & Micotti Da Gloria, E. (2012). In *Antimicrobial potential and chemical composition of agro-industrial wastes*. https://www.researchgate.net/profile/Jose-Prado-Martin/publication/268522606_Antimicrobial_potential_and_chemical_composition_of_agro-industrial_wastes/links/57837a9008aee45b8442c87d/Antimicrobial-potential-and-chemical-composition-of-agro-industrial-wastes.pdf.

Hagerman, A. E., & Butler, L. G. (1978). Protein precipitation method for the quantitative determination of tannins. *Journal of Agricultural and Food Chemistry, 26*(4), 809–812. https://doi.org/10.1021/JF60218A027/ASSET/JF60218A027.FP.PNG_V03

Hano, C., & Tungmunnithum, D. (2020). Plant polyphenols, more than just simple natural antioxidants: Oxidative stress, aging and age-related diseases. *Medicine, 7*(5), 26. https://doi.org/10.3390/MEDICINES7050026

Hartzfeld, P. W., Forkner, R., Hunter, M. D., & Hagerman, A. E. (2002). Determination of hydrolyzable tannins (Gallotannins and Ellagitannins) after reaction with potassium iodate. *Journal of Agricultural and Food Chemistry, 50*(7), 1785–1790. https://doi.org/10.1021/JF0111155

Heiniö, R. L., Liukkonen, K. H., Myllymäki, O., Pihlava, J. M., Adlercreutz, H., Heinonen, S. M., & Poutanen, K. (2008). Quantities of phenolic compounds and their impacts on the perceived flavour attributes of rye grain. *Journal of Cereal Science, 47*(3), 566–575. https://doi.org/10.1016/J.JCS.2007.06.018

Herrera, M. C., & Luque De Castro, M. D. (2005). Ultrasound-assisted extraction of phenolic compounds from strawberries prior to liquid chromatographic separation and photodiode array ultraviolet detection. *Journal of Chromatography. A, 1100*(1), 1–7. https://doi.org/10.1016/J.CHROMA.2005.09.021

Hoang, N. T. T., Golding, J. B., & Wilkes, M. A. (2011). The effect of postharvest 1-MCP treatment and storage atmosphere on "Cripps Pink" apple phenolics and antioxidant activity. *Food Chemistry, 127*(3), 1249–1256. https://doi.org/10.1016/J.FOODCHEM.2011.01.052

Huang, J., & Zhang, Z. (2004). Microwave-assisted extraction of quercetin and acid degradation of its glycosides in Psidium guajava leaves. *Analytical Sciences, 20*(2), 395–397. https://doi.org/10.2116/ANALSCI.20.395

Hurst, W. J., Stanley, B., Glinski, J. A., Davey, M., Payne, M. J., & Stuart, D. A. (2009). Characterization of primary standards for use in the HPLC analysis of the procyanidin content of cocoa and chocolate containing products. *Molecules, 14*(10), 4136–4146. https://doi.org/10.3390/MOLECULES14104136

Huynh, N. T., Van Camp, J., Smagghe, G., & Raes, K. (2014). Improved release and metabolism of flavonoids by steered fermentation processes: A review. *International Journal of Molecular Sciences, 15*(11), Article 19369. https://doi.org/10.3390/IJMS151119369

Ighodaro, O. M., Agunbiade, S. O., & Akintobi, O. (2010). Phytotoxic and anti-microbial activities of flavonoids in ocimum gratissimum. *Life Science Journal, 7*.

Ignat, I., Volf, I., & Popa, V. I. (2011). A critical review of methods for characterisation of polyphenolic compounds in fruits and vegetables. *Food Chemistry, 126*(4), 1821–1835. https://doi.org/10.1016/J.FOODCHEM.2010.12.026

Inoue, K. H., & Hagerman, A. E. (1988). Determination of gallotannin with rhodanine. *Analytical Biochemistry, 169*(2), 363–369. https://doi.org/10.1016/0003-2697(88)90296-5

Jalili, A. (2012). Comparative phenolic profile of Persian walnut (Juglans regia L.) leaves cultivars grown in Iran. *African Journal of Biochemistry Research, 6*(3), 33–38. https://doi.org/10.5897/AJBR11.062

Jandera, P. (2008). Stationary phases for hydrophilic interaction chromatography, their characterization and implementation into multidimensional chromatography concepts. *Journal of Separation Science, 31*(9), 1421–1437. https://doi.org/10.1002/JSSC.200800051

Jha, P., Das, A. J., & Deka, S. C. (2017). Optimization of ultrasound and microwave assisted extractions of polyphenols from black rice (Oryza sativa cv. Poireton) husk. *Journal of Food Science and Technology, 54*(12), 3847–3858. https://doi.org/10.1007/S13197-017-2832-0/TABLES/7

Jiao, G., & Kermanshahi pour, A. (2018). Extraction of anthocyanins from haskap berry pulp using supercritical carbon dioxide: Influence of co-solvent composition and pretreatment. *Lebensmittel-Wissenschaft & Technologie, 98*, 237–244. https://doi.org/10.1016/J.LWT.2018.08.042

Kalili, K. M., & de Villiers, A. (2009). Off-line comprehensive 2-dimensional hydrophilic interaction x reversed phase liquid chromatography analysis of procyanidins. *Journal of Chromatography. A, 1216*(35), 6274–6284. https://doi.org/10.1016/J.CHROMA.2009.06.071

Karasová, G., & Lehotay, J. (2007). Chromatographic determination of derivatives of p-hydroxybenzoic acid in melissa officinalis by HPLC. *Journal of Liquid Chromatography & Related Technologies, 28*(15), 2421–2431. https://doi.org/10.1080/10826070500187624

Kazazi, H., & Rezaei, K. (2010). Effect of various parameters on the selective extraction of main components from hyssop using supercritical fluid extraction (SFE). *Food Science and Technology Research, 15*(6), 645–652. https://doi.org/10.3136/FSTR.15.645

Kelebek, H., & Selli, S. (2011). Determination of volatile, phenolic, organic acid and sugar components in a Turkish cv. Dortyol (Citrus sinensis L. Osbeck) orange juice. *Journal of the Science of Food and Agriculture, 91*(10), 1855–1862. https://doi.org/10.1002/JSFA.4396

Khallouki, F., Haubner, R., Erben, G., Ulrich, C. M., & Owen, R. W. (2012). Phytochemical composition and antioxidant capacity of various botanical parts of the fruits of Prunus × domestica L. from the Lorraine region of Europe. *Food Chemistry, 133*(3), 697–706. https://doi.org/10.1016/J.FOODCHEM.2012.01.071

Khan, M. K., Abert-Vian, M., Fabiano-Tixier, A. S., Dangles, O., & Chemat, F. (2010). Ultrasound-assisted extraction of polyphenols (flavanone glycosides) from orange (Citrus sinensis L.) peel. *Food Chemistry, 119*(2), 851–858. https://doi.org/10.1016/J.FOODCHEM.2009.08.046

Khoddami, A., Wilkes, M. A., & Roberts, T. H. (2013). Techniques for analysis of plant phenolic compounds. *Molecules, 18*(2), 2328–2375. https://doi.org/10.3390/molecules18022328

Klejdus, B., Lojková, L., Lapčík, O., Koblovská, R., Moravcová, J., & Kubáň, V. (2005). Supercritical fluid extraction of isoflavones from biological samples with ultra-fast high-performance liquid chromatography/mass spectrometry. *Journal of Separation Science, 28*(12), 1334–1346. https://doi.org/10.1002/JSSC.200500102

Klejdus, B., Vacek, J., Benešová, L., Kopecký, J., Lapčík, O., & Kubáň, V. (2007). Rapid-resolution HPLC with spectrometric detection for the determination and identification of isoflavones in soy preparations and plant extracts. *Analytical and Bioanalytical Chemistry, 389*(7–8), 2277–2285. https://doi.org/10.1007/S00216-007-1606-3

Krakowska, A., Rafińska, K., Walczak, J., & Buszewski, B. (2018). Enzyme-assisted optimized supercritical fluid extraction to improve Medicago sativa polyphenolics isolation. *Industrial Crops and Products, 124*, 931–940. https://doi.org/10.1016/J.INDCROP.2018.08.004

Kuskoski, E. M., Ríos, J. J., Martín Bueno, J., Fett, R., Troncoso González, A. M., & García Asuero, A. (2012). Capillary gas chromatography-mass spectromety (CGC-MS) analysis and antioxidant activities of phenolic and components of guarana and derivatives. *The Open Analytical Chemistry Journal, 6*(1), 1–8. https://doi.org/10.2174/1874065001206010001

Laborde, J. L., Bouyer, C., Caltagirone, J. P., & Gérard, A. (1998). Acoustic bubble cavitation at low frequencies. *Ultrasonics, 36*(1–5), 589–594. https://doi.org/10.1016/S0041-624X(97)00105-4

Lagaert, S., Beliën, T., & Volckaert, G. (2009). Plant cell walls: Protecting the barrier from degradation by microbial enzymes. *Seminars in Cell and Developmental Biology, 20*(9), 1064–1073. https://doi.org/10.1016/J.SEMCDB.2009.05.008

Landbo, A. K., & Meyer, A. S. (2001). Enzyme-assisted extraction of antioxidative phenols from black currant juice press residues (Ribes nigrum). *Journal of Agricultural and Food Chemistry, 49*(7), 3169–3177. https://doi.org/10.1021/JF001443P

Lapornik, B., Prošek, M., & Wondra, A. G. (2005). Comparison of extracts prepared from plant by-products using different solvents and extraction time. *Journal of Food Engineering, 71*(2), 214–222. https://doi.org/10.1016/J.JFOODENG.2004.10.036

Lee, J., Rennaker, C., & Wrolstad, R. E. (2008). Correlation of two anthocyanin quantification methods: HPLC and spectrophotometric methods. *Food Chemistry, 110*(3), 782–786. https://doi.org/10.1016/J.FOODCHEM.2008.03.010

Li, H., Chen, B., Nie, L., & Yao, S. (2004). Solvent effects on focused microwave assisted extraction of polyphenolic acids from Eucommia ulmodies. *Phytochemical Analysis: Phytochemical Analysis, 15*(5), 306–312. https://doi.org/10.1002/PCA.783

Liazid, A., Palma, M., Brigui, J., & Barroso, C. G. (2007). Investigation on phenolic compounds stability during microwave-assisted extraction. *Journal of Chromatography. A, 1140*(1–2), 29–34. https://doi.org/10.1016/J.CHROMA.2006.11.040

Liggins, J., Bluck, L. J. C., Coward, W. A., & Bingham, S. A. (1998). Extraction and quantification of daidzein and genistein in food. *Analytical Biochemistry, 264*(1), 1–7. https://doi.org/10.1006/ABIO.1998.2825

Liu, Q., Cai, W., & Shao, X. (2008). Determination of seven polyphenols in water by high performance liquid chromatography combined with preconcentration. *Talanta, 77*(2), 679–683. https://doi.org/10.1016/J.TALANTA.2008.07.011

Lopes-Lutz, D., Dettmann, J., Nimalaratne, C., & Schieber, A. (2010). Characterization and quantification of polyphenols in Amazon grape (Pourouma cecropiifolia Martius). *Molecules, 15*(12), 8543–8552. https://doi.org/10.3390/MOLECULES15118543

Lotfi, L., Kalbasi-Ashtari, A., Hamedi, M., & Ghorbani, F. (2015). Effects of enzymatic extraction on anthocyanins yield of saffron tepals (Crocos sativus) along with its color properties and structural stability. *Journal of Food and Drug Analysis, 23*(2), 210–218. https://doi.org/10.1016/J.JFDA.2014.10.011

Lou, Z., Wang, H., Zhu, S., Zhang, M., Gao, Y., Ma, C., & Wang, Z. (2010). Improved extraction and identification by ultra performance liquid chromatography tandem mass spectrometry of phenolic compounds in burdock leaves. *Journal of Chromatography. A, 1217*(16), 2441–2446. https://doi.org/10.1016/J.CHROMA.2009.12.022

Luthria, D. L., & Pastor-Corrales, M. A. (2006). Phenolic acids content of fifteen dry edible bean (Phaseolus vulgaris L.) varieties. *Journal of Food Composition and Analysis, 19* (2–3), 205–211. https://doi.org/10.1016/J.JFCA.2005.09.003

Müller, D., Schantz, M., & Richling, E. (2012). High performance liquid chromatography analysis of anthocyanins in bilberries (Vaccinium myrtillus L.), blueberries (Vaccinium corymbosum L.), and corresponding juices. *Journal of Food Science, 77*(4). https://doi.org/10.1111/J.1750-3841.2011.02605.X

Maran, J. P., Priya, B., & Manikandan, S. (2014). Modeling and optimization of supercritical fluid extraction of anthocyanin and phenolic compounds from Syzygium cumini fruit pulp. *Journal of Food Science and Technology, 51*(9), 1938–1946. https://doi.org/10.1007/S13197-013-1237-Y

Martens, S., & Mithöfer, A. (2005). Flavones and flavone synthases. *Phytochemistry, 66*(20), 2399–2407. https://doi.org/10.1016/J.PHYTOCHEM.2005.07.013

Mason, T. J., Paniwnyk, L., & Lorimer, J. P. (1996). The uses of ultrasound in food technology. *Ultrasonics Sonochemistry, 3*(3), S253–S260. https://doi.org/10.1016/S1350-4177(96)00034-X

Mattila, P., & Hellström, J. (2007). Phenolic acids in potatoes, vegetables, and some of their products. *Journal of Food Composition and Analysis, 20*(3–4), 152–160. https://doi.org/10.1016/J.JFCA.2006.05.007

Mattila, P., & Kumpulainen, J. (2002). Determination of free and total phenolic acids in plant-derived foods by HPLC with diode-array detection. *Journal of Agricultural and Food Chemistry, 50*(13), 3660–3667. https://doi.org/10.1021/JF020028P

Meini, M. R., Cabezudo, I., Boschetti, C. E., & Romanini, D. (2019). Recovery of phenolic antioxidants from Syrah grape pomace through the optimization of an enzymatic extraction process. *Food Chemistry, 283*, 257–264. https://doi.org/10.1016/J.FOODCHEM.2019.01.037

Mendiola, J. A., Herrero, M., Cifuentes, A., & Ibañez, E. (2007). Use of compressed fluids for sample preparation: Food applications. *Journal of Chromatography. A, 1152*(1–2), 234–246. https://doi.org/10.1016/J.CHROMA.2007.02.046

Mercurio, M. D., Dambergs, R. G., Herderich, M. J., & Smith, P. A. (2007). High throughput analysis of red wine and grape phenolicsadaptation and validation of methyl cellulose precipitable tannin assay and modified somers color assay to a rapid 96 well plate format. *Journal of Agricultural and Food Chemistry, 55*(12), 4651–4657. https://doi.org/10.1021/JF063674N

Meyer, A. B. S. (2005). Enzymatic upgrading of antioxidant phenolics in berry juice and in press residues. *Fruit Processing, 6*, 382–387. https://orbit.dtu.dk/en/publications/enzymatic upgrading-of-antioxidant-phenolics-in-berry-juice-and-i.

Može, Š., Polak, T., Gašperlin, L., Koron, D., Vanzo, A., Poklar Ulrih, N., & Abram, V. (2011). Phenolics in Slovenian bilberries (Vaccinium myrtillus L.) and blueberries (Vaccinium corymbosum L.). *Journal of Agricultural and Food Chemistry, 59*(13), 6998–7004. https://doi.org/10.1021/JF200765N

Morata, A., Escott, C., Loira, I., López, C., Palomero, F., & González, C. (2021). Emerging non-thermal technologies for the extraction of grape anthocyanins. *Antioxidants, 10*(12). https://doi.org/10.3390/ANTIOX10121863

Mousavinejad, G., Emam-Djomeh, Z., Rezaei, K., & Khodaparast, M. H. H. (2009). Identification and quantification of phenolic compounds and their effects on antioxidant activity in pomegranate juices of eight Iranian cultivars. *Food Chemistry, 115*(4), 1274–1278. https://doi.org/10.1016/J.FOODCHEM.2009.01.044

Nepote, V., Grosso, N. R., & Guzmán, C. A. (2005). Optimization of extraction of phenolic antioxidants from peanut skins. *Journal of the Science of Food and Agriculture, 85*(1), 33–38. https://doi.org/10.1002/JSFA.1933

Nicoué, E.É., Savard, S., & Belkacemi, K. (2007). Anthocyanins in wild blueberries of Quebec: Extraction and identification. *Journal of Agricultural and Food Chemistry, 55* (14), 5626–5635. https://doi.org/10.1021/JF0703304

Norris, D. (1987). In *Strategic planning, Polaris, and Tomahawk: Technological imperative hypotheses*. https://www.osti.gov/biblio/6909350.

Ouédraogo, J. C. W., Dicko, C., Kini, F. B., Bonzi-Coulibaly, Y. L., & Dey, E. S. (2018). Enhanced extraction of flavonoids from Odontonema strictum leaves with antioxidant activity using supercritical carbon dioxide fluid combined with ethanol. *The Journal of Supercritical Fluids, 131*, 66–71. https://doi.org/10.1016/J.SUPFLU.2017.08.017

Palma, M., Piñeiro, Z., & Barroso, C. G. (2001). Stability of phenolic compounds during extraction with superheated solvents. *Journal of Chromatography. A, 921*(2), 169–174. https://doi.org/10.1016/S0021-9673(01)00882-2

Park, Y. S., Jeon, M. H., Hwang, H. J., Park, M. R., Lee, S. H., Kim, S. G., & Kim, M. (2011). Antioxidant activity and analysis of proanthocyanidins from pine (Pinus densiflora) needles. *Nutrition Research and Practice, 5*(4), 281–287. https://doi.org/10.4162/NRP.2011.5.4.281

Pati, S., Liberatore, M. T., Gambacorta, G., Antonacci, D., & La Notte, E. (2009). Rapid screening for anthocyanins and anthocyanin dimers in crude grape extracts by high performance liquid chromatography coupled with diode array detection and tandem mass spectrometry. *Journal of Chromatography. A, 1216*(18), 3864–3868. https://doi.org/10.1016/J.CHROMA.2009.02.068

Pawar, N., Pai, S., Nimbalkar, M., & Dixit, G. (2011). RP-HPLC analysis of phenolic antioxidant compound 6-gingerol from different ginger cultivars. *Food Chemistry, 126*(3), 1330–1336. https://doi.org/10.1016/J.FOODCHEM.2010.11.090

Pazir, F., Koçak, E., Turan, F., & Ova, G. (2021). Extraction of anthocyanins from grape pomace by using supercritical carbon dioxide. *Journal of Food Processing and Preservation, 45*(8), Article e14950. https://doi.org/10.1111/JFPP.14950

Pereira Do Amaral, F., Napolitano, A., Masullo, M., Campaner Dos Santos, L., Festa, M., Vilegas, W., Pizza, C., & Piacente, S. (2012). HPLC-ESIMS(n) profiling, isolation, structural elucidation, and evaluation of the antioxidant potential of phenolics from Paepalanthus geniculatus. *Journal of Natural Products, 75*(4), 547–556. https://doi.org/10.1021/NP200604K

Piana, F., Ciulu, M., Quirantes-Piné, R., Sanna, G., Segura-Carretero, A., Spano, N., & Mariani, A. (2018). Simple and rapid procedures for the extraction of bioactive compounds from Guayule leaves. *Industrial Crops and Products, 116*, 162–169. https://doi.org/10.1016/J.INDCROP.2018.02.057

Pinelo, M., Arnous, A., & Meyer, A. S. (2006). Upgrading of grape skins: Significance of plant cell-wall structural components and extraction techniques for phenol release. *Trends in Food Science and Technology, 17*(11), 579–590. https://doi.org/10.1016/J.TIFS.2006.05.003

Pinelo, M., Rubilar, M., Jerez, M., Sineiro, J., & Núñez, M. J. (2005). Effect of solvent, temperature, and solvent-to-solid ratio on the total phenolic content and antiradical activity of extracts from different components of grape pomace. *Journal of Agricultural and Food Chemistry, 53*(6), 2111–2117. https://doi.org/10.1021/JF0488110

Pinelo, M., Ruiz-Rodríguez, A., Sineiro, J., Señoráns, F. J., Reglero, G., & Núñez, M. J. (2007). Supercritical fluid and solid-liquid extraction of phenolic antioxidants from grape pomace: A comparative study. *European Food Research and Technology, 226*(1–2), 199–205. https://doi.org/10.1007/S00217-006-0526-3

Pinelo, M., Zornoza, B., & Meyer, A. S. (2008). Selective release of phenols from apple skin: Mass transfer kinetics during solvent and enzyme-assisted extraction. *Separation and Purification Technology, 63*(3), 620–627. https://doi.org/10.1016/J.SEPPUR.2008.07.007

Porter, L. J., Hrstich, L. N., & Chan, B. G. (1985). The conversion of procyanidins and prodelphinidins to cyanidin and delphinidin. *Phytochemistry, 25*(1), 223–230. https://doi.org/10.1016/S0031-9422(00)94533-3

Prasad, K. N., Yang, B., Zhao, M., Ruenroengklin, N., & Jiang, Y. (2009). Application of ultrasonication or high-pressure extraction of flavonoids from litchi fruit pericarp. *Journal of Food Process Engineering, 32*(6), 828–843. https://doi.org/10.1111/J.1745-4530.2008.00247.X

Prior, R. L., Lazarus, S. A., Cao, G., Muccitelli, H., & Hammerstone, J. F. (2001). Identification of procyanidins and anthocyanins in blueberries and cranberries (Vaccinium Spp.) using high-performance liquid chromatography/mass spectrometry. *Journal of Agricultural and Food Chemistry, 49*(3), 1270–1276. https://doi.org/10.1021/JF001211Q

Proestos, C., Boziaris, I. S., Nychas, G. J. E., & Komaitis, M. (2006). Analysis of flavonoids and phenolic acids in Greek aromatic plants: Investigation of their antioxidant capacity and antimicrobial activity. *Food Chemistry, 95*(4), 664–671. https://doi.org/10.1016/J.FOODCHEM.2005.01.049

Proestos, C., & Komaitis, M. (2006). Ultrasonically assisted extraction of phenolic compounds from aromatic plants: Comparison with conventional extraction technics. *Journal of Food Quality, 29*(5), 567–582. https://doi.org/10.1111/J.1745-4557.2006.00096.X

Qin, C., Li, Y., Niu, W., & Ding, Y. (2010). Analysis and characterisation of anthocyanins in mulberry fruit. *Czech Journal of Food Sciences, 28*(2), 117–126. https://www.old-aj.cz/web/cjfs.htm?type=article&id=228_2008-CJFS.

Quideau, S., Deffieux, D., Douat-Casassus, C., & Pouységu, L. (2011). Plant polyphenols: Chemical properties, biological activities, and synthesis. *Angewandte Chemie, 50*(3), 586–621. https://doi.org/10.1002/ANIE.201000044

Ravanfar, R., Tamadon, A. M., & Niakousari, M. (2015). Optimization of ultrasound assisted extraction of anthocyanins from red cabbage using Taguchi design method. *Journal of Food Science and Technology, 52*(12), 8140–8147. https://doi.org/10.1007/S13197-015-1880-6

Reichelt, K. V., Peter, R., Paetz, S., Roloff, M., Ley, J. P., Krammer, G. E., & Engel, K. H. (2010). Characterization of flavor modulating effects in complex mixtures via high temperature liquid chromatography. *Journal of Agricultural and Food Chemistry, 58*(1), 458–464. https://doi.org/10.1021/JF9027552

Ribas-Agustí, A., Gratacós-Cubarsí, M., Sárraga, C., García-Regueiro, J. A., & Castellari, M. (2011). Analysis of eleven phenolic compounds including novel p-coumaroyl derivatives in lettuce (Lactuca sativa L.) by ultra-high-performance liquid chromatography with photodiode array and mass spectrometry detection. *Phytochemical Analysis, 22*(6), 555–563. https://doi.org/10.1002/PCA.1318

Richter, B. E., Jones, B. A., Ezzell, J. L., Porter, N. L., Avdalovic, N., & Pohl, C. (1996). Accelerated solvent extraction: A technique for sample preparation. *Analytical Chemistry, 68*(6), 1033–1039. https://doi.org/10.1021/AC9508199/ASSET/IMAGES/LARGE/AC9508199F00001.JPEG

Robards, K. (2003). Strategies for the determination of bioactive phenols in plants, fruit and vegetables. *Journal of Chromatography. A, 1000*(1–2), 657–691. https://doi.org/10.1016/S0021-9673(03)00058-X

Robbins, R. J. (2003). Phenolic acids in foods: An overview of analytical methodology. *Journal of Agricultural and Food Chemistry, 51*(10), 2866–2887. https://doi.org/10.1021/JF026182T

Roggero, J. P., Archier, P., & Coen, S. (1997). In *Wine: Nutritional and therapeutic benefits.* https://scholar.google.com/scholar_lookup?title=Wine:+Nutritional+and+Therapeutic+Benefits&author=J.-P.+Roggero&author=P.+Archier&author=S.+Coen&publication_year=1997&.

Rostagno, M. A., Palma, M., & Barroso, C. G. (2005). Solid-phase extraction of soy isoflavones. *Journal of Chromatography. A, 1076*(1–2), 110–117. https://doi.org/10.1016/J.CHROMA.2005.04.045

Rostagno, M. A., Palma, M., & Barroso, C. G. (2007). Microwave assisted extraction of soy isoflavones. *Analytica Chimica Acta, 588*(2), 274–282. https://doi.org/10.1016/J.ACA.2007.02.010

Rzeppa, S., Von Bargen, C., Bittner, K., & Humpf, H. U. (2011). Analysis of Flavan-3-ols and procyanidins in food samples by reversed phase high-performance liquid chromatography coupled to electrospray ionization tandem mass spectrometry (RP-HPLC-ESI-MS/MS). *Journal of Agricultural and Food Chemistry, 59*(19), 10594–10603. https://doi.org/10.1021/JF202697J

Salomão, K., Dantas, A. P., Borba, C. M., Campos, L. C., Machado, D. G., Aquino Neto, F. R., & De Castro, S. L. (2004). Chemical composition and microbicidal activity of extracts from Brazilian and Bulgarian propolis. *Letters in Applied Microbiology, 38*(2), 87–92. https://doi.org/10.1111/J.1472-765X.2003.01458.X

Sarneckis, C. J., Dambergs, R. G., Jones, P., Mercurio, M., Herderich, M. J., & Smith, P. A. (2006). Quantification of condensed tannins by precipitation with methyl cellulose: Development and validation of an optimised tool for grape and wine analysis. *Australian Journal of Grape and Wine Research, 12*(1), 39–49. https://doi.org/10.1111/J.1755-0238.2006.TB00042.X

Sati, P., Dhyani, P., Bhatt, I. D., & Pandey, A. (2019). Ginkgo biloba flavonoid glycosides in antimicrobial perspective with reference to extraction method. *Journal of Traditional and Complementary Medicine, 9*(1), 15–23. https://doi.org/10.1016/J.JTCME.2017.10.003

Serea, D., Râpeanu, G., Constantin, O. E., Bahrim, G. E., Stănciuc, N., & Croitoru, C. (2021). Ultrasound and enzymatic assisted extractions of bioactive compounds found in red grape skins Băbească Neagră (Vitis vinifera) variety. *The Annals of the University Dunarea de Jos of Galati. Fascicle VI - Food Technology, 45*(1), 9–25. https://doi.org/10.35219/FOODTECHNOLOGY.2021.1.01

Shadkami, F., Estevez, S., & Helleur, R. (2009). Analysis of catechins and condensed tannins by thermally assisted hydrolysis/methylation-GC/MS and by a novel two step methylation. *Journal of Analytical and Applied Pyrolysis, 85*(1–2), 54–65. https://doi.org/10.1016/J.JAAP.2008.09.001

Shi, J., Nawaz, H., Pohorly, J., Mittal, G., Kakuda, Y., & Jiang, Y. (2007). Extraction of polyphenolics from plant material for functional foods—Engineering and technology. *Food Reviews International, 21*(1), 139–166. https://doi.org/10.1081/FRI-200040606

Siess, M. H., Le Bon, A. M., Canivenc-Lavier, M. C., Amiot, M. J., Sabatier, S., Aubert, S. Y., & Suschetet, M. (1996). Flavonoids of honey and propolis: Characterization and effects on hepatic drug-metabolizing enzymes and benzo[a]pyrene–DNA binding in rats.

Journal of Agricultural and Food Chemistry, 44(8), 2297–2301. https://doi.org/10.1021/JF9504733

Simirgiotis, M. J., Silva, M., Becerra, J., & Schmeda-Hirschmann, G. (2012). Direct characterisation of phenolic antioxidants in infusions from four Mapuche medicinal plants by liquid chromatography with diode array detection (HPLC-DAD) and electrospray ionisation tandem mass spectrometry (HPLC-ESI–MS). *Food Chemistry, 131*(1), 318–327. https://doi.org/10.1016/J.FOODCHEM.2011.07.118

Singanusong, R., Nipornram, S., Tochampa, W., & Rattanatraiwong, P. (2015). Low power ultrasound-assisted extraction of phenolic compounds from mandarin (Citrus reticulata blanco cv. Sainampueng) and lime (citrus aurantifolia) peels and the antioxidant. *Food Analytical Methods, 8*(5), 1112–1123. https://doi.org/10.1007/S12161-014-9992-6/TABLES/3

Solana, M., Mirofci, S., & Bertucco, A. (2016). Production of phenolic and glucosinolate extracts from rocket salad by supercritical fluid extraction: Process design and cost benefits analysis. *Journal of Food Engineering, 168*, 35–41. https://doi.org/10.1016/J.JFOODENG.2015.07.017

Song, L., Liu, P., Yan, Y., Huang, Y., Bai, B., Hou, X., & Zhang, L. (2019). Supercritical CO2 fluid extraction of flavonoid compounds from Xinjiang jujube (Ziziphus jujuba Mill.) leaves and associated biological activities and flavonoid compositions. *Industrial Crops and Products, 139*, Article 111508. https://doi.org/10.1016/J.INDCROP.2019.111508

Sparr Eskilsson, C., & Björklund, E. (2000). Analytical-scale microwave-assisted extraction. *Journal of Chromatography. A, 902*(1), 227–250. https://doi.org/10.1016/S0021-9673(00)00921-3

Stalikas, C. D. (2007). Extraction, separation, and detection methods for phenolic acids and flavonoids. *Journal of Separation Science, 30*(18), 3268–3295. https://doi.org/10.1002/JSSC.200700261

Tomaz, I., Maslov, L., Stupić, D., Preiner, D., Ašperger, D., & Kontić, J. K. (2015). Recovery of flavonoids from grape skins by enzyme-assisted extraction. *Separation Science and Technology, 51*(2), 255–268. https://doi.org/10.1080/01496395.2015.1085881

Tsimogiannis, D., Samiotaki, M., Panayotou, G., & Oreopoulou, V. (2007). Characterization of flavonoid subgroups and hydroxy substitution by HPLC-MS/MS. *Molecules, 12*(3), 593–606. https://doi.org/10.3390/12030593

Um, M., Han, T. H., & Lee, J. W. (2018). Ultrasound-assisted extraction and antioxidant activity of phenolic and flavonoid compounds and ascorbic acid from rugosa rose (Rosa rugosa Thunb.) fruit. *Food Science and Biotechnology, 27*(2), 375–382. https://doi.org/10.1007/S10068-017-0247-3/FIGURES/3

Vinatoru, M. (2001). An overview of the ultrasonically assisted extraction of bioactive principles from herbs. *Ultrasonics Sonochemistry, 8*(3), 303–313. https://doi.org/10.1016/S1350-4177(01)00071-2

Waksmundzka-Hajnos, M., Oniszczuk, A., Szewczyk, K., & Wianowska, D. (2007). Effect of sample-preparation methods on the HPLC quantitation of some phenolic acids in plant materials. *Acta Chromatographica*, (19), 227–237. https://www.infona.pl//resource/bwmeta1.element.baztech-article-BATA-0003-0021.

Wang, L. F., Kim, D. M., & Lee, C. Y. (2000). Effects of heat processing and storage on flavanols and sensory qualities of green tea beverage. *Journal of Agricultural and Food Chemistry, 48*(9), 4227–4232. https://doi.org/10.1021/JF0003597

Wang, L., & Weller, C. L. (2006). Recent advances in extraction of nutraceuticals from plants. *Trends in Food Science and Technology, 17*(6), 300–312. https://doi.org/10.1016/J.TIFS.2005.12.004

Wang, C., & Zuo, Y. (2011). Ultrasound-assisted hydrolysis and gas chromatography-mass spectrometric determination of phenolic compounds in cranberry products. *Food Chemistry, 128*(2), 562–568. https://doi.org/10.1016/J.FOODCHEM.2011.03.066

Wei, S. D., Zhou, H. C., & Lin, Y. M. (2011). Antioxidant activities of fractions of polymeric procyanidins from stem bark of Acacia confusa. *International Journal of Molecular Sciences, 12*(2), 1146–1160. https://doi.org/10.3390/IJMS12021146

Wei, S. D., Zhou, H. C., Lin, Y. M., Liao, M. M., & Chai, W. M. (2010). MALDI-TOF MS analysis of condensed tannins with potent antioxidant activity from the leaf, stem bark and root bark of Acacia confusa. *Molecules, 15*(6), 4369–4381. https://doi.org/10.3390/MOLECULES15064369

Welch, C. R., Wu, Q., & Simon, J. E. (2008). Recent advances in anthocyanin analysis and characterization. *Current Analytical Chemistry, 4*(2), 75–101. https://doi.org/10.2174/157341108784587795

White, B. L., Howard, L. R., & Prior, R. L. (2010). Release of bound procyanidins from cranberry pomace by alkaline hydrolysis. *Journal of Agricultural and Food Chemistry, 58*(13), 7572–7579. https://doi.org/10.1021/JF100700P

Woźniak, L., Marszalek, K., Skapska, S., & Jedrzejczak, R. (2017). The application of supercritical carbon dioxide and ethanol for the extraction of phenolic compounds from chokeberry pomace. *Applied Sciences, 7*(4), 322. https://doi.org/10.3390/APP7040322

Xi, J., & Shouqin, Z. (2007). Antioxidant activity of ethanolic extracts of propolis by high hydrostatic pressure extraction. *International Journal of Food Science and Technology, 42*(11), 1350–1356. https://doi.org/10.1111/J.1365-2621.2006.01339.X

Xu, B. J., & Chang, S. K. C. (2007). A comparative study on phenolic profiles and antioxidant activities of legumes as affected by extraction solvents. *Journal of Food Science, 72*(2). https://doi.org/10.1111/J.1750-3841.2006.00260.X

Xu, D. P., Li, Y., Meng, X., Zhou, T., Zhou, Y., Zheng, J., Zhang, J. J., & Li, H. B. (2017). Natural antioxidants in foods and medicinal plants: Extraction, assessment and resources. *International Journal of Molecular Sciences, 18*(1), 96. https://doi.org/10.3390/IJMS18010096

Xue, H., Xu, H., Wang, X., Shen, L., Liu, H., Liu, C., Qin, Q., Zheng, X., & Li, Q. (2018). Effects of microwave power on extraction kinetic of anthocyanin from blueberry powder considering absorption of microwave energy. *Journal of Food Quality*. https://doi.org/10.1155/2018/9680184

Yedhu Krishnan, R., Neelesh Chandran, M., Vadivel, V., & Rajan, K. S. (2016). Insights on the influence of microwave irradiation on the extraction of flavonoids from Terminalia chebula. *Separation and Purification Technology, 170*, 224–233. https://doi.org/10.1016/J.SEPPUR.2016.06.039

Yiğit, Ü., Turabi Yolaçaner, E., Hamzalıoğlu, A., & Gökmen, V. (2022). Optimization of microwave-assisted extraction of anthocyanins in red cabbage by response surface methodology. *Journal of Food Processing and Preservation, 46*(1), Article e16120. https://doi.org/10.1111/JFPP.16120

You, Q., Wang, B., Chen, F., Huang, Z., Wang, X., & Luo, P. G. (2011). Comparison of anthocyanins and phenolics in organically and conventionally grown blueberries in selected cultivars. *Food Chemistry, 125*(1), 201–208. https://doi.org/10.1016/J.FOODCHEM.2010.08.063

Yu, S., Hongkun, X., Chenghai, L., Chai, L., Xiaolin, S., & Xianzhe, Z. (2016). Comparison of microwave assisted extraction with hot reflux extraction in acquirement and degradation of anthocyanin from powdered blueberry. *International Journal of Agricultural and Biological Engineering, 9*(6), 186–199. https://doi.org/10.25165/IJABE.V9I6.2724

Žiaková, A., Brandšteterová, E., & Blahová, E. (2003). Matrix solid-phase dispersion for the liquid chromatographic determination of phenolic acids in Melissa officinalis. *Journal of Chromatography. A, 983*(1–2), 271–275. https://doi.org/10.1016/S0021-9673(02)01712-0

Zadernowski, R., Czaplicki, S., & Naczk, M. (2009). Phenolic acid profiles of mangosteen fruits (Garcinia mangostana). *Food Chemistry, 112*(3), 685–689. https://doi.org/10.1016/J.FOODCHEM.2008.06.030

Zarena, A. S., & Sankar, K. U. (2012). Phenolic acids, flavonoid profile and antioxidant activity in mangosteen (Garcinia mangostana l.) pericarp. *Journal of Food Biochemistry, 36*(5), 627–633. https://doi.org/10.1111/J.1745-4514.2011.00575.X

Zha, J., Wu, X., Gong, G., & Koffas, M. A. G. (2019). Pathway enzyme engineering for flavonoid production in recombinant microbes. *Metabolic Engineering Communications, 9*, Article e00104. https://doi.org/10.1016/J.MEC.2019.E00104

Zhang, Q. W., Lin, L. G., & Ye, W. C. (2018). Techniques for extraction and isolation of natural products: A comprehensive review. *Chinese Medicine, 13*(1). https://doi.org/10.1186/S13020-018-0177-X

Zhou, Y., Zhao, X., & Huang, H. (2015). Effects of pulsed electric fields on anthocyanin extraction yield of blueberry processing by-products. *Journal of Food Processing and Preservation, 39*(6), 1898–1904. https://doi.org/10.1111/JFPP.12427

Zielinska, M., & Michalska, A. (2016). Microwave-assisted drying of blueberry (Vaccinium corymbosum L.) fruits: Drying kinetics, polyphenols, anthocyanins, antioxidant capacity, colour and texture. *Food Chemistry, 212*, 671–680. https://doi.org/10.1016/J.FOODCHEM.2016.06.003

Zimmermann, B. F., Walch, S. G., Tinzoh, L. N., Stühlinger, W., & Lachenmeier, D. W. (2011). Rapid UHPLC determination of polyphenols in aqueous infusions of Salvia officinalis L. (sage tea). *Journal of Chromatography, B: Analytical Technologies in the Biomedical and Life Sciences, 879*(24), 2459–2464. https://doi.org/10.1016/J.JCHROMB.2011.06.038

Zitka, O., Sochor, J., Rop, O., Skalickova, S., Sobrova, P., Zehnalek, J., Beklova, M., Krska, B., Adam, V., & Kizek, R. (2011). Comparison of various easy-to-use procedures for extraction of phenols from apricot fruits. *Molecules, 16*(4), 2914–2936. https://doi.org/10.3390/MOLECULES16042914

CHAPTER

5 Advancements in high liquid chromatography for plant polyphenol determination

Beatriz Curros[1], Aurora Cittadini[1], José Manuel Lorenzo[1,2]

[1]*Centro Tecnológico de la Carne de Galicia, Ourense, Spain;* [2]*Área de Tecnoloxía dos Alimentos, Facultade de Ciencias, Universidade de Vigo, Ourense, Spain*

1. Introduction

Plant metabolites are categorized into primary and secondary types. Primary metabolites, such as carbohydrates, proteins, and lipids, are essential for cell growth and development. On the other hand, secondary metabolites are substances present in specific cells that play an indirect but crucial role in metabolic processes. Some of these secondary metabolites have bioactive properties that promote beneficial human health. They are called phytochemicals, and phenolic compounds are among them for their significant biological activities (Swallah et al., 2020).

In chemistry, the term "phenol" is defined as an aromatic organic compound consisting of a benzene ring with a hydroxyl group (-OH) attached directly to one of the carbon atoms of the ring. This hydroxyl group can be replaced by one or more substituent groups, thus giving rise to different phenolic compounds. In this sense, the term "polyphenol" usually refers to structures having at least two phenolic rings, irrespective of the number of hydroxyl groups on each ring (Swallah et al., 2020). This leads to the formation of multiple possible structures, resulting in a wide diversity of compounds with molecular masses ranging from small molecules (<100 Da) to large polymers (>30,000 Da) (Lucci et al., 2017).

Although there is no single classification system for phenolic compounds, one of the most used classifications in the literature is based on the number of phenolic rings they contain and the structural elements that link these rings to other rings. Based on this, five basic classes could be distinguished: phenolic acids, flavonoids, stilbenes, lignans, and tannins (Manach et al., 2004). Naturally occurring phenolic acids contain two distinctive carbon structures: hydroxycinnamic and hydroxybenzoic structures. While the basic skeleton remains the same, variations in the number and position of hydroxyl groups on the aromatic rings create diversity among these acids. The basic structure of flavonoids, planar molecules abundant in plants, consists of 15 carbon atoms in a three-ring configuration (C6–C3–C6). Structural diversity in flavonoids arises mainly from differences in hydroxylation, methoxylation,

Plant Polyphenols in the Food and Cosmetic Industries. https://doi.org/10.1016/B978-0-443-19137-4.00005-3

prenylation, or glycosylation patterns. The phenyl rings of the stilbenes are linked by a two-carbon methylene bridge, making their structure (C6–C2–C6) distinguishable. The basic structure of lignans is formed by the phenylpropane group (C6–C3) and is structurally characterized as dimers formed by the coupling of cinnamyl alcohols. Finally, tannins refer to compounds that contain three or more phenol subunits (Lucas Tenório et al., 2022; Stalikas, 2007; Vicente-Zurdo et al., 2024).

The growing interest of scientists and consumers in polyphenols stems from their recognized biological activity and their widespread presence in the diet. These compounds are recognized for their ability to mitigate oxidative stress by neutralizing dangerous reactive oxygen species and chelating metal ions. In this way, polyphenols may help prevent diseases associated with oxidative stress, including cancer, cardiovascular conditions, and neurodegenerative disorders. This potential is attributed to their interactions with signal transduction pathways and cellular receptors that induce adaptive responses (Belščak-Cvitanović et al., 2018; Chiriac et al., 2021; Rudrapal et al., 2022; Vicente-Zurdo et al., 2024; Zhang & Tsao, 2016).

Accurate analytical technology is essential for the analysis of plant polyphenols, so that their use in the food and cosmetic industries can be encouraged (Sun et al., 2022). However, the identification and quantification of these compounds is challenging due to their complex chemical nature and the difficulty of extracting, separating, and purifying them from plant matrices (Blasco et al., 2005). Therefore, sensitive, selective, and precise analytical methods are essential to accurately identify and quantify these compounds simultaneously in plant samples (Vicente-Zurdo et al., 2024). Thus, among the analytical techniques available for the quantification of plant polyphenols, a distinction can be made between those that allow the quantification of total polyphenols and those that allow the qualitative and quantitative analysis of individual polyphenols (Sun et al., 2022). A classification of the analytical methods detailed in this chapter is shown in Fig. 5.1.

It should be noted that prior to analysis, the extraction method should be selected and optimized (Xu et al., 2017). The correct sample preparation is mainly based on the type of plant extract involved, the characteristics of the compounds to be determined, and the selected analytical technique. Maximizing analyte recovery and minimizing interferences is the primary challenge that requires the use of appropriate extraction and clean-up treatments (Sik et al., 2022).

2. Quantification of total polyphenols

Spectrophotometry remains the most used technique for analyzing total polyphenols, as it is simple and cost-effective. However, it only provides an approximate concentration of total phenols above a certain threshold and lacks the ability to quantify individual compounds (Bolat et al., 2024). Moreover, methods based on this technique are susceptible to interference from other nonphenolic compounds in the sample, which can cause overestimation (Sun et al., 2022). Consequently, spectrophotometry techniques are primarily useful for screening purposes, and

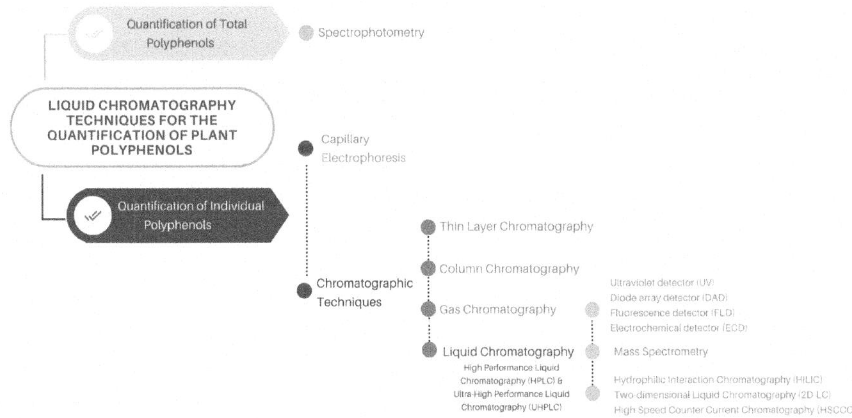

FIGURE 5.1

Liquid chromatograhy techniques for the quantification of plant polyphenols.

quantification is generally reported as total phenolic content (TPC) (Margraf et al., 2015). Therefore, it is essential to recognize that these measurements should be considered only as an "index" (Blasco et al., 2005).

For the spectrophotometric determination of polyphenols, it is necessary to use a chromogenic reagent that reacts with the analytes of interest, causing a colorimetric change, so that the absorbance of the compounds formed can be measured, allowing quantitative analysis (Görög, 2018). One of the most commonly used chromogenic reagents for assessing TPC involves redox reactions (Zhang et al., 2006), such as the Folin–Denis reagent (Folin & Denis, 1912), which operates on the principle that a sample can reduce phosphotungstic–phosphomolybdic acid, resulting in a blue color in its reduced state. This method was later improved by Folin and Ciocalteu (1927) by the addition of lithium salts to the phosphotungstic–phosphomolybdic reagent to prevent the formation of precipitates and turbidity that hindered measurement. Today, this reagent is marketed under the name of Folin–Ciocalteu reagent, although it can also be prepared following the original protocol (Pérez et al., 2023). In this way, spectrophotometric quantification can be carried out under alkaline conditions at 765 nm, the wavelength at which the maximum absorption occurs, using mainly gallic acid as a calibration standard (Echegaray et al., 2021). It should be noted that, although phosphotungstic–phosphomolybdic acid is reduced by the aromatic hydroxyl groups characteristic of plant polyphenolic compounds, the extent to which it is reduced depends on the structural variation of these compounds (especially the number or position of hydroxyl groups). In addition, this reagent can be reduced by many other plant constituents, as well as being affected by the presence of interfering species, such as reducing sugars, organic acids, and so on (Vicente-Zurdo et al., 2024). Therefore, Folin reagent assays cannot accurately quantify differences in the amounts of polyphenols and may be inadequate for comparing phenolic samples from different sources (e.g., species, tissues, etc.) (Appel et al., 2001). Despite these limitations, the spectrophotometric method for quantifying total polyphenols remains a commonly used technique due to its simplicity, cost-effectiveness, and ability to conduct rapid analyses.

An alternative method for analyzing TPC spectrophotometrically is the Prussian Blue method. In this case, the chromogenic reagent is a specific inorganic salt, such as potassium hexacyanoferrate (III) and Fast Blue BB diazonium salt, which reacts with the phenolic compounds in aqueous solution. This reaction produces a blue metallic complex, which can be tracked at 725 nm. This measurement should be carried out within 20 min after mixing all reagents with the sample to ensure color stability (Margraf et al., 2015). Although the Prussian Blue method is faster and more selective than the Folin–Ciocalteu method, it can also be affected by interfering substances. Moreover, the metal complex may precipitate in aqueous conditions, making it necessary to adjust sample dilution before conducting the analysis (Vicente-Zurdo et al., 2024).

It should be noted that some specific polyphenolic subclasses can also be estimated by spectrophotometric methods. These methods are of particular interest for the determination of the very heterogeneous oligomeric or polymeric proanthocyanidins in plant materials, because chromatographic methods are of limited use for their determination (Weber & Passon, 2019). Thus, the vanillin method and the 4-(dimethylamino)-cinnamaldehyde assay are widely used for the quantification of proanthocyanidins (also known as condensed tannins), while for the determination of hydrolyzable tannins, the most widely used methods are based on their reaction with potassium iodate (Bate-Smith assay) and the rhodamine and the sodium nitrate methods (Naczk & Shahidi, 2004). Furthermore, the total flavonoid content can be determined by a sequence of complexation reactions with $AlCl_3$ (Matić et al., 2017).

3. Quantification of individual polyphenols

In contrast to the methods described above for the determination of the total polyphenol content, the determination of these compounds individually requires further pretreatment of the sample and sophisticated equipment for their separation and quantification (Sun et al., 2022). The most commonly used techniques are chromatography and capillary electrophoresis (CE), which can be combined with ultraviolet (UV), fluorometric (FL), mass spectrometry (MS), or electrochemical detection (ECD), with chromatography being the most prevalent technique (Arribas et al., 2012).

Electrophoresis is an analytical separation method that uses an electric field to separate components of a mixture based on their electrophoretic mobility. In this process, only ions are affected by the electric field, while neutral species remain unaffected. CE uses narrow diameter fused silica capillaries so that the migration rate of each compound through the capillary is determined by the strength of the applied electric field. Detection typically occurs near the outlet end of the capillary, where a detector is placed (Chevalier, 2011; Shi et al., 2022). By using this method, it is possible to identify charged low- and medium-molecular-weight compounds with high resolution quickly and efficiently, and with minimal sample and reagent volume requirements (Gatea et al., 2015; Ignat et al., 2011). This technique has several separation modes suitable for a wide range of analytes. The modes mainly used for the separation of polyphenols are capillary zone electrophoresis and micellar electrokinetic chromatography (Arribas et al., 2012). The main drawback of CE in the analysis of phenolic compounds is its low sensitivity as compared to liquid chromatography (LC) (Gatea et al., 2015) and the extended analysis time when using an MS detector due to the large capillary lengths required to couple this type of detector (Pascual-Maté et al., 2018).

On the other hand, chromatographic techniques provide precision and specificity in both the separation and identification of a multitude of compounds in complex matrices. The selection of a particular chromatographic technique depends on the objectives of the study and the analytes of interest. Within the techniques available for the chromatographic determination of polyphenols in plant matrices, gas chromatography (GC) and LC are the most widely used, and even earlier techniques such as thin layer chromatography and column chromatography have been widely used for the identification and purification of phenolic compounds (Ajila et al., 2011).

Currently, LC is the primary method for determining specific polyphenols among the chromatographic techniques (Ajila et al., 2011; Sun et al., 2022). It operates by passing the mobile phase, typically a liquid solvent used to transport the sample, through a column filled with a stationary phase, where each compound interacts in a distinct way. These interactions lead to the separation of the components, depending on their unique physical and chemical characteristics, including size, charge, hydrophobicity, and polarity. A more advanced form of LC is high-performance liquid chromatography (HPLC), which operates under high pressure to increase the flow rate of the mobile phase and improve the separation efficiency. In addition, this technique has also been improved in recent years with the introduction of ultra-high performance liquid chromatography (UHPLC), which has opened new possibilities in the analysis of polyphenols in complex matrices (Sun et al., 2022).

In contrast, GC, employing an inert gaseous mobile phase to carry a sample through a capillary column that contains a stationary phase, presents significant challenges for the quantification of polyphenols (Patil et al., 2024). One of the main concerns results from the low volatility of the phenol compounds due to the ability of the hydroxyl group contained in their structure to form hydrogen bonds, which increases the melting point (Stalikas, 2007). Additionally, given the polar and thermolabile nature of polyphenols, this technique requires a derivatization step (usually through silylation) to mitigate their polarity (Ajila et al., 2011). Moreover, preparation of samples for GC includes the removal of lipids from the extract and liberation of phenolics from ester and glycosidic bonds (Ignat et al., 2011; Naczk & Shahidi, 2004). Despite these drawbacks, GC using a flame ionization detector or coupled to a mass spectrometer has been reported for plant polyphenol quantification, although to a much lesser extent than LC (Bié et al., 2023; Ignat et al., 2011).

4. High-performance liquid chromatography and ultra-high performance liquid chromatography

HPLC offers numerous analytical options for the quantification of plant polyphenols. With respect to the polarity of the mobile and stationary phases, operating in reversed-phase (RP) mode is the most used option. In this mode, a nonpolar stationary phase and a polar mobile phase are employed for the separation of the phenol compounds. Thus, compounds of a more hydrophobic nature will be more retained in the stationary phase (Vicente-Zurdo et al., 2024). The most common stationary phase in RP-HPLC is based on silica particles that have been modified with long hydrocarbon chains, with the octadecyl-bonded C18 column being the most used, and with the shorter alkyl chain C8 (octylsilane) in second place (Žuvela et al., 2019). The column temperature is usually controlled to obtain more reproducible elution times and higher peak resolution, especially when performing long-time analysis (Khoddami et al., 2013; López-fernández et al., 2020).

Regarding the mobile phase, a binary system consisting of an acidified aqueous solution and a polar organic solvent is usually used for the determination of phenolic compounds (Vicente-Zurdo et al., 2024). It should be noted that the pH adjustment of the mobile phase is crucial, as an unstable pH value can lead to the ionization of the phenolic compounds to be analyzed (Bolat et al., 2024), so the pH value should be kept between 2 and 4 (López-fernández et al., 2020). Thus, solvents such as acetonitrile or methanol together with an aqueous solution acidified with acetic acid, phosphoric acid, formic acid, or trifluoracetic acid, to regulate pH, are used as mobile phase to improve chromatographic separation (Chiriac et al., 2021).

Moreover, chromatographic separation is often performed in gradient elution mode because isocratic elution leads to coelution of phenolic compounds (Chiriac et al., 2021; Lucci et al., 2017; Vicente-Zurdo et al., 2024). Thus, in gradient elution mode, the composition of the mobile phase is altered during the chromatographic measurement to facilitate the separation of a mixture of compounds with a wide range of polarities and retention factors (Robards et al., 2004). The gradient program must therefore be managed on a case-by-case basis, depending on the number and chemical properties of the compounds analyzed (Stalikas, 2007).

Conventional HPLC is limited by long analysis time, low resolution, and large solvent consumption (Chiriac et al., 2021). As mentioned above, this technique has been enhanced by the development of UHPLC, which uses shorter columns with smaller particle sizes, resulting in a higher surface area and hence a higher theoretical plate number (Weber & Passon, 2019). In addition, the mobile phase delivery system operates at ultra-high pressures (Vicente-Zurdo et al., 2024). These advancements give UHPLC a greater resolution, higher peak efficiency, faster separation, and less solvent consumption (Lucci et al., 2017; Sun et al., 2022).

Regarding the detectors used in HPLC/UHPLC in the analysis of polyphenols, it should be noted that these compounds present a different degree of absorption in the UV region, given the intrinsic existence of conjugated double and aromatic bonds, so UV detector or diode array detector (DAD) have been used in hundreds of different matrices (Stalikas, 2007; Sun et al., 2022). Both detectors allow the quantification of phenolic compounds by measuring their absorbance at specific wavelengths. The UV detector measures absorbance at a specific wavelength. For routine detection, 280 nm is commonly used because most phenolic compounds show absorption at this wavelength to a greater or lesser extent (Lama-Muñoz & Contreras, 2022). On the contrary, the DAD captures a full spectrum of wavelengths simultaneously, which is recommended for the identification of phenolic compounds, as many of them show several absorption maxima (Carrasco-Pancorbo et al., 2005); for instance, the 270–280 nm range detects benzoic acids, flavonols, and flavanones; 310–330 nm is used for hydroxycinnamic acids and stilbenes; 370–380 nm is optimal for flavones, isoflavones, and flavonols; and 450–550 nm is ideal for anthocyanins (Vicente-Zurdo et al., 2024; Weber & Passon, 2019). On the other hand, while fluorescence detection (FL) provides higher selectivity and sensitivity compared to UV detection, it is limited by the fact that not all phenolic compounds exhibit fluorescence (Carrasco-Pancorbo et al., 2005). However, many of them have fluorescent moieties in their structures, allowing enhanced detection through careful selection of excitation and emission wavelengths (Lama-Muñoz & Contreras, 2022). For instance, flavonols can be detected at an excitation and emission wavelength of 280 and 320 nm, respectively, with minimal interference from other phenolic families. For hydroxycinnamic acids, excitation wavelength (λex) 325 nm and emission wavelength (λem) 375 nm are effective, while λex 375 nm and λem 430 nm are suitable for detecting flavonols, flavones, and isoflavones (Vidal-Casanella et al., 2021). In addition, polyphenols are considered electroactive compounds, so ECD could be used, usually with a coulometric array detector (CoulArray). Nevertheless, these types of detectors generally operate by applying a single detection potential, the equivalent of using a single wavelength in UV detectors, and their combination with gradient elution is often complicated. Furthermore, it can show low responses due to contamination of the electrode. On the contrary, ECD methods may be less affected by interferences due to the sample matrix in some cases (Arribas et al., 2012; Razem et al., 2022).

The main challenge when using the above-mentioned detectors is to achieve low limits of detection and quantification to be sensitive enough to determine polyphenols at trace levels. In addition, determinations with this type of detectors are based solely on retention data with an appropriate standard and on characteristic spectra for analytical identification, when possible. Thus, these detectors may suffer from limitations such as a lack of accuracy due to interferences in complex samples or in the determination of compounds with similar structures. In these cases, MS coupled with LC provides greater sensitivity, precision, and resolution (Sun et al., 2022).

5. Liquid chromatography coupled to mass spectrometry (LC-MS) and liquid chromatography-tandem mass spectrometry (LC-MS/MS)

Advancements in functional component analysis technology, especially LC-MS, have allowed for rapid, high-throughput quantitative analysis with increased sensitivity and precision. In addition, this technology solves the problems presented by the determination of plant polyphenols by traditional spectrophotometry and chromatography (Sun et al., 2022).

Thus, in LC-MS, the sample is first separated by LC, where the components are separated based on their chemical and physical properties. Once the compounds are eluted from the column, they are transformed into the gas phase and introduced into the mass spectrometer, where they are ionized and analyzed according to their mass-to-charge ratio (m/z), so that the mass spectrometer can analyze the ions and any fragment ions produced in the ionization process. The unique fragmentation pattern of each compound, observed in the mass spectrum, serves as a distinctive signature that facilitates accurate identification and structural analysis of molecules. Therefore, LC-MS provides mass information and helps to elucidate the molecular structure, which facilitates the identification of phenolic compounds (Patil et al., 2024; Shi et al., 2022).

Moreover, a second mass spectrometer can be coupled through tandem mass spectrometry (MS/MS). In this way, after ionization and selection of the ion of interest in the mass spectrometer, it is selectively fragmented in a secondary mass analyzer. This allows a more specific and sensitive identification of the individual components in the sample, as it provides information on the structure and composition of the fragment ions. Thus, LC-MS focuses on the identification of components based on their molecular mass, whereas LC-MS/MS provides additional information on the structure and composition of these components by sequential fragmentation of selected ions, which is a major advance in the determination of plant polyphenols, as the structural diversity of these compounds hinders their accurate qualitative and quantitative analysis (Von Bargen et al., 2012).

There are multiple configurations for these couplings, including various ionization sources, which are used to ionize the atoms and molecules before they are separated in the mass analyzer and detected by their mass-to-charge ratio. One of the most common ionization technologies for the determination of polyphenols in complex samples is the electrospray ionization source, which operates at atmospheric pressure and is perfectly compatible with HPLC. However, this determination can suffer from interferences by coeluting compounds, leading to the ionization suppression or enhancement in the ESI source. For this reason, at the sample preparation stage, prior extraction and purification steps are often required, such as solid phase extraction, quenchers, or the use of adsorbent materials, such as primary secondary amines, which are weak anion exchange agents that exhibit good adsorption of polar compounds, including polyphenols. Another option is to dilute the extracts to weaken the matrix effect (Sun et al., 2022). On the other hand, although phenolic compounds can be ionized in both polarities, in the negative ionization mode, the sensitivity of detection is high, but the compounds undergo deprotonation and require higher collision energy for fragmentation (Vicente-Zurdo et al., 2024). It is important to note that because of the positive charge in the anthocyanin structure, this polyphenol subclass is typically studied in a positive ionization mode (Weber & Passon, 2019).

Other atmospheric pressure ionization sources, such as atmospheric pressure chemical ionization or atmospheric pressure photoionization, have also been used. These sources, although less common in polyphenol analysis, are particularly suitable for those compounds that exhibit low ionization efficiencies under ESI conditions (Lucci et al., 2017; Vidal-Casanella et al., 2021). Matrix-assisted laser desorption ionization is another ionization source available for the characterization of polyphenols, especially dealing with polymeric compounds such as proanthocyanidins (Monagas et al., 2010; Pinto et al., 2021).

With respect to mass analyzers, these can be simple or hybrid, such as single quadrupole, triple quadrupole (QqQ), ion trap (IT), or high-resolution mass analyzers like orbital electrostatic IT (Orbitrap mass analyzer), Fourier transform ion cyclotron resonance (FT-ICR), and time-of-flight (TOF). These analyzers differ in the physical principles on which they are based. Thus, ions can be filtered by alternating electrical fields (such as quadrupoles, IT, FT-ICR, or Orbitrap instruments) or separated in a field-free region after acceleration through a fixed accelerating potential (TOF instruments) (Weber & Passon, 2019).

The choice of mass analyzer depends on the type of experiment to be performed. Regarding simple or hybrid mass analyzers, the use of single quadrupoles for the determination of polyphenols can provide an overview of the compounds present in a sample when full scan MS acquisition is performed. However, these instruments are less sensitive than QqQ analyzers, which are highly recommended for quantitative analysis of polyphenols via LC-MS (Lucci et al., 2017). This system consists of two scanning quadrupoles, Q1 and Q3, positioned on both sides of the collision cell (Q2). This common fragmentation mechanism is known as collision-induced dissociation (Dunn, 2011). The instrument offers four monitoring modes: parent ion scan, product ion scan, neutral loss scan, and the highly sensitive selective reaction monitoring (SRM) or multiple reaction monitoring (MRM) mode. SRM or MRM mode operates in a similar way and is primarily used for quantification, while the other modes are employed for analyte identification. In SRM/MRM mode, by adjusting the collision energy for two or more selective ion pairs of each compound, qualitative analysis can be performed, and the most sensitive ion pair is then used for quantitative analysis (Sun et al., 2022).

On the other hand, IT analyzers are commonly used for structural elucidation of target analytes, as they can perform full scan MS and product ion scan MS acquisition modes. These capabilities allow the acquisition of spectra of product ions after multiple analysis steps (MS^n), which are valuable for identifying fragmentation patterns and deducing the analyte's structure. Recently, QTrap mass analyzers, hybrid instruments combining a quadrupole and a linear IT in a similar configuration to QqQ instruments, have grown in popularity, with several studies reporting their use in polyphenol determination (Lucci et al., 2017; Weber & Passon, 2019).

Thus, although simple or hybrid mass analyzers are a recognized technology, the selectivity provided by their unit resolution and the low resolution achievable with QqQ and IT instruments sometimes make it difficult to differentiate between isomeric compounds and to unambiguously assign fragment composition. For this reason, they have been overtaken in recent years by the development of high-resolution mass spectrometry (HRMS), such as FT-ICR MS, TOF MS, or Orbitrap-MS, which have replaced traditional MS (Sun et al., 2022).

6. Liquid chromatography coupled to high-resolution mass spectrometry

HRMS offers many significant advantages for the determination of polyphenols in plant matrices. It achieves high mass resolution and thus high accuracy of mass measurement through multistage scanning. This makes it possible to distinguish even small differences between two compounds, which increases the possibility of unambiguously determining the elemental composition of both known and new compounds in complex matrices. It also allows differentiation between target analytes and other coeluting isobaric compounds, while minimizing sample preparation and reducing the necessity for extensive chromatographic separation (Sun et al., 2022; Vicente-Zurdo et al., 2024).

Currently, TOF and Orbitrap analyzers are the most widely used high-resolution mass analyzers combined with LC for plant polyphenol analysis, while FT-ICR MS is too slow and complex to handle. It should be mentioned that the high mass measurement accuracy of HRMS equipment depends on the stability and accuracy of mass calibration. Unlike TOF systems, which require frequent calibrations, the Orbitrap mass analyzer offers stable calibration for several days. Calibration of the Orbitrap usually uses standard mix solutions suited to the specific mass range, provided by the instrument supplier. In contrast, TOF systems often rely on lock mass correction for higher mass accuracy. This involves the continuous infusion of a user-selected reference compound, such as a well-characterized polyphenol, to correct experimental m/z values against the reference (Lucci et al., 2017).

On the other hand, hybrid mass spectrometers, created by combining quadrupole and IT analyzers such as Q-TOF (quadrupole-time-of-flight), IT-TOF (ion trap-time-of-flight), and Q-Orbitrap (quadrupole-orbitrap), offer an optimal solution by uniting diverse performance characteristics, such as mass resolving power, analysis speed, and a wide dynamic range of mass accuracy, into a single instrument (Lucci et al., 2017; Vidal-Casanella et al., 2021). Compared to Q-TOF, Orbitrap-based systems offer greater sensitivity and more accurate mass measurements in full scan mode, which is crucial for identifying and screening unknown compounds. Thus, Q-Orbitrap technology integrates the high-resolution capabilities of the Orbitrap mass analyzer with the quadrupole's selectivity, making it particularly suitable for analyzing polyphenols in complex matrices (Sun et al., 2022). Among Orbitrap hybrid instruments, those incorporating linear IT technology, like the LTQ-Orbitrap, are also widely used in this field. These instruments enable the screening, identification, and structural characterization of unknown polyphenolic compounds by using exact mass measurements to determine optimal elemental compositions along with precise mass analysis of MS^n product ions in the data-dependent scan (Lucci et al., 2017).

The identification and characterization of plant polyphenols often requires the combined use of HRMS, which provides accurate mass measurements and isotopic pattern analysis, together with MS/MS data, which provide chemical structural information.

7. Other liquid chromatographic techniques

Beyond the techniques discussed previously, alternative liquid chromatographic methods have emerged, expanding the potential for determining polyphenols in plant matrices. Thus, an alternative stationary phase based on hydrophilic interaction chromatography has been developed for polyphenol profiling and is expected to gain popularity due to its good compatibility with MS (Khoddami et al., 2013; Weber & Passon, 2019). This technique uses traditional polar stationary phases in conjunction with mobile phases used in RP, so it is particularly suitable for the determination of the more polar compounds in complex matrices that are weakly retained on RP columns (Pyrzynska & Sentkowska, 2019; Vidal-Casanella et al., 2021) and for oligomeric species (Mir-Cerdà et al., 2023).

Two-dimensional liquid chromatography (2D-LC or LCxLC) is an advanced chromatographic technique that enhances the separation and identification of structurally similar and minor compounds from complex matrices by coupling two distinct liquid chromatographic methods in sequence. Thus, the first dimension separates the sampling using one set of conditions, while a different set is applied in the second dimension. The effluent from the first column is collected in fractions and subsequently injected into the second column for further separation, offering higher peak capacity and resolution by utilizing orthogonal separation mechanisms (Pyrzynska & Sentkowska, 2019; Vidal-Casanella et al., 2021).

High-speed counter-current chromatography (HSCCC) uses a biphasic liquid system for component separation. A centrifugal field enables a liquid stationary phase within an open tube with phase density differences. A notable advantage of

HSCCC is its dual-mode capability: the phases can switch roles during the process, allowing the mobile phase to become stationary and vice versa, ensuring no residual injected material remains in the machine (Ignat et al., 2011).

It is worth noting that beyond their use in the identification and quantification of phenolic compounds, HSCCC and HPLC techniques can also serve as preparative methods. That is, these techniques offer the possibility of isolating and recovering individual phenolic compounds from different extracts after passing through the stationary phase and a nondestructive detector. This allows obtaining reference standards of the different compounds, which is an important tool given their limited availability (Lucas Tenório et al., 2022; Motilva et al., 2013; Weber & Passon, 2019).

8. General data analysis workflow for polyphenol quantification by LC-MS

As mentioned throughout this chapter, LC techniques have been widely used to identify and quantify polyphenols in different matrices, especially in recent years, when coupled to MS. Studies performed with this technique are generally divided into two types. On the one hand, there are targeted analyses that employ available reference standards to identify the phenolic compounds. This type of analysis typically uses sensitive instruments equipped with QqQ in MRM mode. On the other hand, there are nontargeted analyses that are based on acquiring spectrometric data on all compounds that can be ionized and are sufficiently abundant. In this approach, instruments with lower sensitivity but high resolution are generally used, such as Q-Orbitrap or Q-TOF mass analyzers, leading to higher accuracy of the measured m/z values, which is critical to aid identification. Furthermore, with recent advances in MS, the sensitivity of HRMS is improving significantly. Thus, the chemical characteristics of the compounds under study, such as monoisotopic mass, can be compared with those of compounds present in databases. However, given the wide variability of the phenolic compounds, many of them with similar monoisotopic masses, it is essential to have access to data that provides structure-oriented information, which is achieved by coupling a second mass spectrometer through MS/MS. In this regard, a fragmentation of all ions can be helpful, but structural elucidation is significantly improved by having access to selected product ion spectra of precursors (Aceña et al., 2015; Kaufmann, 2014; Oesterle et al., 2023). In addition, the use of full-scan acquisition in combination with data-dependent scanning modes, where two types of scans are recorded simultaneously (e.g., a full-scan followed by an MS/MS ion scan for specific detected signals), offers the key advantage of being able to perform retrospective analysis. This approach allows the identification of unexpected compounds, making use of accurate mass and spectral databases, without having to reanalyze samples (Mir-Cerdà et al., 2023). Although this approach may be sufficient in some cases, full identification of compounds ultimately requires additional assays using candidate standards. Positive identification can be confirmed through complete concordance of chromatographic and spectral data (Lucci et al., 2017).

It should be kept in mind that quantitative analysis requires the use of a calibration curve for each analyte because ionization can vary depending on the molecular structure. Thus, in studies where commercial reference standards are available, the calibration curve is prepared from them. For example, Mizzi et al. (2020) designed a method to identify and quantify, by HPLC-DAD, 15 phenolic compounds common in food matrices but that have similar chemical characteristics and therefore have similar retention times, describing how they solve the problem of peak overlap. On the contrary, when commercial standards are not available, these compounds can be tentatively quantified with respect to other phenolic compounds that are representative of the chemical classes being analyzed (Aceña et al., 2015; Arena et al., 2019; Motilva et al., 2013). Likewise, individual compounds can be isolated and purified for use as reference standards through preparative methods, as indicated above.

The results obtained in these studies produce a large number of data sets that need to be processed, mainly in the case of nontargeted analysis. In recent years, chemometric methods such as principal component analysis and partial least squares-discriminant analysis have been used to obtain relevant information from the results. Chemometric methods are also valuable in the early stages of method development, aiding in the optimization of sample preparation and chromatographic separations (Lucci et al., 2017; Weber & Passon, 2019).

9. Applications of polyphenol quantification by LC techniques in plants

The identification of phenolic compounds naturally present in plant foods allows a better understanding of the different health benefits provided by the consumption of these foods and their derivatives and even allows for assessing their possible application in the pharmaceutical industry (Deschamps et al., 2023; Rathahao-Paris et al., 2016). Numerous studies have been carried out in this regard using LC techniques, such as the qualitative and quantitative characterization of polyphenolic compounds in asparagus using DAD and MS detection (Karaaslan Ayhan & Rosenberg, 2021), in tomato using HPLC-DAD (Xiao et al., 2024), and in *Plantago lanceolata* using LC-MS/MS, as reported by Bahadori et al. (2020).

Furthermore, some studies have focused on the use of metabolomic approaches providing detailed insights into the polyphenol composition of the examined matrix and its bioavailability, bioaccessibility, and interactions with other molecules, supporting the evaluation of its activity and nutraceutical value (Bié et al., 2023; Bozza et al., 2024; Vicente-Zurdo et al., 2024). For example, in the study conducted by

Lucas-González et al. (2023), how different plant-based food matrices affect the bioaccessibility of polyphenols was investigated.

On the other hand, phenolic profile also serves as an important indicator of fruit quality due to its impact on taste, color, and nutritional value. In this sense, quantifying polyphenolic compounds is a standard approach for selecting genotypes, assessing maturity stages, and optimizing storage and processing conditions to produce fresh and processed foods with strong potential for free radical protection (Sánchez-Rangel et al., 2013). For instance, in a study by Castrejón et al. (2008), the TPC and primary phenolic compounds in four raspberry cultivars were analyzed at five distinct ripening stages using HPLC-DAD, suggesting that phenolic profiles are dynamic throughout fruit development and that genetic factors play a crucial role. In a similar manner, Cevallos-Casals et al. (2006), after studying 14 red-fleshed plum and eight peach genotypes, concluded that selecting crops with high phenolic content could enhance functional properties. On the other hand, Deng et al. (2018) investigated how storage conditions affect the phenolic content and antioxidant activity of litchi pericarp. Their findings indicate that storing litchi pericarp at 4 °C is more effective in preserving phenolic compounds compared to room temperature storage.

In addition, understanding variations in polyphenolic profiles can aid in distinguishing geographical origin and ensuring the authenticity of natural products, which can have a substantial economic impact. This process can be addressed by analyzing the data from the point of view of the concentration of polyphenols of interest (profiling approach), as in the work of Arena et al. (2019), who used a 2D-LC method to determine the polyphenolic fraction of *Pistacia vera* extracts of different geographical origin; and in the study of Zhu et al. (2020), who found a significant correlation between the polyphenol profile of *Moringa oleifera* and the sampling area; or the research of Puigventós et al. (2015), who proposed a method for the quantification of 26 polyphenols by LC-ESI-MS/MS in 23 grape-based and cranberry-based natural products and pharmaceutical preparations. Their analysis revealed variations in the concentrations of certain polyphenols, suggesting that phenolic profiles could serve as effective descriptors for authenticating fruit-based products. Alternatively, the authentication process can also be approached from the analysis of so-called fingerprints, that is, the data being analyzed contains mixtures of known and unknown compounds, which may not need to be individually identified or quantified, with the aim of detecting patterns among different samples and identifying significant differences between them (fingerprinting approach) (Castro-Puyana & Herrero, 2013; Lucci et al., 2017). Thus, da Silva Lima et al. (2023) carried out a work to reveal the phenolic fingerprint of eight different genotypes of whole wheat flours by using a nontargeted metabolomic analysis by UHPLC-QTOF-MS. Through multivariate statistical analysis, they successfully differentiated low- and medium-quality flours from high-quality ones.

In recent years, the application of natural antioxidants such as polyphenols has expanded significantly, especially in the food, pharmaceutical, and cosmetic industries. This growth is primarily attributed to their capacity to mitigate oxidative stress damage without the potential risks associated with synthetic antioxidants. While plants remain primary sources, there is a growing interest in identifying alternative sources, particularly among plant-derived by-products from the food industry, following sustainability principles, contributing to waste valorization, reducing environmental impact, and supporting a circular economy (Bhardwaj et al., 2022; Lourenço et al., 2019; Nurzyńska-Wierdak, 2023; Shahidi et al., 2019). Emerging studies emphasize the high polyphenol content in these materials, representing sustainable and cost-effective sources. Thus, for example, Orozco-Flores et al. (2024) identified the polyphenolic profile of apple pomace from the Golden Delicious variety of *Malus domestica*, using HPLC (with UV-Vis standards and spectra) and HPLC-DAD/ESI-MS, concluding that this by-product can be considered a suitable source of phenolic compounds. In a similar manner, Felice et al. (2020) conducted a study focusing on the extraction of phenolic compounds from cocoa by-products, especially beans and husks. They assessed the antioxidant activity, TPC, and polyphenolic composition of these extracts using HPLC-ESI-MS/MS. The findings suggest that cocoa by-products could serve as a valuable ingredient in the food and pharmaceutical industries. It has also been investigated the extraction of polyphenols from thyme (*Thymus serpyllum*) by-products (Jovanović et al., 2022). The research evaluated total polyphenol and flavonoid contents, identified individual polyphenol compounds via HPLC-DAD, and assessed the extract's antioxidant, antimicrobial, and spasmolytic activities. According to the results obtained, extracts from thyme by-products and their encapsulated forms have potential applications as ingredients in food, pharmaceutical, and cosmetic products.

10. Conclusions

Recent advances in chromatographic techniques have significantly enhanced the analysis of plant polyphenols. HPLC, or the improved UHPLC, combined with UV detector or DAD, has proven effective for quantifying high concentrations of individual polyphenols in plants. For low concentrations, LC-MS has become a powerful tool for both qualitative and quantitative analysis, ensuring high sensitivity and selectivity. In particular, QqQ mass analyzers operating in MRM mode have been widely used. In recent years, HRMS techniques, such as Q-TOF and Q-Orbitrap, have gained prominence in plant polyphenol analysis. These technologies offer rapid acquisition rates, excellent full-scan sensitivity, and high mass resolution. In addition, coupling a second mass spectrometer through MS/MS provides structural information by enabling sequential fragmentation of selected ions, facilitating data interpretation based on exact mass and MS/MS spectra.

The ongoing advancement of technology toward increased acquisition speed, resolving power, mass accuracy, and sensitivity will lead to larger data sets and higher-quality results. Data mining, machine learning, the use of predictive models, and other computational approaches will be helpful in the analysis of these results, undoubtedly enabling new applications in polyphenol quantification.

References

Aceña, J., Stampachiacchiere, S., Pérez, S., & Barceló, D. (2015). Advances in liquid chromatography–high-resolution mass spectrometry for quantitative and qualitative environmental analysis. *Analytical and Bioanalytical Chemistry, 407*(21), 6289–6299. https://doi.org/10.1007/s00216-015-8852-6

Ajila, C. M., Brar, S. K., Verma, M., Tyagi, R. D., Godbout, S., & Valéro, J. R. (2011). Extraction and analysis of polyphenols: Recent trends. *Critical Reviews in Biotechnology, 31*(3), 227–249. https://doi.org/10.3109/07388551.2010.513677

Appel, H. M., Govenor, H. L., D'ascenzo, M., Siska, E., & Schultz, J. C. (2001). Limitations of Folin assays of foliar phenolics in ecological studies. *Journal of Chemical Ecology, 27* (4), 761–778. https://doi.org/10.1023/A:1010306103643

Arena, K., Cacciola, F., Mangraviti, D., Zoccali, M., Rigano, F., Marino, N., Dugo, P., & Mondello, L. (2019). Determination of the polyphenolic fraction of Pistacia vera L. kernel extracts by comprehensive two-dimensional liquid chromatography coupled to mass spectrometry detection. *Analytical and Bioanalytical Chemistry, 411*(19), 4819–4829. https://doi.org/10.1007/s00216-019-01649-w

Arribas, A. S., Martínez-Fernández, M., & Chicharro, M. (2012). The role of electroanalytical techniques in analysis of polyphenols in wine. *TrAC, Trends in Analytical Chemistry, 34*, 78–96. https://doi.org/10.1016/j.trac.2011.10.015

Bahadori, M. B., Sarikurkcu, C., Kocak, M. S., Calapoglu, M., Uren, M. C., & Ceylan, O. (2020). Plantago lanceolata as a source of health-beneficial phytochemicals: Phenolics profile and antioxidant capacity. *Food Bioscience, 34*. https://doi.org/10.1016/j.fbio.2020.100536

Belščak-Cvitanović, A., Durgo, K., Huđek, A., Bačun-Družina, V., & Komes, D. (2018). *Overview of polyphenols and their properties polyphenols: Properties, recovery, and applications* (pp. 3–44). Croatia: Elsevier. https://doi.org/10.1016/B978-0-12-813572-3.00001-4

Bhardwaj, K., Najda, A., Sharma, R., Nurzyńska-Wierdak, R., Dhanjal, D. S., Sharma, R., Manickam, S., Kabra, A., Kuča, K., & Bhardwaj, P. (2022). Fruit and vegetable peel-enriched functional foods: Potential avenues and health perspectives. *Evidence-Based Complementary and Alternative Medicine, 2022*. https://doi.org/10.1155/2022/8543881

Bié, J., Sepodes, B., Fernandes, P. C. B., & Ribeiro, M. H. L. (2023). Polyphenols in health and disease: Gut microbiota, bioaccessibility, and bioavailability. *Compounds, 3*(1), 40–72. https://doi.org/10.3390/compounds3010005

Blasco, A. J., Rogerio, M. C., González, M. C., & Escarpa, A. (2005). "Electrochemical index" as a screening method to determine "total polyphenolics" in foods: A proposal. *Analytica Chimica Acta, 539*(1–2), 237–244. https://doi.org/10.1016/j.aca.2005.02.056

Bolat, E., Sarıtas, S., Duman, H., Eker, F., Akdasçi, E., Karav, S., & Witkowska, A. M. (2024). Polyphenols: Secondary metabolites with a biological impression. *Nutrients, 16*(15). https://doi.org/10.3390/nu16152550

Bozza, D., Barboni, D., Spadafora, N. D., Felletti, S., De Luca, C., Nosengo, C., Compagnin, G., Cavazzini, A., & Catani, M. (2024). Untargeted metabolomics approaches for the characterization of cereals and their derived products by means of liquid chromatography coupled to high resolution mass spectrometry. *Journal of Chromatography Open, 6*. https://doi.org/10.1016/j.jcoa.2024.100168

Carrasco-Pancorbo, A., Cerretani, L., Bendini, A., Segura-Carretero, A., Gallina-Toschi, T., & Fernández-Gutiérrez, A. (2005). Analytical determination of polyphenols in olive oils. *Journal of Separation Science, 28*(9–10), 837–858. https://doi.org/10.1002/jssc.200500032

Castrejón, A. D. R., Eichholz, I., Rohn, S., Kroh, L. W., & Huyskens-Keil, S. (2008). Phenolic profile and antioxidant activity of highbush blueberry (Vaccinium corymbosum L.) during fruit maturation and ripening. *Food Chemistry, 109*(3), 564–572. https://doi.org/10.1016/j.foodchem.2008.01.007

Castro-Puyana, M., & Herrero, M. (2013). Metabolomics approaches based on mass spectrometry for food safety, quality and traceability. *TrAC, Trends in Analytical Chemistry, 52*, 74–87. https://doi.org/10.1016/j.trac.2013.05.016

Cevallos-Casals, B. A., Byrne, D., Okie, W. R., & Cisneros-Zevallos, L. (2006). Selecting new peach and plum genotypes rich in phenolic compounds and enhanced functional properties. *Food Chemistry, 96*(2), 273–280. https://doi.org/10.1016/j.foodchem.2005.02.032

Chevalier, F. (2011). Analytical methods: Electrophoresis. In *Encyclopedia of Dairy Sciences* (2nd ed., pp. 185–192). Academic Press.

Chiriac, E. R., Chitescu, C. L., Geană, E. I., Gird, C. E., Socoteanu, R. P., & Boscencu, R. (2021). Advanced analytical approaches for the analysis of polyphenols in plants matrices—A review. *Separations, 8*(5). https://doi.org/10.3390/separations8050065

da Silva Lima, L. R., Barros Santos, M. C., dos Santos D′Almeida, C. T., Cameron, L. C., Gutkoski, L. C., & Ferreira, M. S. L. (2023). Omics data reveals the phenolic fingerprint of Brazilian whole wheat flours of different technological qualities. *Journal of Food Science and Technology, 60*(2), 783–796. https://doi.org/10.1007/s13197-023-05665-8

Deng, M., Deng, Y., Dong, L., Ma, Y., Liu, L., Huang, F., Wei, Z., Zhang, Y., Zhang, M., & Zhang, R. (2018). Effect of storage conditions on phenolic profiles and antioxidant activity of litchi pericarp. *Molecules, 23*(9), 2276. https://doi.org/10.3390/molecules23092276

Deschamps, E., Calabrese, V., Schmitz, I., Hubert-Roux, M., Castagnos, D., & Afonso, C. (2023). Advances in ultra-high-resolution mass spectrometry for pharmaceutical analysis. *Molecules, 28*(5). https://doi.org/10.3390/molecules28052061

Dunn, W. B. (2011). Mass spectrometry in systems biology: An introduction. *Methods in Enzymology, 500*, 15–35. https://doi.org/10.1016/B978-0-12-385118-5.00002-5

Echegaray, N., Pateiro, M., Munekata, P. E. S., Lorenzo, J. M., Chabani, Z., Farag, M. A., & Domínguez, R. (2021). Measurement of antioxidant capacity of meat and meat products: Methods and applications. *Molecules, 26*(13). https://doi.org/10.3390/molecules26133880

Felice, F., Fabiano, A., De Leo, M., Piras, A. M., Beconcini, D., Cesare, M. M., Braca, A., Zambito, Y., & Di Stefano, R. (2020). Antioxidant effect of cocoa by-product and cherry polyphenol extracts: A comparative study. *Antioxidants, 9*(2). https://doi.org/10.3390/antiox9020132

Folin, O., & Ciocalteu, V. (1927). On tyrosine and tryptophane determinations in proteins. *Journal of Biological Chemistry, 73*(2), 627–650. https://doi.org/10.1016/s0021-9258(18)84277-6

Folin, O., & Denis, W. (1912). A new colorimetric method for the determination of vanillin in flavoring extracts. *Journal of Industrial & Engineering Chemistry, 4*(9), 670–672. https://doi.org/10.1021/ie50045a015

Görög, S. (2018). *Ultraviolet-visible spectrophotometry in pharmaceutical analysis* (pp. 1–391). Hungary: CRC Press. https://doi.org/10.1201/9781351077422

Gatea, F., Teodor, E. D., Matei, A. O., Badea, G. I., & Radu, G. L. (2015). Capillary electrophoresis method for 20 polyphenols separation in propolis and plant extracts. *Food Analytical Methods, 8*(5), 1197–1206. https://doi.org/10.1007/s12161-014-0006-5

Ignat, I., Volf, I., & Popa, V. I. (2011). A critical review of methods for characterisation of polyphenolic compounds in fruits and vegetables. *Food Chemistry, 126*(4), 1821–1835. https://doi.org/10.1016/j.foodchem.2010.12.026

Jovanović, A. A., Vajić, U. J. V., Mijin, D. Z., Zdunić, G. M., Šavikin, K. P., Branković, S., Kitić, D., & Bugarski, B. M. (2022). Polyphenol extraction in microwave reactor using by-product of Thymus serpyllum L. and biological potential of the extract. *Journal of Applied Research on Medicinal and Aromatic Plants, 31*. https://doi.org/10.1016/j.jarmap.2022.100417

Karaaslan Ayhan, N., & Rosenberg, E. (2021). Development of comprehensive liquid chromatography with diode array and mass spectrometric detection for the characterization of (poly-)phenolic and flavonoid compounds and application to asparagus. *Food Chemistry, 354*, Article 129518. https://doi.org/10.1016/j.foodchem.2021.129518

Kaufmann, A. (2014). Combining UHPLC and high-resolution MS: A viable approach for the analysis of complex samples? *TrAC, Trends in Analytical Chemistry, 63*, 113–128. https://doi.org/10.1016/j.trac.2014.06.025

Khoddami, A., Wilkes, M. A., & Roberts, T. H. (2013). Techniques for analysis of plant phenolic compounds. *Molecules, 18*(2), 2328–2375. https://doi.org/10.3390/molecules18022328

López-fernández, O., Domínguez, R., Pateiro, M., Munekata, P. E. S., Rocchetti, G., & Lorenzo, J. M. (2020). Determination of polyphenols using liquid chromatography–tandem mass spectrometry technique (LC–MS/MS): A review. *Antioxidants, 9*(6). https://doi.org/10.3390/antiox9060479

Lama-Muñoz, A., & Contreras, M.d. M. (2022). Extraction systems and analytical techniques for food phenolic compounds: A review. *Foods, 11*(22). https://doi.org/10.3390/foods11223671

Lourenço, S. C., Moldão-Martins, M., & Alves, V. D. (2019). Antioxidants of natural plant origins: From sources to food industry applications. *Molecules, 24*(22). https://doi.org/10.3390/molecules24224132

Lucas Tenório, C. J., Assunção Ferreira, M. R., & Lira Soares, L. A. (2022). Recent advances on preparative LC approaches for polyphenol separation and purification: Their sources and main activities. *Trends in Food Science and Technology, 128*, 129–146. https://doi.org/10.1016/j.tifs.2022.08.004

Lucas-González, R., Díez-Riquelme, V., Viuda-Martos, M., Pérez-Álvarez, J.Á., Sánchez-Zapata, E., & Fernández-López, J. (2023). Effect of the food matrix on the (poly)phenol stability of different plant-based meat products and their main ingredients after in vitro gastrointestinal digestion. *Food & Function, 14*(24), 10796–10813. https://doi.org/10.1039/d3fo03265e

Lucci, P., Saurina, J., & Núñez, O. (2017). Trends in LC-MS and LC-HRMS analysis and characterization of polyphenols in food. *TrAC, Trends in Analytical Chemistry, 88*, 1–24. https://doi.org/10.1016/j.trac.2016.12.006

Manach, C., Scalbert, A., Morand, C., Rémésy, C., & Jiménez, L. (2004). Polyphenols: Food sources and bioavailability. *The American Journal of Clinical Nutrition, 79*(5), 727–747. https://doi.org/10.1093/ajcn/79.5.727

Margraf, T., Karnopp, A. R., Rosso, N. D., & Granato, D. (2015). Comparison between Folin-Ciocalteu and Prussian Blue assays to estimate the total phenolic content of juices and teas using 96-well microplates. *Journal of Food Science, 80*(11), Article C2397. https://doi.org/10.1111/1750-3841.13077

Matić, P., Sabljić, M., & Jakobek, L. (2017). Validation of spectrophotometric methods for the determination of total polyphenol and total flavonoid content. *Journal of AOAC International, 100*(6), 1795–1803. https://doi.org/10.5740/jaoacint.17-0066

Mir-Cerdà, A., Nuñez, O., Granados, M., Sentellas, S., & Saurina, J. (2023). An overview of the extraction and characterization of bioactive phenolic compounds from agri-food waste within the framework of circular bioeconomy. *TrAC, Trends in Analytical Chemistry, 161*, Article 116994. https://doi.org/10.1016/j.trac.2023.116994

Mizzi, L., Chatzitzika, C., Gatt, R., & Valdramidis, V. (2020). HPLC analysis of phenolic compounds and flavonoids with overlapping peaks. *Food Technology and Biotechnology, 58*(1), 12–19. https://doi.org/10.17113/ftb.58.01.20.6395

Monagas, M., Quintanilla-López, J. E., Gómez-Cordovés, C., Bartolomé, B., & Lebrón-Aguilar, R. (2010). MALDI-TOF MS analysis of plant proanthocyanidins. *Journal of Pharmaceutical and Biomedical Analysis, 51*(2), 358–372. https://doi.org/10.1016/j.jpba.2009.03.035

Motilva, M. J., Serra, A., & Macià, A. (2013). Analysis of food polyphenols by ultra high-performance liquid chromatography coupled to mass spectrometry: An overview. *Journal of Chromatography A, 1292*, 66–82. https://doi.org/10.1016/j.chroma.2013.01.012

Naczk, M., & Shahidi, F. (2004). Extraction and analysis of phenolics in food. *Journal of Chromatography A, 1054*(1–2), 95–111. https://doi.org/10.1016/j.chroma.2004.08.059

Nurzyńska-Wierdak, R. (2023). Phenolic compounds from new natural sources—Plant genotype and ontogenetic variation. *Molecules, 28*(4). https://doi.org/10.3390/molecules28041731

Oesterle, I., Pristner, M., Berger, S., Wang, M., Verri Hernandes, V., Rompel, A., & Warth, B. (2023). Exposomic biomonitoring of polyphenols by non-targeted analysis and suspect screening. *Analytical Chemistry, 95*(28), 10686–10694. https://doi.org/10.1021/acs.analchem.3c01393

Orozco-Flores, L. A., Salas, E., Rocha-Gutiérrez, B., Peralta-Pérez, M. D. R., González-Sánchez, G., & Ballinas-Casarrubias, L. (2024). Determination of polyphenolic profile of apple pomace (Malus domestica Golden Delicious Variety) by HPLC-MS. *ACS Omega, 9*(1), 196–203. https://doi.org/10.1021/acsomega.3c03218

Pérez, M., Dominguez-López, I., & Lamuela-Raventós, R. M. (2023). The Chemistry behind the Folin-Ciocalteu method for the estimation of (poly)phenol content in food: Total phenolic intake in a Mediterranean dietary pattern. *Journal of Agricultural and Food Chemistry, 71*(46), 17543–17553. https://doi.org/10.1021/acs.jafc.3c04022

Pascual-Maté, A., Osés, S. M., Fernández-Muiño, M. A., & Sancho, M. T. (2018). Analysis of polyphenols in honey: Extraction, separation and quantification procedures. *Separation and Purification Reviews, 47*(2), 142–158. https://doi.org/10.1080/15422119.2017.1354025

Patil, N. D., Bains, A., Sridhar, K., Sharma, M., Dhull, S. B., Goksen, G., Chawla, P., & Inbaraj, B. S. (2024). Recent advances in the analytical methods for quantitative determination of antioxidants in food matrices. *Food Chemistry, 463*. https://doi.org/10.1016/j.foodchem.2024.141348

Pinto, G., De Pascale, S., Aponte, M., Scaloni, A., Addeo, F., & Caira, S. (2021). Polyphenol profiling of chestnut pericarp, integument and curing water extracts to qualify these food by-products as a source of antioxidants. *Molecules, 26*(8), 2335. https://doi.org/10.3390/molecules26082335

Puigventós, L., Navarro, M., Alechaga, É., Núñez, O., Saurina, J., Hernández-Cassou, S., & Puignou, L. (2015). Determination of polyphenolic profiles by liquid chromatography-electrospray-tandem mass spectrometry for the authentication of fruit extracts. *Analytical and Bioanalytical Chemistry, 407*(2), 597–608. https://doi.org/10.1007/s00216-014-8298-2

Pyrzynska, K., & Sentkowska, A. (2019). *Chromatographic analysis of polyphenols* (pp. 353–364). Elsevier BV. https://doi.org/10.1016/b978-0-12-813768-0.00021-9

Rathahao-Paris, E., Alves, S., Junot, C., & Tabet, J. C. (2016). High resolution mass spectrometry for structural identification of metabolites in metabolomics. *Metabolomics, 12*(1), 1–15. https://doi.org/10.1007/s11306-015-0882-8

Razem, M., Ding, Y., Morozova, K., Mazzetto, F., & Scampicchio, M. (2022). Analysis of phenolic compounds in food by coulometric array detector: A review. *Sensors, 22*(19). https://doi.org/10.3390/s22197498

Robards, K., Haddad, P. R., & Jackson, P. E. (2004). High-performance liquid chromatography—Instrumentation and techniques. In *Principles and practice of modern chromatographic methods* (pp. 227–303). Elsevier. https://doi.org/10.1016/b978-0-08-057178-2.50008-x

Rudrapal, M., Khairnar, S. J., Khan, J., Dukhyil, A. B., Ansari, M. A., Alomary, M. N., Alshabrmi, F. M., Palai, S., Deb, P. K., & Devi, R. (2022). Dietary polyphenols and their role in oxidative stress-induced human diseases: Insights into protective effects, antioxidant potentials and mechanism(s) of action. *Frontiers in Pharmacology, 13*. https://doi.org/10.3389/fphar.2022.806470

Sánchez-Rangel, J. C., Benavides, J., Heredia, J. B., Cisneros-Zevallos, L., & Jacobo-Velázquez, D. A. (2013). The Folin-Ciocalteu assay revisited: Improvement of its specificity for total phenolic content determination. *Analytical Methods, 5*(21), 5990–5999. https://doi.org/10.1039/c3ay41125g

Shahidi, F., Vamadevan, V., Oh, W. Y., & Peng, H. (2019). Phenolic compounds in agri-food by-products, their bioavailability and health effects. *Journal of Food Bioactives*, 57–119. https://doi.org/10.31665/jfb.2019.5178

Shi, L., Zhao, W., Yang, Z., Subbiah, V., & Suleria, H. A. R. (2022). Extraction and characterization of phenolic compounds and their potential antioxidant activities. *Environmental Science and Pollution Research, 29*(54), 81112–81129. https://doi.org/10.1007/s11356-022-23337-6

Sik, B., Székelyhidi, R., Lakatos, E., Kapcsándi, V., & Ajtony, Z. (2022). Analytical procedures for determination of phenolics active herbal ingredients in fortified functional foods: An overview. *European Food Research and Technology, 248*(2), 329–344. https://doi.org/10.1007/s00217-021-03908-6

Stalikas, C. D. (2007). Extraction, separation, and detection methods for phenolic acids and flavonoids. *Journal of Separation Science, 30*(18), 3268–3295. https://doi.org/10.1002/jssc.200700261

Sun, M. F., Jiang, C. L., Kong, Y. S., Luo, J. L., Yin, P., & Guo, G. Y. (2022). Recent advances in analytical methods for determination of polyphenols in tea: A comprehensive review. *Foods, 11*(10). https://doi.org/10.3390/foods11101425

Swallah, M. S., Sun, H., Affoh, R., Fu, H., & Yu, H. (2020). Antioxidant potential overviews of secondary metabolites (polyphenols) in fruits. *International Journal of Food Science, 2020*. https://doi.org/10.1155/2020/9081686

Vicente-Zurdo, D., Gómez-Mejía, E., Rosales-Conrado, N., & León-González, M. E. (2024). A comprehensive analytical review of polyphenols: Evaluating neuroprotection in

Alzheimer's disease. *International Journal of Molecular Sciences, 25*(11). https://doi.org/10.3390/ijms25115906

Vidal-Casanella, O., Núñez, O., Granados, M., Saurina, J., & Sentellas, S. (2021). Analytical methods for exploring nutraceuticals based on phenolic acids and polyphenols. *Applied Sciences, 11*(18). https://doi.org/10.3390/app11188276

Von Bargen, C., Hübner, F., Cramer, B., Rzeppa, S., & Humpf, H. U. (2012). Systematic approach for structure elucidation of polyphenolic compounds using a bottom-up approach combining ion trap experiments and accurate mass measurements. *Journal of Agricultural and Food Chemistry, 60*(45), 11274–11282. https://doi.org/10.1021/jf3030369

Weber, F., & Passon, M. (2019). *Characterization and quantification of polyphenols in fruits* (pp. 111 121). Elsevier BV. https://doi.org/10.1016/b978-0-12-813768-0.00007-4

Xiao, X., Liu, F., Sun, M., Tang, Z., Wu, Y., Lyu, J., Khan, K. S., & Yu, J. (2024). Development of a high-performance liquid chromatography method for simultaneous quantification of sixteen polyphenols and application to tomato. *Journal of Chromatography A, 1733*. https://doi.org/10.1016/j.chroma.2024.465254

Xu, C. C., Wang, B., Pu, Y. Q., Tao, J. S., & Zhang, T. (2017). Advances in extraction and analysis of phenolic compounds from plant materials. *Chinese Journal of Natural Medicines, 15*(10), 721–731. https://doi.org/10.1016/S1875-5364(17)30103-6

Žuvela, P., Skoczylas, M., Jay Liu, J., Baczek, T., Kaliszan, R., Wong, M. W., & Buszewski, B. (2019). Column characterization and selection systems in reversed-phase high-performance liquid chromatography. *Chemical Reviews, 119*(6), 3674–3729.

Zhang, H., & Tsao, R. (2016). Dietary polyphenols, oxidative stress and antioxidant and anti-inflammatory effects. *Current Opinion in Food Science, 8*, 33–42. https://doi.org/10.1016/j.cofs.2016.02.002

Zhang, Q., Zhang, J., Shen, J., Silva, A., Dennis, D. A., & Barrow, C. J. (2006). A simple 96-well microplate method for estimation of total polyphenol content in seaweeds. *Journal of Applied Phycology, 18*(3–5), 445–450. https://doi.org/10.1007/s10811-006-9048-4

Zhu, Y., Yin, Q., & Yang, Y. (2020). Comprehensive investigation of Moringa oleifera from different regions by simultaneous determination of 11 polyphenols using UPLC-ESI-MS/MS. *Molecules, 25*(3). https://doi.org/10.3390/molecules25030676

SECTION

Food applications 2

CHAPTER

6 Valorization of polyphenols from citrus by-products for nutraceutical applications

Amine Elbouzidi[1], Mohamed Taibi[2], Mounir Haddou[2], Mohamed Addi[1]

[1]*Laboratoire d'Amélioration des Productions Agricoles, Biotechnologie et Ecophysiologie (LAPABE), Faculty of Sciences, Mohammed Premier University, Oujda, Morocco;* [2]*Laboratory of Bioresources, Biotechnology, and Sustainable Development (L2BD), Faculty of Sciences, Mohammed Premier University, Oujda, Morocco*

1. Introduction

The Citrus genus, which is a prominent global fruit crop, encompasses various species, including sweet oranges (*Citrus sinensis* L.), lemons (*Citrus lemon* L.), pomelos (*Citrus paradisi* Macfad.), and tangerines (*Citrus reticulata* L.), to name a few, each with multiple varieties (Addi et al., 2021). These fruits are known for their nutritional richness, containing vitamins A, C, and E, minerals, dietary fibers, and secondary metabolites such as polyphenols and terpenoids (González-Molina et al., 2010; Sebghatollahi et al., 2022; Zou et al., 2016). The primary phenolic compounds found in citrus fruits are flavonoids and phenolic acids, with the fruit's peel being particularly abundant in antioxidant substances (Kumar et al., 2019; Zibaee et al., 2020). Citrus flavonoids exhibit a wide range of health benefits, including anti-inflammatory, anticancer, antibacterial, and cardiovascular protective properties.

Citrus fruits are a vital component of the Mediterranean diet, encompassing a diverse array of fruits like lemons, oranges, grapefruits, and pomelos, which are widely consumed worldwide (Naureen et al., 2022). From a nutritional standpoint, these fruits are celebrated for their high vitamin C content and the associated antioxidant properties, providing natural protection against oxidative stress and promoting overall health (Liu, 2013; Ma et al., 2020; Zacarías García et al., 2021). However, the significance of citrus fruits extends beyond their vitamin C content. The peel of citrus fruits, which accounts for a substantial portion of the fruit's total weight, is a rich source of bioactive compounds, especially natural antioxidants like flavonoids. Research has uncovered that citrus flavonoids possess not only antioxidant properties but also therapeutic, antiviral, antifungal, and antibacterial effects (Alam et al., 2022; Yi et al., 2017; Zibaee et al., 2020).

The richness of phenolic compounds in citrus peels has attracted the interest of the scientific community, leading to investigations on their potential applications. These compounds are being explored for their role as natural antioxidants, with implications in the food industry as natural preservatives. Additionally, they find

Plant Polyphenols in the Food and Cosmetic Industries. https://doi.org/10.1016/B978-0-443-19137-4.01006-1

applications in the pharmaceutical and cosmetic industries, contributing to the development of health and skincare products (Ramful et al., 2010). This demonstrates the versatility and potential of citrus fruits beyond their refreshing flavors and nutritional value.

Citrus fruits stand as a cornerstone in global fruit cultivation, boasting their significance as one of the most important fruit crops worldwide (Liu et al., 2012). They flourish in a wide range of geographical locations, being cultivated in over 100 countries across tropical, subtropical, and Mediterranean climates (Khan & Kender, 2007). The global impact of citrus fruits is evident in their ranking as the second most produced fruit category on a global scale, yielding over 115 million tons annually. The majority of citrus fruits, including oranges, mandarins, and grapefruits, are typically consumed fresh, accounting for about 75% of their usage. The remaining quarter of citrus fruits undergo processing in the food industry, resulting in the production of citrus juices, citrus-based beverages, jams, marmalades, and dehydrated citrus products (Panwar, Panesar, & Chopra, 2021; Panwar, Saini, et al., 2021). This processing generates significant amounts of waste.

2. Valorization of polyphenols: concept and significance

2.1 Definition and concept of valorization

Food waste is a major global challenge, with approximately one-third of all produced food being lost or wasted throughout the supply chain (Ishangulyyev et al., 2019). This not only leads to wasted resources and economic burdens but also raises environmental concerns related to land use, water consumption, and greenhouse gas emissions. To address this urgent issue, the concept of "valorization" has emerged as a crucial strategy for maximizing food resource utilization and minimizing waste (Socas-Rodríguez et al., 2021). The main objective of valorization is to convert waste into more valuable and useful products, such as materials, chemicals, fuels, and alternative energy sources (Nayak & Bhushan, 2019). This approach aligns with the principles of industrial ecology, which promotes sustainable industrial practices (Xu et al., 2019).

In industrial ecology, the aim is to imitate natural ecosystems, where waste from one organism becomes a resource for another (Korhonen & Snäkin, 2005). Within the realm of food science and technology, valorization involves transforming readily available or underutilized resources into valuable products that have improved functionality and economic viability. This process may include various techniques, such as extraction, purification, modification, and utilization of components found in these resources, contributing to a circular bioeconomy.

2.2 Importance of valorization in reducing food waste

In the field of sustainable agriculture and food processing, there is growing interest in utilizing polyphenols from citrus by-products. Citrus fruits are consumed globally and provide important nutrients and health benefits. However, they also generate significant waste in the form of pith residue, peels, and seeds (Sharma et al., 2019). These waste materials, which are often discarded or underutilized, contain high concentrations of bioactive compounds, particularly polyphenols (Maqbool et al., 2023; Rafiq et al., 2018). These compounds are receiving considerable attention due to their antioxidant, antimicrobial, and functional properties. By harnessing these properties, there are numerous opportunities to enhance the value and reduce the environmental impact of citrus waste (Han et al., 2007).

Traditionally, the focus has been on the primary components of food, such as carbohydrates, proteins, and fats, while overlooking the potential of secondary metabolites like polyphenols. However, there has been a paradigm shift toward the valorization of polyphenols, driven by a growing scientific understanding of their health benefits and diverse functionalities. Unlocking the full potential of polyphenols and utilizing them in various applications is now a primary goal. Polyphenols, which belong to a diverse group of phytochemicals, contribute to the nutritional value and sensory quality of citrus fruits (Mahato et al., 2018). They possess antioxidant activity, which can protect against oxidative stress and prevent chronic diseases like cardiovascular disease and cancer. Some of the most abundant and extensively studied polyphenols found in citrus peels include hesperidin, gallic acid, quercetin, naringenin, and ferulic acid (Al-Ashaal & El-Sheltawy, 2011; Genovese et al., 2014; Ghasemi et al., 2009). These compounds have demonstrated potential health benefits, such as antiinflammatory, anticancer, antidiabetic, and antihypertensive effects. Additionally, they can modulate the gut microbiota and enhance the immune system. Therefore, extracting and utilizing these compounds can contribute to the health and wellness of consumers, while also creating value-added products from citrus waste (Singh et al., 2020).

Furthermore, polyphenols possess functional properties that can be utilized in various food applications. They can enhance the texture, color, and flavor of food products, as well as serve as natural preservatives, thus extending the shelf life of products (Cheynier, 2012). For instance, quercetin and naringenin have been investigated for their potential use in food preservation and as natural disinfectants due to their antimicrobial activity against foodborne pathogens and spoilage microorganisms (Ng et al., 2019; Wang et al., 2016). Ferulic acid has been studied for its potential application in wound healing and as a natural food preservative, owing to its antioxidant and antiinflammatory properties (Bumrungpert et al., 2018).

The valorization of polyphenols from citrus by-products exemplifies innovative approaches in sustainable agriculture, aimed at reducing food waste. This process not only contributes to environmental conservation but also enhances the nutritional and economic value of food products. As the global population continues to grow, the importance of sustainable practices becomes paramount in ensuring food security and effective waste management. Integrating these valuable compounds into the food industry could revolutionize our perception and utilization of what was once considered waste, leading to a more sustainable and resilient food system. The future of food processing and waste management looks promising, with polyphenols playing a crucial role in this transformative journey.

2.3 Sustainability aspects of polyphenol valorization

The sustainable bioeconomy could be significantly improved by valorizing polyphenols derived from food by-products, particularly citrus waste. This transformative approach is crucial for promoting resource conservation, environmental management, and sustainable agriculture. By focusing on underutilized resources like polyphenols, this strategy reduces reliance on virgin resources, minimizing the need for additional crop cultivation or encroachment on intact natural reserves. The valorization of polyphenols exemplifies the principles of a circular economy, enhancing resource efficiency and simultaneously reducing waste accumulation (Zema et al., 2018). Moreover, this strategic reuse of food by-products, notably citrus waste, not only contributes to waste reduction but also plays a key role in mitigating environmental impact, reducing greenhouse gas emissions, and decreasing the ecological footprint associated with waste management. It represents a fundamental shift toward a more sustainable and ecologically balanced exploitation of terrestrial resources. This transition has the potential to improve the nutritional and functional characteristics of food products, offering innovative solutions to global challenges such as food waste and environmental impact (Nieto et al., 2021). Embracing these sustainable practices is crucial for ensuring food security and effective waste management as the world population continues to grow. The promising future of food processing and waste management lies in the valorization of polyphenols, shaping a resilient and sustainable food system.

3. Extraction methods and characterization of polyphenols

Chapter 4 explores traditional and innovative techniques for extracting polyphenols from citrus fruits. Traditional methods like solid–liquid extraction, notably Soxhlet extraction, require careful control of parameters like solvent choice and temperature. However, they face criticism for time and solvent consumption. Innovative methods like pressurized liquid extraction (PLE) and ultrasound-assisted extraction (UAE) offer sustainability and efficiency benefits. Comparison among techniques, including liquid–solid extraction (LSE) and enzyme-assisted extraction (EAE), highlights differences in solvent usage, scalability, and environmental impact (Fig. 6.1). Analytical techniques like Folin–Ciocalteu assay and high-performance liquid chromatography (HPLC) aid in polyphenol characterization, providing insights into composition and concentration, advancing understanding of their role in food and health.

3.1 Traditional extraction techniques

The most commonly used method for extracting polyphenolic compounds from fruit waste is solid–liquid extraction (Oliveira et al., 2016). The choice of solvent and extraction conditions, such as the temperature of extraction, solvent concentration, and the ratio of solvent to solid material, are crucial for maximizing the yield of

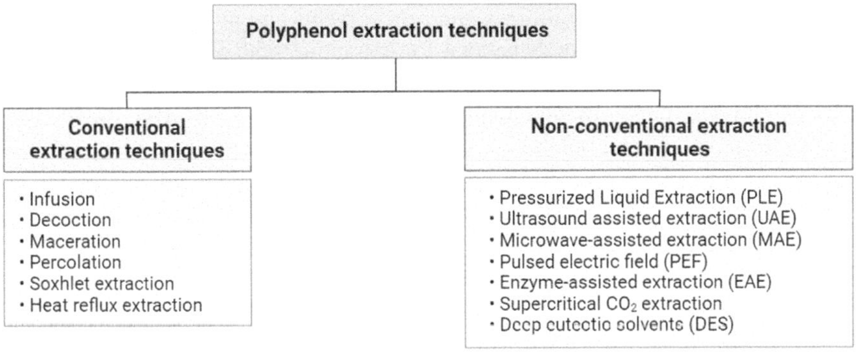

FIGURE 6.1

Various extraction techniques to isolate polyphenolic compounds from food sources, spanning from traditional (conventional techniques) to innovative methods (nonconventional techniques).

polyphenols (Assefa & Keum, 2017). The process begins by grinding the citrus fruits or their waste parts to increase the surface area. Solid–liquid extraction includes methods such as dipping extraction and alcohol extraction (Cai et al., 2021).

Soxhlet extraction is a conventional technique used to extract polyphenolic compounds from citrus fruit (Anoopkumar et al., 2023). This method relies on solvent reflux and siphoning principles for continuous extraction of target analytes from plant tissue (Sander, 2017). The process starts with preparing the sample, which is typically dried and finely ground to increase its surface area. The material is then placed in a Soxhlet extractor. A solvent, such as ethanol or methanol, is chosen for its effective ability to dissolve polyphenols. The solvent is heated, causing it to vaporize. The resulting vapor condenses and percolates through the sample, dissolving the polyphenols. The enriched solvent is then drained back into the distillation flask, and this extraction cycle is repeated several times. After extraction, the solvent is evaporated, often under reduced pressure to preserve the integrity of the polyphenols, resulting in a concentrated extract. Despite being widely used, Soxhlet extraction has been criticized for its lengthy duration and significant solvent consumption (Raynie, 2019). Consequently, researchers have explored alternative methods that offer greater environmental sustainability.

3.2 Innovative extraction methods for polyphenols from citrus fruits

A variety of methods are employed to extract polyphenols from citrus fruits (see Fig. 6.1). These methods aim to improve yield, reduce extraction times, and minimize solvent use, contributing to more efficient and environmentally friendly extraction processes. The choice of method depends on factors such as the type of citrus fruit, the specific polyphenols of interest, and the desired application of the extracted compounds.

PLE is an alternative that utilizes nontoxic solvents and moderate temperatures (Zgórka, 2009), which can be beneficial for extracting both volatile and phenolic compounds from citrus by-products such as orange peel. This technique utilizes solvents at elevated pressures and temperatures to penetrate the fruit matrix and dissolve the targeted compounds (Cai et al., 2021). The high pressure keeps the solvent in a liquid state above its boiling point, improving extraction efficiency by reducing viscosity and surface tension for better penetration into the sample matrix. At the same time, the elevated temperature speeds up solvent diffusion into the solid matrix and desorption of analytes. PLE surpasses traditional methods by extracting polyphenols more quickly and with less solvent, making it environmentally friendly and cost-effective. The process involves multiple cycles of solvent flushing through the sample to ensure thorough extraction. Once extraction is complete, the solvent is separated and can be evaporated, leaving behind concentrated polyphenols. PLE offers advantages such as speed, efficiency, and the ability to preserve the integrity of thermolabile polyphenols due to its short extraction time (Dhua et al., 2022). This method finds application in extracting polyphenols from citrus fruits, as demonstrated by its success in extracting phenolic compounds from orange peel (Barrales et al., 2018). PLE has demonstrated versatility in extracting valuable polyphenols from various citrus fruit by-products, including *C. reticulata* "Chachi" peels, where 70% methanol at 160°C and a pressure of 10.34 MPa proved optimal for extracting major flavones (Feuereisen et al., 2017). These studies highlight the effectiveness of PLE, offering a promising alternative to conventional extraction methods for obtaining polyphenols from citrus fruit by-products.

Deep eutectic solvents (DES) have emerged as highly efficient and environmentally friendly media for extracting polyphenols from citrus fruits. These solvents are composed of two or more compounds with a lower melting point than that of the individual components, allowing them to remain in liquid form at room temperature (Zhou et al., 2023). This characteristic makes them advantageous for extraction purposes. The extraction process involves mixing citrus fruit material with DES and stirring to facilitate the dissolution of polyphenols into the solvent. DES offers several benefits, including lower toxicity, biodegradability, and the ability to dissolve a wide range of polar and nonpolar compounds (Kalogiouri et al., 2022). Furthermore, DES can be tailored to target specific types of polyphenols by adjusting their composition (Oliveira et al., 2022). This optimization of the extraction process ensures maximum yield and purity. Choline chloride-based DES, when paired with glycerol or ethylene glycol, has proven to be particularly effective, showcasing high extraction efficiency and supporting sustainability by utilizing food waste biomass and reducing reliance on conventional organic solvents (Heidari et al., 2020). After extraction, the polyphenol-rich DES phase can be separated from the fruit matrix and further processed to isolate the desired compounds (Sridhar et al., 2021).

Supercritical CO_2 extraction is an efficient technique used to extract polyphenols from citrus fruits (Raveendran et al., 2005). This method utilizes carbon dioxide at supercritical conditions, where it exhibits both liquid and gas properties, acting as a solvent (Paes et al., 2014). The process involves placing citrus fruit material into an extraction vessel and introducing supercritical CO_2. The CO_2 penetrates the fruit matrix and dissolves polyphenols due to its high solvation power. One advantage of this method is the ability to adjust CO_2's solvating properties through temperature and pressure, allowing for selective compound extraction. After the extraction, depressurization causes the CO_2 to return to its gaseous state, releasing the extracted polyphenols. The recycled gas leaves behind a pure extract without any solvent residues. The environmentally friendly nature of CO_2, being nontoxic, nonflammable, and readily available, contributes to the appeal of this method. Supercritical CO_2 extraction is especially beneficial for extracting heat-sensitive polyphenols, as it operates at lower temperatures compared to traditional methods and avoids the use of toxic organic solvents (Paes et al., 2014). This ensures safer extracts for consumption and reduces the environmental impact. In the case of citrus fruits, research has shown the potential for optimizing supercritical CO_2 extraction to enhance polyphenol yield and purity. Particularly, the addition of ethanol as an entrainer has been found to be effective. A study demonstrated that using 95% ethanol as an entrainer significantly improved extraction efficiency, increasing polyphenol yield by 2.45 times and purity by 15.49% compared to extractions without an entrainer. The optimal conditions for this process include a solid/liquid ratio of 1:2.6 (g/mL), an extraction pressure of 37 MPa, and a temperature of 51°C (Palma & Taylor, 1999). Supercritical CO_2 extraction, often enhanced by cosolvents like ethanol, proves to be a promising and environmentally friendly alternative for extracting polyphenols from citrus fruits and other plant materials (Garcia-Salas et al., 2010).

The extraction of polyphenols from citrus fruits using UAE is a technique that has been explored in various studies to optimize the recovery of these bioactive compounds (Khan et al., 2010). UAE is a process that utilizes ultrasonic waves to create cavitation bubbles in a solvent (Bucur et al., 2023). When these bubbles collapse, they enhance the penetration of the solvent into the plant tissues, thereby improving the extraction of desired compounds. Numerous studies have demonstrated that UAE significantly increases the yield of polyphenols compared to conventional extraction methods. For example, when UAE was used on orange peels, it enhanced the extraction of polyphenols by 62.5% compared to the conventional solid–liquid extraction method (Carrera et al., 2012). Similarly, when phenolic compounds were extracted from *Satsuma mandarin* peels using UAE, higher yields of phenolic acids and flavanone glycosides were obtained compared to the maceration extraction method (Ma et al., 2008). To maximize extraction efficiency, it is crucial to optimize the UAE parameters such as temperature, ultrasonic power, and extraction time. For instance, the optimal conditions for extracting flavanone glycosides from orange peel were determined to be a temperature of 40°C, a sonication power of 150W, and a 4:1 ethanol:water ratio (Boukroufa et al., 2015). Furthermore, the use of UAE has been associated with a reduction in extraction time and energy consumption (Pradal et al., 2016).

Microwave-assisted extraction (MAE) is a technique widely used to extract polyphenolic compounds from various plants, including citrus fruits. This method involves using microwave radiation to heat up the solvent and the plant material, which helps release the bioactive compounds, including polyphenols (Murugesan et al., 2024). In the field of citrus applications, MAE has proven to be effective in extracting polyphenols from the peels of *C. sinensis*, *C. reticulata* (mandarin), and the by-products of *C. paradisi* Macfad. (Boukroufa et al., 2015; Ciğeroğlu et al., 2021; Nayak et al., 2015). The optimization of MAE parameters is crucial to maximizing the yield of polyphenols. Factors such as microwave power, extraction time, solvent-to-solid ratio, and solvent composition significantly influence the extraction efficiency. For instance, optimal conditions for extracting polyphenols from *C. sinensis* peels involved using a solution of 51% acetone in water, a microwave intensity of 500 W, an extraction time of 122 s, and a solvent-to-material ratio of 25 mL/g (Boukroufa et al., 2015). Similarly, for *C. reticulata* peels, the optimal conditions included a microwave intensity of 152 W, an extraction time of 49 s, a liquid-to-solid ratio of 16, and a methanol concentration of 66% (Nayak et al., 2015). These conditions have been found to enhance the total phenolic content (TPC) and total antioxidant activity (TAA) of the extracts, surpassing those obtained using conventional, ultrasound-assisted, and accelerated solvent extraction techniques (Boukroufa et al., 2015). In addition to improving polyphenol yields, MAE also offers the advantages of reduced time and energy consumption (Li et al., 2012).

Pulsed electric field (PEF) technology is increasingly recognized as an effective nonthermal method for improving the extraction of polyphenols from various botanical sources, particularly citrus fruits (El Kantar et al., 2018). The PEF extraction process involves applying short, high-voltage electric fields to botanical tissues, which causes the cell membranes to become permeable and allows the release of polyphenols from within the cells (El Darra et al., 2013). Studies indicate that PEF treatment significantly increases the extraction of juice and polyphenolic compounds from citrus fruits. For instance, applying PEF at electric field intensities of 3 kV/cm and 10 kV/cm has been found to increase the juice yield from oranges, pomelos, and lemons by 25%, 37%, and 59%, respectively (El Kantar et al., 2018). Additionally, using higher electric field strengths on orange peels has been associated with a higher extraction of polyphenols, reaching a maximum of 22 mg GAE/g Dry Matter (DM) (El Kantar et al., 2018). Similarly, varying the electric field strengths (1, 3, 5, and 7 kV/cm) during PEF treatment of orange peel has resulted in increased recovery rates of total polyphenols and flavonoids, with the most significant cellular disintegration observed at a treatment duration of 60 μS (El Darra et al., 2013). The optimization of PEF parameters, including electric field intensity, exposure duration, and pulse frequency, is crucial for maximizing extraction efficiency. For example, treating lemon peel residues with PEF at an intensity of 7 kV/cm has been shown to improve polyphenol extraction efficiency by 300% (Peiró et al., 2019). The advantages of PEF technology go beyond increased extraction yields and include the preservation of antioxidant properties and the avoidance of organic solvents, making it an environmentally friendly alternative to traditional extraction methods (Agcam et al., 2014; Barbosa-Pereira et al., 2018).

EAE is an innovative method that is gaining attention for its efficient extraction of polyphenols from various plant materials, including citrus fruits. This technique utilizes specific enzymes to break down plant cell walls, enhancing the release of phenolic compounds (Ricarte et al., 2020). The extraction of polyphenols from citrus fruits using EAE has proven to be effective due to the enzymes' ability to target and degrade complex polysaccharides in cell walls that bind phenolic compounds (Cai et al., 2021). Various enzymes, such as pectinase, cellulase, tannase, and β-glucosidase, have been investigated for their potential to improve the release and bioconversion of phenolics from citrus juice by-products (Ruviaro et al., 2019). These enzymes facilitate the extraction of phenolics and promote their hydrolysis from sugar residues, resulting in an enhanced phenolic profile and higher antioxidant activity. Specifically, β-glucosidase has been identified as optimal for producing hesperetin and naringenin under certain conditions, demonstrating the specificity and efficiency of EAE (Ruviaro et al., 2019). The utilization of cellulase, pectinase, and hemicellulase has facilitated an enhancement in both the yield and quality of polyphenolic extracts, while concurrently reducing the extraction time and temperature. For instance, employing cellulase in the extraction of polyphenols from grapefruit resulted in a yield of 6.5%, as opposed to 3.8% obtained through Soxhlet extraction. Furthermore, EAE has proven efficacious in preserving the antioxidant and antiinflammatory properties inherent in polyphenolic extracts derived from citrus fruits (Cai et al., 2021). Given these features, EAE can serve as a pretreatment technology in combination with other extraction methods or as a primary extraction method using multiple enzymes to extract the desired compound.

3.3 Comparison of extraction methods

Extraction techniques differ in their underlying principles, operational conditions, and extraction efficiencies. Table 6.1 provides a comparative analysis of the advantages and disadvantages of several prominent extraction methods, including LSE, PLE, DES, supercritical CO_2 extraction, UAE, MAE, PEF, and EAE. The selection of an appropriate extraction technique is crucial in ensuring efficient and effective isolation of bioactive compounds from citrus fruit by-products. Each method offers distinct advantages and limitations, necessitating careful consideration of factors such as the nature of the target compounds, sample characteristics, environmental impact, cost-effectiveness, and scalability.

3.4 Characterization of polyphenols: analytical techniques for polyphenol characterization

The Folin–Ciocalteu assay is a widely used spectrophotometric method for the quantitative determination of TPC in samples containing polyphenols. The principle behind this method involves the reduction of the Folin–Ciocalteu reagent, a mixture of phosphomolybdic and phosphotungstic acids, by polyphenols under alkaline conditions, leading to the formation of a blue complex that can be measured spectrophotometrically (Ainsworth & Gillespie, 2007). While the Folin–Ciocalteu assay provides a rapid and simple means of estimating the TPC, it has some limitations.

Table 6.1 Comparison of polyphenol extraction methods.

Method	Advantages	Disadvantages	References
Liquid–solid extraction (LSE)	- Simplicity and accessibility, - Ability to extract various phenolic compounds, - Suitable for limited resource settings.	- High solvent consumption, - Long extraction times, - Low extraction yields, - Exposure risk to organic vapors, - Potential for degradation of target compounds.	(Cai et al., 2021; Jahromi, 2019)
Pressurized liquid extraction (PLE)	- Efficient extraction of polyphenolic compounds, - Reduced solvent usage, - Shorter extraction times.	- Costly advanced instrumentation required. - Risk of degrading heat-sensitive polyphenols.	Sentkowska et al. (2024)
Deep eutectic solvents (DES)	- Low toxicity, - Biodegradability. - Ability to dissolve a wide range of compounds, - High extraction efficiency and quality.	- Complex preparation of DES, - Solvents may need to be tailored, - Recovery of polyphenols can be challenging, - Scalability issues for industrial applications.	(Dheyab et al., 2021; Zhou et al., 2023)
Supercritical CO_2	- Nontoxicity, - Nonflammability, - Environmental friendliness, - Operates at low temperatures, - Preserving bioactivity, - High selectivity, efficiency, and high-purity extracts.	- High cost of equipment, - Complexity of the process, - Challenge of optimizing extraction parameters.	Sridhar et al. (2021)
Ultrasound-assisted extraction (UAE)	- Higher extraction efficiency and quality, - More efficient than conventional methods, - Allows for the extraction of a wide range of phenolic compounds.	- Complexity of the procedure, - Need for specialized equipment.	Jahromi (2019)

Table 6.1 Comparison of polyphenol extraction methods.—*cont'd*

Method	Advantages	Disadvantages	References
Microwave-assisted extraction (MAE)	- Significant reduction in extraction time, - Improved yield of polyphenols, - Suitable for thermolabile substances.	- High capital cost and maintenance, - Potential risk of degradation of target compounds.	Sparr Eskilsson and Björklund (2000)
Pulsed electric field (PEF)	- Environmentally friendly, - Nonthermal treatment preserves quality, - Suitable for scaling up.	- Efficiency depends on solvent nature, - Need for extensive research on solvents, - Complexity and need for specialized equipment.	(Athanasiadis et al., 2022; Peiró et al., 2019)
Enzyme-assisted extraction (EAE)	- Operates under mild conditions, - Selective extraction of compounds, - Preserves integrity of heat-sensitive polyphenols.	- Potential cost of enzymes, - Longer reaction times, - Scaling up challenges.	Jovanovic et al. (2017)

The assay is not specific to polyphenols and can also react with other reducing agents present in the sample, leading to potential overestimation of the TPC. Additionally, the response of the assay can vary depending on the structure and chemical properties of the individual polyphenols present in the sample (Georgé et al., 2005). Despite these limitations, the Folin–Ciocalteu assay remains a valuable tool for the initial screening and characterization of polyphenol-rich samples. It is often used in combination with other analytical techniques, such as HPLC or liquid chromatography–mass spectrometry (LC-MS), to gain more detailed information about the individual polyphenolic compounds present and their concentrations.

Infrared spectroscopy, particularly Fourier transform infrared (FT-IR) spectroscopy, has been demonstrated as a valuable tool for the high-throughput screening and quantification of bioactive compounds in citrus fruits. The technique's efficacy in discriminating between different citrus fruit groups and predicting the concentration of total carotenoids, flavonoids, and phenolic compounds is highlighted by its ability to differentiate citrus fruits into distinct clusters that correspond to their taxonomic relationships (Song et al., 2018). FT-IR spectroscopy relies on the unique vibrational signatures of chemical bonds present in molecules, allowing for the identification and quantification of various functional groups and compounds. In the context of citrus fruits, FT-IR can provide valuable insights into the polyphenolic profile by detecting and analyzing the characteristic absorption bands associated with phenolic rings, hydroxyl groups, and other structural features of polyphenols. The process typically involves extracting the polyphenolic compounds from the citrus fruit sample using suitable solvents, followed by the preparation of a thin film or potassium bromide (KBr) pellet for analysis (HuiShan et al., 2006). The FT-IR spectrometer then irradiates the sample with infrared radiation, and the resulting absorption or transmission spectrum is recorded. This spectrum serves as a fingerprint, revealing the presence and relative abundance of various polyphenolic compounds based on their distinctive absorption patterns (Sun et al., 2022).

HPLC is an analytical technique widely used for the characterization and quantification of polyphenols. HPLC allows for the separation, identification, and quantification of individual polyphenolic compounds within complex mixtures. The technique involves injecting a sample into a column packed with a stationary phase, typically a silica-based material or a polymer-based material, and then passing a mobile phase (a liquid solvent mixture) through the column. The polyphenols in the sample interact differently with the stationary phase based on their chemical properties, resulting in different retention times as they travel through the column (Shui & Leong, 2002). The separated compounds are then detected by various detectors, such as UV-visible detectors, diode array detectors, or mass spectrometers, which provide information about their identity and concentration (Lamuela-Raventós et al., 2014). The choice of detector depends on the specific polyphenols being analyzed and the desired level of sensitivity and structural information. Optimization of HPLC methods, including the choice of mobile phase, gradient program, and column type, is crucial for achieving efficient separation and accurate quantification of polyphenols. The use of reversed-phase HPLC (RP-HPLC) is common due to its effectiveness in separating a wide range of polyphenolic structures, from simple phenolic acids to complex flavonoids (Waterhouse et al., 1999). Additionally, comprehensive two-dimensional liquid chromatography (LC $\times$ LC) has emerged as a technique that provides higher peak capacities than one-dimensional separations, which is particularly beneficial when analyzing polyphenols in foodstuffs with very complex matrices (Cacciola et al., 2017).

HPLC is suitable for determining most polyphenols, but achieving the required high sensitivity for trace polyphenols is challenging with UV-visible detectors or diode array detectors (Sun et al., 2022). LC–MS/MS is a powerful analytical technique that combines the separation capabilities of HPLC with the highly sensitive and selective detection of a tandem mass spectrometer (López-Fernández et al., 2020). In LC–MS/MS analysis of polyphenols, the sample is first separated by HPLC based on the polarity and hydrophobicity of the compounds. The eluted compounds are then introduced into the mass spectrometer, where they undergo ionization, typically through electrospray ionization (ESI) or atmospheric pressure chemical ionization (APCI). The ionized molecules are subsequently subjected to two stages of mass analysis, known as tandem mass spectrometry (MS/MS). In the first stage of MS/MS, precursor ions of specific mass-to-charge ratios (*m/z*) corresponding to the polyphenols of interest are selectively isolated. These precursor ions are then fragmented through collision-induced dissociation (CID), generating characteristic product ions. The second stage of MS/MS involves the detection and analysis of these product ions, which provide structural information and allow for unambiguous identification of the polyphenols. LC–MS/MS is a powerful technique for polyphenol characterization due to its selectivity, sensitivity, and adaptability to different sample types and analytical requirements. Its application ranges from simple and rapid quantification methods to more comprehensive analyses involving sample preparation and optimization of chromatographic conditions to achieve accurate and precise measurements of polyphenols in various matrices (Redeuil et al., 2009; Sapozhnikova, 2014; Schoedl et al., 2011).

Nuclear magnetic resonance (NMR) spectroscopy has proven to be an invaluable tool in the characterization of polyphenols. This nondestructive technique provides detailed structural information about the polyphenolic compounds present in a sample, making it invaluable for the identification and elucidation of their chemical structures. In NMR spectroscopy, the sample is placed in a strong magnetic field, and the nuclei of certain atoms, such as hydrogen (^{1}H) or carbon (^{13}C), are excited by radio frequency pulses. The resonance frequencies of these nuclei depend on their chemical environment, which is influenced by the presence of nearby functional groups, substituents, and stereochemical arrangements (Kim et al., 2008). For polyphenol characterization, both ^{1}H-NMR and ^{13}C-NMR spectroscopy are commonly used. ^{1}H-NMR provides information about the number and chemical environment of hydrogen atoms present in the molecule, while ^{13}C-NMR reveals details about the carbon skeleton and the connectivity of the different carbon atoms (Hikichi & Yasuda, 1987). NMR spectroscopy allows for the identification of various structural features of polyphenols, such as the position and number of hydroxyl groups, the degree of polymerization, the presence of glycosidic linkages, and the stereochemistry of the compounds (Wishart, 2008).

4. Antioxidant and antimicrobial properties of citrus by-product polyphenols

Polyphenols derived from citrus by-products, notably flavonoids and phenolic acids, exhibit antioxidant properties primarily through mechanisms of free radical neutralization (Addi et al., 2021; Dzah et al., 2020; Kaurinovic & Vastag, 2019). These compounds act as donors of hydrogen or electrophilic groups, thereby stabilizing oxygenated free radicals and mitigating oxidative damage to biological macromolecules such as lipids, proteins, and DNA (Addi et al., 2021; Dangles, 2012; Pisoschi et al., 2021). Consequently, polyphenols contribute to the preservation of cellular structural integrity and genome stability (Lewandowska et al., 2016; Zagoskina et al., 2023). Moreover, polyphenols possess the ability to modulate the activity of endogenous antioxidant enzymes, including superoxide dismutase and catalase (Du et al., 2007). These enzymes are pivotal in cellular defense against oxidative stress, facilitating the breakdown of reactive oxygen species (ROS) into inert byproducts. Thus, by regulating the expression and functionality of these enzymes, polyphenols bolster intracellular defense mechanisms against the harmful impacts of oxidative stress (Elsayed Azab et al., 2019; Rudrapal et al., 2022).

Several methods are used to assess the antioxidant capacity of polyphenols derived from citrus by-products, each with its own advantages and limitations (Barbosa et al., 2021). ORAC (oxygen radical absorbance capacity), ABTS (2,2′-azino-bis (3-ethylbenzothiazoline 6-sulfonate)), FRAP (ferric reducing antioxidant power), and DPPH (2,2-diphenyl-1-picrylhydrazyl) tests are also widely used (Chedea & Pop, 2019; Dudonné et al., 2009). These in vitro assays provide measurements of the capacity of polyphenols to neutralize free radicals, thus offering indications of their antioxidant potential (Dudonné et al., 2009).

Among the abundant polyphenols in citrus and its by-products are compounds such as naringenin, naringin, hesperidin, and rutin (Khan et al., 2014). These polyphenols have been extensively studied for their antioxidant capacity and beneficial effects on human health (Addi et al., 2021).

Naringenin is a flavanone flavonoid that can exist in two forms: its glycosidic form, called naringin, and its aglycone form, naringenin (Alam et al., 2014). Studies have demonstrated the remarkable antioxidant capacity of these compounds (Barbosa et al., 2018). For instance, a study by Cavia-Saiz et al. (2010) showed that naringin exhibits a lower capacity to inhibit ABTS radical generation compared to naringenin (27.1 ± 0.4 μmol L^{-1} vs. 7.9 ± 0.2 μmol L^{-1}). The aglycone demonstrates higher antioxidant efficacy than the glycoside according to the FRAP test. Both flavanones dose-dependently reduce ferric ions. Moreover, the aglycone exhibits superior superoxide radical scavenging activity, especially in the presence of xanthine oxidase, compared to the Nitroblue Tetrazolium (NBT) method, with an IC_{50} of 4.4 μmol L^{-1} for the aglycone and 94.7 μmol L^{-1} for naringenin (Cavia-Saiz et al., 2010). Another study reported by Bacanlı et al. (2015) evaluated the antioxidant capacity of naringin using ABTS methods and revealed IC_{50} values of 9026 μM. Another study reported by K. Ouali et al. (2007) demonstrated that hesperidin, a flavonoid from citrus by-products, can protect and prevent embryos from oxidative stress and regenerate β cells in the endocrine pancreas in experimental diabetes. Furthermore, Yang et al. (2008) reported that rutin can lead to a reduction in oxidative stress. Rutin showed strong DPPH radical scavenging activity. At a concentration of 0.05 mg/mL, it exhibited 90.4% inhibition. These findings suggest that polyphenols present in citrus by-products can significantly contribute to the prevention of oxidative stress-related diseases and the promotion of optimal health. They also underscore the importance of including citrus by-products in the diet to benefit from their antioxidant properties.

Alongside their antioxidant properties, polyphenols in citrus by-products also demonstrate antimicrobial properties that may play a beneficial role in the prevention and treatment of microbial infections (Lamine et al., 2021; Liu et al., 2021). These compounds interfere with the growth and survival of pathogenic microorganisms by altering the permeability of their cell membranes, inhibiting their vital enzymes, or disrupting their metabolic processes (Álvarez-Martínez et al., 2020). In addition, polyphenols have the ability to modulate the host's immune response, enhancing its ability to fight infection (Bernini & Velotti, 2021).

Several studies have investigated the antimicrobial properties of these compounds against various microorganisms. For example, research conducted by Veiko et al. (2023) demonstrated the antimicrobial activity of naringenin and naringin against *Staphylococcus aureus*, with minimum inhibitory concentrations (MICs) of 125 μg/mL and 1000 μg/mL, respectively. Additionally, Mundlia et al. (2019) showed that naringenin, when conjugated with pectin, exhibited significantly higher antimicrobial action against *S. aureus*, *Staphylococcus epidermidis*, and *Pseudomonas aeruginosa*, while comparable antimicrobial activity was observed against *Escherichia coli* and *Bacillus subtilis*. Hesperidin has also been evaluated for its antimicrobial potential. Studies by Choi et al. (2022) revealed the inhibitory effects of hesperidin against certain bacterial strains. It effectively inhibited bacterial growth, with MIC values of 125 μg/mL against *S. aureus*, 250 μg/mL against *B. cereus*, 500 μg/mL against *E. coli*, and *P. aeruginosa* (Choi et al., 2022).

Overall, the combination of these antioxidant and antimicrobial properties gives polyphenols derived from citrus by-products significant potential in health promotion and disease control (Lamine et al., 2021). Their use in various food and nutraceutical products offers promising prospects for the prevention and treatment of disorders linked to oxidative stress and microbial infections.

5. Challenges in the valorization of polyphenols derived from citrus by-products

The natural variability of polyphenol content in citrus by-products can be influenced by many factors, including citrus variety, stage of ripening, environmental conditions, and agronomic practices (Costanzo et al., 2022; M'hiri et al., 2017). This variability can make it difficult to produce standardized products with consistent levels of polyphenols, requiring additional efforts in raw material selection and quality control (Bansal et al., 2016; Hafting et al., 2015).

Scaling up the production of polyphenol extracts from citrus by-products for industrial application can present technical, economic, and logistical challenges (Freitas et al., 2021; Taghian Dinani & van der Goot, 2023). Extraction processes need to be adapted to handle large quantities of raw material while maintaining extract efficiency and quality (Gil-Martín et al., 2022; Yammine et al., 2018). In addition, waste management and environmental sustainability issues must be taken into account to ensure responsible production (Dhua et al., 2022; Doria et al., 2021; Dzah et al., 2020).

The marketing of nutraceutical products based on citrus by-product polyphenols is subject to strict regulation in many jurisdictions (Galieni et al., 2020).

Manufacturers must comply with food safety, good manufacturing practice, and product labeling standards (Mittal et al., 2024). In addition, they must provide robust scientific evidence of the efficacy and safety of their products to obtain regulatory approval (Varzakas et al., 2016).

6. Opportunities for innovation

Advances in eco-friendly extraction methods aim to reduce the environmental footprint of citrus polyphenol extraction processes (Satari & Karimi, 2018). Approaches such as green solvent extraction, the use of supercritical fluids, and extraction assisted by nonthermal technologies such as ultrasound and microwaves are being studied to improve the efficiency and sustainability of extraction processes (Picot-Allain et al., 2021; Shen et al., 2023).

New nutraceutical formulations aim to improve the bioavailability, stability, and efficacy of citrus by-product polyphenols (Lauro et al., 2015). Techniques such as nanoencapsulation, the use of innovative carrier matrices, and synergistic combination with other bioactive compounds are being explored to maximize the health benefits of finished products (Pateiro et al., 2021).

In addition to traditional applications in dietary supplements and health products, polyphenols from citrus by-products open the way to new emerging markets and applications (Barreca et al., 2020). For example, they can be used in the formulation of antiaging cosmetics and skin care products, or in the development of food packaging materials based on bioactive polymers, opening up new business opportunities (Messinese et al., 2024).

The valorization of polyphenols from citrus by-products for nutraceutical applications offers promising potential, but requires a multidisciplinary approach integrating basic research, technological innovation, and regulatory compliance. Continued efforts in this area can lead to the creation of safer, more effective, and more sustainable food and health products for consumers worldwide.

References

Addi, M., Elbouzidi, A., Abid, M., Tungmunnithum, D., Ahmed, E., & Hano, C. (2021). An overview of bioactive flavonoids from citrus fruits. *Applied Sciences, 12*(1), 29.

Agcam, E., Akyildiz, A., & Evrendilek, G. A. (2014). Comparison of phenolic compounds of orange juice processed by pulsed electric fields (PEF) and conventional thermal pasteurisation. *Food Chemistry, 143*, 354–361. https://doi.org/10.1016/j.foodchem.2013.07.115

Ainsworth, E. A., & Gillespie, K. M. (2007). Estimation of total phenolic content and other oxidation substrates in plant tissues using Folin-Ciocalteu reagent. *Nature Protocols, 2*(4), 875–877. https://doi.org/10.1038/nprot.2007.102

Al-Ashaal, H. A., & El-Sheltawy, S. T. (2011). Antioxidant capacity of hesperidin from Citrus peel using electron spin resonance and cytotoxic activity against human carcinoma cell lines. *Pharmaceutical Biology, 49*(3), 276–282. https://doi.org/10.3109/13880209.2010.509734

Alam, F., Mohammadin, K., Shafique, Z., Tayyeba Amjad, S., & Asad, M. H. H.bin (2022). Citrus flavonoids as potential therapeutic agents: A review. *Phytotherapy Research, 36*(4), 1417–1441.

Alam, M. A., Subhan, N., Rahman, M. M., Uddin, S. J., Reza, H. M., & Sarker, S. D. (2014). Effect of citrus flavonoids, naringin and naringenin, on metabolic syndrome and their mechanisms of action. *Advances in Nutrition, 5*(4), 404–417. https://doi.org/10.3945/an.113.005603

Álvarez-Martínez, F. J., Barrajón-Catalán, E., Encinar, J. A., Rodríguez-Díaz, J. C., & Micol, V. (2020). Antimicrobial capacity of plant polyphenols against gram-positive bacteria: A comprehensive review. *Current Medicinal Chemistry, 27*(15), 2576–2606. https://doi.org/10.2174/0929867325666181008115650

Anoopkumar, A. N., Aneesh, E. M., Sirohi, R., Tarafdar, A., Kuriakose, L. L., Surendhar, A., Madhavan, A., Kumar, V., Awasthi, M. K., & Binod, P. (2023). Bioactives from citrus food waste: Types, extraction technologies and application. *Journal of Food Science & Technology, 61*, 1–15.

Assefa, A. D., & Keum, Y. S. (2017). Effect of extraction solvent and various drying methods on polyphenol content and antioxidant activities of yuzu (Citrus junos Sieb ex Tanaka). *Journal of Food Measurement and Characterization, 11*(2), 576–585. https://doi.org/10.1007/s11694-016-9425-x

Athanasiadis, V., Pappas, V. M., Palaiogiannis, D., Chatzimitakos, T., Bozinou, E., Makris, D. P., & Lalas, S. I. (2022). Pulsed electric field-based extraction of total polyphenols from Sideritis raiseri using hydroethanolic mixtures. *Oxygen, 2*(2), 91–98. https://doi.org/10.3390/oxygen2020008

Bacanlı, M., Ahmet Basaran, A., & Basaran, N. (2015). The antioxidant and antigenotoxic properties of citrus phenolics limonene and naringin. *Food and Chemical Toxicology, 81*, 160–170. https://doi.org/10.1016/j.fct.2015.04.015

Bansal, G., Suthar, N., Kaur, J., & Jain, A. (2016). Stability testing of herbal drugs: Challenges, regulatory compliance and perspectives. *Phytotherapy Research, 30*(7), 1046–1058. https://doi.org/10.1002/ptr.5618

Barbosa, C. H., Andrade, M. A., Séndon, R., Silva, A. S., Ramos, F., Vilarinho, F., Khwaldia, K., & Barbosa-Pereira, L. (2021). Industrial fruits by-products and their antioxidant profile: Can they be exploited for industrial food applications? *Foods, 10*(2). https://doi.org/10.3390/foods10020272

Barbosa, P.d. P. M., Ruviaro, A. R., & Macedo, G. A. (2018). Comparison of different Brazilian citrus by-products as source of natural antioxidants. *Food Science and Biotechnology, 27*(5), 1301–1309. https://doi.org/10.1007/s10068-018-0383-4

Barbosa-Pereira, L., Guglielmetti, A., & Zeppa, G. (2018). Pulsed electric field assisted extraction of bioactive compounds from cocoa bean shell and coffee silverskin. *Food and Bioprocess Technology, 11*(4), 818–835. https://doi.org/10.1007/s11947-017-2045-6

Barrales, F. M., Silveira, P., Barbosa, P.d. P. M., Ruviaro, A. R., Paulino, B. N., Pastore, G. M., Macedo, G. A., & Martinez, J. (2018). Recovery of phenolic compounds from citrus by-products using pressurized liquids — An application to orange peel. *Food and Bioproducts Processing, 112*, 9–21. https://doi.org/10.1016/j.fbp.2018.08.006

Barreca, D., Mandalari, G., Calderaro, A., Smeriglio, A., Trombetta, D., Felice, M. R., & Gattuso, G. (2020). Citrus flavones: An update on sources, biological functions, and health promoting properties. *Plants, 9*(3). https://doi.org/10.3390/plants9030288

Bernini, R., & Velotti, F. (2021). Natural polyphenols as immunomodulators to rescue immune response homeostasis: Quercetin as a research model against severe COVID-19. *Molecules, 26*, 5803.

Boukroufa, M., Boutekedjiret, C., Petigny, L., Rakotomanomana, N., & Chemat, F. (2015). Bio-refinery of orange peels waste: A new concept based on integrated green and solvent free extraction processes using ultrasound and microwave techniques to obtain essential oil, polyphenols and pectin. *Ultrasonics Sonochemistry, 24*, 72–79. https://doi.org/10.1016/j.ultsonch.2014.11.015

Bucur, M. P., Radulescu, M. C., Radu, G. L., & Bucur, B. (2023). Cavitation-effect-based treatments and extractions for superior fruit and milk valorisation. *Molecules, 28*(12). https://doi.org/10.3390/molecules28124677

Bumrungpert, A., Lilitchan, S., Tuntipopipat, S., Tirawanchai, N., & Komindr, S. (2018). Ferulic acid supplementation improves lipid profiles, oxidative stress, and inflammatory status in hyperlipidemic subjects: A randomized, double-blind, placebo-controlled clinical trial. *Nutrients, 10*(6), 713. https://doi.org/10.3390/nu10060713

Cacciola, F., Farnetti, S., Dugo, P., Marriott, P. J., & Mondello, L. (2017). Comprehensive two-dimensional liquid chromatography for polyphenol analysis in foodstuffs. *Journal of Separation Science, 40*(1), 7–24. https://doi.org/10.1002/jssc.201600704

Cai, H., You, S., Xu, Z., Li, Z., Guo, J., Ren, Z., & Fu, C. (2021). Novel extraction methods and potential applications of polyphenols in fruit waste: A review. *Journal of Food Measurement and Characterization, 15*(4), 3250–3261. https://doi.org/10.1007/s11694-021-00901-1

Carrera, C., Ruiz-Rodríguez, A., Palma, M., & Barroso, C. G. (2012). Ultrasound assisted extraction of phenolic compounds from grapes. *Analytica Chimica Acta, 732*, 100–104. https://doi.org/10.1016/j.aca.2011.11.032

Cavia-Saiz, M., Busto, M. D., Pilar-Izquierdo, M. C., Ortega, N., Perez-Mateos, M., & Muñiz, P. (2010). Antioxidant properties, radical scavenging activity and biomolecule protection capacity of flavonoid naringenin and its glycoside naringin: A comparative study. *Journal of the Science of Food and Agriculture, 90*(7), 1238–1244. https://doi.org/10.1002/jsfa.3959Spain

Chedea, V. S., & Pop, R. M. (2019). *Total polyphenols content and antioxidant DPPH assays on biological samples* (pp. 169–183). Elsevier BV. https://doi.org/10.1016/b978-0-12-813768-0.00011-6

Cheynier, V. (2012). Phenolic compounds: From plants to foods. *Phytochemistry Reviews, 11*(2–3), 153–177. https://doi.org/10.1007/s11101-012-9242-8

Choi, S. S., Lee, S. H., & Lee, K. A. (2022). A comparative study of hesperetin, hesperidin and hesperidin glucoside: Antioxidant, anti-inflammatory, and antibacterial activities in vitro. *Antioxidants, 11*(8). https://doi.org/10.3390/antiox11081618

Ciğeroğlu, Z., Bayramoğlu, M., Kırbaslar, S.İ., & Sahin, S. (2021). Comparison of microwave-assisted techniques for the extraction of antioxidants from Citrus paradisi Macf. biowastes. *Journal of Food Science and Technology, 58*(3), 1190–1198. https://doi.org/10.1007/s13197-020-04632-x

Costanzo, G., Vitale, E., Iesce, M. R., Naviglio, D., Amoresano, A., Fontanarosa, C., Spinelli, M., Ciaravolo, M., & Arena, C. (2022). Antioxidant properties of pulp, peel and seeds of Phlegrean mandarin (Citrus reticulata Blanco) at different stages of fruit ripening. *Antioxidants, 11*(2). https://doi.org/10.3390/antiox11020187

Dangles, O. (2012). Antioxidant activity of plant phenols: Chemical mechanisms and biological significance. *Current Organic Chemistry, 16*(6), 692–714. https://doi.org/10.2174/138527212799957995

Dheyab, A. S., Bakar, M. F. A., Alomar, M., Sabran, S. F., Hanafi, A. F. M., & Mohamad, A. (2021). Deep eutectic solvents (DESs) as green extraction media of beneficial bioactive phytochemicals. *Separations, 8*(10). https://doi.org/10.3390/SEPARATIONS8100176

Dhua, S., Kumar, K., Sharanagat, V. S., & Nema, P. K. (2022). Bioactive compounds and its optimization from food waste: Review on novel extraction techniques. *Nutrition & Food Science, 52*(8), 1270–1288. https://doi.org/10.1108/NFS-12-2021-0373

Doria, E., Boncompagni, E., Marra, A., Dossena, M., Verri, M., & Buonocore, D. (2021). Polyphenols extraction from vegetable wastes using a green and sustainable method. *Frontiers in Sustainable Food Systems, 5*. https://doi.org/10.3389/fsufs.2021.690399

Du, Y., Guo, H., & Lou, H. (2007). Grape seed polyphenols protect cardiac cells from apoptosis via induction of endogenous antioxidant enzymes. *Journal of Agricultural and Food Chemistry, 55*(5), 1695–1701. https://doi.org/10.1021/jf063071b

Dudonné, S., Vitrac, X., Coutiére, P., Woillez, M., & Mérillon, J. M. (2009). Comparative study of antioxidant properties and total phenolic content of 30 plant extracts of industrial interest using DPPH, ABTS, FRAP, SOD, and ORAC assays. *Journal of Agricultural and Food Chemistry, 57*(5), 1768–1774. https://doi.org/10.1021/jf803011r

Dzah, C. S., Duan, Y., Zhang, H., Serwah Boateng, N. A., & Ma, H. (2020). Latest developments in polyphenol recovery and purification from plant by-products: A review. *Trends in Food Science and Technology, 99*, 375–388. https://doi.org/10.1016/j.tifs.2020.03.003

El Darra, N., Grimi, N., Vorobiev, E., Louka, N., & Maroun, R. (2013). Extraction of polyphenols from red grape pomace assisted by pulsed ohmic heating. *Food and Bioprocess Technology, 6*(5), 1281–1289. https://doi.org/10.1007/s11947-012-0869-7

El Kantar, S., Boussetta, N., Lebovka, N., Foucart, F., Rajha, H. N., Maroun, R. G., Louka, N., & Vorobiev, E. (2018). Pulsed electric field treatment of citrus fruits: Improvement of juice and polyphenols extraction. *Innovative Food Science and Emerging Technologies, 46*, 153–161. https://doi.org/10.1016/j.ifset.2017.09.024

Elsayed Azab, A., Adwas, A. A., Elsayed, A. S. I., Adwas, A. A., Elsayed, A. S. I., & Quwaydir, F. A. (2019). Oxidative stress and antioxidant mechanisms in human body. *Journal of Applied Biotechnology & Bioengineering, 6*(1), 43–47. https://doi.org/10.15406/jabb.2019.06.00173

Feuereisen, M. M., Gamero Barraza, M., Zimmermann, B. F., Schieber, A., & Schulze-Kaysers, N. (2017). Pressurized liquid extraction of anthocyanins and biflavonoids from Schinus terebinthifolius Raddi: A multivariate optimization. *Food Chemistry, 214*, 564–571. https://doi.org/10.1016/j.foodchem.2016.07.002

Freitas, L. C., Barbosa, J. R., da Costa, A. L. C., Bezerra, F. W. F., Pinto, R. H. H., & Carvalho Junior, R. N.d. (2021). From waste to sustainable industry: How can agro-industrial wastes help in the development of new products? *Resources, Conservation and Recycling, 169*. https://doi.org/10.1016/j.resconrec.2021.105466

Galieni, A., Falcinelli, B., Stagnari, F., Datti, A., & Benincasa, P. (2020). Sprouts and microgreens: Trends, opportunities, and horizons for novel research. *Agronomy, 10*(9), 1424. https://doi.org/10.3390/agronomy10091424

Garcia-Salas, P., Morales-Soto, A., Segura-Carretero, A., & Fernández-Gutiérrez, A. (2010). Phenolic-compound-extraction systems for fruit and vegetable samples. *Molecules, 15* (12), 8813–8826. https://doi.org/10.3390/molecules15128813Spain

Genovese, S., Fiorito, S., Locatelli, M., Carlucci, G., & Epifano, F. (2014). Analysis of biologically active oxyprenylated ferulic acid derivatives in citrus fruits. *Plant Foods for Human Nutrition, 69*(3), 255–260. https://doi.org/10.1007/s11130-014-0427-8

Georgé, S., Brat, P., Alter, P., & Amiot, M. J. (2005). Rapid determination of polyphenols and vitamin C in plant-derived products. *Journal of Agricultural and Food Chemistry, 53*(5), 1370–1373. https://doi.org/10.1021/jf048396b

Ghasemi, K., Ghasemi, Y., & Ebrahimzadeh, M. A. (2009). Antioxidant activity, phenol and flavonoid contents of 13 citrus species peels and tissues. *Pakistan Journal of Pharmaceutical Sciences, 22*(3), 277–281.

Gil-Martín, E., Forbes-Hernández, T., Romero, A., Cianciosi, D., Giampieri, F., & Battino, M. (2022). Influence of the extraction method on the recovery of bioactive phenolic compounds from food industry by-products. *Food Chemistry, 378*, Article 131918. https://doi.org/10.1016/j.foodchem.2021.131918

González-Molina, E., Domínguez-Perles, R., Moreno, D. A., & García-Viguera, C. (2010). Natural bioactive compounds of Citrus limon for food and health. *Journal of Pharmaceutical and Biomedical Analysis, 51*(2), 327–345.

Hafting, J. T., Craigie, J. S., Stengel, D. B., Loureiro, R. R., Buschmann, A. H., Yarish, C., Edwards, M. D., & Critchley, A. T. (2015). Prospects and challenges for industrial production of seaweed bioactives. *Journal of Phycology, 51*(5), 821–837. https://doi.org/10.1111/jpy.12326Canada

Han, X., Shen, T., & Lou, H. (2007). Dietary polyphenols and their biological significance. *International Journal of Molecular Sciences, 8*(9), 950–988. https://doi.org/10.3390/i8090950China

Heidari, H., Ghanbari-Rad, S., & Habibi, E. (2020). Optimization deep eutectic solvent-based ultrasound-assisted liquid-liquid microextraction by using the desirability function approach for extraction and preconcentration of organophosphorus pesticides from fruit juice samples. *Journal of Food Composition and Analysis, 87*, Article 103389. https://doi.org/10.1016/j.jfca.2019.103389

Hikichi, K., & Yasuda, M. (1987). Two dimensional ^1H and ^{13}C nuclear magnetic resonance studies of polyvinyl alcohol. *Polymer Journal, 19*(9), 1003–1012. https://doi.org/10.1295/polymj.19.1003

HuiShan, H. R., YiBin, X. P., & HaiYan, T. H. H. Q. (2006). Application Fourier transform near infrared spectrometer in rapid estimation of soluble solids content of intact citrus fruits. *Journal of Zhejiang University - Science B, 7*(10), 794–799.

Ishangulyyev, R., Kim, S., & Lee, S. H. (2019). Understanding food loss and waste-why are we losing and wasting food? *Foods, 8*(8). https://doi.org/10.3390/foods8080297

Jahromi, S. G. (2019). Extraction techniques of phenolic compounds from plants. *Plant Physiological Aspects of Phenolic Compounds*, 1–18.

Jovanovic, A., Petrovic, P., Đordjevic, V., Zdunic, G., Savikin, K., & Bugarski, B. (2017). Polyphenols extraction from plant sources. *Lekovite sirovine, 37*, 45–49. https://doi.org/10.5937/leksir1737045j

Kalogiouri, N. P., Palaiologou, E., Papadakis, E. N., Makris, D. P., Biliaderis, C. G., & Mourtzinos, I. (2022). Insights on the impact of deep eutectic solvents on the composition of the extracts from lemon (Citrus Limon L.) peels analyzed by a novel RP-LC–QTOF-MS/MS method. *European Food Research and Technology, 248*, 2913–2927.

Kaurinovic, B., & Vastag, D. (2019). *Flavonoids and phenolic acids as potential natural antioxidants*. IntechOpen. https://doi.org/10.5772/intechopen.83731

Khan, M. K., Abert-Vian, M., Fabiano-Tixier, A. S., Dangles, O., & Chemat, F. (2010). Ultrasound-assisted extraction of polyphenols (flavanone glycosides) from orange (Citrus sinensis L.) peel. *Food Chemistry, 119*(2), 851–858. https://doi.org/10.1016/j.foodchem.2009.08.046

Khan, M. K., Huma, Z. E., & Dangles, O. (2014). A comprehensive review on flavanones, the major citrus polyphenols. *Journal of Food Composition and Analysis, 33*(1), 85–104. https://doi.org/10.1016/j.jfca.2013.11.004

Khan, I. A., & Kender, W. J. (2007). *Citrus breeding: Introduction and objectives citrus genetics, breeding and biotechnology* (pp. 1–8). Pakistan: CABI Publishing.

Kim, J., Gracz, H. S., Roberts, G. W., & Kiserow, D. J. (2008). Spectroscopic analysis of poly (bisphenol A carbonate) using high resolution 13C and 1H NMR. *Polymer, 49*(2), 394–404. https://doi.org/10.1016/j.polymer.2007.11.046

Korhonen, J., & Snäkin, J. P. (2005). Analysing the evolution of industrial ecosystems: Concepts and application. *Ecological Economics, 52*(2), 169–186. https://doi.org/10.1016/j.ecolecon.2004.07.016

Kumar, M., Dahuja, A., Sachdev, A., Kaur, C., Varghese, E., Saha, S., & Sairam, K. V. S. S. (2019). Valorisation of black carrot pomace: Microwave assisted extraction of bioactive phytoceuticals and antioxidant activity using Box–Behnken design. *Journal of Food Science and Technology, 56*, 995–1007.

Lamine, M., Gargouri, M., Rahali, F. Z., & Mliki, A. (2021). Recovering and characterizing phenolic compounds from citrus by-product: A way towards agriculture of subsistence and sustainable bioeconomy. *Waste and Biomass Valorization, 12*(8), 4721–4731. https://doi.org/10.1007/s12649-020-01306-9

Lamuela-Raventós, R. M., Vallverdú-Queralt, A., Jáuregui, O., Martínez-Huélamo, M., & Quifer-Rada, P. (2014). Improved characterization of polyphenols using liquid chromatography. In *Polyphenols in plants: Isolation, purification and extract preparation* (pp. 261–292). Spain: Elsevier Inc.. https://doi.org/10.1016/B978-0-12-397934-6.00014-0

Lauro, M. R., Crasci, L., Carbone, C., Aquino, R. P., Panico, A. M., & Puglisi, G. (2015). Encapsulation of a citrus by-product extract: Development, characterization and stability studies of a nutraceutical with antioxidant and metalloproteinases inhibitory activity. *LWT - Food Science and Technology, 62*(1), 169–176. https://doi.org/10.1016/j.lwt.2015.01.017

Lewandowska, H., Kalinowska, M., Lewandowski, W., Stępkowski, T. M., & Brzóska, K. (2016). The role of natural polyphenols in cell signaling and cytoprotection against cancer development. *The Journal of Nutritional Biochemistry, 32*, 1–19. https://doi.org/10.1016/j.jnutbio.2015.11.006

Li, H., Deng, Z., Wu, T., Liu, R., Loewen, S., & Tsao, R. (2012). Microwave-assisted extraction of phenolics with maximal antioxidant activities in tomatoes. *Food Chemistry, 130*(4), 928–936. https://doi.org/10.1016/j.foodchem.2011.08.019

Liu, R. H. (2013). Health-promoting components of fruits and vegetables in the diet. *Advances in Nutrition, 4*(3), 384S.

Liu, Y., Benohoud, M., Galani Yamdeu, J. H., Gong, Y. Y., & Orfila, C. (2021). Green extraction of polyphenols from citrus peel by-products and their antifungal activity against Aspergillus flavus. *Food Chemistry X, 12*. https://doi.org/10.1016/j.fochx.2021.100144

Liu, Y. Q., Heying, E., & Tanumihardjo, S. A. (2012). History, global distribution, and nutritional importance of citrus fruits. *Comprehensive Reviews in Food Science and Food Safety, 11*(6), 530–545.

López-Fernández, O., Domínguez, R., Pateiro, M., Munekata, P. E. S., Rocchetti, G., & Lorenzo, J. M. (2020). Determination of polyphenols using liquid chromatography-tandem mass spectrometry technique (LC-MS/MS): A review. *Antioxidants, 9*(6), 479.

Ma, Y. Q., Ye, X. Q., Fang, Z. X., Chen, J. C., Xu, G. H., & Liu, D. H. (2008). Phenolic compounds and antioxidant activity of extracts from ultrasonic treatment of Satsuma mandarin (Citrus unshiu Marc.) peels. *Journal of Agricultural and Food Chemistry, 56*(14), 5682–5690. https://doi.org/10.1021/jf072474o

Ma, G., Zhang, L., Sugiura, M., & Kato, M. (2020). *Citrus and health the genus citrus* (pp. 495–511). Elsevier.

Mahato, N., Sharma, K., Sinha, M., & Cho, M. H. (2018). Citrus waste derived nutra-/pharmaceuticals for health benefits: Current trends and future perspectives. *Journal of Functional Foods, 40*, 307–316. https://doi.org/10.1016/j.jff.2017.11.015

Maqbool, Z., Khalid, W., Atiq, H. T., Koraqi, H., Javaid, Z., Alhag, S. K., Al-Shuraym, L. A., Bader, D. M. D., Almarzuq, M., Afifi, M., & AL-Farga, A. (2023). Citrus waste as source of bioactive compounds: Extraction and utilization in health and food industry. *Molecules, 28*(4). https://doi.org/10.3390/molecules28041636

Messinese, E., Pitirollo, O., Grimaldi, M., Milanese, D., Sciancalepore, C., & Cavazza, A. (2024). By-products as sustainable source of bioactive compounds for potential application in the field of food and new materials for packaging development. *Food and Bioprocess Technology, 17*(3), 606–627. https://doi.org/10.1007/s11947-023-03158-2

M'hiri, N., Ioannou, I., Ghoul, M., & Mihoubi Boudhrioua, N. (2017). Phytochemical characteristics of citrus peel and effect of conventional and nonconventional processing on phenolic compounds: A review. *Food Reviews International, 33*(6), 587–619. https://doi.org/10.1080/87559129.2016.1196489

Mittal, R. K., Mishra, R., Sharma, V., & Purohit, P. (2024). Bioactive exploration in functional foods: Unlocking nature's treasures. *Current Pharmaceutical Biotechnology, 25*, 1419–1435.

Mundlia, J., Ahuja, M., Kumar, P., & Pillay, V. (2019). Improved antioxidant, antimicrobial and anticancer activity of naringenin on conjugation with pectin. *Biotechnology, 9*, 1–14.

Murugesan, S., Maran, P., Venkatesan, M., & Alexander, R. A. (2024). Microwave assisted extraction of polyphenols from Pithecellobium dulce Benth fruit peels and evaluation of its anticancer and antioxidant activity. *Waste and Biomass Valorization, 15*(2), 841–855. https://doi.org/10.1007/s12649-023-02183-8

Naureen, Z., Dhuli, K., Donato, K., Aquilanti, B., Velluti, V., Matera, G., Iaconelli, A., & Bertelli, M. (2022). Foods of the Mediterranean diet: Citrus, cucumber and grape. *Journal of Preventive Medicine and Hygiene, 63*(2 Suppl. 3), E21.

Nayak, A., & Bhushan, B. (2019). An overview of the recent trends on the waste valorization techniques for food wastes. *Journal of Environmental Management, 233*, 352–370. https://doi.org/10.1016/j.jenvman.2018.12.041

Nayak, B., Dahmoune, F., Moussi, K., Remini, H., Dairi, S., Aoun, O., & Khodir, M. (2015). Comparison of microwave, ultrasound and accelerated-assisted solvent extraction for recovery of polyphenols from Citrus sinensis peels. *Food Chemistry, 187*, 507–516. https://doi.org/10.1016/j.foodchem.2015.04.081

Ng, K. R., Lyu, X., Mark, R., & Chen, W. N. (2019). Antimicrobial and antioxidant activities of phenolic metabolites from flavonoid-producing yeast: Potential as natural food preservatives. *Food Chemistry, 270*, 123–129. https://doi.org/10.1016/j.foodchem.2018.07.077

Nieto, G., Fernández-López, J., Pérez-Álvarez, J. A., Peñalver, R., Ros-Berruezo, G., & Viuda-Martos, M. (2021). Valorization of citrus co-products: Recovery of bioactive compounds and application in meat and meat products. *Plants, 10*(6), 1069. https://doi.org/10.3390/plants10061069

Oliveira, G. D. A. R., De Oliveira, A. E., Da Conceição, E. C., & Leles, M. I. G. (2016). Multi-response optimization of an extraction procedure of carnosol and rosmarinic and carnosic acids from rosemary. *Food Chemistry, 211*, 465–473. https://doi.org/10.1016/j.foodchem.2016.05.042

Oliveira, J. A. R., Menezes, de P., Barbosa, P., & Macêdo, G. A. (2022). High concentrate flavonoids extract from citrus pomace using enzymatic and deep eutectic solvents extraction. *Foods, 11*, 3205.

Ouali, K., Trea, F., Toumi, L., Bairi, A., Maurel, D., & Guellati, M.-A. (2007). L'hespéridine, un antioxydant flavonoïde qui diminue le stress oxydatif et prévient les malformations fœtales au cours du diabète gestationnel expérimental. *Phytothérapie, 5*(4), 204–209. https://doi.org/10.1007/s10298-007-0253-3

Paes, J., Dotta, R., Barbero, G. F., & Martínez, J. (2014). Extraction of phenolic compounds and anthocyanins from blueberry (Vaccinium myrtillus L.) residues using supercritical CO2 and pressurized liquids. *Journal of Supercritical Fluids, 95*, 8–16. https://doi.org/10.1016/j.supflu.2014.07.025

Palma, M., & Taylor, L. T. (1999). Extraction of polyphenolic compounds from grape seeds with near critical carbon dioxide. *Journal of Chromatography A, 849*(1), 117–124. https://doi.org/10.1016/S0021-9673(99)00569-5

Panwar, D., Panesar, P. S., & Chopra, H. K. (2021). Recent trends on the valorization strategies for the management of citrus by-products. *Food Reviews International, 37*(1), 91–120. https://doi.org/10.1080/87559129.2019.1695834

Panwar, D., Saini, A., Panesar, P. S., & Chopra, H. K. (2021). Unraveling the scientific perspectives of citrus by-products utilization: Progress towards circular economy. *Trends in Food Science and Technology, 111*, 549–562. https://doi.org/10.1016/j.tifs.2021.03.018

Pateiro, M., Gómez, B., Munekata, P. E. S., Barba, F. J., Putnik, P., Kovačević, D. B., & Lorenzo, J. M. (2021). Nanoencapsulation of promising bioactive compounds to improve their absorption, stability, functionality and the appearance of the final food products. *Molecules, 26*(6). https://doi.org/10.3390/molecules26061547

Peiró, S., Luengo, E., Segovia, F., Raso, J., & Almajano, M. P. (2019). Improving polyphenol extraction from lemon residues by pulsed electric fields. *Spain Waste and Biomass Valorization, 10*(4), 889–897. https://doi.org/10.1007/s12649-017-0116-6

Picot-Allain, C., Mahomoodally, M. F., Ak, G., & Zengin, G. (2021). Conventional versus green extraction techniques — A comparative perspective. *Current Opinion in Food Science, 40*, 144–156. https://doi.org/10.1016/j.cofs.2021.02.009

Pisoschi, A. M., Pop, A., Iordache, F., Stanca, L., Predoi, G., & Serban, A. I. (2021). Oxidative stress mitigation by antioxidants - An overview on their chemistry and influences on health status. *European Journal of Medicinal Chemistry, 209*. https://doi.org/10.1016/j.ejmech.2020.112891

Pradal, D., Vauchel, P., Decossin, S., Dhulster, P., & Dimitrov, K. (2016). Kinetics of ultrasound-assisted extraction of antioxidant polyphenols from food by-products: Extraction and energy consumption optimization. *Ultrasonics Sonochemistry, 32*, 137–146. https://doi.org/10.1016/j.ultsonch.2016.03.001

Rafiq, S., Kaul, R., Sofi, S. A., Bashir, N., Nazir, F., & Ahmad Nayik, G. (2018). Citrus peel as a source of functional ingredient: A review. *Journal of the Saudi Society of Agricultural Sciences, 17*(4), 351–358. https://doi.org/10.1016/j.jssas.2016.07.006

Ramful, D., Bahorun, T., Bourdon, E., Tarnus, E., & Aruoma, O. I. (2010). Bioactive phenolics and antioxidant propensity of flavedo extracts of Mauritian citrus fruits: Potential prophylactic ingredients for functional foods application. *Toxicology, 278*(1), 75–87.

Raveendran, P., Ikushima, Y., & Wallen, S. L. (2005). Polar attributes of supercritical carbon dioxide. *Accounts of Chemical Research, 38*(6), 478–485. https://doi.org/10.1021/ar040082m

Raynie, D. E. (2019). Looking at the past to understand the future: Soxhlet extraction. *LC-GC North America, 37*(8), 510–513.

Redeuil, K., Bertholet, R., Kussmann, M., Steiling, H., Rezzi, S., & Nagy, K. (2009). Quantification of flavan-3-ols and phenolic acids in milk-based food products by reversed-phase liquid chromatography-tandem mass spectrometry. *Journal of Chromatography A, 1216*(47), 8362–8370. https://doi.org/10.1016/j.chroma.2009.09.056

Ricarte, G. N., Coelho, M. A. Z., Marrucho, I. M., & Ribeiro, B. D. (2020). Enzyme-assisted extraction of carotenoids and phenolic compounds from sunflower wastes using green solvents. *3 Biotech, 10*(9). https://doi.org/10.1007/s13205-020-02393-0

Rudrapal, M., Khairnar, S. J., Khan, J., Dukhyil, A. B., Ansari, M. A., Alomary, M. N., Alshabrmi, F. M., Palai, S., Deb, P. K., & Devi, R. (2022). Dietary polyphenols and their role in oxidative stress-induced human diseases: Insights into protective effects, antioxidant potentials and mechanism(s) of action. *Frontiers in Pharmacology, 13*. https://doi.org/10.3389/fphar.2022.806470

Ruviaro, A. R., Barbosa, P.d. P. M., & Macedo, G. A. (2019). Enzyme-assisted biotransformation increases hesperetin content in citrus juice by-products. *Food Research International, 124*, 213–221. https://doi.org/10.1016/j.foodres.2018.05.004

Sander, L. C. (2017). Soxhlet extractions. *Journal of Research of the National Institute of Standards and Technology, 122*, 1.

Sapozhnikova, Y. (2014). Development of liquid chromatography-tandem mass spectrometry method for analysis of polyphenolic compounds in liquid samples of grape juice, green tea and coffee. *Food Chemistry, 150*, 87–93. https://doi.org/10.1016/j.foodchem.2013.10.131

Satari, B., & Karimi, K. (2018). Citrus processing wastes: Environmental impacts, recent advances, and future perspectives in total valorization. *Resources, Conservation and Recycling, 129*, 153–167. https://doi.org/10.1016/j.resconrec.2017.10.032

Schoedl, K., Forneck, A., Sulyok, M., & Schuhmacher, R. (2011). Optimization, in-house validation, and application of a liquid chromatography-tandem mass spectrometry (LC-MS/MS)-based method for the quantification of selected polyphenolic compounds in leaves of grapevine (Vitis vinifera L.). *Journal of Agricultural and Food Chemistry, 59*(20), 10787–10794. https://doi.org/10.1021/jf202753g

Sebghatollahi, Z., Ghanadian, M., Agarwal, P., Seyedhosseini Ghaheh, H., Mahato, N., Yogesh, R., & Hossein Hejazi, S. (2022). Citrus flavonoids: Biological activities, implementation in skin health, and topical applications: A review. *ACS Food Science & Technology, 2*(9), 1417–1432.

Sentkowska, A., Ivanova-Petropulos, V., & Pyrzynska, K. (2024). What can be done to get more—Extraction of phenolic compounds from plant materials. *Food Analytical Methods, 17*, 594–610.

Sharma, K., Mahato, N., & Lee, Y. R. (2019). Extraction, characterization and biological activity of citrus flavonoids. *Reviews in Chemical Engineering, 35*(2), 265–284. https://doi.org/10.1515/revce-2017-0027

Shen, L., Pang, S., Zhong, M., Sun, Y., Abdul, Q., Liu, Y., Rashid, A., Xu, B., Liang, Q., Ma, H., & Ren, X. (2023). A comprehensive review of ultrasonic assisted extraction

(UAE) for bioactive components: Principles, advantages, equipment, and combined technologies. *Ultrasonics Sonochemistry, 101*, Article 106646. https://doi.org/10.1016/j.ultsonch.2023.106646

Shui, G., & Leong, L. P. (2002). Separation and determination of organic acids and phenolic compounds in fruit juices and drinks by high-performance liquid chromatography. *Journal of Chromatography A, 977*(1), 89–96. https://doi.org/10.1016/S0021-9673(02)01345-6

Singh, B., Singh, J. P., Kaur, A., & Singh, N. (2020). Phenolic composition, antioxidant potential and health benefits of citrus peel. *Food Research International, 132*, Article 109114. https://doi.org/10.1016/j.foodres.2020.109114

Socas-Rodríguez, B., Álvarez-Rivera, G., Valdés, A., Ibáñez, E., & Cifuentes, A. (2021). Food by-products and food wastes: Are they safe enough for their valorization? *Trends in Food Science and Technology, 114*, 133 147. https://doi.org/10.1016/j.tifs.2021.05.002

Song, S. Y., Kim, C. H., Im, S. J., & Kim, I. J. (2018). Discrimination of citrus fruits using FT-IR fingerprinting by quantitative prediction of bioactive compounds. *Food Science and Biotechnology, 27*(2), 367–374. https://doi.org/10.1007/s10068-017-0263-3

Sparr Eskilsson, C., & Björklund, E. (2000). Analytical-scale microwave-assisted extraction. *Journal of Chromatography A, 902*(1), 227–250. https://doi.org/10.1016/S0021-9673(00)00921-3

Sridhar, A., Ponnuchamy, M., Kumar, P. S., Kapoor, A., Vo, D. V. N., & Prabhakar, S. (2021). Techniques and modeling of polyphenol extraction from food: A review. *Environmental Chemistry Letters, 19*(4), 3409–3443. https://doi.org/10.1007/s10311-021-01217-8

Sun, M. F., Jiang, C. L., Kong, Y. S., Luo, J. L., Yin, P., & Guo, G. Y. (2022). Recent advances in analytical methods for determination of polyphenols in tea: A comprehensive review. *Foods, 11*(10). https://doi.org/10.3390/foods11101425

Taghian Dinani, S., & van der Goot, A. J. (2023). Challenges and solutions of extracting value-added ingredients from fruit and vegetable by-products: A review. *Critical Reviews in Food Science and Nutrition, 63*(25), 7749–7771. https://doi.org/10.1080/10408398.2022.2049692

Varzakas, T., George, Z., & Verpoort, F. (2016). Plant food residues as a source of nutraceuticals and functional foods. *Foods, 5*(4), 88. https://doi.org/10.3390/foods5040088

Veiko, A. G., Olchowik-Grabarek, E., Sekowski, S., Roszkowska, A., Lapshina, E. A., Dobrzynska, I., Zamaraeva, M., & Zavodnik, I. B. (2023). Antimicrobial activity of quercetin, naringenin and catechin: Flavonoids inhibit Staphylococcus aureus-induced hemolysis and modify membranes of bacteria and erythrocytes. *Molecules, 28*(3). https://doi.org/10.3390/molecules28031252

Wang, W., Sun, C., Mao, L., Ma, P., Liu, F., Yang, J., & Gao, Y. (2016). The biological activities, chemical stability, metabolism and delivery systems of quercetin: A review. *Trends in Food Science and Technology, 56*, 21–38. https://doi.org/10.1016/j.tifs.2016.07.004

Waterhouse, A. L., Price, S. F., & Mccord, J. D. (1999). Reversed-phase high-performance liquid chromatography methods for analysis of wine polyphenols. *Methods in Enzymology, 299*, 113–121. https://doi.org/10.1016/S0076-6879(99)99014-6

Wishart, D. S. (2008). Quantitative metabolomics using NMR. *TrAC, Trends in Analytical Chemistry, 27*(3), 228–237. https://doi.org/10.1016/j.trac.2007.12.001

Xu, C., Nasrollahzadeh, M., Selva, M., Issaabadi, Z., & Luque, R. (2019). Waste-to-wealth: Biowaste valorization into valuable bio(nano)materials. *Chemical Society Reviews, 48*(18), 4791–4822. https://doi.org/10.1039/c8cs00543e

Yammine, S., Brianceau, S., Manteau, S., Turk, M., Ghidossi, R., Vorobiev, E., & Mietton-Peuchot, M. (2018). Extraction and purification of high added value compounds from by-products of the winemaking chain using alternative/nonconventional processes/technologies. *Critical Reviews in Food Science and Nutrition, 58*(8), 1375–1390. https://doi.org/10.1080/10408398.2016.1259982

Yang, J., Guo, J., & Yuan, J. (2008). In vitro antioxidant properties of rutin. *LWT - Food Science and Technology, 41*(6), 1060–1066. https://doi.org/10.1016/j.lwt.2007.06.010

Yi, L., Ma, S., & Ren, D. (2017). Phytochemistry and bioactivity of Citrus flavonoids: A focus on antioxidant, anti-inflammatory, anticancer and cardiovascular protection activities. *Phytochemistry Reviews, 16*, 479–511.

Zacarías-García, J., Rey, F., Gil, J.-V., Rodrigo, M. J., & Zacarías, L. (2021). Antioxidant capacity in fruit of Citrus cultivars with marked differences in pulp coloration: Contribution of carotenoids and vitamin C. *Food Science and Technology International, 27*(3), 210–222.

Zagoskina, N. V., Zubova, M. Y., Nechaeva, T. L., Kazantseva, V. V., Goncharuk, E. A., Katanskaya, V. M., Baranova, E. N., & Aksenova, M. A. (2023). Polyphenols in plants: Structure, biosynthesis, abiotic stress regulation, and practical applications (review). *Federation International Journal of Molecular Sciences, 24*(18). https://doi.org/10.3390/ijms241813874

Zema, D. A., Calabrò, P. S., Folino, A., Tamburino, V., Zappia, G., & Zimbone, S. M. (2018). Valorisation of citrus processing waste: A review. *Waste Management, 80*, 252–273. https://doi.org/10.1016/j.wasman.2018.09.024

Zgórka, G. (2009). Pressurized liquid extraction versus other extraction techniques in micropreparative isolation of pharmacologically active isoflavones from Trifolium L. species. *Talanta, 79*(1), 46–53. https://doi.org/10.1016/j.talanta.2009.03.011

Zhou, M., Fakayode, O. A., & Li, H. (2023). Green extraction of polyphenols via deep eutectic solvents and assisted technologies from agri-food by-products. *Molecules, 28*(19). https://doi.org/10.3390/molecules28196852

Zibaee, E., Kamalian, S., Tajvar, M., Amiri, M. S., Ramezani, M., Moghadam, A. T., Emami, S. A., & Sahebkar, A. (2020). Citrus species: A review of traditional uses, phytochemistry and pharmacology. *Current Pharmaceutical Design, 26*(1), 44–97.

Zou, Z., Xi, W., Hu, Y., Nie, C., & Zhou, Z. (2016). Antioxidant activity of citrus fruits. *Food Chemistry, 196*, 885–896.

CHAPTER

Bee products for nutraceutical applications

7

Hassan Laaroussi[1, 6], Asmae El Ghouizi[1, 2], Meryem Bakour[1, 2], Driss Ousaaid[1, 3], Soumaya Touzani[4] and Badiaa Lyoussi[5]

[1]*Laboratory of Biotechnology, Conservation, and Valorization of Bioresources (BCVB), Research Unit: Api-Phytotherapy, Physiology, Environment, and Health, Department of Biology, Faculty of Sciences Dhar Mehraz, Sidi Mohamed Ben Abdellah University, Fez, Morocco;* [2]*The Higher Institute of Nursing Professions and Health Techniques (ISPITS), Fez, Morocco;* [3]*Laboratory of Drug Sciences, Faculty of Medicine, Pharmacy, and Dental Medicine, Sidi Mohamed Ben Abdellah University, Fez, Morocco;* [4]*Laboratory of Mini-Invasive Surgery, Robotics, Artificial Intelligence and Educational Innovations, Faculty of Medicine Pharmacy and Dentistry, University Sidi Mohamed Ben Abdellah, Fez, Morocco;* [5]*Laboratory of Natural Substances, Pharmacology, Environment, Modeling, Health, and Quality of Life (SNAMOPEQ), Department of Biology, Faculty of Sciences Dhar Mehraz, Sidi Mohamed Ben Abdellah University, Fez, Morocco;* [6]*Centre of Biological Engineering (CEB), University of Minho (UMINHO), Braga, Portugal*

1. Introduction

Over the past decades, the establishment of bio-organic products as sustainable sources of a wide range of high-value-added bioactive molecules has supported their growing demand and justified the growing research in this area. This fact, coupled with consumer awareness and their increasing demand for healthier lifestyles based on organic, safer, and nutritious functional food products encouraged the investigation of using natural antioxidant compounds, such as flavonoids, phenolic acids, stilbens, lignans, vitamins (C, E, A, K), and minerals (zinc, selenium, iodine, etc.) present in plants and their by-products as primary and secondary metabolites (Aziz et al., 2022; Novais et al., 2022).

Several biological matrices have been largely investigated regarding their content of micro/macro-antioxidant components and their functional effect in cosmetic, pharmaceutical, and nutraceutical industries, specifically fruits, agro-food by-products, aromatic medicinal plants, algae, and microalgae (Socas-Rodríguez et al., 2021). Recently, special attention has been given to the most produced and consumed bee-hive products worldwide, namely, honey, propolis, bee pollen, and bee bread, from which a high amount of secondary metabolites and other bio-valuable molecules has been recovered (Ali et al., 2010; Kowalczuk et al., 2023; Moskwa et al., 2023).

Nowadays, the design of more effective and alternative methods for the transformation of products and fortification of foodstuffs with a simultaneous reduction in

Plant Polyphenols in the Food and Cosmetic Industries. https://doi.org/10.1016/B978-0-443-19137-4.00008-9

environmental footprint and energy expenditure is recognized as the biggest challenge in today's industrial progress (Shabir et al., 2023). Antioxidant compounds, mainly phenolic molecules, have been used not only for medicinal and pharmaceutical purposes but also as fortified ingredients and techno-functional agents to improve the quality and functionality of end-food products/foodstuffs, contributing to a sustainable development based on circular economy (CE) and zero waste (ZW) concepts (Kieliszek et al., 2023).

For this purpose, this book chapter intends to discuss the potential application of clean processes and innovative technologies for the recovery of antioxidant compounds from beehive products and highlight their recent advanced incorporation as functional ingredients in the nutraceutical sector.

2. Bee-hive products: a sustainable source of bio-valuable molecules

Beehive products such as bee pollen, bee bread, honey, and propolis are among the most interesting sources of several important nutrients and bioactive compounds, making them functional products of excellence not only for bees but also for humans (Kolayli & Keskin, 2020).

In the hive, bee pollen and bee bread were produced as follows: firstly, bees gather pollen grains from flowers, combine them with nectar and enzymes, carry the mixture to the hive as pellets, and finally store it in wax-sealed cells inside the hive. This process creates bee pollen. Following that, lactic fermentation is carried out to produce bee bread (Bakour et al., 2022b).

Bee products are considered natural, which fulfill the basic requirements to be safe and efficacious against many human diseases. Since the Stone Age, bee products have been coming to the forefront as natural medications and have been widely discussed and utilized by different civilizations. Being one of the most popular natural products and the oldest foods covering nutritional needs. In the last few decades, several breakthrough discoveries have unveiled the complexity and chemical composition of bee products, which make them complete food.

This part provides an in-depth summary of the macro and micronutrients of bee pollen, bee bread, propolis, and honey. These products are highly dependent on the environmental conditions, especially the botanical origin and geographic location (Tsuruda et al., 2021). The macro and micronutrients constitute important constituents that reflect the quality and biological properties of bee products.

The chemical composition of bee products is highly related to several factors, such as floral origin, ecoclimatic conditions, nutrition, health status of bees, harvesting techniques, processing, and storage (Mohammed, 2022).

Bee nutrition constitutes a key determinant in the macro- and microelement content of honey. Regarding this impact, the scientists found that bees fed with different proportions of persimmon sap produced significant variability in different physicochemical characteristics of honey (Pourakbari et al., 2012). In addition, feeding bees with different syrups affects the enzymes, vitamins, macro and micro-elements concentrations of honey (Collison, 2016; Guler et al., 2017).

The response of bees to multiple factors could also be the principal factor to explain the variability of the physicochemical characteristics of honey. The study of the molecular response of bees against infections revealed that numerous genes are implicated in the process, including immune genes (Doublet et al., 2017). Glucose oxidase and serine protease were found downregulated after *Apis mellifera* infection, which can affect the chemical characteristics of bee products (Doublet et al., 2017).

2.1 Macronutrients in bee products

2.1.1 Proteins and amino acids

Bee secretions and floral (nectar, honeydew, and pollen grains) are typically used to explain where the amino acids in bee bread and bee pollen come from Bakour et al. (2022a). According to World Health Organization and United Nations University (2007), amino acid requirements in diet are as follows: histidine (15 mg per gram of proteins), isoleucine (30 mg per gram of proteins), leucine (59 mg per gram of proteins), lysine (45 mg per gram of proteins), methionine (16 mg per gram of proteins), threonine (23 mg per gram of proteins), tryptophan (6 mg per gram of proteins), valine (23 mg per gram of proteins), phenylalanine + tyrosine (38 mg per gram of proteins).

The maintenance of cell structure, the synthesis of hormones and enzymes, biochemical catalysis, waste removal, and the initiation of apoptosis are all tasks that proteins are physiologically involved in LaPelusa and Kaushik, (2021). For this reason, consuming protein is crucial to a healthy diet for people. Plants and animals can both provide enough protein, but animal proteins are highly associated with health threats (Bessada et al., 2019). This brings up the issue of food security and the challenge of finding sustainable and ecologically viable high-nutritional foods with alternative protein sources. In this regard, plant proteins (such as soya and fava beans, chia and quinoa seeds, almonds, and nuts) are increasingly being employed as a cost-effective alternative to animal protein in the human diet, as well as a functional ingredient in the food industry (Sá et al., 2020), and as a result, plant-based diets are becoming globally more popular.

2.1.2 Carbohydrates

Besides proteins, carbohydrates are considered the main nutrients for bees in the hive. Carbohydrates make up a large portion of bee pollen (13%–55%) and bee bread (24%–35%), respectively (Campos et al., 2008). Depending on their nature, dietary carbohydrates have been acknowledged to be health beneficial for many disorders such as cancer, diabetes, and cardiovascular diseases (Goldenberg et al., 2021; Mann et al., 2007), and they are required in the diet of physically active people (Kanter, 2018).

As primer metabolites in plants, carbohydrates are mainly found in bee pollen with an average of 15%–55%, and they are mainly composed of polysaccharides (starch) and cell wall materials (Campos et al., 2008).

2.1.3 Fiber

Dietary fibers are compounds of plant origin, consisting of nonstarch complex carbohydrates and lignin. These elements are not digestible in the small intestine of mammals due to the absence of specific enzymes. Dietary fibers are compounds of plant origin, consisting of nonstarch complex carbohydrates and lignin. These elements are not digestible in the small intestine of mammals due to the absence of specific enzymes. Fibers have no energy value (Turner & Lupton, 2011). In bee pollen, the outer coat of the pollen grain, which is composed of cellulose and sporopollenin, is the main source of the crude fiber (Negrão et al., 2014). The amount of fiber content in bee pollen varied according to the botanical origin and seasonal variation (Al-Kahtani & Taha, 2021; Anjos et al., 2017). The bee pollen samples originating from Poland, Romania, Spain, Hungary, Colombia, Bulgaria, India, and China showed an average content of 14.5 $\pm$ 3.5%. Brazilian bee pollen contains a range from 12.79 to 20.30 g/100g. In addition, it was found that Oak and Hardy kiwi bee pollens have a content of 4.2 $\pm$ 0.53% and 3.1 $\pm$ 0.38% of fiber, respectively. Portuguese bee pollens have a range of crude fibers oscillating from 1.23% to 5.45%. According to Brazilian legislation, bee pollen should contain a 2% minimum content of crude fiber in bee pollen (Bárbara et al., 2015).

Similarly, bee bread was rich in crude fiber. In a study conducted by Aylanc et al. (2023), it was revealed that bee bread samples from Portugal have a range of content of 3.8–4.8 g/100g.

Recent evidence has proved that dietary fiber intake has many health benefits, such as the protection against cardiovascular diseases, the reduction in the prevalence of ischemic strokes, the decrease in systolic and diastolic blood pressure, the prevention of obesity and diabetes (Anderson et al., 2009; Lattimer & Haub, 2010).

2.1.4 Fatty acids

The term "fatty acid" refers to an organic acid that has at least one carboxyl group (–C (=O) OH, –COOH, or –CO2H) and a long carbon chain. The links of which can be double bonds in the case of unsaturated fatty acids or single bonds in the case of saturated fatty acids. The primary constituents of dietary fats are FAs, which are typically derived from triglycerides and phospholipids. Saturated fatty acids (SFAs), monounsaturated fatty acids (MUFAs), and polyunsaturated fatty acids (PUFAs) are all included depending on the number of double bonds (Chen & Liu, 2020).

As consumers, we are required to consume fatty acids for the growth, good/proper functioning of our body, and even to prevent diseases, especially cancer and cardiovascular dysfunction (Abdelhamid et al., 2018; Ilag, 2018). This highlights the importance and suggests incorporating fatty acids from organic and sustainable sources into the daily human diet.

2.2 Micronutrients in bee products

2.2.1 Vitamins

Vitamins are organic compounds involved in a wide range of health advantages, such as bone and teeth formation, tissue maintenance, and acting as cofactors for numerous enzymatic reactions (Godswill et al., 2020).

2.2.2 Minerals

Foods contain mineral elements in a variety of chemical forms. Compounds, complexes, and free ions are among these forms (Damodaran et al., 2007). Minerals are necessary for the skeleton's structural integrity as well as the soft tissues' regulatory processes, which include the transmission of neuromuscular signals, blood coagulation, oxygen transport, and enzyme activities (National Research Council, 1989).

3. Eco-friendly process to recover bioactive compounds from bee-hive products

The recovery of functional and bio-valuable extracts/ingredients or purified molecules from foods, plants, and plant-derived resources is a leading strategy and essential step to promote and target the industrial use of natural and biological matrices for subsequent application in cosmetic, pharmaceutical, and nutraceutical industries.

It is well recognized that the use of nonconventional/organic solvents in the extraction and the fractioning process of selected bio-ingredients might possess a negative impact on the environment and cause harmful effects on the consumer's health because the total elimination and removal of the utilized hazardous solvents from the final extract cannot be ensured, they may pass to the food matrix where they will be used (Duan et al., 2016).

Usually, numerous petroleum solvents have been used for the recovery of micro/macro-antioxidant phytochemicals and other high-added value molecules from medicinal plants, agro-industrial by-products, marine sources, and other functional foods (Torres-Valenzuela et al., 2020).

Since conventional solvents with different polarities are required for the selective extraction and isolation process of target phyto-compounds of different chemical structures, the choice of solvents should be based on the polarity and chemical nature of the biomolecules to be extracted, which increases their selectivity toward a chemical group of molecules or even a precise molecule. For instance, hydrosoluble antioxidant molecules are easily extracted using water, ethanol, and methanol as polar solvents and/or propanol and acetone alone or mixed with water (Ferreira-Santos, Genisheva, et al., 2020). However, these organic solvents are generally considered inefficient since they take a long time to purify or even to extract the desired molecules and require higher amounts of solvents, which leads to a large amount of hazardous waste generation.

Subsequently, to decrease the environmental footprint and to avoid the harmful effects of derivatives and residues of organic solvents in the foodstuff, the search for new safer and more effective extraction processes, including ecological solvents (biodegradable, recyclable, nonvolatile, and nontoxic solvents), and green technologies remains necessary.

Nowadays, processing and food industries are undergoing a real revolution as a response to consumers' expects and changes legislations toward more eco-friendly agricultural systems and green solutions to extract and preserve/enhance food quality, contributing to more sustainable and circular economic development based on the "zero waste concept," Which consents to a rational combination of healthy, environmental, economic, and social benefits (Sagar et al., 2018). In this context, green extraction techniques offer significant advantages in the recovery of bioactive extracts by improving extraction efficiency, reducing the use of hazardous solvents, and minimizing energy consumption and environmental impact (Fig. 7.1).

The reduction and minimization of the use of toxic solvents in extractions is now considered one of the priorities of the policy of the European Union. In this view, the EU policy from 2010 to 2050 targets the reduction in the use of toxic solvents in the

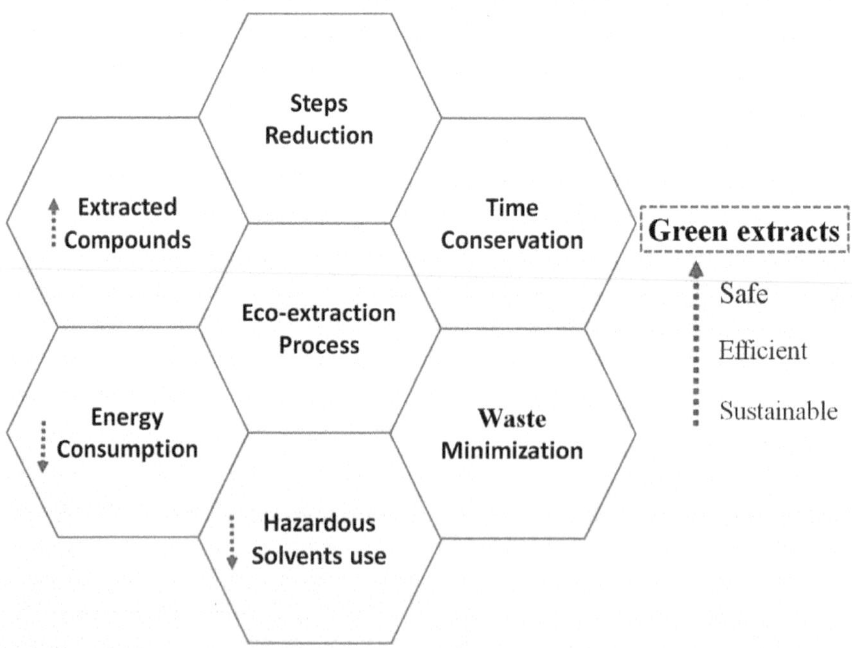

FIGURE 7.1

Advantage of green extraction techniques in the process of recovering bioactive extracts.

extraction process of bio-compounds destined for human consumption (Bubalo et al., 2018). Accordingly, the selection of eco-solvents is generally based on their safety (skin and respiratory system absorptions, toxicity, mutagenicity, carcinogenicity, and other toxic effects), process safety (volatility, explosiveness, flammability, mass transfer), environmental impact (persistence, water contamination, ecotoxicity, ozone layer depletion), and process sustainability (solvent recycling ability). Meaning that green solvents have to be chemically and physically stable, and recyclable with low volatility. Moreover, a suitable solvent should also be effective in producing high-quality extracts and maintaining the bio-properties of the extracted components to meet both technological and economic demands (Murador et al., 2019).

For this reason, nonconventional safe solvents (including water, deep eutectic solvents, carbon dioxide, and ethanol) have been recommended to be used in modern extraction and green process technologies for the recovery of a variety of target molecules of different chemical classes from different plants and plant by-products such as bee hive products. These include pressurized technologies like supercritical fluid extraction (SFE) and subcritical water extraction (SWE), as well as microwave-assisted extraction (MAE), ultrasound-assisted extraction (UAE), ohmic heating-assisted extraction (OHAE), and other extraction processes (Table 7.1).

In the following section, the above-mentioned clean processes and innovative technologies used to recover phenolic compounds from bee-hive products will be discussed.

3.1 Polyphenols as the main targeted bio-compounds

Due to the above-mentioned advantages, green processes are progressively used for the recovery of bio-valuable molecules. These technologies have been applied for antioxidant molecules' extraction, especially phenolic compounds, from a wide variety of foods, plant-based foods, agro-foods, and their by-products. From the same point of view, nonconventional methods used for polyphenols recovery from bee-hive products have been recently examined (Table 7.1).

Polyphenols are multifunctional chemical substances present in a variety of aquatic and terrestrial bio-resources such as algae, fruits, cereals, vegetables, and medicinal plants, among others. Currently, 4000 flavonoids have been identified among more than 8000 phenolic compounds that are known. Secondary metabolites known as polyphenols are crucial for the protection and development of plants. They also contribute to some sensorial properties such as astringency, bitterness, and color.

From a chemical structure point of view, the presence of hydroxyl groups and benzene ring(s) is a common and shared characteristic of all phenolic molecules. In addition to the carbon side chain, the positioning of hydroxyl groups on the benzene ring(s) gives rise to a variable number of phenolic molecules (Ferreira-Santos, Zanuso, et al., 2020). According to their chemical structure, flavonoids and nonflavonoid molecules are regrouped into two main groups, namely lipophilic and

Table 7.1 Green technologies to recover bioactive compounds from bee-hive products.

Propolis			
Extraction methods	**Solvents/Conditions**	**Extracted compounds**	**References**
Ultrasound-assisted extraction	Ethanol-water (70–30,v/v) 20 kHz/60 °C/15 min	Apigenin (10.50 mg/g), caffeic acid (16.50 mg/g), chrysin (126.71 mg/g), p-coumaric acid (212.94 mg/g), galangin (17.99 mg/g), isorhamnetin (17.99 mg/g), kaempferol (10.50 mg/g), luteolin (5.25 mg/g), myricetin (194.94 mg/g), pinocembrin (13.50 mg/g), rutin (59.23 mg/g), quercetin (9.75 mg/g).	Oroian et al. (2020)
	Ethanol-water (80–20, v/v) 40 kHz/30 °C/30–60 min	30 min: TPC (97.96 mg GAE/mL of extract), TFC (67.50 μg QE/mL of extract); 45 min: TPC (97.56 mg GAE/mL of extract), TFC (41.58 μg QE/mL of extract), 60 min: TPC (80.2 mg GAE/mL of extract), TFC (25.09 μg QE/mL of extract).	Zainal et al. (2022)
	Ethanol-water (70–30, v/v) 40 kHz/60 °C/30 min	TAA (732 mg GAE/100g), TPC (2218 mg GAE/100g)	Medina-Jaramillo et al. (2022)
	Ethanol-water (99–1, v/v) 40 kHz/50 °C/120 min	TPC (612.14 mg GAE/g), FRAP (534.39 μM ferrous sulfate./g), DPPH (72.37 μmol TEAC/g)	Cavalaro et al. (2020)
	Ethanol-water (60–40, v/v) 40 kHz/220 W/53°C/16 min	Benzaldehyde (0.41%), caffeic acid (3.03%), chrysin (0.87%), p-coumaric acid (2.11%), ferulic acid (1.94%), benzoic acid (1.88%), galangin (2.04%), rutin (1.69%), quercetin (1.92%), isorhamnetin (1.52%), kaempferol (0.96%).	Ding et al. (2019)
	Ethanol-water (70–30, v/v) 20 kHz/210 W/30 min	Extraction yield (11.86%), TPC (98.74 mg GAE/g), TFC (15.07 mg QE/g), chrysin, vanillin, p-cinnamic acid, ferulic acid, caffeic acid, benzoic acid, benzene acetic acid, 4-hydroxybenzoic acid, 3- methoxy-4-pentoxybenzoic acid, D-glucuronic acid, malic acid, and malonic acid.	Pobiega et al. (2019)
Ultrasound-assisted extraction-low pressure	Ethanol-water (80–20, v/v) 60 kHz/75 °C/30min	Quercetin (2.24 mg/g), gallic Acid (0.44 mg/g), formononetin (6.66 mg/g), kaempferol (0.78 mg/g), p-coumaric acid (0.10 mg/g), caffeic acid (3.02 mg/g), catechin (1.16 mg/g), epicatechin (11.49 mg/g), rutin (0.48 mg/g).	Teixeira et al. (2023)
Supercritical fluid extraction	CO2/50 °C; 150 bar/60min	Caffeic acid (4.1 g/100g), ferulic acid (2.1 g/100g), p-coumaric acid (36.8 g/100g), CAPE (2.8 g/100g), cinnamic acid (3.8 g/100g), quercetin (3.2 g/100g), kaempferol (13.2 g/100g), galangin (8.3 g/100g), myricetin (1.3 g/100g), pynobanksin (8.8 g/100g), chrysin (12.7 g/100g).	Fachri et al. (2020)
	CO2/50 °C; 33.7 MPa/60min	TPC (28.95 μg GAE/mg), FRAP (0.48 μmolFe2+/mg), p-coumaric acid (1.16 μg/mg), pynobanksin (1.39 μg/mg), chrysin (2.03 μg/mg), pinocemrin (36.63 μg/mg), galangine (8.08 μg/mg), pinostrobin (79.56 μg/mg).	Kuś et al. (2018)

Supercritical/high pressure–assisted extraction	Ethanol-water (80–20, v/v) 60 kHz/350 bar/50 °C/20min	Quercetin (2.06 mg/g), gallic Acid (2.15 mg/g), formononetin (8.39 mg/g), kaempferol (0.34 mg/g) p-coumaric acid (7.87 mg/g), caffeic acid (0.23 mg/g), catechin (0.66 mg/g), epicatechin (0.51 mg/g), rutin (4.17 mg/g).	T. D. Teixeira et al. (2023)
Microwave-assisted extraction	Ethanol-Water (70:30, v/v) 140 W/30 °C/1 min	Apigenin (9.97 mg/g), caffeic acid (15.67 mg/g), chrysin (120.35 mg/g), p-coumaric acid (202.24 mg/g), galangin (17.09 mg/g), isorhamnetin (17.09 mg/g), kaempferol (9.97 mg/g), luteolin (4.98 mg/g), myricetin (185.15 mg/g), pinocembrin (13.50 mg/g), rutin (56.26 mg/g), quercetin (9.26 mg/g).	Oroian et al. (2020)
	Ethanol-Water (70:30, v/v) 700 W/10 S	TPC (1.218 mg GAE/L), tartaric acid esters (356 mg QE/L), total flavonols (55 mg CAE/L), antioxidant activity (84%)	Yıldırım (2022)
	Water 700 W/10 S	TPC (96 mg GAE/L), tartaric acid esters (135 mg QE/L), total flavonols (98 mg CAE/L), antioxidant activity (79%)	
	Ethanol-Water (70:30, v/v) 700 W/30 S	TPC (1.536 mg GAE/L), tartaric acid esters (413 mg QE/L), total flavonols (114 mg CAE/L), antioxidant activity (37%)	
	Water 700 W/30 S	TPC (1.36 mg GAE/L), tartaric acid esters (133 mg QE/L), total flavonols (53 mg CAE/L), antioxidant activity (27%)	
	Water 800 W/5 min	Vanillin (6.00 mg/g), p-coumaric acid (10.5 mg/g), ferulic acid (7.6 mg/g), vanillic acid (3.7 mg/g), caffeic acid (0.6 mg/g).	Juodeikaitė et al. (2022)
	Ethanol 700 W/10 s	DPPH (IC50 = 1.77 mg/mL), TPC (205.07 mg/mL)	Mojica et al. (2021)
	Ethanol-water (80:20, v/v) 2450 kHz/300 W/106°C 15 min	Total phenolic acids (81.3 mg/g), total flavones (17.3 mg/g), total flavonols (17.2 mg/g), total flavanones (16.3 mg/g), total dihydroflavonols (30.1 mg/g), total flavonoids (81.6 mg/g), total phenolics (172.2 mg/g), caffeic acid (3.4 mg/g), p-coumaric acid (8.6 mg/g), ferulic acid (6.2 mg/g), isoferulic acid (1.8 mg/g), 3,4- dimethyl-caffeic acid (5.2 mg/g), quercetin (0.9 mg/g), cinnamic acid (1.6 mg/g), p inobanksin-5-methyl-ether (4.0 mg/g), quercetin- 3-methyl-ether (1.2 mg/g), apigenin (1.0 mg/g), pinobanksin (4.1 mg/g), kaempferol (1.2 mg/g), isorhamnetin (1.4 mg/g), luteolin (0.8 mg/g), cinnamylidene acetic acid (6.9 mg/g), caffeic acid (4.3 mg/g), chrysin (13.2 mg/g), pinocembrin (16.3 mg/g), galangin (8.1 mg/g), caffeic acid phenylethyl ester (6.6 mg/g).	Pellati et al. (2013)
Ohmic heating-assisted extraction	Distilled water 16.6 KPa/100 V/58°C/30min	TPC (41.59 mg GAE/g), TFC (15.11 mg QE/g)	Kartika and Lastriyanto (2020)
Ohmic heating-assisted extraction	Ethanol-water (70:30, v/v), 16.6 KPa/220 V/37°C/30 min	TPC (45.572 mg GAE/g), TFC (15.19 mg QE/g)	
Batch-type extraction at elevated temperatures	Choline chloride:propylene glycol (1:1, n/n)/50°C/3 h	Artepillin C (2.66 mg/mL)	Funari et al. (2019)

Continued

Table 7.1 Green technologies to recover bioactive compounds from bee-hive products.—*cont'd*

Propolis			
Extraction methods	**Solvents/Conditions**	**Extracted compounds**	**References**
	Choline chloride:propylene glycol (1:2, n/n)/50°C/3 h	Artepillin C (2.23 mg/mL)	
	EtOH:water 7:3 (v/v)/50°C/3 h	Artepillin C (2.13 mg/mL)	
	Choline chloride: DL-Lactic acid; water 1:2:2 (mol/mol/mol)/50°C/3 h	Artepillin C (1.86 mg/mL)	
	L-lysine:water 1:10 (w/w)/50°C/3 h	Artepillin C (1.06 mg/mL)	
	CC:Gly:water 1:2:1 (mol/mol)/50°C/3 h	Artepillin C (0.55 mg/mL)	
High-pressure-assisted extraction	Ethanol:water, (75:25, v/v)/500 MPa/1 min	TPC (290.4 mg GAE/g), TFC (230.4 mg QE/g)	Jun (2006)
Bee pollen			
Ultrasound-assisted extraction	Choline chloride:1,2- propanediol (1:1 M ratio 30% aqueous solution)/35 min	Quercetin (0.383 μg/mg), naringenin (0.048 μg/mg), kaempferol (0.383 μg/mg), isorhamnetin (3.149 μg/mg)	Meng et al. (2018)
	Choline chloride: Lactic acid (1:1 M ratio) 90 W/15 min	Gallic acid (170.24 μg/100 g), protocatechuic acid (88.76 μg/100 g), ssalicyclic acid (13.67μg/100 g), caffeic acid (222.47μg/100 g), syringic acid (nd), chlorogenic acid (70.39 μg/100 g), verbascoside (nd), p-coumaric acid (168.20 μg/100 g), rutin (335.33 μg/100 g), trans ferulic acid (27.33 μg/100 g), hesperidin (5.27 μg/100 g), ethyl gallate (nd), indole-3-acetic acid (59.77 μg/100 g), myricetin (3138.09 μg/100 g), quercetin (2387.18 μg/100 g), lutolein (151.53 μg/100 g), naringenin (22.42 μg/100 g), kaempferol (13,882.01 μg/100 g), isorhamnetin (374.60 μg/100 g), apigenin (198.16 μg/100 g); total (22,457.01 μg/100 g).	
	Choline chloride: lactic acid (1:1 M ratio) 180 W/30 min	Gallic acid (134.47 μg/100 g), protocatechuic acid (123.75 μg/100 g), salicylic acid (19.98 μg/100 g), caffeic acid (244.92 μg/100 g), syringic acid (nd), chlorogenic acid (69.03 μg/100 g), verbascoside (nd), p-coumaric acid (250.92 μg/100 g), rutin (506.92 μg/100 g), trans ferulic acid (145.71 μg/100 g), hesperidin (16.99 μg/100 g), ethyl gallate (nd), Indole-3-acetic acid (94.26 μg/100 g), Myricetin (3416.63 μg/100 g), Quercetin (2975.79 μg/100 g), Lutolein (188.65 μg/100 g), Naringenin (80.30 μg/100 g), Kaempferol (17,834.48 μg/100 g), Isorhamnetin (562.70 μg/100 g), Apigenin (268.86 μg/100 g); Total (26,934.34 μg/100 g)	Çelik et al. (2022)

	Choline chloride: lactic acid (1:1 M ratio) 90 W/15 min	L-tryptophan (3.21 mg/kg), taurine (14.93 mg/kg), L-phenylalanine (65.17 mg/kg), L-tyrosine (43.70 mg/kg), L-leucine (51.63 mg/kg), L-isoleucine (11.55 mg/kg), gamma-aminobutyric acid (79.85 mg/kg), L-methionine (7.08 mg/kg), L-glutamic acid (237.62 mg/kg), beta-alanine (369.82 mg/kg), L-valine (55.13 mg/kg), ethanolamine (0.57 mg/kg), L-alanine (224.57 mg/kg), L-threonine (55.86 mg/kg), L-serine (128.20 mg/kg), L-glycine (nd), L-aspartic acid (1998.30 mg/kg), L-asparagine (nd), Trans-4-hydroxy L-proline (nd), L-proline (nd mg/kg), L-glutamine (1.43 mg/kg), L-citrulline (60.89 mg/kg), L-arginine (203.04 mg/kg), L-histidine (289.74 mg/kg), L- ornithine (9.55 mg/kg), L-lysine (9.55 mg/kg); total (17,161.74 mg/kg).	
	Choline chloride: lactic acid (1:1 M ratio) 180 W/30 min	L-tryptophan (2.93 mg/kg), taurine (12.46 mg/kg), L-phenylalanine (99.36 mg/kg), L-tyrosine (49.61 mg/kg), L-leucine (59.64 mg/kg), L-isoleucine (10.25 mg/kg), gamma-aminobutyric acid (109.09 mg/kg), L-methionine (9.76 mg/kg), L-glutamic acid (350.43 mg/kg), beta-alanine (1043.42 mg/kg), L-valine (71.61 mg/kg), ethanolamine (0.35 mg/kg), L-alanine (284.89 mg/kg), L-threonine (62.38 mg/kg), L-serine (293.26 mg/kg), L-glycine (nd), L-aspartic acid (1137.35 mg/kg), L-asparagine (nd), trans-4-hydroxy L-proline (nd), L-proline (15,387.87 mg/kg), L-glutamine (nd), L-citrulline (1.84 mg/kg), L-arginine (295.57 mg/kg), L-histidine (403.49 mg/kg), L-ornithine (16.08 mg/kg), L-lysine (138.45 mg/kg); total (19,952.83 mg/kg).	
Microwave-assisted extraction	Ethanol: Water (80:20, v/v)/25 °C 20 min	Quercetin-diglucoside (0.38 mg/g), methyl herbacetin-O-dihexoside (0.12 mg/g), quercetin-O-hexosyl-O-rutinoside (0.04 mg/g), methyl herbacetin-3-O-rutinoside (nd), kaempferol-O-derivative (1.05 mg/g), kaempferol-3-O-rutinoside (nd), quercetin-3-O-glucoside (nd), luteolin (1.10 mg/g).	Aylanc et al. (2022)
	Water/50 kHz/30 min	4-Hydroxy benzoic acid, abscisic acid, apigenin, cafeic acid, catechin, chlorogenic acid, cinnamic acid, chrysin, coumaric acid, epigenin, ferulic acid, galic acic, kaempferol, luteolin, pinocembrin, pinocembrin, resveratrol, rutin, sinapic acid, syringic acid, vanillic acid.	Muñoz et al. (2020)
	Ultrapure water/80 kHz/20 °C/30 min	Quercetin-diglucoside (0.38 mg/g), methyl herbacetin-O-dihexoside (0.13 mg/g), quercetin-O-hexosyl-O-rutinoside (0.05 mg/g), methyl herbacetin-3-O-rutinoside (nd), kaempferol-O-derivative (1.49 mg/g), kaempferol-3-O-rutinoside (nd), quercetin-3-O- glucoside (nd), luteolin (1.54 mg/g).	Aylanc et al. (2022)
	Ethanol-water (65:45. v/v)/500 W/25 °C/30 min	Rutin (nq), quercetin (1.37 mg/g), kaempferol (23.44 mg/g), isohamnetin (nq).	Lv et al. (2015)

Continued

Table 7.1 Green technologies to recover bioactive compounds from bee-hive products.—*cont'd*

Propolis			
Extraction methods	**Solvents/Conditions**	**Extracted compounds**	**References**
High hydrostatic pressure-assisted extraction	Ethanol: Water (50:50, v/v)/20 MPa/10 min	Quercetin, kaempferol, isorhamnetin, 13′-carboxy-alpha-tocopherol, cinnzeylanine, melilotoside C, sophoranone, brassicoside, hordatine B, mulberrofuran T, artocarpin, pebrellin.	Tuoheti et al. (2020)
Supercritical fluid extraction	CO_2:Ethanol (10:90, v/v); 31.3 °C; 25 MPa/15min	–	Arena et al. (2022)
Supercritical fluid extraction	CO_2/50°c; 30 MPa/90 min	Myristic acid (0.80 mg/g), pentadecanoic acid (1.32 mg/g), palmitic acid (44.78 mg/g), palmitoleic acid (0.38 mg/g), margaric acid (1.71 mg/g), stearic acid (2.84 mg/g), oleic acid (18.93 mg/g), linnoleic acid (14.87 mg/g), α-linolenic acid (7.32 mg/g), behenic acid (1.43 mg/g), saturated fatty acid (52.89 mg/g), monounsaturated fatty acid (19.31 mg/g), polyunsaturated fatty acid (22.19 mg/g), sterols (1.39 mg/g).	Xu et al. (2011)
Bee bread			
Ultrasound-assisted extraction	Ethanol-water (60:40, v/v) 45 kHz/65 °C/25min	Gallic acid (nd), protocatechuic acid (nd), p-hydroxybenzoic acid (nd), vanillic acid (nd), caffeic acid (0.10 mg/L), chlorogenic acid (nd), p-coumaric acid (0.11 mg/L), rosmarinic acid (0.23 mg/L), myricetin (3.15 mg/L), luteolin (1.17 mg/L), quercetin (0.06 mg/L), kaempferol (31.25 mg/L)	Dranca et al. (2020)
Microwave-assisted extraction	HNO_3:HCl (10:2, v/v)/20 min 180 °C/800 W	TPC (19.63–21.85 mg RUE/10 g), TFC (7.88–15.67 mg RUE/10 g), radical scavenging activity (17.23–27.40 mg RUE/10 g)	Adaškevičiūtė et al. (2019)
Honey			
Ultrasound-assisted extraction	Water/15 min/40 °C	Cannabidiolic acid, cannabigerolic acid, tetrahydrocannabinol, tetrahydrocannabinolic acid	Brighenti et al. (2019)
	Water/500 W/20kHz/20 min/25 °C	Catechin (4.63 μg/g), naringenin (12.79 μg/g), cinnamic acid (9.74 μg/g), kaempferol (nd), chlorogenic acid (nd), quercetin (19.84 μg/g), myricetin (nd), catechol (11.99 μg/g), ferulic acid (9.18 μg/g), gentisic acid (31.03 μg/g), caffeic acid (nd), gallic acid (183.92 μg/g), total (283.18 μg/g).	Makowicz et al. (2018)

Microwave-assisted extraction	Acidified water (HCl, pH = 2)/10 min/25 °C	Rhamnetin (1.594 μg/g), naringenin (3.492 μg/g), quercetin (4.851 μg/g), rutin (2.070 μg/g), quercetin (<0.002 μg/g), hesperidin (<0.001 μg/g), vanillic acid (12.69 μg/g), p-coumaric acid (46.51 μg/g), chlorogenic acid (53.23 μg/g), protocatechuic acid (30.33 μg/g), syringic acid (3.932 μg/g).	Biesaga and Pyrzyńska (2013)
	Aqueous acidified: ethanol (40:60, v/v) 10 min/25 °C	Rhamnetin (1.929 μg/g), naringenin (3.872 μg/g), quercetin (10.20 μg/g), rutin (7.133 μg/g), quercetrin (1.862 μg/g), hesperidin (6.893 μg/g), vanillic acid (15.85 μg/g), p-coumaric acid (275.1 μg/g), chlorogenic acid (81.82 μg/g), protocatechuic acid (56.41 μg/g), syringic acid (4.531 μg/g).	
	Acidified water (HCl, pH = 2)/160 W/10 min/ 25 (°C)	Rhamnetin (6.571 μg/g), naringenin (1.113 μg/g), quercetin (6.941 μg/g), rutin (2.670 μg/g), quercetrin (0.981 μg/g), hesperidin (<0.001 μg/g), vanillic acid (13.95 μg/g), p-coumaric acid (91.91 μg/g), chlorogenic acid (45.62 μg/g), protocatechuic acid (35.51 μg/g), syringic acid (4.221 μg/g).	
	Aqueous acidified:ethanol (40:60, v/v)/ 160 W/10 min/25	Rhamnetin (4.396 μg/g), naringenin (3.030 μg/g), quercetin (13.042 μg/g), rutin (4.494 μg/g), quercetrin (1.727 μg/g), hesperidin (4.881 μg/g), vanillic acid (25.42 μg/g), p-coumaric acid (250.1 μg/g), chlorogenic acid (73.61 μg/g), protocatechuic acid (61.58 μg/g), syringic acid (4.842 μg/g).	

hydrophilic antioxidants. For this reason, the recovery of these compounds from natural sources required the choice of suitable solvents and appropriate processes, green technologies, for example, to better recover and preserve the biological properties of the extracted components.

In recent years, scientists have been interested in extracting polyphenols and other high- value molecules from propolis, bee pollen, bee bread, and honey, as the main produced and consumed bee-hive products using different emergent technologies, including ultrasound-assisted technology (Table 7.1). For the recovery of these biomolecules, ultrasound (US) is usually used to assist the extraction process (a process known as ultrasound-assisted extraction). UAE is used both to produce natural extracts and for analytical uses. US assistance allows to intensify the extraction and reduces the extraction duration, energy consumption, and environmental impacts generated during the extraction process.

In the last decade, ultrasound-assisted extraction has been frequently used to recover biomolecules from functional foods and other natural sources, including propolis, bee pollen, bee bread, and other beehive products. UAE uses mechanical waves having a frequency higher than 20 kHz and selective solvents to extract the target bio-compounds (Okolie et al., 2019).

Passing through a flexible medium, US waves successively produce, as a result of longitudinal particle displacement, refraction, and compression phases, which in turn raise the solvent penetration into the vegetal cells, inducing the breakdown of cell walls and thus enhancing the release of intracellular constituents (extracts) into the extraction medium (Bouras et al., 2015). In the opposite of conventional techniques, UAE is an alternative leading technology with multiple advantages such as low solvent consumption, less processing time/shorter extraction duration, high yield recovery, the preservation of final product quality, and low extraction temperature. Moreover, the process's simplicity, low cost, and lower environmental impacts encourage the industrial use of this emerging technology as an efficient, eco-friendly process. Despite the advantages and characteristics of this technology, it is associated with some defects and drawbacks, including the degradation of extracted bio-compounds, particularly when applying higher ultrasound frequency waves for a long period. Indeed, US waves could alter chemical reactions, leading to the possible production of free radicals, oxidation, and hydrolysis of extracted thermolabile molecules (Dimitrov et al., 2022).

For instance, the UAE has been used to obtain extracts rich in phenolic compounds from propolis (Oroian et al., 2020). In this study, the application of ultrasound waves (20 kHz), using a mixture of ethanol and water (70–30,v/v) as eco-solvents at 60 °C for//15 min leads to high recovery of individual phenolic compounds compared with a conventional maceration technique, increasing their quantity as well as the displayed antioxidant activity of the whole extract. From the qualitative point of view, both extraction techniques recovered 12 phenolic components belonging to different chemical groups, namely Flavons (Apigenin, Luteolin, Chrysin), Flavonols (galangin, kaempferol, isorhamnetin, quercetin, and rutin), flavanones (pinocembrin), flavanol (myricetin), and hydroxycinnamic acids

(Caffeic acid and p-Coumaric acid). Moreover, X and coworkers evaluated the impact of extraction time on total phenolic compounds (TPC) and total flavonoid compounds (TFC) recovery from propolis (Zainal et al., 2022). Obtained data showed that the application of US (40 kHz), using ethanol and water (80–20, v/v) as solvents at 30°C, decreased the recovery of phenolic compounds (TPC and TFC) when the extraction process was longer. More precisely, at 30 min: TPC and TFC values were 97.96 mg GAE/mL of extract and 67.50 μg QE/mL of extract, respectively. However, at 60 min, values were decreased to 80.2 mg GAE/mL of extract and 25.09 μg QE/mL of extract for TPC and TFC, respectively. This could be due to the negative impact (the possible degradation of extracted bio-compounds), particularly when applying higher ultrasound frequency waves for a long extraction time.

In the same context, bee pollen and bee bread were also subjected to UAE (Aylanc et al., 2022; Dranca et al., 2020; Meng et al., 2018; Muñoz et al., 2020; Çelik et al., 2022). In a recent study, Saffet and collaborators investigated the effect of the UAE method and deep eutectic solvents (lactic acid and choline chloride) on the recovery of several bioactive molecules from bee pollen. In this regard, the effects of different process variables such as sonication time (15, 30, and 45 min), the molar ratio of the DES (1, 1.5, and 2), and ultrasonic waves power (90, 135, and 180 W) on individual phenolic compounds individual and total organic acids, were explored by response surface methodology (RSM) (Çelik et al., 2022). Accordingly, the authors documented that gallic acid, protocatechuic acid, salicylic acid, caffeic acid, chlorogenic acid, p-coumaric acid, rutin, trans ferulic acid, hesperidin, indole-3-acetic acid, myricetin, quercetin, luteolin, naringenin, kaempferol, isorhamnetin, and apigenin were the most quantified phenolic compounds. However, syringic acid, verbascoside, and ethyl gallate were not detected in all extraction conditions. In addition, kaempferol, quercetin, and myricetin were recovered at higher concentrations using deep eutectic solvents (DESs) as compared to the maceration technique using ethanol as a control solvent. Moreover, the authors showed that ultrasonic waves with a power of 180W, a sonication time of 45 min, and a choline chloride/lactic acid (1:1 M ratio) are the optimal extraction conditions. Although the advantage they showed for obtaining bio-functional ingredients, till now, there is no study investigating NADES as nonconventional solvents for the extraction and recovery of individual phenolic compounds and other bioactive antioxidant molecules from bee bread.

Typically, natural deep eutectic solvents (NADES) are recognized as emerging green solvents derived from abundant, cheaper, safe, and biodegradable natural compounds that are present in our daily diet, such as sugars, amines, carboxylic acids, and choline polyalcohol. NADEs show very good physicochemical characteristics based on their high solubilization degree, adaptable viscosity, liquid state at different temperatures, and low volatility among other characteristics. Which fully fit the principles of green chemistry. Although they can be modified to have increased affinity for the target compounds, the Federal Drug Administration (FDA) does not yet regulate NADES since there is little knowledge regarding the biological activity and toxicity of the extracted materials, which restricts their industrial use, especially in the pharmaceutical industry (Bubalo et al., 2018; Martins et al., 2017). Moreover, the high boiling point characteristic of these eco-solvents (affecting their separation from the final mixture) might limit their industrial applications. In addition to the UAE, other extraction technologies have also been used to extract polyphenols as value-added molecules from bee products. Evaluating propolis as a raw material, Mircea and coworkers identified and quantified 12 phenolic compounds extracted from Romanian propolis using an ethanol-water mixture (ethanol-water (70:30, v/v)) as solvents and reported that Microwave-Assisted Extraction (140 W) produces antioxidant-rich extracts with substantial polyphenols content since it can attain high temperatures, which is a critical factor during bio-compounds extraction process, as most of these processes are temperature dependent. Accordingly, the antioxidant extract obtained at 60 °C is rich in flavonoids and nonflavonoid molecules (phenolic acids), namely: apigenin, caffeic acid, chrysin, p-coumaric acid, galangin, isorhamnetin, kaempferol, luteolin, myricetin, pinocembrin, rutin, and quercetin. In another work, x et al. investigated the effect of eco-solvents (water alone or mixed with ethanol) and the extraction time for the recovery of phenolic compounds from propolis using MAE (Yıldırım, 2022). Results show that the highest TPC, tartaric acid content, and total flavonols content were recorded in an ethanol-water mixture (70:30, v/v) and 30 s as extraction conditions. However, the lowest values were referred to MAE using water alone at an extraction time of 10 s. Owing to its viability, affordability, safety, and environmentally friendly characteristics, Water is widely used as the cleanest solvent for the extraction of various functional compounds from different biological matrices, including bee products. Despite these advantageous characteristics, it is not a suitable/appropriate solvent for the recovery of nonpolar and less polar substances, which supports the obtained results (Yıldırım, 2022).

Bee bread and bee pollen have also been subjected to MAE to extract phenolic compounds. For instance, Adaškevičiūtė et al. (2019) have examined the potential of MAE for the recovery of total polyphenols and total flavonoids content using water as a solvent and found high amounts of several individual phenolic acids, flavan-3-ols, and flavonols in the obtained extract.

Various phenolic compounds and other micro/macro-antioxidant molecules are extracted from bee pollen and honey. Applying MAE, Aylanc et al. (2022) extracted luteolin, quercetin-glucoside, methyl herbacetin-O-dihexoside, quercetin-O-hexosyl-O-rutinoside, kaempferol-O-derivative, using ethanol-water mixtures (80:20, v/v) from bee pollen. Other studies on MAE for polyphenols recovery from bee products have been summarized in the present chapter (Table 7.1). Organic honey is another functional food very consumed all over the world. It is well known for its high nutritional and medicinal properties. Recently, water or acidified water has been recognized as the main co-solvent used to better extract phenolic compounds from organic honey. Magdalena and collaborators (Biesaga & Pyrzyńska, 2013) studied the phenolic composition of organic honey extracted by UAE and MAE techniques. Obtained results showed that the application of ultrasound waves, using acidified water (HCl, pH = 2) as a solvent at 25 °C during 10 min, increased the recovery of several flavonoids and phenolic acids specifically, naringenin, quercetin, rutin, quercetin, hesperidin, vanillic acid, p-coumaric acid, chlorogenic acid, protocatechuic acid, syringic acid compared with MAE, except of Rhamnetin which is better recovered by MAE than UAE.

Moreover, novel and emergent technologies such as Ohmic heating extraction (OHE), supercritical fluid extraction (SFE), and high-pressure-assisted extraction (HPAE) have been adopted for the recovery of phenolic compounds from bee products (Kartika & Lastriyanto, 2020; Tuoheti et al., 2020; Xu et al., 2011). Recently, Annisa et al. (Kartika & Lastriyanto, 2020) examined the effect of ohmic heating pretreatment and distilled water or ethanol-water mixture (70:30, v/v) of propolis on TPC and TFC. Obtained results documented that a high applied voltage allows for better extraction of both flavonoids and nonflavonoid phenolic compounds. In general, this nonpulsed electro-technology might enhance the recovery of high-added-value compounds of natural origins by promoting the formation of pores (electro-permeabilization) in cell membranes and thus facilitating and increasing the solvent(s)-biomolecules surface interaction (Teixeira, 2018). In addition to its reliability as a novel extraction technology, OHE decreases the environmental impacts caused by the extraction process of conventional techniques by reducing waste generation and water consumption (Chemat et al., 2017). Moreover, enhanced solvent extraction or pressurized solvent extraction, commonly known as high-pressure-assisted extraction, has been broadly applied to recover valuable antioxidant components with potential industrial applications from different medicinal plants, functional foods, and agro-industrial waste (Fierascu et al., 2019). Recently, Tuhanguli and collaborators applied HPAE technique during 10 min at 30 °C and 20 MPa conditions to extract antioxidant organic compounds such as quercetin, kaempferol, Isorhamnetin, 13′-Carboxy-alpha-tocopherol, cinnzeylanine, melilotoside C, sophoranone, brassicoside, hordatine B, mulberrofuran T, artocarpin, pebrellin from bee pollen using ethanol-water (50:50, v/v) as liquid solvents (Rocha et al., 2018). In specific conditions, the possibility of using one or a mixture of organic solvents changes the polarity of the extraction fluid, allowing for more selective extraction of the desired components, such as phenolic compounds. Globally, to be considered clean/green, a recovered extract or compounds, especially those intended directly or indirectly (incorporated in foodstuffs) for human consumption must meet several quality criteria and go on hand with specific guidelines: it should be natural, of high physical and chemical quality with potential functional properties (e.g., antibacterial, antifungal, antioxidant, color, flavor, and foaming properties), with less environmental footprint. To achieve these requirements, the given extracts must be pasteurized and sterilized to denature undesirable microorganisms. In addition, extracts may also contain pollutants that are typically transferred from polluted soils or the environment to plant materials (nectar, resin, or/and pollen) during bee-hive products production/storage such as heavy metals, pesticides, polycyclic aromatic hydrocarbons (PAHs), and many other contaminants.

3.2 Environmental life cycle assessment of phenolic compound recovery process

The Environmental Life Cycle Assessment (E-LCA) is a strong and crucial quantitative concept for decision-making in a variety of industries since it offers information and data on product development and helps in creating strategic planning policies to identify and quantify the environmental impacts of products and used processes, starting from the raw material (for example, samples and solvents, among others), until the final product (extracts, antioxidants, and other bio-compounds, residues, etc., see Fig. 7.2) (Barba et al., 2019; Barjoveanu et al., 2020). It is critical, in this field, to determine the viability of various extraction methods, which requires the analysis and quantification of the consumed energy, the used reagents and solvents, the needed time, the recycling process, and the organization of the end product to provide proof-of-concept and pinpoint implementation strategies.

Since bee-hive products are one of the most naturally occurring resources that can be converted into other goods with a high added value, the incorporation of their valuable ingredients, especially polyphenols, is crucial not only for health care (cosmetic, nutraceutical, and pharmaceutical industries) but also for the basic bio-circular economy (Bakour et al., 2021).

According to the International Standards Organization (ISO) 14,040 and 14,044 (ISO, 2006a; 2006b), the ELCA targets the following objectives: collect and list the relevant inputs and outputs from the product system, evaluate possible potential environmental issues related to each production system, interpret the data from

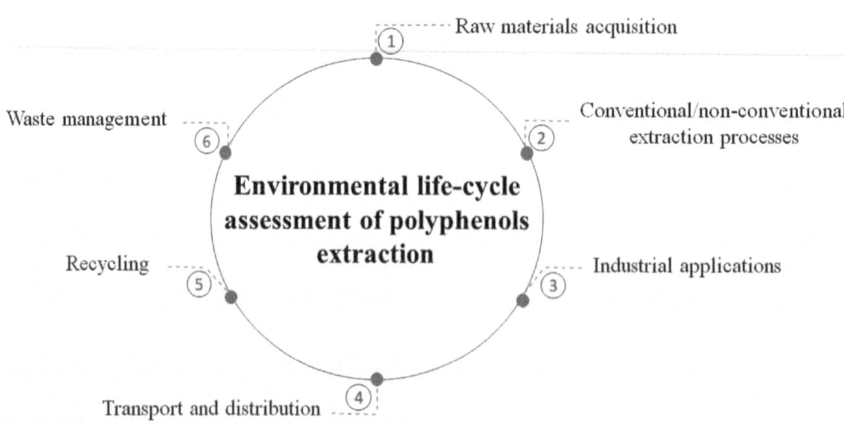

FIGURE 7.2

Environmental life-cycle assessment of polyphenols extraction.

the inventory analysis, and conduct scale assessment and expected impact of the study's goals. To date, there is no study approaching the ecological footprint and life cycle assessment to explore the environmental impacts of polyphenols extraction from different bee-hive products; nevertheless, there are some general scenarios that must be taken into consideration during the bio-active molecules' recovery process from bee-products too. These include the use of hazardous solvents, their emissions, and the required energy for their recycling.

These scientific claims imply that the application of E-LCA as a tool for quantitative scale in the recovery process of polyphenols and other valuable ingredients not only assesses the social demand and required process for obtaining the final product, but also makes it possible to quantify how the generated co-products or involved techniques affect the renewable and nonrenewable resources, climate change, and the environment as a whole. Allowing for a better valorization of different bio-resources while maintaining environmental sustainability under the concept of a "zero-waste value chain" (Vega et al., 2021).

3.3 Advanced applications of bee products as natural biochemicals in the food industry

To understand the role of bioactive compounds of bee products in food processing and their application in the food industry, it is necessary to emphasize/establish the relationship between natural products' biomolecules and the food industry. Emphasis on the different plant phytochemical compounds used in the food industry.

Throughout history, natural products (which include any substance produced by living organisms) have been utilized to enhance various food characteristics. Plant extracts or plant mixtures have been widely used in food preparations for their aromatic, preservative, and coloring properties. Besides their therapeutic properties, largely due to the abundance of secondary metabolites they contain, natural products have played a significant part in food processing.

Given the importance of several additive classes in food manufacturing, with a great majority being of synthetic origin, their use has been remarkably increased. Nowadays, and considering the safety concerns, the food industry is actively engaged in seeking natural alternatives. These compounds are intended to serve as potential natural food ingredients or bio-based additives, trying to satisfy the increased customer expectations and prioritize natural sources for a safer and more sustainable approach.

In this framework, numerous studies have been carried out on major bio compounds (proteins, carbohydrates, lipids) and minor bio compounds such as antioxidants, vitamins, or minerals and their beneficial use as preservative agents of food organoleptic characteristics, flavoring agents, fragrances, and functional additives. These natural compounds could exert an effect on both the food matrix and the organism that consumes it (González-Manzano & Dueñas, 2021).

Plant-derived compounds such as plant extracts and essential oils or plant by-products generated from food processing are widely used in the food industry due to their remarkable bioactive properties, especially antimicrobial activity (Gyawali & Ibrahim, 2014). For instance, essential oils obtained from spices and herbs have gained widespread use as preservatives in various food products, including fruits, vegetables, juices, milk, cheese, beef, chicken, sausages, and fish. Their potential as natural and sustainable additives has garnered increasing interest in the field of food processing (Li et al., 2022; Saeed et al., 2022). Due to rising environmental regulations and increased public focus on food safety and health, essential oils are gaining attention as shelf-life enhancers due to their antibacterial and antifungal properties and are being researched as natural food preservatives (Atki et al., 2019; Liu et al., 2021; Moghadam et al., 2023; Somrani et al., 2020; Zhang & Wang, 2019).

In this context, F. Youssef and her coworkers demonstrated the antimicrobial, antiviral, antimycobacterial, anti-*Helicobacter pylori*, antitrypanosomal, and antioxidant activities of *C. indicum* and *C. morifolium,* suggesting that their essential oils could be used in different food products as natural preservatives possessing antioxidant potential (Youssef et al., 2020). Another work carried out by J. Mota et al. has demonstrated that the inclusion of 10% of sweet lupin protein extract in formulations improves the nutritional value and quality of cookies (Mota et al., 2020). Furthermore, olive leaf extract has been examined for its preservative properties on retailed poultry meat quality; results showed that olive leaf extract exhibits potential utility as a dual-function agent, serving as a natural antioxidant and an antimicrobial preservative for chilled poultry meat held at refrigerated temperature (Saleh et al., 2020). The outcomes of another previous investigation showed that Gum *Cordia,* which is a natural polysaccharide easily extracted from the mature *Cordia* fruit, can be used in the formulation of cornstarch-based products as a substitute for chemical and enzymatic modifications in terms of cost-effectiveness, safety, reduced time consumption, and minimal process adjustments needed (Hussain et al., 2020).

3.3.1 Bee-hive products in the food industry

At present, the way food is produced and consumed is constantly evolving from mass production to standardized methods. This change is driven by increasing consumer demands, the focus on convenience in daily life, and the awareness that better nutrition can enhance health and overall well-being and prevent nutrition-related diseases. In response to these trends, the incorporation of bioactive components into food products is steadily growing as a means to combat nutrient deficiency. Consequently, there has been an emergence of "functional foods," denoting food products fortified by essential nutrients and bioactive compounds, such as lipids, proteins, and antioxidants, which hold significant importance for human health (Fernández-García et al., 2009; Galanakis, 2021).

Bee products, particularly honey, pollen, bee bread, and propolis, are foods that not only fulfill basic nutritional requirements but also have a beneficial impact that positively affects human health. They are recognized as "functional food" when integrated into other food products to enhance their nutritional value or when consumed independently due to their natural and abundant nutritional content and substantial bioactive constituents such as proteins, simple sugars, essential amino acids, monounsaturated fatty acids, and antioxidant compounds. These features strengthen immunity, help the body fight actively against bacteria, stimulate quality tissue regeneration, and consequently protect overall body health and treat it (Giampieri et al., 2022; Yucel et al., 2017). This section will include an overview of the functionality of bee products and their application in the food industry.

4. Honey as a functional food

As a natural functional food, honey is the main bee-derived product that has been valued for its nutritional and medicinal properties throughout various studies. Honey primarily contains water and sugars derived from fructose and glucose, as well as nutraceutical elements. Overall, honey is rich in various acids, including gluconic, succinic, malic, acetic, citric, and butyric acids. Eighteen free amino acids, with proline being the most abundant. In terms of vitamins, honey contains small amounts of ascorbic acid, niacin, pantothenic acid, riboflavin, and thiamine, which are not significant from a nutritional perspective. The mineral content of honey includes potassium, sodium, calcium, magnesium, chlorine, iron, copper, manganese, phosphorus, sulfur, and silica, but the levels can vary depending on the nectar source. In addition, honey contains small amounts of vegetable substances that contribute to its aroma and taste (Zammit Young & Blundell, 2023). Despite its complex composition, honey can be easily digested and assimilated.

Honey possesses the capacity to harbor live probiotics that are transferred from the guts of honey bees during honey production. Consequently, honey could serve as a valuable source of potential probiotics in the food industry (Esawy et al., 2012). The probiotics derived from the honey bee gut consisted of a wide range of microorganisms, which included *Bifidobacterium*, lactic acid bacteria (LAB), a subgroup of LAB known as fructophilic lactic acid bacteria (FLAB), as well as yeasts and other bacterial types like *Bacillus* spp (Mustar & Ibrahim, 2022). In a study carried out by Olofsson et al., 13 lactic acid bacteria were found in significant concentrations in fresh honey sourced from the honeybee stomach (*Apis mellifera*) (Olofsson et al., 2016).

Several clinical studies have provided evidence supporting the beneficial impact of probiotics on human health, such as gastrointestinal conditions, including irritable bowel syndrome, gastrointestinal disorders, Helicobacter eradication, inflammatory bowel disease, and diarrheas, as well as on allergic diseases like atopic dermatitis. In addition, numerous clinical studies have demonstrated the effectiveness of probiotics in treating obesity, insulin resistance syndrome, type 2 diabetes, and nonalcoholic fatty liver disease. Moreover, probiotics have been shown to have positive effects on human health by boosting the body's immune system through immunomodulation and preventing certain types of cancer or managing cancer-related side effects (Markowiak & Śliżewska, 2017).

In addition to its probiotic composition, honey plays a crucial role as a prebiotic to keep probiotic bacteria alive. Prebiotics are acquired through the consumption of functional foods containing oligosaccharides that foster the growth of beneficial bifidobacteria while inhibiting the proliferation of harmful bacteria. This process ensures the optimal protection of the intestinal flora (Mitsuoka, 2014). Prebiotics consist of undigested carbohydrates such as lactulose, lactitol, oligosaccharides, and inulin. These undigested oligosaccharides are naturally present in fruits, vegetables, milk, and honey. Inulin is a mixture of fructose polymers found in nature as stored carbohydrates, with a varying polymerization degree ranging from 2 to 60 degrees. The units with a lower polymerization degree are known as fructooligosaccharides or oligofructose. Studies indicated that consuming fructo-oligosaccharides daily has a bifidogenic effect; they specifically encourage the growth of beneficial microorganisms like *Lactobacillus* and *Bifidobacterium* (Güleç & Sarper, 2022; Sabater-Molina et al., 2009).

Within this framework, honey has been extensively studied for the benefic effect of its probiotics and prebiotics. Indeed, the impact of honey oligosaccharides on the bacterial population in the human gut intestinal tract has been demonstrated by numerous studies, which revealed that honey with higher levels of oligosaccharides led to a substantial increase in the growth of beneficial bacteria. Hence, the dietary application of honey as a probiotic source and nondigestible oligosaccharides could maintain a positive microbial balance toward a more favorable bacterial community (Mohan et al., 2017). Mazruei Arani, N. et al. have proven that consuming honey probiotics for 12 weeks among patients with diabetic nephropathy showed positive outcomes in terms of insulin metabolism, total-/HDL-cholesterol levels, serum hs-CRP levels, and plasma MDA levels (Mazruei Arani et al., 2019).

5. Application of honey in the food industry

By considering the new challenges faced by the industrialized world, studies are now prioritizing the improvement of scientific knowledge regarding alternative approaches to attain optimal nutrition, which leads to the emergence of enriched processed foods with natural bioactive products as a viable solution to offer essential micronutrients and health-promoting substances. In this context, honey is widely used in food processing because of its physicochemical properties and bioactive composition.

5.1 Honey in bakery

Honey contains a high percentage of sugars such as fructose and glucose (80%–90%), which makes it advantageous for enhancing browning and retaining moisture in baked products. Honey liquid or honey powder can be marketed directly to the consumer or used as a functional ingredient in beverages, sauces, additives, nuts, or culinary preparations (« Full Text PDF », 2023). The study conducted by S.

Sathivel and coworkers suggested that the use of honey powder produced with retrograded starch offers the advantage of being easier to handle and evenly dispersed in products when compared to liquid honey. Moreover, the high heat stability of retrograded starch makes it suitable for baked or high-temperature cooked foods. The study indicates that bread made with either honey or honey powder shared similar qualities with bread baked using sucrose (Sathivel, 2013). Q. Tong et al. stated that honey powder is a valuable addition to bakery product formulations as it enhances their nutritional, sensory, and preservation qualities. It significantly influences the rheological properties of the dough, promoting fermentation, improving mixing tolerance, and contributing to crust formation in bread. They suggested that incorporating honey powder in baked products can preserve them fresh and moist, and prevent their deterioration (Tong et al., 2010). In another work, the mineral content of traditional bread enriched with 15% honey was determined by Inductively Coupled Plasma Atomic Emission Spectrometry (ICP-AES), and results suggested that the mineral profile of a honey-enriched bread showed a calcium and iron content up to 12.53% and 35.34% more, respectively (Juhaimi et al., 2016).

In the same frame, several patents have been filed regarding the use of honey in bread making. In China, Zhejiang Special Education Career Academy registered a patent introducing a novel preparation technique for making low-sugar wormwood honey soft bread. The bread's ingredients include wormwood powder, honey, strong flour, salt, yeast, egg, and butter. By incorporating specific raw materials such as wormwood and natural honey in the preparation process, the bread gains a unique flavor and increased flavone content while reducing sugar content. In addition, the technique helps enhance the utilization of certain metal ions, such as calcium, iron, and magnesium, by decomposing part of the phytic acid in the bread. A new dough soaking technology is introduced during the dough stirring process to address issues related to the use of nonflour functional auxiliary materials, which can result in hard texture, inadequate rising, and insufficient elasticity. The formula is scientifically balanced, considering nutrition, taste, color, and wormwood's strong flavor, catering to diverse regional taste preferences (解殿伟, 2018). Another Brazilian patent was filed for a formulation and preparation method for creating gluten-free honey bread with a chocolate coating. It pertains to the domain of dough formulations and bakery products, utilizing liquid ingredients like vegetable fat, honey, glucose syrup, and water, along with dry ingredients such as rice flour, potato starch, sugar, cassava starch, and dehydrated whole egg, which are mixed, homogenized, and baked (Brueckheimer, 2021).

5.2 Honey as a natural sweetener

Honey, which dates back to ancient times, is likely the oldest sweetener known to mankind. Throughout human history, honey held significant importance as a carbohydrate source and served as the only available sweetener until the 17th century, when sugar production started (Bogdanov et al., 2008). Honey is a natural sweetener that has the potential to substitute artificial sweeteners in beverages, bakery foods, dairy desserts, and candies without compromising health safety. Honey also has the capacity to enhance the acceptability of beverages and create an instant energy drink with natural ingredients (González-Montemayor et al., 2019). In a study conducted by S. Sharma et al., honey was used to prepare an acceptable ready-to-serve RTS drink of lemon juice and to evaluate its quality characteristics for 6 months under different storage conditions. The study demonstrated that the honey lemon ready-to-serve beverage had higher ascorbic acid content and antioxidant activity compared with sugar lemon RTS at the fresh stage. In addition, honey lemon RTS received superior sensory scores for color, flavor, taste, and overall acceptability even after 6 months of refrigerated storage, with minimal changes in quality and no microbial growth, suggesting that honey could be used as a completely natural sweetener in beverages since it proves to be storage-stable and suitable for commercialization (Sharma et al., 2016).

In this connection, J. Janik, Susan, and coworkers have filed a patent application for capsules containing honey or alternative sweeteners designed for beverage usage. The capsules are composed of materials that can either dissolve in fluids or remain intact and edible when added to beverages. Furthermore, these capsules can also incorporate nutritional supplements or other additives along with honey (Janik et al., 2007). Another Chinese patent was filed about a preparation method for honey grapefruit tea used for relieving fatigue. The mixture undergoes homogeneous processing and sterilization, resulting in the creation of fatigue-eliminating honey citron tea (刘本君, 2018). Mohammad a.m.s.h. Alzemi has registered a patent for a honey gelatin product used as a sweetener that consists of a teaspoon of filtered alfalfa honey combined with a solidified animal gelatin cube made from bovine gelatin, forming a cubic-shaped lozenge. The final product is made to be used for sweetening hot beverages such as tea and coffee (Alzemi, 2015). A separate patent was registered for a honey food bar containing 40% or more fat designed to mimic intermittent fasting and consists of a blend of macadamia nuts, honey, pecans, almonds, almond butter, poop, inulin, flax seeds, coconut oil, cocoa derivatives, and coconut flour. These honey food bars usually contain 100–300 kcal, with at least 40% of the calories derived from fat and 10%–30% from carbohydrates. The protein content is kept below 10 g (Longo, 2021).

5.3 Honey in dairy and fermented products

Dairy products are economically the most important food sector all over the world, and they are well-liked due to their elevated nutritional content and beneficial properties that contribute to improved health, including the alleviation of various health conditions. They are among the most important foods used in our daily diet and can be a suitable option to deliver bioactives to the body. Nowadays, there is increasing interest in incorporating honey into dairy products due to its role as a natural sweetener and its ability to induce various changes in numerous characteristics such as bacterial survival, nutritional composition, bio-ingredient content, antioxidant activity, and sensory properties. M. Mohamed Ismail et al. have shown that fortifying stirred yogurt (Rayeb milk) with 15% barley flour and 4% honey increased curd tension and water-holding capacity and decreased coagulation time and susceptibility to syneresis. The values of carbohydrate, total solids, dietary fiber, ash, total protein, water-soluble nitrogen, total volatile fatty acids, unsaturated fatty acids, oleic, linoleic, α-linolenic acids, and antioxidant activity were higher in Rayeb milk supplemented with barley flour and honey than in the control. That results in an acceptable functional fermented dairy product with potential health benefits (Ismail et al., 2018b). Another Egyptian study aimed to produce a functional fermented dairy product, specifically Rayeb milk, that possesses the nutritional and health benefits associated with goat's milk, dried dates, honey, and probiotics. Their chemical composition, starter bacteria populations, and sensory qualities were evaluated after 14 days of storage. Results showed that supplementing goat's milk with Tamr and honey resulted in reduced concentrations of saturated fatty acids in Rayeb milk. Also, Rayeb milk enriched with Tamr and honey exhibited higher levels of carbohydrates, total solids, dietary fiber, ash, total protein (TP), unsaturated fatty acids, omega-3, omega-6, omega-9 fatty acids, and antioxidant activity compared with the control. In addition, the presence of Tamr and honey significantly increased the populations of probiotic bacteria (*L. acidophilus and Bifidobacterium*), surpassing the recommended level for bifidobacteria as a probiotic. They suggested that the addition of Tamr and honey greatly enhanced the sensory characteristics of Rayeb milk (Ismail et al., 2018a). Another Egyptian study investigated the possibility of increasing the nutritional and health values of bio-yogurt by adding honey and the possibility of using honey as a prebiotic for yogurt cultures. Results showed that the inclusion of honey has led to an increase in the total solids content of the product, consequently enhancing its overall nutritional value. The combination of honey with yogurt milk significantly activated Bifidobacteria, indicating that honey can serve as both a sweetener and a prebiotic in the production of bio-yogurt. Organoleptic evaluations of the yogurt demonstrated no disparity in color and appearance, but notable differences were observed in terms of body, texture, and flavor. Results also showed that the addition of honey substantially improved the sensory evaluation scores of the resulting yogurt (Ammar et al., 2015). In the same context, a Brazilian study aimed to investigate the impact of incorporating honey on the technological, physicochemical, and sensory qualities of goat yogurt enriched with probiotic *Lactobacillus acidophilus* during refrigerated storage. Their outcomes demonstrated the successful integration of probiotic *L. acidophilus* and honey into a novel goat dairy product. This new product exhibited satisfactory nutritional and sensory quality, and its potential functional properties added value to the market (Machado et al., 2017).

Honey has been used in the fermentation process of alcoholic beverages. By way of example, a study was conducted by E. Getachew Fentie et al. to create a direct fermentation system for Tej, the traditional Ethiopian honey wine. The researchers isolated fermentative microbial strains from Tej samples and conducted thorough screening to identify suitable candidates. Results showed that the identified strains belonged to the Saccharomycetaceae and Lactobacillaceae families and demonstrated excellent tolerance to osmotic stress and a lower pH environment. This finding indicates that a mixture of Saccharomyces and Lactobacillus strains isolated from fermented honey wine can be utilized as a starter culture to produce Ethiopian honey Tej while preserving its major quality characteristics (Fentie et al., 2022).

Numerous patents have been registered for using honey in dairy products. In that regard, Chinese inventors have registered patents that relate to a kind of honey acidophilus milk and its preparation method that uses honey, Bee Pollen, milk, and beneficial bacteria as raw materials.

The method involves adding different beneficial bacteria in sequence and fermenting the mixture to prepare honey acidophilus milk, resulting in a completely natural and green nutraceutical yogurt without any added essence, pigment, or preservative. The inventors suggested that their invention helps the consumer to improve body immunity, delay aging, and provide health benefits such as skincare and liver protection. The "prebiotics" in honey and Bee Pollen promote the growth of probiotics during Yogurt fermentation, resulting in a more thorough fermentation and a finer and smoother mouthfeel (史云峰, 2016). In the same framework, a Russian inventor filed a patent for the method of production of functional ice cream by adding honey as an ingredient. These technologies enable the production of a product with great storage stability, higher food value, and a pleasant sour milk taste with a blooming honey tone (ПрисяЖная and Грибанова, 2016). Bee pollen (BP) is produced as a result of bees agglutinating pollen grains with salivary secretions, nectar, and/or honey.

6. Bee pollen and bee bread as functional and superfoods

Bee pollen is produced as a result of agglutinating pollen grains with bees' salivary secretions, nectar, and/or honey. After being stored in honeybee comb cells, bee pollen undergoes a process where it is mixed with digestive enzymes, honey, and organic acids secreted by the bees. This mixture then undergoes lactic acid fermentation, catalyzed by bacteria such as *Pseudomonas* spp. and *Lactobacillus* spp., as well as yeasts like *Saccharomyces* spp., under anaerobic conditions, leading to the transformation of bee pollen to bee bread. During the conversion process, various microorganisms, including lactic acid bacteria, aerobic bacteria, and yeasts, start to grow using pollen nutrients and cause a drop in pH (Aylanc et al., 2021). Bee pollen and bee bread are a valuable repository of bioactive natural compounds, including minerals, protein, carbohydrates, vitamins (A, C, and E), and lipids, including fatty acids such as Omega-3 and Omega-6, as well as bioactive micronutrients including vitamins and minerals (Bakour et al., 2022; El Ghouizi et al., 2023). In addition to its nutritional content, bee pollen offers an abundant reserve of potent antioxidant compounds such as resveratrol, quercetin, kaempferol, cinnamic acid, and caffeic acids (Bakour et al., 2022; El Ghouizi et al., 2023). Bee pollen is used in diets as supplementary nutrition due to its beneficial effects for the treatment of microbial infections or inflammation (Komosinska-Vassev et al., 2015). Bee bread and bee pollen are valuable and health-promoting food products that possess not only a high nutritional value but also play a significant role in metabolic regulation processes; therefore, they are employed to solve issues like children's poor appetite, developmental delays, and malnutrition in both children and elderly people. In addition, bee pollen and bee bread are recommended for use during recovery periods after surgeries and for individuals engaged in strenuous physical and mental activities. They also exhibit antibacterial effects and can help alleviate the adverse effects of antibiotic treatments (Kieliszek et al., 2018; Komosinska-Vassev et al., 2015). In terms of definition, the functional properties of food components or ingredients go beyond their nutritional value and play a significant role in their application in food products. These properties affect the final product's appearance, taste, texture, and overall acceptance. Bee pollen, being a complex colloidal system, contains various components like biopolymers, oil droplets, gaseous bubbles, and lipid crystals, whose interactions on the surface are vital in determining the functional traits of bee pollen (Thakur & Nanda, 2020). Bee pollen not only provides nutritional benefits but also possesses specific physicochemical composition and techno-functional properties, including water, protein, and lipid content, as well as protein solubility, carbohydrate solubility, and emulsifying ability. These properties make it suitable for various food applications. Furthermore, its oil-absorption capacity enables it to retain flavors and enhance mouthfeel, while also contributing to the consistency and structure of food products (Algethami et al., 2022). A recent study carried out by H. Laaroussi et al. found that bee pollen extracts contain 15 phenolic compounds from various families, such as flavons, flavonols, flavanones, flavan-3-ols, hydroxybenzoic acids, hydroxycinnamic acids, and stilbenes, which contribute to its strong antioxidant capacity and ability to inhibit digestive enzymes. Authors suggested that bee pollen has the potential as a valuable nutritional, bioactive, and antifoaming ingredient offering a natural alternative to synthetic products in the food industry (Laaroussi et al., 2022).

7. Application of bee pollen and bee bread in food manufacturing

7.1 In dairy fermented products

Based on their functional characteristics, bee pollen and bee bread can be used as natural food ingredients in a variety of food products. Literature data indicated that bee pollen and bee bread serve as a natural source of probiotics, housing a diverse array of microorganisms, including bacteria and fungi, which play a role in their production through bee pollen lactic fermentation. It has been found that bee bread contains Lactobacillus bacteria and fructophilic lactic acid bacteria, which offer a valuable reservoir of compounds with significant techno-functional properties. These compounds have practical applications in the food industry as natural preservatives and functional cultures for large-scale production of fermented foods (Baky et al., 2023). Presently, the food industry faces difficulties in preserving the survival of bifidobacteria in fermented dairy food because of the bacteria's high nutrient requirements and the occurrence of proteolytic and glycolytic reactions (Camacho-Bernal et al., 2021). In that respect, a recent study was conducted to explore the potential application of bee pollen, bee bread, and royal jelly as fortifiers in functional fermented dairy products. The effect of these ingredients on starter cultures and the overall physicochemical and sensory attributes of the products was assessed. Results of this study indicated that Bee pollen and bee bread did not show adverse effects on starter cultures or organoleptic properties, and suggested that incorporating bee product-fortified fermented dairy into daily consumption could provide valuable nutrients, given their high nutritional value and quality protein content. In addition, the products exhibit potent in vitro antioxidant capacities and proteolytic activities, all while maintaining favorable sensory attributes (Darwish et al., 2023). Other studies showed that adding bee pollen to yogurt production triggers significant increases in oil and protein content and exhibits positive effects on both physicochemical and sensory characteristics (Pop et al., 2015; Zlatev et al., 2018). Moreover, pollen increased probiotic features when pollen was added to milk instead of yogurt, in contrast to the sensory attributes, which were adversely affected (Yerlikaya, 2014).

In this context, there have been several patents filed that describe the use of bee pollen and bee bread in food products. A German patent was registered for the described foodstuff that contains a variety of ingredients such as dry fruits, cereals and seeds, nuts and seeds, plant oils, natural flavorings and spices, vitamins, minerals, honey, stabilizers, fruit juices, milk products, bee products (bee pollen, bee bread, royal jelly, and propolis). The final product delivers revitalizing energy to the body, refills depleted energy reserves after physical activity, and positively impacts overall health and well-being (Gleich, 2012). Another invention involved a highly nutritional buccal tablet made of 60%–90% natural bee bread that belongs to the bee product category (江武军 et al., 2023). Meanwhile, bee pollen has been the target of multiple patents since it is a highly nutritive product. A recent invention related to the healthcare beverage industry has placed particular emphasis on fermented sports drinks made of bee pollen and their method of preparation. These inventions aimed to obtain a muscle-increasing beverage with high nutritive value. And obtain better taste (张政 et al., 2022; 冯彬, 2017).

Based on the antioxidant properties and biomolecule composition, many patents were published concerning bee pollen alone or combined with other ingredients and intended for applications in therapy, as well as for use as food supplements or functional foods to promote human health (林军庭, 2014; 赖晔 et al., 2018).

7.2 Bee pollen and bee bread in bakery products

Biscuits are well-liked bakery items due to their convenient consumption, extended shelf life, and diverse flavors and textures. They are composed of flour, fat/oil, sugar, water, and chemical leavening agents like baking soda. Consumers have expressed concerns about the safety of these ingredients, which has led to numerous studies over the years to improve their nutritional value.

R. Végh and coworkers have conducted a study on biscuits enriched with monofloral bee pollens of different origins (rapeseed, phacelia, and sunflower), considering their nutritional characteristics, antioxidant properties, color, texture, sensory profile, customer preferences, and purchasing behavior. Results of this study reported that biscuits containing pollens of sunflowers were the most preferred by customers and recommended to be used for developing pollen-enriched foods in the future (Végh et al., 2023). Another study explored the incorporation of bee pollen with different percentages as a functional ingredient in cookies. Physicochemical and sensory attributes, along with determining the total phenolic content and antioxidant activity of the different fractions of the enriched cookies, were assessed. Results showed that bee pollen addition decreased the carbohydrate content of the cookies and increased ash, total protein, and total fat content, spread ratio, browning index, and hardness of the cookies. This indicated that the inclusion of bee pollen can enhance the quality aspects and functional attributes of the cookies (Dundar, 2022). However, in the health food industry, bee bread is less accessible compared with pollen, as it is more challenging to extract from the hive. However, due to its more potent composition compared with pollen, bee bread can be consumed in smaller quantities or for a shorter duration to achieve the desired health benefits (Kieliszek et al., 2023).

Be pollen has been employed as a principal ingredient in the preparation of bee pollen-based cookies. The invention described a method for producing a bee pollen cookie using bee pollen and plain flour as its main ingredients. The final cookies presented potential advantages in terms of cancer prevention and resistance, while also enhancing the immune response of the body. Its suitability for consumption extends to individuals of all age groups (程妮 et al., 2012).

8. Incorporation of propolis in the food system

Owing to its antioxidant and antimicrobial characteristics, propolis has attracted the attention of many researchers in numerous scientific fields with a wide range of industrial applications. Considering its physicochemical, nutritive, and techno-functional properties, Propolis is increasingly being used as a functional ingredient, which is driven by a current consumer trend toward a combination of "organic products" and a "healthy lifestyle." Today, consumers are looking for fresh food, barely processed with fewer synthetic preservatives, supplemented with natural molecules/extracts with sensory characteristics and health benefits. This new trend widens the consumer market and creates new business opportunities for propolis, involving its incorporation as a functional and techno-functional agent in the food industry.

Recently, propolis has been used in the food industry as a dietary supplement and active ingredient, conserving/improving the quality of end-food products, enhancing health, and preventing disease. In general, this natural product has a worldwide reputation as a powerful bio-antioxidant ingredient in foodstuffs, especially in fish and meat products; smooth foods, juice, yogurt, milk, and other dairy products (Tumbarski et al., 2022).

8.1 Preservation of fish and meat products

Meat and meat-derived foods are well known as major sources of fats, proteins, amino acids, and oligo-elements required for the proper functioning of our body. These products are extremely susceptible to microbial contamination and often act as vectors for a variety of food spoilage bacteria and foodborne pathogens. In addition to bacterial and fungal spoilage, some meat products like sausage and x are also vulnerable to lipid oxidation, which increases the danger of microbial spoilage, which increases the risk of food intoxication. In this regard, propolis or its active molecules can be used widely in the biopreservation of meat products as a natural antimicrobial and antioxidant agent. The antimicrobial properties and antioxidant potential of propolis extracts in the biopreservation of meat products have already been documented. Ali and coworkers evaluated the antioxidant and antimicrobial activities of propolis extract (0.6%) in a freshly prepared oriental sausage. Accordingly, propolis extract prevented lipid oxidation in fresh sausage and inhibited yeast mold lipolytic proteolytic proliferation. Moreover, propolis-treated sausages treated with propolis extract showed longer shelf life (21 days) as compared to the control sausages, which had a maximum shelf life of 12 days (Ali et al., 2010).

To evaluate the effect of encapsulated propolis extract (EPE) on the shelf life of sausages, Safaei and Roosta Azad (2020) examined the antibacterial effect of EPE and documented that the encapsulated propolis extract had a high inhibitory activity against P. *aeroginosa* and S. *aureus*, indicating that EPE can provide powerful antibacterial activity and might increase shelf life and long-term storage of sausage. In addition to the microbial aspect of meat, lipid peroxidation is among the most critical issues related to meat quality. In fact, lipid oxidation deteriorates the sensory quality and alters the physicochemical characteristics of meat.

Therefore, controlling the lipid oxidation status of meat and meat-derived products with the active packaging method containing antioxidant substances will positively contribute to the bio-preservation of the product to enhance the shelf life of meat sausages and improve their functional properties.

The biopreservation effect of propolis is most often related to its rich content of macro- and micro-antioxidant components, in which flavonoids and phenolic acids are recognized as major active molecules with great interest.

Aquaculture is susceptible to many bacterial contaminations, especially when grown in uncontrolled conditions. Infections in aquaculture lead to a drop in production rate and increased mortality, which have not only health issues but also have a heavy economic impact (Hatha et al., 2005). This required specific conditions and treatments of fish farms. However, the intensified use of antibiotics in aquaculture leads to antibiotic resistance in fish as well as in humans as final consumers. For instance, the emergence of resistance was specifically observed for *Pasteurella piscida, Aeromonas salmonicida, Yersinia ruckeri, A. hydrophila, V. anguillarum, Vibrio salmonicida, E. tarda, and Edwardsiella icttaluri.*

In general, most fish contaminants are extremely dangerous and can spread diseases to people through zoonotic or food-borne transmission. Therefore, it is imperative to search for valuable bio-antimicrobials from sustainable sources to combat fish pathogenic diseases. As a natural product, propolis brings together these characteristics. In this sense, Abd-El-Rhman and co-authors examined the effect of propolis extract (PE) against *Aeromonas hydrophila*. Results documented that PE boosts immunity and increases the resistance of *Oreochromis niloticus* against *Aeromonas hydrophila* (Abd El Rhman, 2009). Likewise, Tukmechi et al. (2010) studied the effect of propolis extract (PE) against common pathogenic bacteria in fish, namely *Aeromonas hydrophila*, *Yersinia ruckeri*, and *Streptococcus iniae*. Accordingly, the examined extract exhibited significant antimicrobial activity against the studied strains.

8.2 Possible issues and future challenges

Although eco-friendly extraction techniques possess potential and growing interest in functional and bio-valuable rich-antioxidant extracts recovery from functional foods, scientific information on using these processes is limited to case-by-case approaches. Despite the high number of publications and investigations targeting the eco-friendly techniques assisted extraction for the recovery of bio-antioxidant molecules from different biomass, including bee-hive products, industrial applications of MAE, UAE, and another innovative and emergent process for micro/macro-antioxidant extraction and polyphenols recovery from sustainable sources are quite rare and still limited to the production of high added value components from wastes and agro-food by-products.

In fact, the limited comprehensive data on individual phenolic components' absorption and bioavailability in vitro systems and in vivo organisms, and their chemical instability toward, for example, solvents, oxygen, light, enzymes, pH, and temperature changes, are recently recognized among the major issues and challenges. Moreover, the nonselective extraction, possible degradation of thermolabile compounds, and high cost of most sophisticated techniques are also disadvantages that limit their industrial scale-up (Santos et al., 2021).

Despite the drawbacks mentioned above, the increased consumer demand for a healthier lifestyle and safe food products with high nutraceutical and pharmaceutical interest justifies the continuous search for new, safer, and alternative sources of functional nutrients promoting a sustainable and healthy diet. The application of extracts from bee-hive products in the food industry is a leading strategy to enrich and fortify the target foodstuffs with different bio-valuable ingredients. However, this critical step required detailed registrations and legal regulations. Recently, legal regulations concerning the incorporation of honey, bee pollen, and propolis as food supplements have been approved in the European Union, Canada, Brazil, Korea, China, Japan,

Australia, and the United States (Berretta et al., 2017). Unfortunately, thus far, even with the high efforts made by the scientific community, there is no Moroccan standard or regulation that manages the quality of local bee products. This makes it necessary to define the quality norms and criteria standards first before moving on to fortifying foodstuffs based on extracts from bee-hive products.

9. Conclusion and perspectives

Beehive products, mainly honey, propolis, bee pollen, and bee bread, are an exhaustible source of biomolecules with high industrial interest. Trying to meet the high expectation levels and the growing demands of consumers toward a healthier lifestyle, intending to consume foods with fewer synthetic preservatives, the current trend involves the incorporation of beehive products, especially propolis and bee pollen, as natural preservative agents and techno-functional ingredients in food manufacturing. The use of bee hive products or their antioxidant-rich extracts in bread, yogurt, meat, and fish increases the nutritional value of the end-product, improves its techno-functional characteristics as well as boosts its antioxidant properties, and thus increases its shelf life. However, from the industrial point of view, there are some critical limitations and hindrances related to the use of bee hive products in the nutraceutical sector, in which the major obstacle is related to their wide variability and complexity in terms of chemical composition, which highlights the need for the typification and standardization of these biological matrices before their safe incorporation into liquid or solid foodstuffs.

To keep up with the evolution of food technology and considering the consumers' awareness, it can be stated that the use of standardized apiary products and their safe extracts as nutraceutical agents can lead to the creation and promotion of "clean-labeled" functional food.

References

Abd-El-Rhman, A. M. M. (2009). Antagonism of Aeromonas hydrophila by propolis and its effect on the performance of Nile tilapia, Oreochromis niloticus. *Fish & Shellfish Immunology, 27*(3), 454–459.

Abdelhamid, A. S., Martin, N., Bridges, C., Brainard, J. S., Wang, X., Brown, T. J., Hanson, S., Jimoh, O. F., Ajabnoor, S. M., Ho Deane, K., Song, F., & Hooper, L. (2018). Polyunsaturated fatty acids for the primary and secondary prevention of cardiovascular disease. *Cochrane Database of Systematic Reviews*, (7).

Adaškevičiūtė, V., Kaškonienė, V., Kaškonas, P., Barčauskaitė, K., & Maruška, A. (2019). Comparison of physicochemical properties of bee pollen with other bee products. *Biomolecules, 9*(12), 819.

Al-Kahtani, S., & Taha, E.-K. A. (2021). Seasonal variations in nutritional composition of honeybee pollen loads. *Journal of the Kansas Entomological Society, 93*(2), 105–112.

Algethami, J. S., Abd El-Wahed, A. A., Elashal, M. H., Ahmed, H. R., Elshafiey, E. H., Omar, E. M., Naggar, Y. Al, Algethami, A. F., Shou, Q., Alsharif, S. M., Xu, B., Shehata, A. A., Guo, Z., Khalifa, S. A. M., Wang, K., & El-Seedi, H. R. (2022). Bee pollen: Clinical trials and patent applications. *Nutrients, 14*(14), 2858. https://doi.org/10.3390/nu14142858

Ali, F. H., Kassem, G. M., & Atta-Alla, O. A. (2010). Propolis as a natural decontaminant and antioxidant in fresh oriental sausage. *Veterinaria Italiana, 46*(2), 167–172.

Alzemi, M.a m s h (2015). In *Honey cube. United States US20150004291A1, filed 27 juin 2013 and issued 1 janvier 2015*. https://patents.google.com/patent/US20150004291A1/en?q=(honey+sweetener)&oq=honey+sweetener&page=1.

Ammar, M. A., El-Tahra, M. M., Ismail, A., El-Wahab, E., Khalil, & Eid, M. Z. (2015). Impact of fortification with honey on some properties of bio- yoghurt. *Journal of Microbiology, Biotechnology and Food Sciences, 4*(6), 503–508. https://doi.org/10.15414/jmbfs.2015.4.6.503-508

Anderson, J. W., Baird, P., Davis, R. H., Jr., Ferreri, S., Knudtson, M., Koraym, A., Waters, V., & Williams, C. L. (2009). Health benefits of dietary fiber. *Nutrition Reviews, 67*(4), 188–205. https://doi.org/10.1111/j.1753-4887.2009.00189.x

Anjos, O., Santos, A. J. A., Dias, T., & Estevinho, L. M. (2017). Application of FTIR-ATR spectroscopy on the bee pollen characterization. *Journal of Apicultural Research, 56*(3), 210–218. https://doi.org/10.1080/00218839.2017.1289657

Arena, R., Medeiros, A. C. L.de, Secci, G., Mancini, S., Manuguerra, S., Bovera, F., Santulli, A., Parisi, G., Messina, C. M., & Piccolo, G. (2022). Effects of dietary supplementation with honeybee pollen and its supercritical fluid extract on immune response and fillet's quality of farmed gilthead seabream (Sparus aurata). *Animals, 12*(6), 675.

Atki, El, Yassine, I., El Aouam, F., Kamari, A., Taroq, K., Nayme, M., Timinouni, B., Lyoussi, A., & Abdellaoui. (2019). Antibacterial activity of cinnamon essential oils and their synergistic potential with antibiotics. *Journal of Advanced Pharmaceutical Technology & Research, 10*(2), 2019.

Aylanc, V., Ertosun, S., Russo-Almeida, P., Falcão, S. I., & Vilas-Boas, M. (2022). Performance of green and conventional techniques for the optimal extraction of bioactive compounds in bee pollen. *International Journal of Food Science and Technology, 57*(6), 3490–3502.

Aylanc, V., Falcão, S. I., Ertosun, S., & Vilas-Boas, M. (2021). From the hive to the table: Nutrition value, digestibility and bioavailability of the dietary phytochemicals present in the bee pollen and bee bread. *Trends in Food Science & Technology, 109*, 464–481. https://doi.org/10.1016/j.tifs.2021.01.042

Aylanc, V., Falcão, S. I., & Vilas-Boas, M. (2023). Bee pollen and bee bread nutritional potential: Chemical composition and macronutrient digestibility under in vitro gastrointestinal system. *Food Chemistry, 413*, Article 135597.

Aziz, A., Noreen, S., Khalid, W., Mubarik, F., Niazi, A. K., Ali, C. M. G., Lima, W. S., Alansari, Areej, A., & Eskandrani. (2022). Extraction of bioactive compounds from different vegetable sprouts and their potential role in the formulation of functional foods against various disorders: A literature-based review. *Molecules, 27*(21), 7320.

Bárbara, M., Machado, C., Sodré, G., Dias, L., Estevinho, L., & de Carvalho, C. (2015). Microbiological assessment, nutritional characterization and phenolic compounds of bee pollen from Mellipona mandacaia Smith, 1983. *Molecules, 20*(7), 12525–12544. https://doi.org/10.3390/molecules200712525

Bakour, M., Laaroussi, H., Elaraj, T., & Lyoussi, B. (2021). The beekeeping state and inventory of mellifero-medicinal plants in the North-Central of Morocco. *The Scientific World Journal, 2021*.

Bakour, M., Laaroussi, H., Ousaaid, D., El Ghouizi, A., Es-Safi, I., Hamza, M., & Badiaa, L. (2022a). Bee bread as a promising source of bioactive molecules and functional properties: An up-to-date review. *Antibiotics, 11*(2), 203.

Bakour, M., Laaroussi, H., Ousaaid, D., El Ghouizi, A., Es-Safi, I., Mechchate, H., & Lyoussi, B. (2022). Bee bread as a promising source of bioactive molecules and functional properties: An up-to-date review. *Antibiotics, 11*(2), 203. https://doi.org/10.3390/antibiotics11020203

Bakour, M., Laaroussi, H., Ousaaid, D., El Ghouizi, A., Es-Safi, I., Mechchate, H., & Lyoussi, B. (2022b). New insights into potential beneficial effects of bioactive compounds of bee products in boosting immunity to fight COVID-19 pandemic: Focus on zinc and polyphenols. *Nutrients, 14*(5), 942.

Baky, M. H., Abouelela, M. B., Wang, K., & Farag, M. A. (2023). Bee pollen and bread as a super-food: A comparative review of their metabolome composition and quality assessment in the context of best recovery conditions. *Molecules, 28*(2), 715. https://doi.org/10.3390/molecules28020715

Barba, F. J., Rosello Soto, E., Brncic, M., & Lorenzo Rodriquez, Jose M. (2019). *Green extraction and valorization of by-products from food processing*. CRC Press.

Barjoveanu, G., Pătrăuțanu, O.-A., Teodosiu, C., & Volf, I. (2020). Life cycle assessment of polyphenols extraction processes from waste biomass. *Scientific Reports, 10*(1), Article 13632.

Berretta, A. A., Arruda, C., Felipe, G. M., Baptista, N., Nascimento, A. P., Marquele-Oliveira, F., Hori, J. I., Barud, H., Damaso, B., & César, R. (2017). Functional properties of Brazilian propolis: From chemical composition until the market. *Superfood and Functional Food-An Overview of Their Processing and Utilization, 4*, 2017.

Bessada, S. M. F., Barreira, J. C. M., & Oliveira, M. B. P. P. (2019). Pulses and food security: Dietary protein, digestibility, bioactive and functional properties. *Trends in Food Science & Technology, 93*, 53–68.

Biesaga, M., & Pyrzyńska, K. (2013). Stability of bioactive polyphenols from honey during different extraction methods. *Food Chemistry, 136*(1), 46–54.

Bogdanov, S., Jurendic, T., Sieber, R., & Gallmann, P. (2008). Honey for nutrition and health: A review. *Journal of the American College of Nutrition, 27*(6), 677–689. https://doi.org/10.1080/07315724.2008.10719745

Bouras, M., Chadni, M., Barba, F. J., Grimi, N., Bals, O., & Vorobiev, E. (2015). Optimization of microwave-assisted extraction of polyphenols from Quercus bark. *Industrial Crops and Products, 77*, 590–601. https://doi.org/10.1016/j.indcrop.2015.09.018

Brighenti, V., Licata, M., Pedrazzi, T., Maran, D., Bertelli, D., Pellati, F., & Benvenuti, S. (2019). Development of a new method for the analysis of cannabinoids in honey by means of high-performance liquid chromatography coupled with electrospray ionisation-tandem mass spectrometry detection. *Journal of Chromatography A, 1597*, 179–186.

Brueckheimer, P. L. (2021). In *Composição e processo de preparo para obtenção de pão de mel sem glúten com cobertura de chocolate. BR102020000291A2, filed 7 janvier 2020 et issued 13 juillet 2021*. https://patents.google.com/patent/BR102020000291A2/en?q=(honey+bread)&oq=honey+bre ad.

Bubalo, M., Cvjetko, S., Radojčić Vidović, I., Redovniković, S., & Jokić. (2018). New perspective in extraction of plant biologically active compounds by green solvents. *Food and Bioproducts Processing, 109*, 52–73.

Çelik, S., Kutlu, N., Gerçek, Y. C., Bayram, S., Ravi, P., & Nesrin, E. B. (2022). Optimization of ultrasonic extraction of nutraceutical and pharmaceutical compounds from bee pollen with deep eutectic solvents using response surface methodology. *Foods, 11*(22), 3652.

Camacho-Bernal, G. I., Cruz-Cansino, N.del S., Ramírez-Moreno, E., Delgado-Olivares, L., Zafra-Rojas, Q. Y., Castañeda-Ovando, A., & Suárez-Jacobo, Á. (2021). Addition of bee products in diverse food sources: Functional and physicochemical properties. *Applied Sciences, 11*(17), 8156. https://doi.org/10.3390/app11178156

Campos, M. G. R., Bogdanov, S., Almeida-Muradian, L. B.de, Szczesna, T., Mancebo, Y., Frigerio, C., & Ferreira, F. (2008). Pollen composition and standardisation of analytical methods. *Journal of Apicultural Research, 47*(2), 154–161. https://doi.org/10.1080/00218839.2008.11101443

Cavalaro, R. I., de Freitas Fabricio, L. F., & de Souza Vieira, T. M. F. (2020). Ultrasound-assisted extraction of antioxidants from Baccharis dracunculifolia and green propolis. *Processes, 8*(12), 1530.

Chemat, F., Rombaut, N., Meullemiestre, A., Turk, M., Perino, S., Fabiano-Tixier, A.-S., & Abert-Vian, M. (2017). Review of green food processing techniques. Preservation, transformation, and extraction. *Innovative Food Science & Emerging Technologies, 41*, 357–377.

Chen, J., & Liu, H. (2020). Nutritional indices for assessing fatty acids: A mini-review. *International Journal of Molecular Sciences, 21*(16), 5695.

Collison, C. (2016). A closer look: Feeding sugar syrup/HMI. In *Bee culture*.

Damodaran, S., Parkin, K. L., & Fennema, O. R. (2007). *Fennema's food chemistry*. CRC press.

Darwish, A. M. G., Abd El-Wahed, A. A., Shehata, M. G., El-Seedi, H. R., Masry, S. H. D., Khalifa, S. A. M., Mahfouz, H. M., & El-Sohaimy, S. A. (2023). Chemical profiling and nutritional evaluation of bee pollen, bee bread, and royal jelly and their role in functional fermented dairy products. *Molecules, 28*(1), 227. https://doi.org/10.3390/molecules28010227

Dimitrov, K., Pradal, D., Vauchel, P., Fabiano-Tixier, A.-S., & Chemat, F. (2022). Ultrasounds technology to recover polyphenols. In *Technologies to recover polyphenols from agrofood by-products and wastes* (pp. 201–223). Elsevier.

Ding, Q., Wu-Chen, R. A., Wu, Q., Jiang, H., Zhang, T., Luo, L., Ma, H., Ma, S., & He, R. (2019). Kinetics of ultrasound-assisted extraction of flavonoids and triterpenes and structure characterization of Chinese northeast black bee propolis. *Chiang Mai Journal of Science, 46*(1), 72–92.

Doublet, V., Poeschl, Y., Gogol-Döring, A., Alaux, C., Annoscia, D., Aurori, C., Barribeau, S. M., Bedoya-Reina, O. C., Mark, J. F., Brown, J. C., & Bull. (2017). Unity in defence: Honeybee workers exhibit conserved molecular responses to diverse pathogens. *BioMed Central, 18*(1), 2017.

Dranca, F., Ursachi, F., & Oroian, M. (2020). Bee bread: Physicochemical characterization and phenolic content extraction optimization. *Foods, 9*(10), 1358.

Duan, L., Dou, L.-L., Guo, L., Li, P., & Liu, E.-H. (2016). Comprehensive evaluation of deep eutectic solvents in extraction of bioactive natural products. *ACS Sustainable Chemistry & Engineering, 4*(4), 2405–2411. https://doi.org/10.1021/acssuschemeng.6b00091

Dundar, A. N. (2022). Total phenolic and antioxidant bioaccessibilities of cookies enriched with bee pollen. *Journal of Food Processing and Preservation, 46*(6), Article e16085. https://doi.org/10.1111/jfpp.16085

El Ghouizi, A., Bakour, M., Laaroussi, H., Ousaaid, D., El Menyiy, N., Hano, C., & Lyoussi, B. (2023). Bee pollen as functional food: Insights into its composition and therapeutic properties. *Antioxidants, 12*(3), 557. https://doi.org/10.3390/antiox12030557

Esawy, M. A., Ghada, E. A., Awad, E. F., Ahmed, E. N., Danial, N., & Mansour. (2012). Evaluation of honey as new reservoir for probiotic bacteria. *Advances in Food Sciences, 34*, 72–81.

Fachri, B. A., Sari, P., Yuwanti, S., & Subroto, E. (2020). Experimental study and modeling on supercritical CO2 extraction of Indonesian raw propolis using response surface method: Influence of pressure, temperature and CO2 mass flowrate on extraction yield. *Chemical Engineering Research and Design, 153*, 452–462.

Fentie, E. G., Jeong, M., Emire, S. A., Dessalegn Demsash, H., Kim, M.-C., Lim, K., & Shin, J.-H. (2022). Development of mixed starter culture for the fermentation of Ethiopian honey wine, tej. *Scientific Reports, 12*(1), Article 13431. https://doi.org/10.1038/s41598-022-17594-1

Fernández-García, E., Carvajal-Lérida, I., & Pérez-Gálvez, A. (2009). In vitro bioaccessibility assessment as a prediction tool of nutritional efficiency. *Nutrition Research, 29*(11), 751–760.

Ferreira-Santos, P., Genisheva, Z., Botelho, C., Santos, J., Ramos, C., Teixeira, J. A., & Rocha, C. M. R. (2020). Unravelling the biological potential of pinus pinaster bark extracts. *Antioxidants, 9*(4), 334.

Ferreira-Santos, P., Zanuso, E., Genisheva, Z., Rocha, C. M. R., & Teixeira, J. A. (2020). Green and sustainable valorization of bioactive phenolic compounds from pinus by-products. *Molecules, 25*(12), 2931.

Fierascu, R. C., Fierascu, I., Avramescu, S. M., & Sieniawska, E. (2019). Recovery of natural antioxidants from agro-industrial side streams through advanced extraction techniques. *Molecules, 24*(23), 4212. https://doi.org/10.3390/molecules24234212

Funari, C. S., Sutton, A. T., Lajarim Carneiro, R., Fraige, K., Cavalheiro, A. J., Silva Bolzani, V.da, Hilder, E. F., & Dario Arrua, R. (2019). Natural deep eutectic solvents and aqueous solutions as an alternative extraction media for propolis. *Food Research International, 125*, Article 108559.

Full Text PDF. (2023). In *Consulté le juillet 26*. https://www.researchgate.net/profile/Buesra-Bildir/publication/370375534_Potential_Use_of_Honey_Powder_in_Food_Products/links/64 4cbc4a809a53502137fd3c/Potential-Use-of-Honey-Powder-in-Food-Products.pdf.

Güleç, H., & Sarper, F. (2022). Honey as a functional food; its culinary status. *Turkish Journal of Agriculture - Food Science and Technology, 10*(8), 1426–1433. https://doi.org/10.24925/turjaf.v10i8.1426-1433.5163

Galanakis, C. M. (2021). Functionality of food components and emerging technologies. *Foods, 10*(1), 128. https://doi.org/10.3390/foods10010128

Giampieri, F., Quiles, J. L., Cianciosi, D., Forbes-Hernández, T. Y., Orantes-Bermejo, F. J., Alvarez-Suarez, J. M., & Maurizio, B. (2022). Bee products: An emblematic example of underutilized sources of bioactive compounds. *Journal of Agricultural and Food Chemistry, 70*(23), 6833–6848. https://doi.org/10.1021/acs.jafc.1c05822

Gleich, A. (2012). In *Nahrungsmittel. Germany DE102011101102A1, filed 10 mai 2011 and issued 15 novembre 2012*. https://patents.google.com/patent/DE102011101102A1/en?q=(bee+bread)&oq=bee+bread+.

Godswill, A. G., Victory Somtochukwu, I., Ikechukwu, A. O., & Chinelo Kate, E. (2020). Health benefits of micronutrients (vitamins and minerals) and their associated deficiency diseases: A systematic review. *International Journal of Food Sciences, 3*(1), 1–32. https://doi.org/10.47604/ijf.1024

Goldenberg, J. Z., Day, A., Grant, D., Brinkworth, J., Sato, S., Yamada, T., Jönsson, J., Beardsley, A. J., Johnson, L., Thabane, et, Bradley, C., & Johnston. (2021). Efficacy and safety of low and very low carbohydrate diets for type 2 diabetes remission: Systematic review and meta-analysis of published and unpublished randomized trial data. *BMJ, 372*.

González-Manzano, S., & Dueñas, M. (2021). Applications of natural products in food. *Foods, 10*(2), 300. https://doi.org/10.3390/foods10020300

González-Montemayor, Á.-M., Flores-Gallegos, A. C., Serrato-Villegas, L. E., López-Pérez, M. G., Montañez-Sáenz, J. C., & Rodríguez-Herrera, R. (2019). Honey and syrups: Healthy and natural sweeteners with functional properties. In Natural Beverages, *édité par Alexandru Mihai Grumezescu et Alina Maria Holban* (pp. 143–177). Academic Press. https://doi.org/10.1016/B978-0-12-816689-5.00006-7

Guler, A., Garipoglu, A. V., Onder, H., Biyik, S., Kocaokutgen, H., Deniz, & Ekinci. (2017). Comparing biochemical properties of pure and adulterated honeys produced by feeding honeybees (Apis mellifera L.) colonies with different levels of industrial commercial sugars. *Kafkas Üniversitesi Veteriner Fakültesi Dergisi., 23*(2), 2017.

Gyawali, R., & Ibrahim, S. A. (2014). Natural products as antimicrobial agents. *Food Control, 46*(décembre), 412–429. https://doi.org/10.1016/j.foodcont.2014.05.047

Hatha, M., Vivekanandhan, A. A., & Julie Joice, G. (2005). Antibiotic resistance pattern of motile aeromonads from farm raised fresh water fish. *International Journal of Food Microbiology, 98*(2), 131–134.

Hussain, S., Mohamed, A. A., Mohamed, S. A., Ibraheem, M. A., Qasem, A. A. A., Ali Shahzad, S., & Ababtain, I. A. (2020). Use of gum cordia (Cordia myxa) as a natural starch modifier; effect on pasting, thermal, textural, and rheological properties of corn starch. *Foods, 9*(7), 909. https://doi.org/10.3390/foods9070909

Ilag, L. L. (2018). Are long-chain polyunsaturated fatty acids the link between the immune system and the microbiome towards modulating cancer? *Medicine, 5*(3), 102.

Ismail, M., Mohamed, H., & Esraa, M. E. (2018a). Quality of Rayeb milk fortified with Tamr and honey. *British Food Journal, 120*(2), 499–514. https://doi.org/10.1108/BFJ-04-2017-0259

Ismail, M. M., Mohamed, F. H., & Esraa, M. E. (2018b). Using goat's milk, barley flour, honey, and probiotic to manufacture of functional dairy product. *Probiotics and Antimicrobial Proteins, 10*(4), 677–691. https://doi.org/10.1007/s12602-017-9316-4

ISO. (2006a). *Environmental management: Life cycle assessment; principles and framework*. ISO.

ISO. (2006b). *Environmental management: Life cycle assessment; requirements and guidelines. 14044*. Geneva, Switzerland: ISO.

Janik, J., Janik, S., & Carlone, D. (2007). Capsule filled with honey. In *United States US20070071808A1, filed 29 novembre 2006 and issued 29 Mars 2007*. https://patents.google.com/patent/US20070071808A1/en?q=(honey+sweetener)&oq=honey+sweetener.

Juhaimi, A. L. F., Ghafoor, K., Babiker, E. E., Musa Özcan, M., & Harmankaya, M. (2016). *Mineral contents of traditional breads enriched with floral honey*. NISCAIR-CSIR.

Jun, X. (2006). Comparison of antioxidant activity of ethanolic extracts of propolis obtained by different extraction methods. *Canadian Journal of Chemical Engineering, 84*(4), 447–451.

Juodeikaitė, D., Žilius, M., & Briedis, V. (2022). Preparation of aqueous propolis extracts applying microwave-assisted extraction. *Processes, 10*(7), 1330.

Kanter, M. (2018). High-quality carbohydrates and physical performance: Expert panel report. *Nutrition Today, 53*(1), 35.

Kartika, A. A., & Lastriyanto, A. (2020). Propolis extraction using vacuum resistive heating method. *Journal of Physics: Conference Series, 1665*, Article 012009.

Kieliszek, M., Piwowarek, K., Kot, A. M., Błażejak, S., Chlebowska-Śmigiel, A., & Wolska, I. (2018). Pollen and bee bread as new health-oriented products: A review. *Trends in Food Science & Technology, 71*, 170–180. https://doi.org/10.1016/j.tifs.2017.10.021

Kieliszek, M., Piwowarek, K., Kot, A. M., Wojtczuk, M., Roszko, M., Bryła, M., & Petkoska, A. T. (2023). Recent advances and opportunities related to the use of bee products in food processing. *Food Science and Nutrition, 11*(8), 4372–4397. https://doi.org/10.1002/fsn3.3411

Kolayli, S., & Keskin, M. (2020). Natural bee products and their apitherapeutic applications. *Studies in Natural Products Chemistry, 66*, 175–196.

Komosinska-Vassev, K., Olczyk, P., Kaźmierczak, J., Mencner, L., & Olczyk, K. (2015). Bee pollen: Chemical composition and therapeutic application. *Evidence-Based Complementary and Alternative Medicine*, 297425. https://doi.org/10.1155/2015/297425

Kowalczuk, I., Gębski, J., Stangierska, D., Agata, & Szymańska. (2023). Determinants of honey and other bee products use for culinary, cosmetic, and medical purposes. *Nutrients, 15* (3), 737.

Kuś, P., Jerković, I., Jakovljević, M., & Jokić, S. (2018). Extraction of bioactive phenolics from black poplar (Populus nigra L.) buds by supercritical CO 2 and its optimization by response surface methodology. *Journal of Pharmaceutical and Biomedical Analysis, 152*, 128–136. https://doi.org/10.1016/j.jpba.2018.01.046

Laaroussi, H., Ferreira-Santos, P., Genisheva, Z., Bakour, M., Ousaaid, D., Asmae, El G., Teixeira, J. A., & Lyoussi, B. (2022). Unveiling the Techno- functional and bioactive properties of bee pollen as an added-value food ingredient. *Food Chemistry*, Article 134958. https://doi.org/10.1016/j.foodchem.2022.134958

LaPelusa, A., & Kaushik, R. (2021). Physiology, proteins. In *StatPearls*. StatPearls Publishing.

Lattimer, J. M., & Haub, M. D. (2010). Effects of dietary fiber and its components on metabolic health. *Nutrients, 2*(12), 1266–1289. https://doi.org/10.3390/nu2121266

Li, Y.-xin, Erhunmwunsee, F., Liu, M., Yang, K., Zheng, W., & Tian, J. (2022). Antimicrobial mechanisms of spice essential oils and application in food industry. *Food Chemistry, 382*, Article 132312. https://doi.org/10.1016/j.foodchem.2022.132312

Liu, X., Chen, L., Kang, Y., He, D., Yang, B., & Wu, K. (2021). Cinnamon essential oil nanoemulsions by high-pressure homogenization: Formulation, stability, and antimicrobial activity. *Lebensmittel-Wissenschaft & Technologie, 147*, Article 111660.

Longo, V. D. (2021). barra alimentar para simulação de jejum intermitente. In *BR112021008998A2, filed 12 novembre 2019 and issued 10 août 2021.* https://patents.google.com/patent/BR112021008998A2/en?q=(honey+food)&oq=honey+food &page=2.

Lv, H., Wang, X., He, Y., Wang, H., Yourui, & Suo. (2015). Identification and quantification of flavonoid aglycones in rape bee pollen from Qinghai- Tibetan Plateau by HPLC-DAD-APCI/MS. *Journal of Food Composition and Analysis, 38*, 49–54.

Machado, T. A. D. G., Gomes de Oliveira, M. E., Campos, M. I. F., Paloma Oliveira Antonino de Assis, Leite de Souza, E., Madruga, M. S., Pacheco, M. T. B., Pintado, M. M. E., & Rita de Cássia Ramos do Egypto Queiroga. (2017). Impact of honey on quality characteristics of goat yogurt containing probiotic Lactobacillus acidophilus. *Lebensmittel-Wissenschaft & Technologie, 80*, 221–229. https://doi.org/10.1016/j.lwt.2017.02.013

Makowicz, E., Jasicka-Misiak, I., Teper, D., & Kafarski, P. (2018). HPTLC fingerprinting—Rapid method for the differentiation of honeys of different botanical origin based on the composition of the lipophilic fractions. *Molecules, 23*(7), 1811.

Mann, J., Cummings, H. N. E., Key, Liu, G., Riccardi, C., Summerbell, U., & Van Dam, V. (2007). FAO/WHO scientific update on carbohydrates in human nutrition: Conclusions. *European Journal of Clinical Nutrition*, (1), 2007.

Markowiak, P., & Śliżewska, K. (2017). Effects of probiotics, prebiotics, and synbiotics on human health. *Nutrients, 9*(9), 1021. https://doi.org/10.3390/nu9091021

Martins, P. L. G., Rafaela Braga, A., & Vera de Rosso, V. (2017). Can ionic liquid solvents be applied in the food industry? *Trends in Food Science & Technology, 66*, 117–124.

Mazruei Arani, N., Emam-Djomeh, Z., Tavakolipour, H., Sharafati-Chaleshtori, R., Soleimani, A., & Asemi, Z. (2019). The effects of probiotic honey consumption on metabolic status in patients with diabetic nephropathy: A randomized, double-blind, controlled trial. *Probiotics and Antimicrobial Proteins, 11*(4), 1195–1201. https://doi.org/10.1007/s12602-018-9468-x

Medina-Jaramillo, C., Carvajal-Díaz, L. M., & López-Córdoba, A. (2022). Propolis from native stingless bees: Ultrasound-assisted extraction. *Vitae, 29*(2).

Meng, Z., Zhao, J., Duan, H., Guan, Y., & Zhao, L. (2018). Green and efficient extraction of four bioactive flavonoids from pollen typhae by ultrasound-assisted deep eutectic solvents extraction. *Journal of Pharmaceutical and Biomedical Analysis, 161*, 246–253. https://doi.org/10.1016/j.jpba.2018.08.048

Mitsuoka, T. (2014). Development of functional foods. *Bioscience of Microbiota, Food and Health, 33*(3), 117–128. https://doi.org/10.12938/bmfh.33.117

Moghadam, Z. A., Hosseini, H., Hadian, Z., Asgari, B., Mirmoghtadaie, L., Mohammadi, A., Shamloo, E., & Seyed Javadi, N. H. (2023). Evaluation of the antifungal activity of cinnamon, clove, thymes, zataria multiflora, cumin and caraway essential oils against ochratoxigenic aspergillus ochraceus. *Journal of Pharmaceutical Research International, 21*. https://journaljpri.com/index.php/JPRI/article/view/1055.

Mohammed, M. E. A. (2022). Factors affecting the physicochemical properties and chemical composition of bee's honey. *Food Reviews International, 38*(6), 1330–1341. https://doi.org/10.1080/87559129.2020.1810701

Mohan, A., Quek, S.-Y., Gutierrez-Maddox, N., Gao, Y., & Quan and, S. (2017). Effect of honey in improving the gut microbial balance. *Food Quality and Safety, 1*(2), 107–115. https://doi.org/10.1093/fqsafe/fyx015

Mojica, E.-R. E., Charland, J. M., Deocaris, C. C., & Micor, J. R. L. (2021). Potential and performance of accelerated solvent extraction (ASE) in obtaining bioactive compounds from bee propolis as compared with other extraction methods. *Journal of Tropical Life Science, 11*(2).

Moskwa, J., Naliwajko, S. K., Dobiecka, D., & Socha, K. (2023). Bee products and colorectal cancer—Active components and mechanism of action. *Nutrients, 15*(7), 1614.

Mota, J., Lima, A., Ferreira, R. B., & Raymundo, A. (2020). Lupin seed protein extract can efficiently enrich the physical properties of cookies prepared with alternative flours. *Foods, 9*(8), 1064. https://doi.org/10.3390/foods9081064

Muñoz, E., Velásquez, P., Rodriguez, K., Montenegro, G., & Giordano, A. (2020). Influence of Brassica campestris and Galega officinalis on antioxidant activity of bee pollen. *Revista Brasileira de Farmacognosia, 30*(3), 444–449. https://doi.org/10.1007/s43450-020-00065-x

Murador, D. C., de Souza Mesquita, L. M., Vannuchi, N., Braga, A. R. C., & Rosso, V. V de (2019). Bioavailability and biological effects of bioactive compounds extracted with natural deep eutectic solvents and ionic liquids: Advantages over conventional organic solvents. *Current Opinion in Food Science, 26*, 25–34. https://doi.org/10.1016/j.cofs.2019.03.002

Mustar, S., & Ibrahim, N. (2022). A sweeter pill to swallow: A review of honey bees and honey as a source of probiotic and prebiotic products. *Foods, 11*(14), 2102. https://doi.org/10.3390/foods11142102

National Research Council. (1989). *Diet and health: Implications for reducing chronic disease risk*. National Academies Press.

Negrão, A. F., Lidia, M. R. C. B., & Orsi, R. O. (2014). Influence of the collection season on production, size, and chemical composition of bee pollen produced by Apis mellifera L. *Journal of Apicultural Science, 58*(2), 5.

Novais, C., Molina, A. K., Abreu, R. M. V., Santo-Buelga, C., Ferreira, I. C. F. R., Pereira, C., & Barros, L. (2022). Natural food colorants and preservatives: A review, a demand, and a challenge. *Journal of Agricultural and Food Chemistry, 70*(9), 2789–2805.

Okolie, C. L., Akanbi, T. O., Mason, B., Udenigwe, C. C., & Aryee, A. N. A. (2019). Influence of conventional and recent extraction technologies on physicochemical properties of bioactive macromolecules from natural sources: A review. *Food Research International, 116*, 827–839. https://doi.org/10.1016/j.foodres.2018.09.018

Olofsson, T. C., Èile, B., Pawel, M., Christina, L., Lennart, L., & Vásquez, A. (2016). Lactic acid bacterial symbionts in honeybees–An unknown key to honey's antimicrobial and therapeutic activities. *International Wound Journal, 13*(5), 668–679.

Oroian, M., Dranca, F., & Ursachi, F. (2020). Comparative evaluation of maceration, microwave and ultrasonic-assisted extraction of phenolic compounds from propolis. *Journal of Food Science and Technology, 57*, 70–78.

Pellati, F., Pio Prencipe, F., Bertelli, D., & Benvenuti, S. (2013). An efficient chemical analysis of phenolic acids and flavonoids in raw propolis by microwave- assisted extraction combined with high-performance liquid chromatography using the fused- core technology. *Journal of Pharmaceutical and Biomedical Analysis, 81*, 126–132.

Pobiega, K., Kraśniewska, K., Derewiaka, D., & Gniewosz, M.I. (2019). Comparison of the antimicrobial activity of propolis extracts obtained by means of various extraction methods. *Journal of Food Science and Technology, 56*(12), 5386–5395.

Pop, C., Vlaic, R., Fărcas, A., Liana, S., Delia, G., Semeniuc, C., & Ancuța. (2015). Influence of pollen, chia seeds and cranberries addition on the physical and probiotics characteristics of yogurt. *Bulletin UASVM Food Science and Technology., 72*(1), 2015.

Pourakbari, Y., Ali, R., & Ahmad, G. (2012). Effect of Persimmon Sap nutrition on behavioral properties in honey bee and physicochemical properties in honey of colony of Apis mellifera. *Academic Journal of Entomology, 5*(3), 147–150.

Rocha, C. M. R., Genisheva, Z., Ferreira-Santos, P., Rodrigues, R., Vicente, A. A., Teixeira, J. A., & Pereira, R. N. (2018). Electric field-based technologies for valorization of bioresources. *Bioresource Technology, 254*, 325–339. https://doi.org/10.1016/j.biortech.2018.01.068

Sá, A.G.A., Y.M.F. Moreno and B.A.M. Carciofi. 2020. Plant proteins as high-quality nutritional source for human diet. Trends in Food Science & Technology 97. Elsevier: 170-184.

Sabater-Molina, M., Larqué, E., Torrella, F., & Zamora, S. (2009). Dietary fructooligosaccharides and potential benefits on health. *Journal of Physiology and Biochemistry, 65*(3), 315–328. https://doi.org/10.1007/bf03180584

Saeed, K., Pasha, I., Chughtai, M. F. J., Ali, Z., Bukhari, H., & Zuhair, M. (2022). Application of essential oils in food industry: Challenges and innovation. *Journal of Essential Oil Research, 34*(2), 97–110. https://doi.org/10.1080/10412905.2022.2029776

Safaei, M., & Roosta Azad, R. (2020). Preparation and characterization of poly-lactic acid based films containing propolis ethanolic extract to be used in dry meat sausage packaging. *Journal of Food Science and Technology, 57*, 1242–1250.

Sagar, N. A., Pareek, S., Sharma, S., Yahia, E. M., & Lobo, M. G. (2018). Fruit and vegetable waste: Bioactive compounds, their extraction, and possible utilization. *Comprehensive Reviews in Food Science and Food Safety, 17*(3), 512 531. https://doi.org/10.1111/1541-4337.12330

Saleh, E., Eldin Morshdy, A., El-Manakhly, E., Al-Rashed, S., Hetta, H. F., Jeandet, P., Yahia, R., El-Saber Batiha, G., & Ali, E. (2020). Effects of olive leaf extracts as natural preservative on retailed poultry meat quality. *Foods, 9*(8), 1017. https://doi.org/10.3390/foods9081017

Santos, P. M., Ferreira, Z., Genisheva, B., Cristina, M. R., Rocha, J. A., & Teixeira. (2021). *Valorization of natural antioxidants for nutritional and health applications* (p. 2021).

Sathivel, S. (2013). Application of honey powder in bread and its effect on bread characteristics. *Journal of Food Processing & Technology, 04*(11). https://doi.org/10.4172/2157-7110.1000279

Shabir, I., Dash, K. K., Dar, A. H., Pandey, V. K., Fayaz, U., Srivastava, S., & Nisha, R. (2023). Carbon footprints evaluation for sustainable food processing system development: A comprehensive review. *Future Foods, 7*, Article 100215.

Sharma, S., Vaidya, D., & Rana, N. (2016). Honey as natural sweetener in lemon ready-to-serve drink. *International Journal of Bio-Resource and Stress Management, 7*, 320–325.

Socas-Rodríguez, B., Álvarez-Rivera, G., Valdés, A., Ibáñez, E., & Cifuentes, A. (2021). Food by-products and food wastes: Are they safe enough for their valorization? *Trends in Food Science & Technology, 114*, 133–147. https://doi.org/10.1016/j.tifs.2021.05.002

Somrani, M., Inglés, M.-C., Debbabi, H., Abidi, F., & Palop, A. (2020). Garlic, onion, and cinnamon essential oil anti-biofilms' effect against listeria monocytogenes. *Foods, 9*(5), 567. https://doi.org/10.3390/foods9050567

Teixeira, J. A. (2018). Electric field-based technologies for valorization of bioresources. *Bioresource Technology, 254*, 325–339.

Teixeira, T. D., Machado, B. A. S., Barreto, G.de A., Pereira Dos Anjos, J., Leal, I. L., Nascimento, R. Q., Hodel, K. V. S., & Umsza-Guez, M. A. (2023). Extraction of antioxidant compounds from Brazilian green propolis using ultrasound-assisted associated with low- and high-pressure extraction methods. *Molecules, 28*(5), 2338.

Thakur, M., & Nanda, V. (2020). Composition and functionality of bee pollen: A review. *Trends in Food Science & Technology, 98*(avril), 82–106. https://doi.org/10.1016/j.tifs.2020.02.001

Tong, Q., Zhang, X., Wu, F., Tong, J., Zhang, P., Jing, & Zhang. (2010). Effect of honey powder on dough rheology and bread quality. *Food Research International, 43*(9), 2284–2288. https://doi.org/10.1016/j.foodres.2010.08.002

Torres-Valenzuela, L. S., Ballesteros-Gómez, A., & Rubio, S. (2020). Green solvents for the extraction of high added-value compounds from agri-food waste. *Food Engineering Reviews, 12*, 83–100.

Tsuruda, J. M., Chakrabarti, P., & Sagili, R. R. (2021). Honey bee nutrition. *Veterinary Clinics: Food Animal Practice, 37*(3), Article 505-19. https://doi.org/10.1016/j.cvfa.2021.06.006

Tukmechi, A., Ownagh, A., & Ali, M. (2010). In vitro antibacterial activities of ethanol extract of iranian propolis (EEIP) against fish pathogenic bacteria (Aeromonas hydrophila, Yersinia ruckeri & Streptococcus iniae). *Brazilian Journal of Microbiology, 41*, 1086–1092.

Tumbarski, Y., Topuzova, M., & Todorova, M. (2022). Food industry applications of propolis: A review. *Journal of Hygienic Engineering & Design, 40*.

Tuoheti, T., Rasheed, H. A., Meng, L., & sheng Dong, M. (2020). High hydrostatic pressure enhances the anti-prostate cancer activity of lotus bee pollen via increased metabolites. *Journal of Ethnopharmacology, 261*, Article 113057.

Turner, N. D., & Lupton, J. R. (2011). Dietary fiber. *Advances in Nutrition, 2*(2), 151–152.

Végh, R., Csóka, M., Stefanovits-Bányai, É., Juhász, R., & Sipos, L. (2023). Biscuits enriched with monofloral bee pollens: Nutritional properties, techno-functional parameters, sensory profile, and consumer preference. *Foods, 12*(1), 18. https://doi.org/10.3390/foods12010018

Vega, G. C., Sohn, J., Voogt, J., Birkved, M., Olsen, S. I., & Nilsson, A. E. (2021). Insights from combining techno-economic and life cycle assessment–A case study of polyphenol extraction from red wine pomace. *Resources, Conservation and Recycling, 167*, Article 105318.

World Health Organization and United Nations University. (2007). In, *935. Protein and amino acid requirements in human nutrition*. World Health Organization.

Xu, X., Dong, J., Mu, X., & Sun, L. (2011). Supercritical CO2 extraction of oil, carotenoids, squalene and sterols from lotus (Nelumbo nucifera Gaertn) bee pollen. *Food and Bioproducts Processing, 89*(1), 47–52. https://doi.org/10.1016/j.fbp.2010.03.003

Yıldırım, H. K. (2022). Assessment of propolis treated by different extraction methods. *Brazilian Archives of Biology and Technology, 65*.

Yerlikaya, O. (2014). Effect of bee pollen supplement on antimicrobial, chemical, rheological, sensorial properties and probiotic viability of fermented milk beverages. *Mljekarstvo/Dairy, 64*(4).

Youssef, F. S., Eid, S. Y., Alshammari, E., Ashour, M. L., Wink, M., & El-Readi, M. Z. (2020). Chrysanthemum indicum and Chrysanthemum morifolium: Chemical composition of their essential oils and their potential use as natural preservatives with antimicrobial and antioxidant activities. *Foods, 9*(10), 1460. https://doi.org/10.3390/foods9101460

Yucel, B., Topal, E., & Kosoglu, M. (2017). *Bee products as functional food*. InTech. https://doi.org/10.5772/65477

Zainal, W. N. H. W., Azian, N. A. A. M., Saleha Albar, S., & Ainin, S. R. (2022). Effects of extraction method, solvent and time on the bioactive compounds and antioxidant activity of Tetrigona apicalis Malaysian propolis. *Journal of Apicultural Research, 61*(2), 264–270.

Zammit Young, G.-W., & Blundell, R. (2023). A review on the phytochemical composition and health applications of honey. *Heliyon, 9*(2), Article e12507. https://doi.org/10.1016/j.heliyon.2022.e12507

Zhang, H., & Wang, J. (2019). Constituents of the essential oils of garlic and citronella and their vapor-phase inhibition mechanism against *S.aureus. Food Science and Technology Research, 25*(1), 65–74. https://doi.org/10.3136/fstr.25.65

Zlatev, Z., Taneva, I., Baycheva, S., Mitko, & Petev. (2018). A comparative analysis of physico-chemical indicators and sensory characteristics of yogurt with added honey and bee pollen. *Bulgarian Journal of Agricultural Science, 24*(1), 2018.

ПрисяЖная С. П. С.Л. Грибанова. 2016. Способ производства МороЖеного функционального назначения. RU2598636C1, filed 6 mai 2015 and issued 27 septembre 2016. https://patents.google.com/patent/RU2598636C1/en?q=(honey+dairy+product)&oq=honey+d airy+product&page=2.

冯彬. 2017. 一种蜂花粉蜂蜜发酵饮料制备方法. China CN107156580A, filed 26 juin 2017 and issued 15 septembre 2017. https://patents.google.com/patent/CN107156580A/en?q=(bee+pollen+fermented)&oq=bee+p ollen+fermented.

刘本君. 2018. 一种消除疲劳的蜂蜜柚子茶制备方法. China CN108617819A, filed 26 juillet 2018 and issued 9 octobre 2018. https://patents.google.com/patent/CN108617819A/en?q=(honey+sweetener)&oq=honey+sweetener.

史云峰. 2016. 蜂蜜酸奶及其制备方法. China CN106135410A, filed 1 août 2016 and issued 23 novembre 2016. https://patents.google.com/patent/CN106135410A/en?q=(honey+dairy+product)&oq=honey+ dairy+product.

崔波 张政. 陶海腾. 李渐鹏 于滨. 刘桂梅. 一种蜂花粉发酵运动饮料及其制备方法. China CN115336693A, filed 25 août 2022 . (2022), https://patents.google.com/patent/CN115336693A/en?q=(bee+pollen+fermented)&oq=bee+p ollen+fermented.

林军庭. 2014. 一种蜂花粉保健品及制备方法. China CN104083464A, filed 23 juin 2014 and issued 8 octobre 2014. https://patents.google.com/patent/CN104083464A/en?q=(bee+pollen+health)&oq=bee+polle n+health+.

席芳贵 江武军, 张进明, 张串联, 夏晓翠, 刘锋, 李倩, et al., 一种天然蜂粮含片及其生产方法. China CN113662971B, filed 30 août 2021. (2023) https://patents.google.com/patent/CN113662971B/en?q=(bee+bread)&oq=bee+bread+&page=1.

程妮. 高慧 and 曹炜. 2012. 蜂花粉饼干的制备方法. China CN102763701A, filed 3 juillet 2012 and issued 7 novembre 2012. https://patents.google.com/patent/CN102763701A/en?q=(bee+pollen+biscuit)&oq=bee+polle n+biscuit.

解殿伟. 2018. 一种低糖艾草蜂蜜面包及其制备方法. China CN107711987A, filed 14 novembre 2017 and issued 23 février 2018. https://patents.google.com/patent/CN107711987A/en?q=(honey+bread)&oq=honey+bread.. China CN105010927B.

赖秋萍 赖晔, 丘爱芳 张方珍, 刘功良 (2018) Filed 31 juillet 2015 and issued 7 août 2018. https://patents.google.com/patent/CN105010927B/en?q=(bee+pollen+antioxidant)&oq=bee+ pollen+antioxidant.

SECTION

3

Cosmetic applications

CHAPTER

Plant polyphenols as natural antioxidants for cosmetic applications

8

Magdalena Anna Malinowska[1], Marta Sharafan[2], Karolina Jafernik[3], Marta Klimek-Szczykutowicz[4], Halina Ekiert[5], Elżbieta Sikora[1] and Agnieszka Szopa[5]

[1]*Department of Organic Chemistry and Technology, Faculty of Chemical Engineering and Technology, Cracow University of Technology Cracow, Poland;* [2]*Cracow University of Technology, CUT Doctoral School, Faculty of Chemical Engineering and Technology, Department of Organic Chemistry and Technology, Cracow, Poland;* [3]*Department of Medical Cosmetology, Faculty of Medicine and Health Sciences, Andrzej Frycz Modrzewski Krakow University, Krakow, Poland;* [4]*Department of Pharmaceutical Sciences, Collegium Medicum, Jan Kochanowski University, Kielce, Poland;* [5]*Department of Medicinal Plant and Mushroom Biotechnology, Jagiellonian University Medical College, Kraków, Poland*

1. Introduction

The growing awareness of the harmful effects of external factors and aging on skin health has led to an increased demand for effective and safe antioxidants in cosmetic formulations. Antioxidants play a crucial role in neutralizing reactive oxygen species (ROS) and protecting the skin from oxidative damage, thus preserving its youthful appearance (Chiu & Kimball, 2003; Michalak, 2022). In recent years, there has been a growing interest in using the potential of natural compounds as antioxidants, especially phenolic acids and flavonoids of natural origin (Ferrier et al., 2022; Hoang et al., 2021; Kubica et al., 2017).

Phenolic compounds, including flavonoids and phenolic acids, are secondary metabolites found in various plant sources, including fruits, vegetables, grains, and medicinal herbs. Their diverse chemical structures and biological activities have attracted significant attention from researchers and industries alike (Hoang et al., 2021; Kubica et al., 2022). Flavonoids, a subcategory of phenolic compounds, cover a wide range of biologically active molecules, such as flavonols, flavones, flavan-3-ols, flavanonols, flavanones, isoflavones, and anthocyanins, which are known for their potential to reduce oxidative stress (Arct & Pytkowska, 2008; Liga et al., 2023; Parashar et al., 2023; Šamec et al., 2021). The very important group of plant antioxidants is phenolic acids, divided into two subcategories: hydroxycinnamic acids and hydroxybenzoic acids (Goleniowski et al., 2013; Kumar & Goel, 2019; Shahidi & Yeo, 2018) (Fig. 8.1).

Plant Polyphenols in the Food and Cosmetic Industries. https://doi.org/10.1016/B978-0-443-19137-4.01008-5

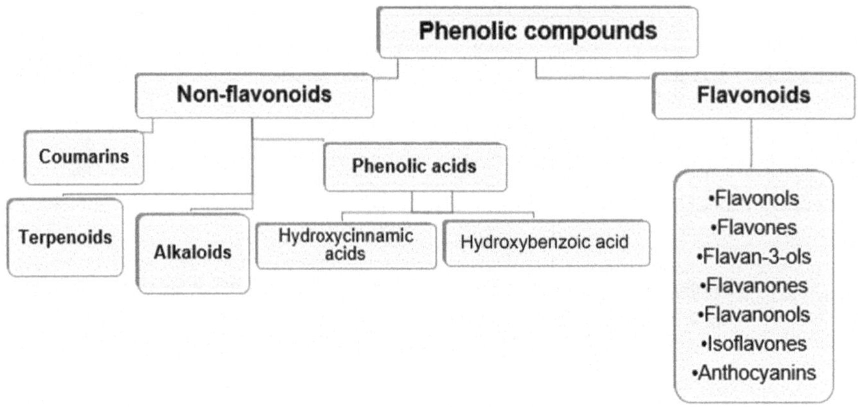

FIGURE 8.1

Phenolic compounds—general division.

Numerous studies have demonstrated the potent antioxidant properties of phenolic acids and flavonoids, attributing their efficacy to their ability to scavenge free radicals and inhibit lipid peroxidation (Malinowska et al., 2020; Toiu et al., 2017). Moreover, these natural antioxidants have shown antiinflammatory, antimicrobial, and photoprotective effects, further contributing to their suitability for cosmetic applications (Leyva-López et al., 2016; Saraf & Kaur, 2010; Toiu et al., 2019).

The primary advantage of incorporating these natural compounds into cosmetic products is their ability to neutralize harmful free radicals generated by UV radiation, pollution, and other environmental factors. By reducing oxidative stress, flavonoids and phenolic acids help prevent premature aging, collagen degradation, and DNA damage, leading to improved skin texture and reduced fine lines and wrinkles (Heim et al., 2002; Zeb, 2020). Importantly, flavonoids as well as phenolic acids exhibit antiinflammatory properties, which can help soothe and calm irritated or sensitive skin. This antiinflammatory action contributes to a more even complexion and reduced redness, making them valuable ingredients in products targeted at addressing skin conditions like rosacea and acne (Kiokias & Oreopoulou, 2021; Simmler et al., 2013). Another cosmetic target of using flavonoids of natural origin and phenolic acids is to brighten the skin and ensure its radiance. It has been proven that plant extracts rich in these compounds have been shown to inhibit melanin production and promote a more even skin tone. These ingredients are often used to target hyperpigmentation and dark spots and promote a radiant complexion (Liu-Smith & Meyskens, 2016; Panzella & Napolitano, 2019).

The stimulation of collagen production in human skin is also a significant challenge that may be supported by the application of plant polyphenols. This can lead to improved skin texture and a reduction in the appearance of sagging skin. Some flavonoids stimulate collagen synthesis in human fibroblasts, promoting elasticity and firmness (Mandrone et al., 2018; Stipcevic et al., 2006). Many studies confirm that plant extracts rich in polyphenols can protect collagen and elastin by inhibiting collagenase and elastase, helping to preserve the integrity of structural skin proteins (Majeed et al., 2020; Thring et al., 2009; Yingchutrakul et al., 2021).

Despite their numerous skin-beneficial properties, there are still many challenges that need to be addressed when using plant extracts rich in flavonoids and phenolic acids in cosmetics. These compounds can be sensitive to light, heat, and air, leading to potential stability issues in cosmetic formulations. Ensuring the stability and efficacy of these compounds throughout the product's shelf life can pose formulation challenges for manufacturers (Pinto et al., 2021). Moreover, the efficacy of antioxidant compounds can vary depending on factors such as concentration, formulation, and individual skin characteristics, which is, of course, related to the ability to penetrate the *stratum corneum*. Achieving consistent and predictable results may require careful formulation and testing (Dal Belo et al., 2009). Finally, the use of phenolic acids and flavonoids in cosmetics may be subject to regulatory guidelines and restrictions. Manufacturers need to ensure compliance with relevant regulations to ensure the safety and efficacy of their products.

In conclusion, antioxidant compounds of plant origin play a crucial role in cosmetic products by providing antioxidant protection, antiinflammatory effects, skin brightening, and collagen production. However, challenges such as stability issues, potential sensitivity reactions, variable efficacy, and regulatory considerations need to be carefully addressed during product development. Despite these challenges, the growing demand for natural and effective skincare solutions underscores the importance of harnessing the potential of antioxidant compounds in cosmetic formulations.

This chapter aims to collect the latest scientific data (from 2010 to 2023) regarding activities supporting the use of selected key active compounds from the group of flavonoids (rutin, quercetin, luteolin, apigenin, kaempferol, hesperidin) and phenolic acids (ferulic, rosmarinic, ellagic, salicylic, caffeic acids) in cosmetology. Moreover, their use in modern cosmetic production was characterized. An interesting aspect of the work is the presentation of the latest innovative research on the possibilities of producing these compounds through plant biotechnology methods.

2. Antioxidant mechanisms of phenolic acids and flavonoids

Flavonoids and phenolic acids contribute to plant protection against various environmental stresses. These compounds have gained considerable attention for their crucial role in maintaining cellular redox balance and mitigating oxidative stress-induced damage (Goleniowski et al., 2013; Jena et al., 2023; Pizzino et al., 2017).

The antioxidative mechanisms of flavonoids and phenolic acids are multifaceted and include both direct scavenging of reactive oxygen species (ROS) and modulation of cellular antioxidant defense systems (Kubica et al., 2020; Miceli et al., 2017). One of the primary mechanisms through which these compounds exert their antioxidative effects is by directly scavenging ROS, including superoxide anion radicals, hydrogen peroxide, and hydroxyl radicals. Flavonoids, as well as phenolic acids, with their distinct polyphenolic structures, possess multiple hydroxyl groups that enable them to effectively neutralize ROS through electron donation, thereby preventing oxidative damage to cellular biomolecules such as lipids, proteins, and nucleic acids (Seleem et al., 2017).

Hydrogen atom transfer (HAT), single electron transfer (SET), sequential proton loss electron transfer (SPLET), and proton loss (PL) are fundamental antioxidative mechanisms through which flavonoids and phenolic acids exert their protective effects against oxidative stress and cellular damage. These mechanisms involve intricate interactions between these bioactive compounds and ROS, enabling them to neutralize harmful radicals and maintain cellular redox balance (Georgiev et al., 2014).

In the HAT mechanism, flavonoids and phenolic acids act as potent hydrogen donors. When a ROS, such as a lipid peroxyl radical, approaches a cellular component, it seeks to steal a hydrogen atom to stabilize itself. Flavonoids and phenolic acids readily donate hydrogen atoms from their hydroxyl groups, effectively quenching the ROS and converting it into a less damaging molecule. This process not only prevents oxidative damage to cellular structures but also regenerates the antioxidant capacity of these compounds, allowing them to engage in multiple antioxidative cycles (Rosiak et al., 2022).

On the contrary, single-electron transfer (SET) involves the transfer of a single electron from a flavonoid or phenolic acid molecule to a ROS, neutralizing its reactivity. This mechanism is particularly effective against radicals that are difficult to quench through HAT due to their unique electronic structure. Flavonoids and phenolic acids, with their conjugated double bonds and aromatic rings, possess a surplus of electrons that they can donate to stabilize radicals. This electron donation disrupts the chain reaction of radical propagation, preventing oxidative damage to cellular components (Malacarne et al., 2011).

According to the density functional theory (DFT) theoretical calculations, in both the gas and benzene phases, the initial preference of all flavonoids is to engage in continuous di-hydrogen atom transfer (HAT) reactions, originating from the B ring OH groups. This leads to the capture of two free radicals, resulting in the creation of stable quinones for ampelopsin (dihydromyricetin, 3,3′,4′,5,5′,7-hexahydroxyflavanone), taxifolin (dihydroquercetin, (2R,3R)-3,3′,4′,5,7-pentahydroxyflavanone), dihydroorobol (dihydro derivative of orobol) (Johnson & Schmidt, 2004), and eriodictyol (3,4,5,7-tetrahydroxyflavanone). For hesperetin ((2S)-5,7-dihydroxy-2-(3-hydroxy-4-methoxyphenyl)-2,3-dihydrochromen-4-one), this process results in the formation of benzodioxole. The third free radical is captured through HAT in the gas phase. In the benzene phase, ampelopsin also shows a propensity for employing HAT to capture the third free radical, while the other flavonoids utilize the sequential proton loss electron transfer (SPLET) mechanism. In the aqueous phase, the studied flavonoids initially undergo consecutive proton loss (PL) reactions from all OH groups, leading to the formation of polyanions. This sequence of multiple PL reactions commences from the 7–OH group (Zheng et al., 2022).

All the mechanisms highlight the versatility of flavonoids and phenolic acids in combating oxidative stress. These compounds exhibit selective reactivity toward different types of radicals, allowing them to efficiently neutralize a wide range of ROS. In addition, their antioxidative effects extend to metal ions that contribute to ROS formation. Flavonoids and phenolic acids possess metal-chelating properties, which inhibit metal-ion-mediated ROS generation, further reducing oxidative stress (Messaadia et al., 2020).

Furthermore, flavan-3-ols, like catechin, as well as phenolic acids, are known to enhance the activities of endogenous antioxidant enzymes, such as superoxide dismutase (SOD), catalase (CAT), and glutathione peroxidase (GPx). These compounds can upregulate the expression and activity of these enzymes by modulating signaling pathways, including the nuclear factor erythroid 2-related factor 2 (Nrf2) pathway (Simos et al., 2012). Activation of Nrf2 leads to increased transcription of antioxidant response element (ARE)-containing genes, including those encoding antioxidant enzymes and phase II detoxification enzymes. This coordinated upregulation enhances the cellular defense against oxidative stress, enabling cells to better cope with ROS-induced damage (Karakas et al., 2016).

Moreover, flavonoids and phenolic acids exhibit metal chelating properties, which contribute to their antioxidative effects. By binding to transition metal ions such as iron and copper, these compounds inhibit the Fenton reaction, a process that generates highly reactive hydroxyl radicals in the presence of metal ions and hydrogen peroxide. By sequestering these metal ions, flavonoids and phenolic acids reduce the potential for hydroxyl radical formation and subsequent oxidative damage (Ebrahimzadeh et al., 2008; Morel et al., 1993).

In conclusion, flavonoids and phenolic acids represent a diverse group of natural compounds with potent antioxidative mechanisms that contribute to cellular redox homeostasis and protection against oxidative stress-induced damage. Their ability to directly scavenge ROS, enhance antioxidant enzyme activities, chelate metal ions, and modulate inflammatory pathways collectively underpins their role in maintaining cellular health and mitigating oxidative stress-related pathologies. As research in this field continues to evolve, a deeper understanding of the specific mechanisms through which flavonoids and phenolic acids exert their antioxidative effects holds promise for the development of novel therapeutic strategies aimed at combating oxidative stress-associated diseases.

3. Phenolic acids and flavonoids classification and biosynthetic pathways

3.1 Phenolic acids

Phenolic acids (phenolcarboxylic acids) are a class of phenolic secondary metabolites present in a wide range of plant-based foods. Their primary role in plant cells is associated with enzyme activity, allelopathy, photosynthesis, and protein synthesis (Goleniowski et al., 2013; Kumar & Goel, 2019; Shahidi & Yeo, 2018).

Phenolic acids usually occur in plant cells as esters or glycosides and may be conjugated with flavonoids, hydroxyfatty acids, alcohols, or glycosides. They may also be linked to structural components of the plant, for example, cellulose and proteins through the ester, ether, or acetal bonds. Phenolic acids can be divided into two main subgroups: hydroxycinnamic acids and hydroxybenzoic acids (Goleniowski et al., 2013; Kumar & Goel, 2019; Shahidi & Yeo, 2018). Hydroxycinnamic acids include, among others, *p*-coumaric acid, ferulic acid, sinapic acid, caffeic acid, chlorogenic acid, and rosmarinic acid. Hydroxybenzoic acids include, among others, benzoic acid, protocatechuic acid, syringic acid, veratric acid, and salicylic, gallic (and its derivative—ellagic acid), vanillic, *p*-hydroxybenzoic, 2,5-dihydroxybenzoic, 3,4-dihydroxybenzoic, 2,3-dihydroxybenzoic and 3,5-dihydroxybenzoic acids, which are derived from benzoic acid (Abramovič, 2015; Goleniowski et al., 2013; Kumar & Goel, 2019; Lorigooini et al., 2020; Shahidi & Yeo, 2018). The chemical structures of selected hydroxycinnamic and hydroxybenzoic acids are presented in Fig. 8.2.

3.2 Flavonoids

Flavonoids are a large class of plant secondary metabolites with health-beneficial properties, synthesized via the phenylpropanoid pathway. They act as plant pigments, phytoalexins, allelopathic and antimicrobial compounds (Panche et al., 2016; Shahidi & Yeo, 2018).

The basic chemical structure of flavonoids consists of two benzene rings (A and B) linked by a heterocyclic pyran/pyrone ring (C). They may occur in both free and glycoside forms. Flavonoids can be divided into seven subgroups: flavonols (quercetin, rutin, myricetin, morin, kaempferol), flavones (apigenin, baicalein, tangeretin, rhoifolin), flavan-3-ols (catechin, epicatechin, epigallocatechin, gallocatechin), flavanones (naringin, naringenin, hesperetin, hesperidin, eriodictyol), flavanonols (aromadedrin, engeletin, taxifolin), isoflavones (daidzein, daidzin, genistein, genistin, glycitein), and anthocyanins (cyanidin, malvidin, peonidin, delphinidin) (Fig. 8.3; Shahidi & Yeo, 2018; Al Mamary, 2016; Šamec et al., 2021).

3.3 Biosynthesis of phenolic compounds

Phenolic compounds can be produced by shikimic acid and phenylpropanoid pathways (Figs. 8.4 and 8.5). The shikimic acid pathway involves two substrates—phosphoenolpyruvic acid (PEP) and erythrose-4-phosphate (E4P), which react and give 3-deoxy-D-arabino-heptulosonic acid 7-phosphate (DAHP). Then, 3-dehydroquinate (DHQ) is generated through the elimination of phosphate from DAHP by 3-dehydroquinate synthase (DHQS). Next, 3-dehydroquinate dehydratase (DHQD) dehydrates 3-dehydroquinate and converts it to 3-dehydroshikimate (DHS). Shikimate-5-dehydrogenase (SD) catalyzes the nicotinamide adenine dinucleotide phosphate (NADP)-dependent reduction of 3-dehydroshikimate to shikimate. Shikimate is

HYDROXYCINNAMIC ACIDS	HYDROXYBENZOIC ACIDS
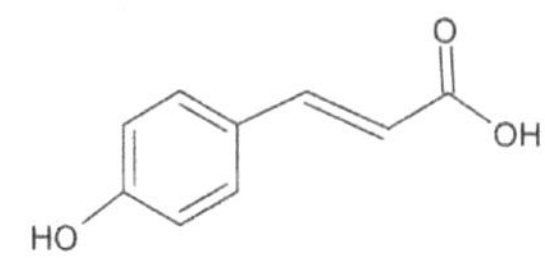	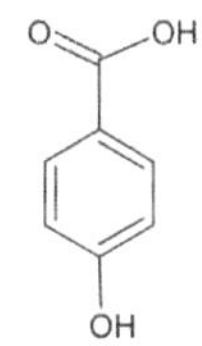
p-Coumaric acid	*p*-Hydroxybenzoic acid
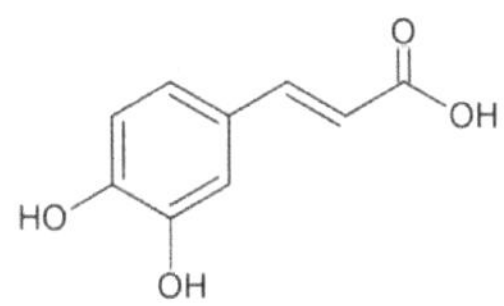	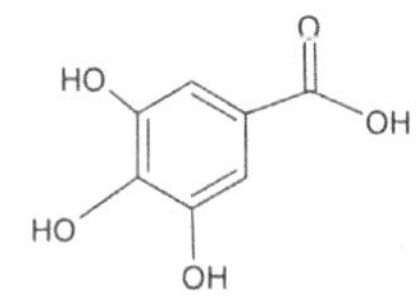
Caffeic acid	Gallic acid
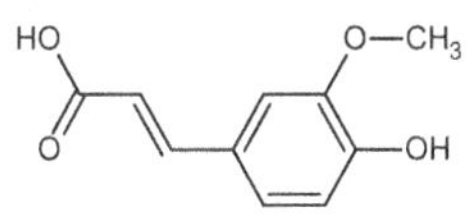	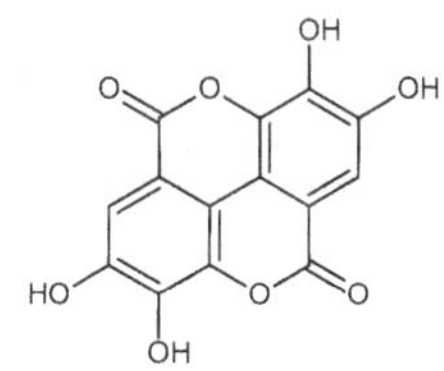
Ferulic acid	Ellagic acid
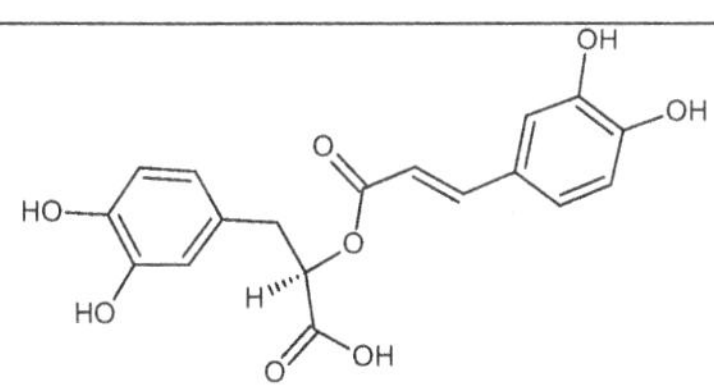	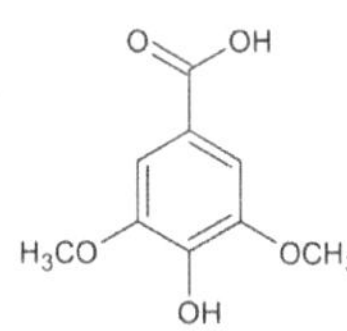
Rosmarinic acid	Syringic acid

FIGURE 8.2

Chemical structures of the selected phenolic acids.

then converted to shikimate 3-phosphate (S3P) by shikimate kinase (SK). Next, 5-enolpyruvylshikimic acid 3-phosphate (EPSP) is generated by adding phosphoenol pyruvate (PEP) to shikimate 3-phosphate. Lastly, chorismate synthase (CS) generates chorismic acid by trans-1,4 elimination of 5-enolpyruvylshikimate 3-phosphate. Chorismic acid is a key metabolite for aromatic acids: L-phenylalanine (L-Phe), L-tyrosine

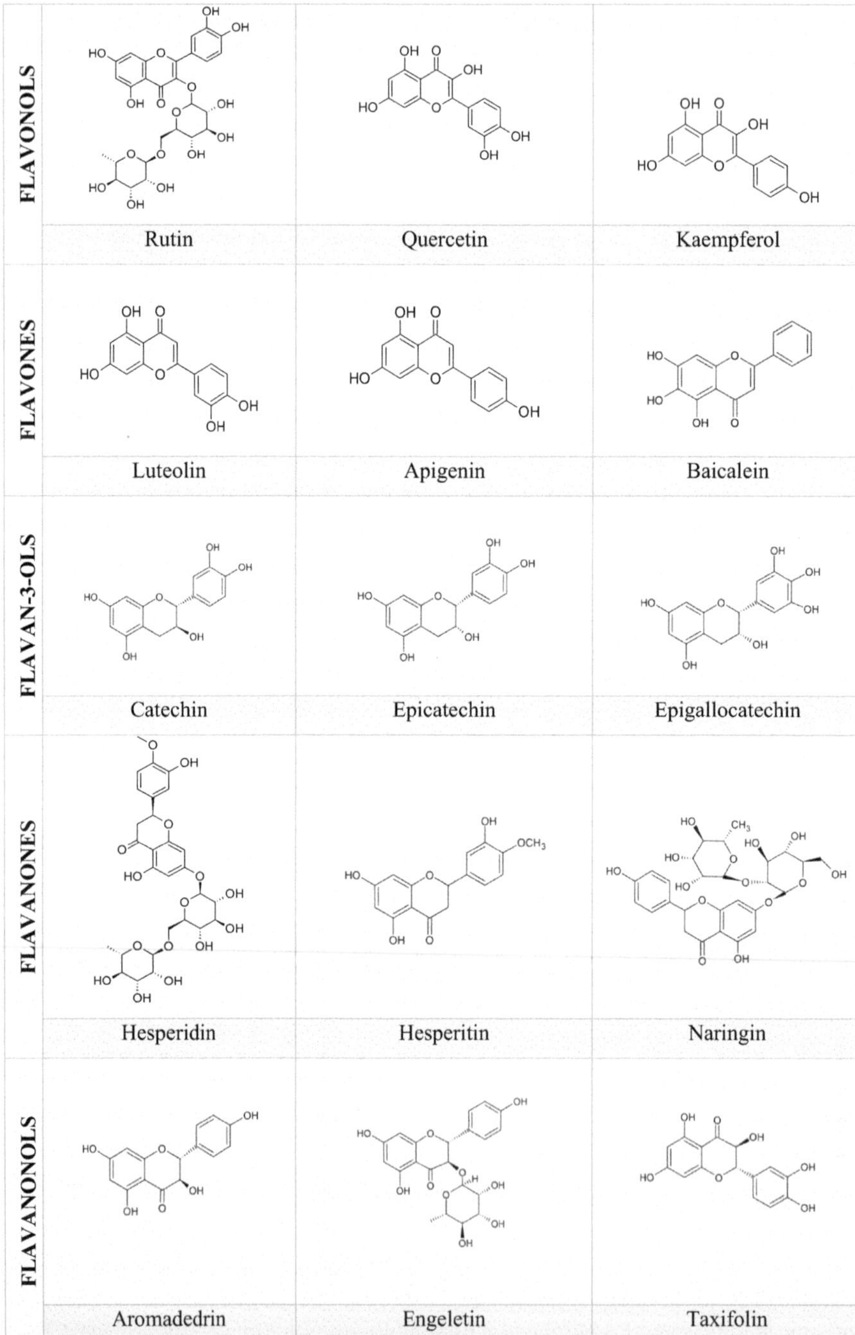

FIGURE 8.3

Flavonoid classification with examples of the selected compounds.

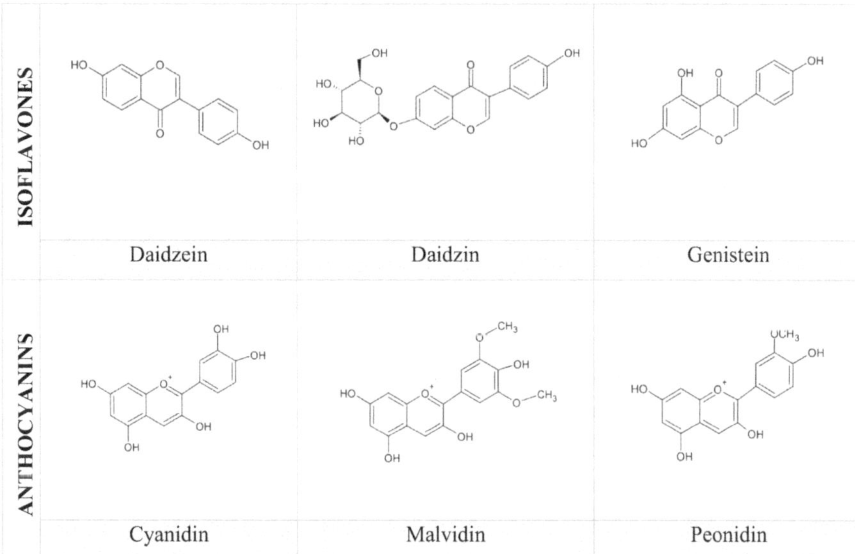

FIGURE 8.3 cont'd.

(L-Tyr), and L-tryptophan (L-Trp) (Francenia Santos-Sánchez et al., 2019; Potts, 1941; Al Mamary, 2016).

Phenylpropanoid pathway starts from L-phenylalanine deamination under the influence of phenylalanine ammonia-lyase (PAL) and gives cinnamic acid, from which hydroxybenzoic acids are derived. However, some of the hydroxybenzoic acids are produced during the shikimate pathway, for example, gallic acid, which is directly generated from dehydroshikimic acid (Goleniowski et al., 2013; Kalinowska et al., 2014; Al Mamary, 2016). Cinnamic acid is hydroxylated to *p*-coumaric acid by the action of cinnamate 4-hydroxylase (C4H). It is also formed from tyrosine under the influence of tyrosine ammonia lyase (TAL). *p*-Coumaric acid hydroxylation subsequently leads to the formation of caffeic acid. Next, methylation of caffeic acid gives ferulic acid. Hydroxylation of ferulic acid leads to the formation of 5-hydroxyferulic acid, from which sinapic acid is produced after *O*-methylation of the same hydroxyl group. The last step in the phenylpropanoid pathway is the conversion of *p*-coumaric acid to *p*-coumaroyl-CoA catalyzed by coumarate-CoA ligase (4CL) (Kalinowska et al., 2014; Potts, 1941; Vermerris & Nicholson, 2006) (Fig. 8.5).

Flavonoids are also produced through the phenylpropanoid pathway. However, their biosynthesis, in addition, involves *p*-coumaroyl-CoA condensation with three malonyl-CoA molecules initiated by chalcone synthase (CHS). It gives rise to chalcone, which produces scaffolds from which flavonoids are derived (Falcone Ferreyra et al., 2012) (Fig. 8.6). Chalcone isomerase catalyzes the cyclization of chalcone to flavanones, from which flavones are biosynthesized by flavone synthase,

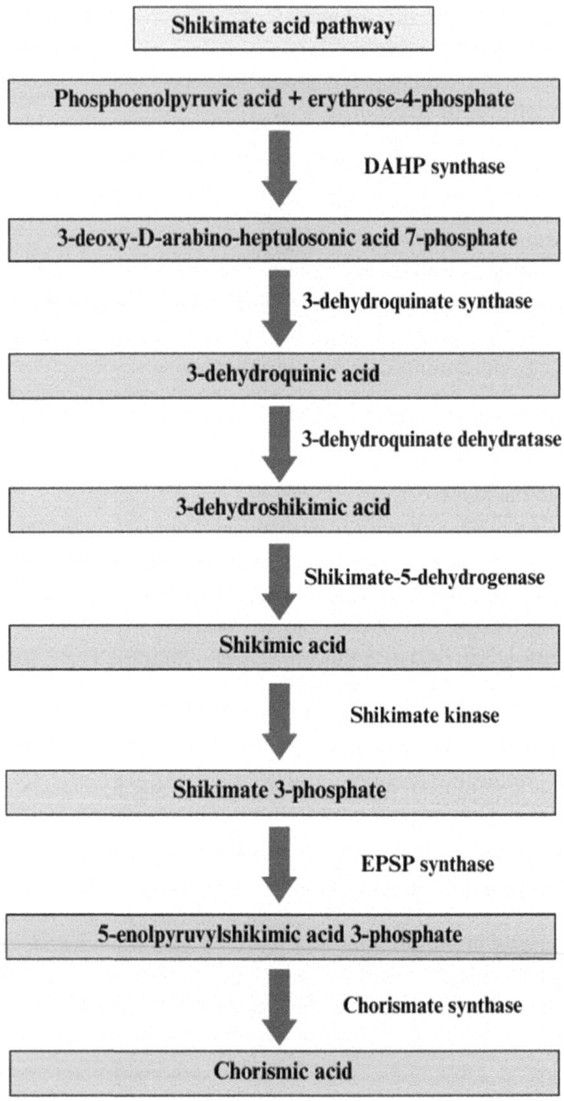

FIGURE 8.4

Shikimic acid pathway of phenolic compounds. Enzymes involved: 3-deoxy-D-arabino-heptulosonic acid 7-phosphate (DAHP), 3-dehydroquinate synthase (DHQS), 3-dehydroquinate dehydratase (DHQD), shikimate-5-dehydrogenase (SD), shikimate kinase (SK), 5-enolpyruvylshikimic acid 3-phosphate (EPSP), chorismate synthase (CS).

dihydroflavonols by flavanone 3β-hydroxylase, and isoflavones by isoflavone synthase. Dihydroflavonols give rise to flavonols by flavonol synthase and leucoanthocyanidins by dihydroflavonol reductase. Anthocyanidins are produced from

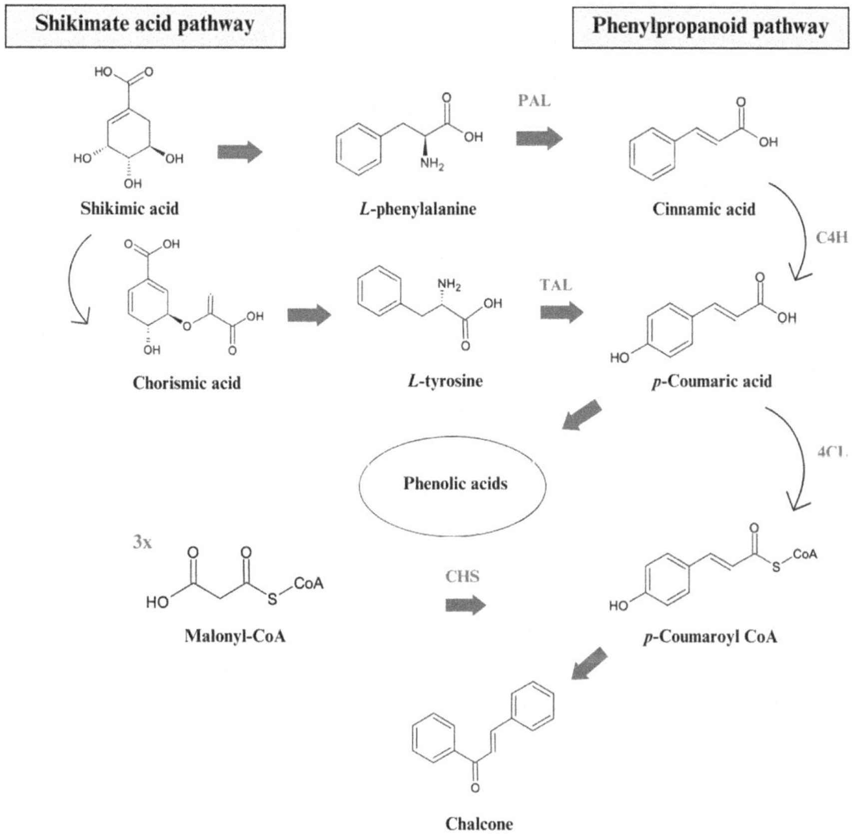

FIGURE 8.5

Schematic representation of the shikimate and phenylpropanoid pathways leading to the biosynthesis of phenolic acids and chalcone

leucoanthocyanidins by anthocyanidin synthase. Anthocyanidins give rise to 2,3-*cis*-(−)-flavan-3-ols by anthocyanidin reductase; however, 2,3-*trans*-(−)-flavan-3-ols are biosynthesized from leucoanthocyanidins by leucoanthocyanidin reductase (Dias et al., 2021; Zhu et al., 2015; Ejiofor & Chikodili Igbokwe, 2021 (Fig. 8.6).

4. Important flavonoids for cosmetic applications

Flavonoids are widely used in cosmetic products owing to their broad spectrum of biological activity. They exhibit numerous properties beneficial for cosmetic applications, including antiinflammatory, antiaging, skin-lightening, and antimicrobial effects (Patil et al., 2013; Jo et al., 2022; Kalita & Das, 2018; Memar et al., 2023;

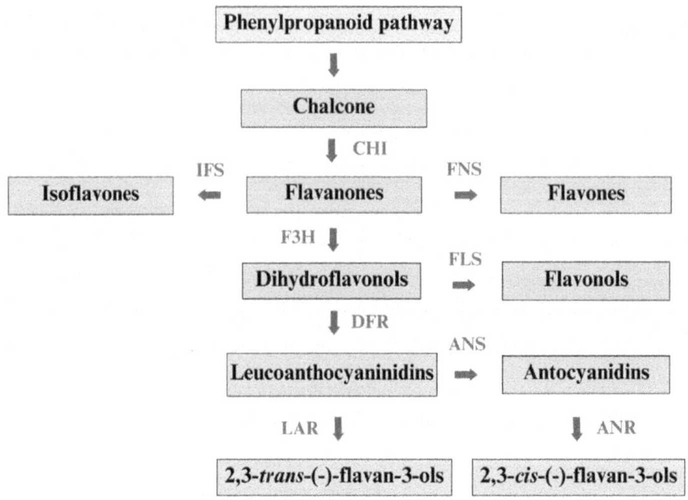

FIGURE 8.6

Biosynthesis of flavonoids. Enzymes involved: chalcone isomerase (CHI), flavone synthase (FNS), isoflavone synthase (IFS), flavanone 3β-hydroxylase (F3H), flavonol synthase (FLS), dihydroflavonols reductase (DFR), anthocyanidin synthase (ANS), anthocyanidin reductase (ANR), leucoanthocyanidin reductase (LAR).

Xia et al., 2014; Lee et al., 2018; Novotná et al., 2023; Bahadır Acıkara et al., 2019) (Fig. 8.7).

Flavonoids are recognized as important antioxidants in the plant kingdom. Among the most extensively studied are rutin, quercetin, luteolin, apigenin, kaempferol, and hesperidin. The antioxidant effect of flavonoids is mainly based on neutralizing free radicals such as $O^{2\bullet-}$ and $OH^{\bullet}$, increased the expression levels of endogenous antioxidant enzymes and protective effect against DNA, tissues and protein damage caused by internal and external factors (Ahmad et al., 2012; Arafa et al., 2009; Balakrishnan & Menon, 2007; Chen, Gu, et al., 2010; Chen, Ye, et al., 2010; Chen, Park, et al., 2017; El-Sayed et al., 2008; Hosseinimehr et al., 2009; Kamaraj et al., 2011; Lobo et al., 2010; Patil et al., 2013; Pradeep et al., 2012; Sahu et al., 2013). Furthermore, flavonoids have been shown to protect ultraviolet B (UVB)-induced damage by scavenging ROS and strengthening the cell membrane and mitochondria against ROS-induced damage (Santhanam et al., 2018).

Awareness of the biological properties of flavonoids has sparked growing interest in flavonoid applications in medicine, pharmacy, cosmetology, and the food and nutraceutical industries. One of the main limitations blocking the wider use of flavonoids is their low bioavailability and solubility, poor absorption, and rapid metabolism, which is why research focuses on these directions. One of the most promising strategies is the development and application of nanotechnology (Liga et al., 2023).

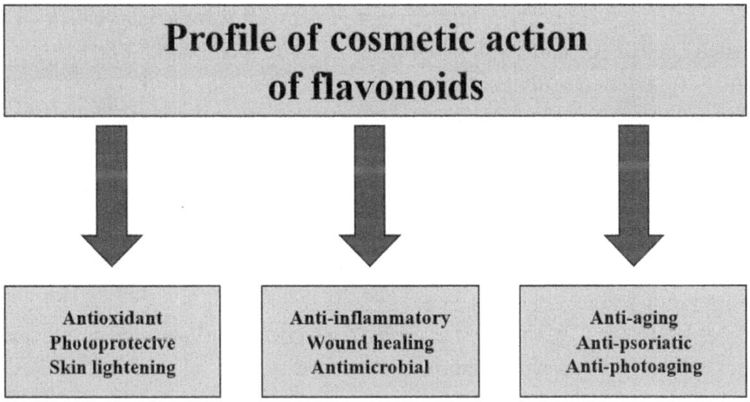

FIGURE 8.7

Profile of action of flavonoids.

Selected flavonoids that are widely used in the cosmetics industry, the mechanism of antioxidant action, other biological activities, and examples of use in cosmetics are presented in detail below (Fig. 8.3).

4.1 Rutin

4.1.1 Chemical structure and plant resources

Rutin (i.e., rutoside) (Fig. 8.8) is a common secondary metabolite found in the plant kingdom. It is a flavonoid glycoside also known as quercetin-3-*O*-rutinoside, rutoside, sophorin, and it's sometimes included in the so-called vitamin P. Its chemical name is 2-(3,4-dihydroxyphenyl)-5,7-dihydroxy-3-[α-L-rhamnopyranosyl-(1 → 6)-β-D-glucopyranosyloxy]-4H-chromen-4-one. Rutin is in the form of

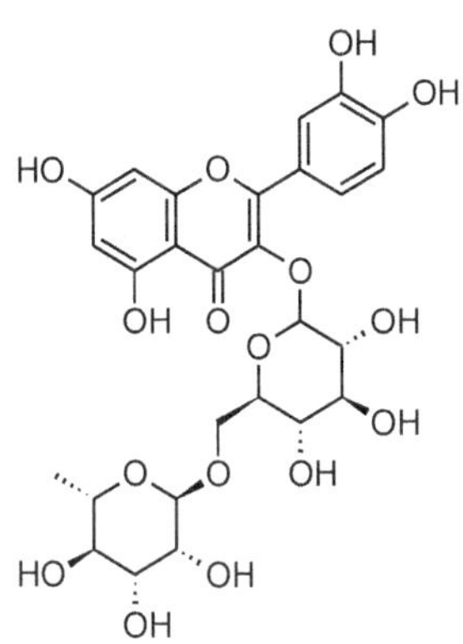

FIGURE 8.8

Chemical structure of rutin.

a yellow powder that is poorly soluble in water, but well soluble in pyridine. In addition, it melts at 125°C, and its pKa value is 6.17 (Yong et al., 2020).

Rutin is a flavonoid widely distributed in vegetables and fruits, as well as many medicinal plants. Its name comes from *Ruta graveolens*, which is rich in this compound. This compound is found in numerous plant species, including *Carpobrotus edulis, Camellia sinensis, Citrus sinensis, Fagopyrum esculentum, Ficus carica, Malus domestica,* and *Sophora japonica.* The first commercial source of rutin is considered to be flowers of green buckwheat (*Fagopyrum esculentum*), which contain 3%–5% rutin, and was used as the first commercial source of rutin (Auchter, 1945; Farha et al., 2022; Semwal et al., 2021).

According to the CosIng database (2026), eight rutin derivatives are used in cosmetics, such as darutoside, disodium rutinyl disulfate, glucosylrutin, methoxy-PEG-7 rutinyl succinate, narirutin, nicotiflorin (kaempferol-3-rutinoside), rutinyl succinate, and troxerutin. Darutoside is a diterpenoid compound with significant antiinflammatory activity (Wang & Sun, 2022). Glucosylrutin is a potent antioxidant with a high epidermal bioavailability (Wolber et al., 2004). Narirutin is a compound with strong antioxidant and anticancer properties (Pandey et al., 2024). Troxerutin (vitamin P4) is a derivative of rutin found in tea, coffee, cereal grains, and a number of vegetables and fruits. It is believed to have antioxidant and antiinflammatory, antidiabetic, and anticancer properties (Ahmadi et al., 2021).

4.1.1.1 Review of the latest scientific studies focused on the cosmetological applications

4.1.1.1.1 Antioxidant properties. Rutin is considered one of the most effective antioxidants, often used as a comparative sample in antioxidant research (Lobo et al., 2010).

Studies by Patil et al. (2013) showed a protective effect in mice against gamma radiation. Studies have confirmed the oxidative stress-relieving effect by neutralizing free radicals such as $O^{2\bullet-}$ and $OH^{\bullet}$ (Patil et al., 2013). In vivo studies confirmed the antioxidant effect of rutin in Cd^{+} $EtOH^{-}$ induced test oxidative stress in rats. Administration of doses at concentrations of 50 and 100 mg/kg p.o. after 2 weeks, it reduced the induced oxidative stress and improved sperm quality in rats (Abarikwu et al., 2017). The results demonstrated antioxidant activity using the DPPH (1,1-diphenyl-2-picrylhydrazyl) method and an increase in SOD activity in the myocytes of newborn rats and H9C2 cells (embryonic rat cardiomyocytes) in an ex vivo study. The treatment also increases LVDP (left ventricular developed pressure) and $\pm$ dP/dtmax (left ventricular end-diastolic pressure) (Patil et al., 2013). The conducted studies suggest that rutin may protect against ischemia/reperfusion injury by improving the cardiovascular health factor (Bhandary et al., 2012). Recent studies by Li et al. (2022) have focused on a more soluble form of rutin—sodium rutin. Administration of rutin sodium at a concentration of 0.2 mg/mL to male mice increased lifespan by 10%. In addition, it also alleviated the pathological changes associated with aging and improved the behavior of aging mice. Rutin sodium supplementation altered whole-body metabolism in mice, which showed increased energy expenditure and a lower respiratory quotient. Transcriptomic analysis showed that rutin sodium affects the expression of metabolic genes. Metabolomic analysis demonstrated that rutin sodium reduces hepatic steatosis through increased β-oxidation of lipids. Treatment with rutin sodium increased the level of autophagy both in vivo and in vitro. Inhibition of autophagy partially abolished the effect of sodium rutin on lipolysis in hepatoblastoma (HepG2) cells.

4.1.1.2 Antiphotoaging effect

Li et al. (2021) demonstrated the antiphotoaging potential of rutin bydeveloping a novel lyophilized rutin nanocrystal to improve its saturation solubility. The lyophilized nanocrystals were subsequently re-dispersed in a carbopol gel to formulate a rutin nanocrystal target gel (NC gel) to enhance transdermal delivery efficiency. These results confirmed that the NC gel is an ideal carrier for the topical application of rutin. NC gel also demonstrated strong free radical scavenging activity (DPPH test). The skin photoaging model was performed on Bagg's albino (BALB/c) mice. Pharmacodynamic results in vivo studies have shown that NC gel can effectively prevent skin photoaging and tissue damage caused by UV radiation.

4.1.1.3 Photoprotective activity

Moraes et al. (2017) conducted a normal stability test with formulations containing rutin succinate with or without sunscreen for 90 days. The combination of rutin succinate with chemical filters improved the functionality of the formulation, as it resulted in a more stable formulation that retained the effectiveness of the added sunscreens.

Kim et al. (2013) showed inhibition of the inflammatory effect induced by UVB radiation. The study was conducted on the JB6P+ mouse epidermal (JB6 P+) cell line. Rutin inhibited UVB-induced cyclooxygenase-2 (COX-2) expression at both the protein and transcript levels. UVB-induced activation of nuclear factor (NF)-κB and activator protein-1 was inhibited in a dose-dependent manner by rutin treatment (25, 50, and 100 μM). Western blot data revealed that rutin attenuated UVB-induced phosphorylation of extracellular regulated kinase (ERK), mitogen-activated protein kinase kinase 4 (MKK4), c-Jun-N-terminal kinase (JNK), mitogen-activated protein kinase kinase 3 (MKK3), and p38, but not mitogen-activated kinase kinase (MEK).

Oliveira et al. (2016) investigated the effect of rutin-entrapped gelatin nanoparticles for antioxidant activity and sun protection. The nanoparticles with rutin increased the antioxidant activity by 74% compared with the free rutin solution and also increased the sun protection factor (SPF) by 48%. In addition, they obtained good results in skin compatibility tests. The results demonstrated that rutin nanoparticles can be a safe SPF enhancer in sunscreens.

Kamel and Mostafa (2015) demonstrated that preparations with rutin nanocarriers increased antioxidant activity and also increased SPF protection against UV radiation.

Tomazelli et al. (2018) confirmed the antioxidant and photoprotective effects of rutin in the studies. A sunscreen formulation containing 0.1% (w/w) rutin, 3.0% (w/w) butylmethoxydibenzoylmethane, and 8.0% (w/w) octyldimethyl PABA (2-ethylhexyl 4-(dimethylamino)benzoate) was tested. The DPPH test confirmed the antioxidant properties of rutin, revealing an approximately 40% increase in radical scavenging potential in the presence of the bioactive compound. Rutin combined with UV filters significantly increased the clinical SPF by approximately 70% compared with the free formula.

The studies conducted by Peres et al. (2016) have proven antioxidant activity, in vitro photoprotective efficacy, photostability, and in vivo skin tolerance of rutin. Preparations with rutin were compatible with human skin and showed an antioxidant potential 75% higher than samples containing only UVB filters. In addition, rutin increased the wavelengths of the preparations, showing photoprotective enhancement, especially in the UVA range.

Gęgotek et al. (2017) confirmed the antioxidant and antiinflammatory effects in fibroblasts exposed to UVA and UVB radiation. Rutin reduced the UV-induced inflammatory response and ROS production and increased the activity/level of antioxidants SOD, glutathione peroxidase (GSH-Px), vitamin E, glutathione (GSH), and thioredoxin (Trx). The results demonstrated that rutin prevented changes in the levels of lipid mediators: malondialdehyde (MDA), 4-hydroxynonenal (4-HNE), and endocannabinoids, as well as the expression of endocannabinoid receptors CB1/2, vanilloid receptor 1(VR1), and G-protein coupled receptor 55 (GPR55). In addition, protein modification was inhibited (especially the formation of tyrosine derivatives), and the level of pro-apoptotic markers—caspase-3 (CASP3) and cytochrome c (Cyt-c) was reduced.

Choi, Kundu, et al. (2014) investigated the effects of rutin on UVB-induced inflammation in mouse skin in vivo. Topical application of rutin to the back skin of hairless HR-1 female mice 30 min before UVB irradiation reduced epidermal hyperplasia. Significant inhibition of UVB-induced COX-2 and induced nitric oxide synthase (iNOS) expression has also been demonstrated. Rutin inhibited DNA binding of activator protein-1 (AP-1) and phosphorylation of signal transducer and activator of transcription-3 (STAT3) in UVB-exposed mouse skin. The study confirmed the reduction in rutin-induced UVB phosphorylation of p38 mitogen-activated protein (MAP) kinase and JNK kinase.

4.1.1.3.1 Skin lightening activity. Jo et al. (2022) investigated the tyrosinase and melanin inhibitory effects of rutin. The tyrosinase inhibitory activity of supercritical water (SCW)-treated rutin ranged from 21.72% to 60.05% at 1 mg/mL. The ethyl acetate fraction showed 14.91% melanin-inhibiting activity at 10 μg/mL. The rates of inhibition of protein expression by melanocyte-inducing transcription factor (MITF), tyrosinase, human tyrosinase-related protein 1/2 (TRP-1/2) in the ethyl acetate fractions were 14.05%, 72%, 93.05% and 53.44% at 10 μg/mL, respectively.

4.1.1.3.2 Antihair loss effect. Carelli et al. (2012) studied a special combination of zeaxanthin, spermidine, and rutin for protection against apoptosis of human dermal papilla cells. Incubation of dermal follicular papilla cells (HFDPC) with staurosporine resulted in apoptosis that was completely inhibited by exposure to rutin (2.2 μM), spermidine (1 μM) and zeaxanthin (80 μM). The study found that the best effects were shown when the compounds were used together rather than separately. Moreover, treatment preserved the expression of antiapoptotic molecules such as B-cell lymphoma 2 (Bcl-2), MAP-kinases, and their phosphorylated forms.

4.1.1.3.3 Antiallergic effect. Choi and Kim (2013) demonstrated a reduction in the symptoms of atopic dermatitis and allergic contact dermatitis. The studies were performed on a model of atopic dermatitis in BALB/c mice by repeated local exposure of the ears to an extract of house dust mites (*Dermatophagoides farinae*) and 2,4-dinitrochlorobenzene. In the model of allergic contact dermatitis, a local lymph node test sensitized to 2,4-dinitrofluorobenzene was used. Topical application of rutin reduced serum immunoglobulin E (IgE), inhibited ear mast cell infiltration, and reduced serum histamine levels. Rutin in the allergic contact dermatitis model suppressed lymphocyte proliferation, serum immunoglobulin G2a (IgG2a) levels, and expression of interferon gamma (INF-γ), interleukins (IL-4, IL-5, IL-10, IL-17), and tumor necrosis factor-α (TNF-α).

4.1.1.3.4 Antiinflammatory activity. Kalita and Das (2018) conducted research on the effect of phospholipid complex in polymer matrix for sustained delivery of rutin via the skin for the treatment of acute and chronic inflammatory diseases. The results demonstrated that rutin in a phospholipid complex has a greater penetrating ability than free rutin. Permeability tests were performed on a damaged rat paw and human cadaver skin. The antiinflammatory efficacy of the optimized RNP patch formulation was carried out using the carrageenan-induced hind paw edema method in rats. The rutin patch significantly inhibited inflammation (91.4%) compared with standard diclofenac sodium gel (39.58%) after 24 h.

Tian et al. (2019) demonstated that rutin has the potential to treat vascular inflammation by reducing TNF-α production and nuclear factor kappa B (NF-κB) activation. Studies have also shown a reduction in inflammatory markers such as IL-6, mitogen-activated protein kinase (MAPK), NF-κB, i-NOS, and COX-2. At concentrations of 50 and 100 μM, it showed antiinflammatory activity against lipopolysaccharide (LPS)-activated RAW 264.7 cells by reducing phagocytic activity. In addition, it reduced the levels of NO, TNF-α, IL-1β, and IL-6 in RAW 264.7 cells.

Jeong et al. (2018) showed that treatment with rutin may be useful in allergic inflammatory diseases because it reduces the proliferation of human mast cells (HMC-1) without showing cytotoxicity. Studies have confirmed the effect of reducing mouse double minute 2 (MDM2) and STAT6 levels, along with IL-13 and Bcl2 expression in thymic stroma lymphopoietin-stimulated cells.

Studies by Choi et al. (2013) confirmed the inhibitory effect of B[α]PDE-induced COX-2 expression in JB6 P + mouse epidermal cells by reducing the RAF/MEK/ERK and Akt signaling pathways at 40 μM. This study established that rutin reduces the activity of epidermal growth factor receptor (EGFR) kinase as an ATP-competitive inhibitor.

4.1.1.3.5 Antidiabetic activity. The studies conducted by Doostkam et al. (2021) demonstrated a protective effect against diabetic auditory neuropathy in diabetics.

Diabetic rats were treated with rutin (at doses of 50, 100 and 150 mg/kg, respectively). Rutin administration improved hearing threshold, wave latency, wave morphology, the volume and number of spiral ganglion neurons, as well as superoxide dismutase (SOD) activity, while reducing malondialdehyde (MDA) levels in rats.

Calzada et al. (2017) demonstrated that oral administration of rutin (30 mg/kg) reduced postprandial glucose levels in rats with alloxan-induced diabetes compared with acarbose.

Ganesan et al. (2020) confirmed the antidiabetic effect of rutin in alloxan-induced diabetic rats. Orally administered rutin at a dose of 100 mg/kg daily for 28 days prevented the formation of ketone bodies in the urine and lowered creatinine, urea, triglycerides and cholesterol. In addition, studies have confirmed the protective effect of rutin against diabetic ketoacidosis and fibrosis, as it improves the expression of genes associated with metabolic acidosis, including aquaporins (AQP2/AQP3) and vasopressin V2 receptor (V2R).

4.1.1.3.6 Antiobesity activity. Choi and Kim (2013) demonstrated an antiobesity effect in mice on a high-fat diet that were given rutin at doses of 5 and 50 mg/kg per day. Administration of rutin did not cause weight gain in mice fed a high-fat diet after 28 days; total cholesterol levels decreased, while expression of peroxisome proliferator-activated receptor γ (PPARγ) and CCAAT-enhancer-binding protein α (C/EBPα) mRNA in hepatocytes was suppressed.

4.1.1.3.7 Antibacterial activity. Memar et al. (2023) showed the antibacterial activity of rutin nanocrystals. Antibacterial activity was assessed by determining the minimum inhibitory concentration (MIC) using the microdilution technique. Antibacterial activity against *Pseudomonas aeruginosa*, *Staphylococcus aureus,* and *Streptococcus mutans* was demonstrated.

4.1.1.3.8 Practical utilization in cosmetology. Rutin and its derivatives are widely used in the cosmetics industry. According to the CosIng database, nine forms of rutin are used in cosmetic products. According to CosIng database (2026), these compounds exhibited antioxidant, skin conditioning, surfactant, skin protecting, hair conditioning, and UV absorber properties. Rutin is an ingredient of cosmetic products marketed worldwide. It is used for facial, body, and hair care products, including creams, serums, and masks. Examples of cosmetic products containing rutin are presented in Table 8.1.

4.2 Quercetin

4.2.1 Chemical structure and plant resources

Quercetin (3,3′,4′,5,7-pentahydroxyflavone) (Fig. 8.9) consists of two benzene rings connected by a heterocyclic pyrone ring. Its name is related to *Quercetum* (oak forest), named after *Quercus* in 1857. It has the form of a yellow powder with a bitter taste, is insoluble in water, hardly soluble in alcohol, but is well soluble in glacial acetic acid (Baghel et al., 2012). It is found mainly in onions, red wine, apples, green tea, blueberries, as well as in plants with valuable medicinal properties, such as

Table 8.1 Examples of cosmetic preparations containing flavonoids in their composition.

Producer	Country of origin	Trade name	Cosmetic form	The form of rutin in the composition of the cosmetics (INCI)	Properties of the cosmetics according to the producer
Rutin and its derivatives					
Lumene (www.lumene.com)	Finland	Bright eyes under eye concealer	Concelear	Darutoside	Reduces the appearance of dark undereye circles and helps to reduce puffiness for a visibly brighter, rejuvenated eye area
Fiorga laboratoires (www.filorga.com)	France	Time-filler absolute eye correction cream	Cream	Darutoside	Smoothing eye contour
Yves Rocher (www.yves-rocher.pl)		Sensitive Végétal intense soothing moisturizing care	Serum	Rutin	Soothes, strengthens, and protects sensitive skin
Doctor Babor (www.babor.com)	Germany	Pro-growth factor EGF	Serum	Rutin	Activate natural skin renewal and essential rejuvenation mechanisms
Eucerin (www.eucerin.com)		Pigment control sun fluid SPF 50+	Suncreen	Glucosylrutin	Prevents sun-induced hyperpigmentation
KLAPP cosmetics (www.klapp-skincare.com)		Eye care stri-pexan	Cream	Rutin	Antiaging eye cream
Nivea (www.nivea.com)		Urban skin detox mask hidratante	Mask	Glucosylrutin	Moisturizing the skin
Korres (www.korres.com)	Greece	Golden Krocus Hydra-Filler plumping cream	Cream	Troxerutin	Antioxidant for all skin types
Collistar (www.collistar.com)	Italy	Intensive antistretch marks body cream	Cream	Rutin	Preventing and reducing stretch marks
Rilastil (www.rilastil.com)		Soothing micellar solution	Micellar water	Methoxy-PEG-7 rutinyl succinate	No-rinse makeup remover
Dermedics (www.dermedics-asia.com)	Malaysia	Youth Expert Ultimate Corrective Serum	Serum	Rutin	Regeneration of the damaged extracellular matrix and deactivation of inflammation

Continued

Table 8.1 Examples of cosmetic preparations containing flavonoids in their composition.—*cont'd*

Producer	Country of origin	Trade name	Cosmetic form	The form of rutin in the composition of the cosmetics (INCI)	Properties of the cosmetics according to the producer
Rutin and its derivatives					
Nanowhite (www.nanowhite.com)		Intensive brightening serum	Serum	Rutin	Helps reduce stubborn spots by up to 95%. Brightens up skin tone and improves clarity
Bandi professional (www.bandi.pl)	Poland	Cleansing milk remover with liposome complex	Cleansing milk	Disodium rutinyl disulfate	Cleansing
Bielenda professional (www.bielenda.com)		Supremelab S.O.S. Strengthening capillary skin face cream SPF 15	Cream	Troxerutin	The cream is an ideal support for sensitive capillaries, prone to irritation.
Dermedic (www.dermedic.pl)		Redness calm antiaging night cream	Cream	Troxerutin	Soothes redness and irritation, increases the level of hydration,
Fitomed (www.fitomed.pl)		Serum lukrecja gładka	Serum	Rutin	Moisturizes, smoothes the skin, and gives a feeling of satin skin
Yoskine (www.yoskine.com)		Asayake pure	Micellar water	Rutin	Remove make-up and skin impurities
ESPA (www.espaskincare.com)	United Kingdom	Lift & Firm eye moisturizer	Cream	Disodium rutinyl disulfate	Improve skin hydration and enhance firmness and elasticity
Mary Kay (www.marykay.com)	USA	Timewise replenishing Serum + C	Serum	Rutin	Antioxidants and antiirritants to leave skin feeling soft and smooth
PCA skin (www.pcaskin.com)		Ideal complex eye cream	Cream	Darutoside	Firms and lifts the skin around the eye
Perricone MD (www.perriconemd.com)		Neuropeptide smoothing facial conformer	Serum	Rutin	Helps minimize the appearance of fine lines and wrinkles and improves skin tone and radiance

Neogen (www.neogen.com)	South Korea	Oxygen AHA bubble peeling mask	Mask	Rutin	Cleanses, smooth, and glowing
ESPA (www.espaskincare.com)	United Kingdom	Lift & Firm eye moisturizer	Cream	Disodium rutinyl disulfate	Improve skin hydration and enhance firmness and elasticity
Mary Kay (www.marykay.com)	USA	Timewise replenishing Serum + C	Serum	Rutin	Antioxidants and antiirritants to leave skin feeling soft and smooth
PCA skin (www.pcaskin.com)		Ideal complex eye cream	Cream	Darutoside	Firms and lifts the skin around the eye
Perricone MD (www.perriconemd.com)		Neuropeptide smoothing facial conformer	Serum	Rutin	Helps minimize the appearance of fine lines and wrinkles and improves skin tone and radiance
Quercetin and its derivatives					
Garnier (www.garnier.com)	France	Ambre solaire sensitive face sun cream mist SPF 50	Cream	Isoquercetin	Helps protect against long-term damage from the sun and premature skin aging
Biopeptix (www.biopeptix.com)	Israel	Contour eye lift	Cream	Quercetin caprylate	Reduces dark circles and puffy eyes
Epionce (www.epionce.com)	Japan	Renewal facial cream	Cream	Quercetin	Helps reduce the appearance of fine lines, wrinkles, and other signs of skin aging
Dr Irena Eris (www.drirenaeris.com)	Poland	Tokyo Lift antiwrinkle protecting & illuminating serum	Serum	Quercetin	Illuminating and protecting day & night serum
Sesderma (www.sesderma.com)	Spain	Factor G Renew Rejuvenating Gel Cream	Cream	Quercetin	Antiaging, activates skin regeneration
Dermalosophy (www.dermalosophy.co.il)	Switzerland	Total repair Forte	Cream	Quercetin caprylate	Renewing the skin
Sisley (www.sisley.com)	United Kingdom	Ecological compound	Serum	Isoquercetin	Helps skin fight environmental stress and restores the skin to a more toned, supple state
Dr Dennis Gross (www.drdennisgross.com)	USA	Vitamin C lactic oil-free radiant moisturizer	Serum	Quercetin	Restores radiance and evens the skin tone

Continued

Table 8.1 Examples of cosmetic preparations containing flavonoids in their composition.—*cont'd*

Producer	Country of origin	Trade name	Cosmetic form	The form of rutin in the composition of the cosmetics (INCI)	Properties of the cosmetics according to the producer
Rutin and its derivatives					
Teami (www.teamiblends.com)		Overnight sleep mask	Mask	Dihydroquercetin	Moisturizing
Sothys (www.sothys.fr)		[W.]™ + brightening fluid	Cream	Quercetin caprylate	Helps boost clarity and protects the skin from free-radical damage
Luteolin					
Aiken (www.aiken.com.my)	Malaysia	Prebiotic & probiotic shampoo anti hair fall	Shampoo	Luteolin	Helps strengthen the scalp's natural defense system for healthier & stronger hair
		Prebiotic & probiotic shampoo antidandruff	Shampoo	Luteolin	Antiirritation properties to soothe the scalp
Skin actives scientific (www.skinactives.com)	USA	Skin brightening cream	Cream	Luteolin	Helps reduce spots of hyperpigmentation
		Ultra calming cream	Cream	Luteolin	Moisturizes and soothes redness and irritation
Apigenin					
Kevin Murphy (www.kevinmurphy.com)	Australia	Body. Mass	Serum	Apigenin	Leave in plumping conditioning treatment
Bioderma (www.bioderma.com)	France	Cicabio Arnica+	Cream	Apigenin	Reduces the appearance of bruises, knocks, and bumps
Sephora (www.sephora.com)		Waterproof eye makeup remover	Makeup remover	Apigenin	Make-up remover suitable for contact lens wearers and those with sensitive eyes
Catrice (www.catrice.eu)	Germany	Lash brow boost elixir	Serum	Apigenin	Activates the growth of eyelashes and eyebrows
Lakme (www.lakme.com)	India	K'therapy active prevention shampoo	Shampoo	Apigenin	Prevents hair loss
Regaliz (www.regalizindia.com)		Folirich hair growth serum	Serum	Apigenin	Helps in hair growth, prevents follicle aging, hair fall, and increases hair strength

Dermedic (www.dermedic.com)	Poland	Capilarte serum treatment	Serum	Apigenin	Stimulates the growth of new, properly structured hair shafts
Marti Derm (www.martiderm.com)	Spain	Hair system anti-hair-loss ampoules	Ampoules	Apigenin	Slcw down the hair loss process and stimulate the growth of new hair
Camille Rose (www.camillerose.com)	United Kingdom	Rejuva-drops come back!	Serum	Apigenin	Effect on the regrowth of hair loss
MD Hair (www.md-factor.com)	USA	Customized hair care serum	Serum	Apigenin	Support scalp health and stimulate hair follicles to switch from resting to the active growth phase
Kaempferol and its derivatives					
Patyka (www.patyka.com)	France	Instant wrinkle Filler	Cream	Kaempferol	Smooths and fills
Comfort zone (www.world.comfortzoneskin.com)	Italy	Body active cream	Cream	Kaempferol	Toning and antioxidant action
Chantarelle (www.chantarelle.pl)	Poland	Nano serum C peptide with alpine Floral stem cell	Serum	Tiliroside	Effectively stimulates the renewal of dermal stem cells, clearly rejuvenating skin and improving its condition
AmorePacific (www.ap-beauty.com)	South Korea	Moisture-bound rejuvenating serum	Serum	Kaempferol	Moisturizing
Iope (www.iope.com)		Bio Eye Cream Youth Complete	Cream	Kaempferol	Specifically designed for the delicate skin around the eyes, absorbing quickly for long-lasting nourishment
Zelens (www.zelens.com)	United Kingcom	Hyaluronic acid complex serum drops	Serum	Kaempferol	Moisturizing
Naturium (www.naturium.com)	USA	Quadruple hyaluronic acid serum 5%	Serum	Kaempferol	Moisturizing

Continued

Table 8.1 Examples of cosmetic preparations containing flavonoids in their composition.—*cont'd*

Producer	Country of origin	Trade name	Cosmetic form	The form of rutin in the composition of the cosmetics (INCI)	Properties of the cosmetics according to the producer
			Rutin and its derivatives		
Hesperidin and its derivatives					
Dermedic (www.dermedic.com)	Poland	Oilage antiaging concentrated antiwrinkle eye cream	Cream	Glucosyl hesperidin	Eye cream for sensitive skin with signs of aging
Redblocker (www.redblocker.pl)		Local treatment serum	Serum	Hesperidin	Helps to eliminate defects visible on the skin: redness, visible capillaries, local bruising, or discolorations
Lineages (www.lineagenaturalcosmetics.ca)	South Korea	Eye sleeping mask	Mask	Hesperidin	Helps to soothe and maintain deep hydration while you sleep
Numbuzin (www.numbuzin.com)		No. 5 daily multi-vitamin cream	Cream	Hesperidin	Moisturizing
Isadora (www.isadora.com)	Sweden	Nature enhanced length mascara black	Mascara	Hesperidin	A lengthening mascara
Insta Natural (www.instanatural.com)	USA	Dead Sea mud mask	Mask	Hesperidin	Absorbs excess oil, dirt, and toxins from the skin.
Mad Hippie (www.madhippie.com)		Hydrating nutrient mist	Serum	Hesperidin	Protect against photodamage
Skin actives scientific (www.skinactives.com)		Capillary repair cream	Cream	Hesperidin	Helps reduce the appearance of varicose veins and capillaries

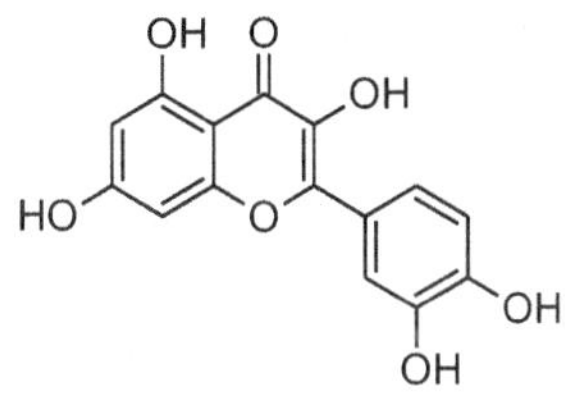

FIGURE 8.9

Chemical structure of quercetin.

Hypericum perforatum, *Ginkgo biloba,* and others (Almeida et al., 2018; Anand David et al., 2016; Ding et al., 2021; Shi et al., 2021).

According to the CosIng database (2026), quercetin can be found in eight derivatives in cosmetics. Derivatives may exhibit skin conditioning and antioxidant properties, act as humectants, and exert a protective effect on the skin.

4.2.1.1 Review of the latest scientific studies focused on the cosmetological applications

4.2.1.1.1 Antioxidant properties Quercetin is the subject of many studies on its antioxidant activity. Studies have shown that its antioxidant activity is mainly related to the effect on GSH, enzymatic activity, signal transduction pathways, and ROS. Through these mechanisms, quercetin helps maintain the oxidative balance.

Animal studies demonstrated that quercetin induces the synthesis of GSH, which catalyzes the decomposition of H_2O_2 (produced by superoxide dismutase to neutralize free radicals) to nontoxic H_2O (Granado-Serrano et al., 2012; Kobori et al., 2015). Kinaci et al. (2012) confirmed that quercetin treatment in renal ischemia/reperfusion (I/R) rats enhanced the antioxidant capacity by increasing GSH levels.

Ademosun et al. (2016) showed that the –OH groups on the side phenyl ring of quercetin have a stronger inhibitory effect on enzymes with oxidative properties, such as acetylcholinesterase (AChE) and butyrylcholinesterase (BChE).

Adedara et al. (2017) demonstrated the antioxidant effect of quercetin in studies in rats subchronically treated with manganese alone at 15 mg/kg body weight or orally co-treated with quercetin at 10 and 20 mg/kg body weight for 45 consecutive days. Quercetin attenuated manganese-induced decreases in antioxidant enzyme activity and increases in acetylcholinesterase activity, hydrogen peroxide production, and lipid peroxidation levels in the hypothalamus and brain in rats.

The studies conducted by Chen, Park, et al. (2017) demonstrated that pretreatment with quercetin significantly increased the expression levels of endogenous antioxidant enzymes such as Cu/Zn SOD, Mn SOD, CAT, and GSH peroxidase in hippocampal CA1 pyramidal neurons of animals suffering from ischemic injury.

The studies conducted by Akkoyun et al. (2016) confirmed the action of quercetin in preventing heart damage by scavenging free radicals. Quercetin administered to rats with LPS-induced endotoxemia increased SOD and CAT levels and decreased MDA levels, compared with controls.

Manca et al. (2014) investigated the effect of incorporated quercetin in glycerosomes on their protective effect against oxidative stress-induced skin damage. Quercetin incorporated into liposomal and glycerosomal nanoformulations showed a strong ability to scavenge free radicals (DPPH test) and protect human keratinocytes in vitro against hydrogen peroxide damage. Moreover, quercetin-loaded vesicles were avidly taken up by keratinocytes in vitro. Overall, results indicate that quercetin incorporated into liposomal and glycerosomal nanoformulations improves keratinocyte protection against oxidative stress damage.

4.2.1.1.2 Antipsoriatic effect The studies conducted by Chen, Lu, et al. (2017) confirmed the supportive effect of quercetin in psoriasis. Studies have shown that the use of quercetin reduces the expression of NF-κB-inducing kinase (NIK) and NF-κB, including IkappaB kinase (IKK) and transcription factor (RelB), and increases the expression of tumor necrosis factor receptor associated-factor 3 (TRAF3). In addition, quercetin also increases the activity of GSH, CAT, and SOD and reduces the level of MDA in skin tissues induced by imiquimod (IMQ).

4.2.1.1.3 Protective effect against UVB radiation Studies demonstrated that quercetin prevents UVB-induced damage by removing ROS and strengthening the cell membrane and mitochondria against ROS-induced damage. In addition, it also inhibits cell membrane motility and mitochondrial membrane depolarization. Therefore, consuming quercetin also inhibits this imbalance and is used to prevent UVB-induced skin damage (Casagrande et al., 2006; Zhu et al., 2017).

4.2.1.1.4 Stimulating-melanogenesis activity Nagata et al. (2004) investigated the effects of quercetin on melanogenesis in human melanoma cells (HMVII) and in normal human epidermal melanocytes (nHEM) in the absence of UV radiation. After adding quercetin to the culture medium, the melanin content in HMVII and in human normal epidermal melanocytes. Quercetin administration for 7 days at a concentration of 20 μM for HMVII cells or 1 μM for nHEM cells increased the melanin content 7-fold. Tyrosinase activity was also increased, up to 61.8-fold higher than the control.

Takekoshi et al. (2013) investigated the effect of quercetin on melanin production in the tissues of hair follicles from the buccal region of C3H/HeN Jel mice. Studies have shown that hair follicle tissue synthesized greater amounts of melanin when administered quercetin than control tissue, with the amount depending on the concentration of added quercetin. Moreover, TRP-2, which is a melanogenic enzyme, increased in a concentration-dependent manner after quercetin treatment, whereas its mRNA expression was not affected. These results suggest that quercetin may have potential application in the treatment of vitiligo and acquired pigment loss, such as graying of the hair.

4.2.1.1.5 Wound healing activity Irfan et al. (2022) investigated the effect of wound acceleration using human umbilical cord mesenchymal stem cells (hUC-MSCs) and treated with 20 mM quercetin or rutin in an in vitro scratch test and an in vivo cold burn wound model developed by directly exposing the skin of the rat's back to liquid nitrogen. The scratch test showed increased cell migration toward wound closure in the treated groups compared with the control. A comparison of the treatment groups showed that the burn wound treated with quercetin had significantly greater healing potential than the MSCs treated with rutin. The results of histological examinations showed increased regeneration of skin layers together with hair follicles in the quercetin group.

4.2.1.1.6 Antiinflammatory activity García-Mediavilla et al. (2007) demonstrated the anti-inflammatory effect of quercetin in studies on the human hepatocyte-derived cell line. In in vitro studies have shown that quercetin significantly decreases levels of inflammatory mediators such as NO synthase, COX-2, and C-reactive protein (CRP).

Quercetin has also been studied in rats against arthritis. Studies have shown the inhibitory effect of quercetin on acute and chronic inflammation (Guardia et al., 2001; Mamani-Matsuda et al., 2006). In addition, quercetin in a 2-month supplementation in a dose of 500 mg decreases the level of CRP in amateur athletes (Askari et al., 2012).

4.2.1.1.7 Antiallergic activity Karuppagounder et al. (2015) investigated the effects of quercetin on skin lesions, high-mobility group box (HMGB)1 cascade signaling, and inflammation in an atopic dermatitis (AD) mouse model. After AD induction, quercetin (50 mg/kg, p.o.) was administered daily for 2 weeks. Studies have shown that quercetin weakened the development of AD-like skin lesions, inhibited hyperkeratosis, parakeratosis, acanthosis, mast cells, and infiltration of inflammatory cells. In addition, administration of quercetin impacted on down-regulated cytoplasmic HMGB1, RAGE, nuclear p-NF-κB, ERK 1/2, COX-2, TNF-α, IL-1β, IL-2Rα, interferon-γ (IFN-γ), and IL-4 and up-regulated nuclear Nrf2.

Sagit et al. (2017) investigated the effect of quercetin in an experimental rat model of allergic rhinitis. A significant increase was detected in posttreatment total IgE levels in the control group. However, there was no increase in IgE in the group treated with quercetin. On histopathological evaluation, it was observed that findings of allergic rhinitis were suppressed in the quercetin group when compared to the control group. In immunohistochemical evaluation, it was detected that COX-2 and vasoactive intestinal peptide (VIP) expressions were weaker in the quercetin group compared with the control group.

4.2.1.1.8 Antimicrobial activity Shu et al. (2011) demonstrated antibacterial activity of quercetin against strains such as *Streptococcus mutans* and *Streptococcus sobrinus*, with MIC values ranging from 1 to 8 mg/mL. In addition, Jaisinghani (2017) demonstrated the activity of quercetin against *Staphylococcus aureus* (MIC 20 mg/mL).

Studies have shown that quercetin has dose-dependent antiviral effects against *herpes* viruses (HSV-1 and HSV-2) in cell culture (Lee et al., 2017; Lyu et al., 2005). In addition, studies show that quercetin inhibits respiratory viruses in cultured cells and exerts protective effects on the lungs (Kumar et al., 2003). Furthermore, quercetin has demonstrated potential therapeutic benefits in the early stage of COVID-19 (Kalil et al., 2021; Khan et al., 2022).

4.2.1.1.9 Practical utilization in cosmetology Quercetin and its derivatives are widely used in the cosmetics industry. According to the CosIng database (2026), ten forms of quercetin are used in cosmetic products (Table 8.2). According to Cos-Ing, quercetin exhibit antioxidant, humectant, skin conditioning, and skin protecting properties. Quercetin is widely used as an ingredient in cosmetic products worldwide, including formulations intended for facial, body, and hair care, such as creams, serums, and masks. Examples of cosmetics containing quercetin are presented in Table 8.1.

4.3 Luteolin

4.3.1 Chemical structure and plant resources

Luteolin, commonly named 3′, 4′, 5, 7-tetrahydroxyflavone (Fig. 8.10), is one of the most widespread flavones. Luteolin is a common flavonoid that occurs in vegetables and fruits such as parsley, celery, perilla leaf, chamomile tea, and green pepper. The molar mass of this compound is 286.24 g/mol. Luteolin is soluble in organic solvents such as ethanol, dimethyl sulfoxide (DMSO), and dimethylformamide (DMF). Luteolin is sparingly soluble in aqueous buffers (Lo et al., 2021).

Table 8.2 Forms of quercetin and its derivatives and their use in cosmetic preparations according to CosIng database.

The form	Activity
Dihydroquercetin	Antioxidant
Dihydroquercetin dithiolanepentyl ether	Antioxidant Humectant Skin conditioning Skin conditioning—emollient
Dihydroquercetin glucoside	Antioxidant Skin conditioning
Dihydroquercetin pentacaprylate	Antioxidant Humectant Skin conditioning—emollient
Dihydroquercetin tetracaprylate	Antioxidant Humectant Skin conditioning Skin conditioning—emollient
Isoquercetin	Antioxidant
Quercetagetin	Skin protecting
Quercetin	Antioxidant Skin conditioning
Quercetin caprylate	Antioxidant Skin conditioning
Quercetin dimethyl ether glucoside	Skin conditioning Skin protecting

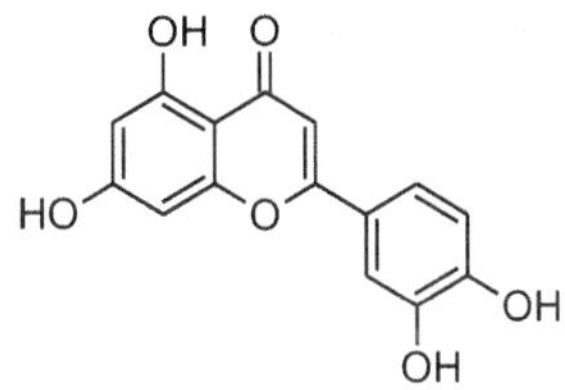

FIGURE 8.10

Chemical structure of luteolin.

4.3.1.1 Review of the latest scientific studies focused on the cosmetological applications

4.3.1.1.1 Antioxidant properties Choi, Islam, et al. (2014) investigated the effect of C-glycosylation at different positions of luteolin; the structure-activity relationships of luteolin and a pair of isomeric C-glycosylated derivatives, orientin and isoorientin, were evaluated. Studies have been conducted on the effects of C-glycosylation on the antioxidant, anti-Alzheimer's disease (AD), antidiabetic and antiinflammatory effects of luteolin and its two C-glycosides via in vitro assays of peroxynitrite (ONOO(−)), total ROS, NO, DPPH, aldose reductase, protein tyrosine phosphatase 1B (PTP1B), acetylcholinesterase (AChE), butyrylcholinesterase (BChE), and β-site amyloid precursor cleaving enzyme 1 (BACE1), and cellular assays of NO production and inducible iNOS/COX-2 expression in LPS-stimulated RAW 264.7 cell. Modification/transformation of luteolin structures using various methods can significantly improve their antioxidant activity. Luteolin was the most potent inhibitor of ROS generation. In addition, luteolin showed the most potent anti-AD activity as determined by its inhibition of AChE, BChE, and BACE1. Luteolin also inhibited NO production and iNOS protein expression in LPS-stimulated macrophages.

Aruwa et al. (2021) also investigated the method of enzymatic transformation of luteolin using fungal laccase from *Trametes pubescens,* which rendered a dimer, namely m/z 569. This transformation moderately enhanced the antioxidant activity of luteolin.

4.3.1.1.2 Antiphotoaging effect Mu et al. (2021) investigated the role of luteolin in UVB-induced photoaging. In the in vivo photoaging model, the dorsal skin of rats was irradiated with UVB. In vitro studies were performed on human dermal fibroblasts (HDFs) exposed to UVB irradiation. In vivo studies showed that luteolin reduced erythema and wrinkle formation. In vitro studies showed that luteolin inhibited the UVB-induced decrease in cell viability. In both in vitro and in vivo models, luteolin reduced oxidative stress and MMP activation and increased collagen expression. The findings suggest that luteolin protects skin cells against UVB-induced aging through the SIRT3/ROS/MAPK axis and may be a promising agent for preventing UVB-induced photoaging.

Lim et al. (2013) conducted studies on the effect of luteolin on UVB-induced photoaging using human keratinocytes (HaCaT) and SKH-1 hairless mice. In vitro studies demonstrated that luteolin inhibited UVB-induced matrix metalloproteinase-1(MMP-1) expression in HaCaT cells. It also suppressed the UVB-induced activation of activator protein-1 (AP-1), a transcription factor that regulates the MMP-1 promoter, as well as the expression of c-Fos (proto-oncogene) and c-Jun, the major component of the AP-1 complex. In vivo luteolin inhibited UVB-induced wrinkle formation and MMP-13 expression in mouse skin.

4.3.1.1.3 Photoprotective activity Abbas et al. (2022) performed in vitro studies of luteolin enclosed in an integrated lecithin-bile salt system called bilosomes on skin permeability and reducing skin damage caused by UV radiation. Studies have shown that luteolin delivered in this form is safe and soothes skin damage caused by UV radiation.

Yan et al. (2016) investigated the protective effect of luteolin against ultraviolet (UVA) radiation. The study was carried out on human skin fibroblasts (HSF). Studies have shown that after incubation with micromolar concentrations of luteolin reduced ROS production and attenuated autophagy. Western blot analysis showed that the expression levels of hypoxia-inducible factor 1-alpha (HIF-1α), microtubule-associated protein 1A/1B-light chain 3 (LC3)-II, and autophagy-related protein—Beclin 1 gradually decreased in UVA-irradiated HSF after luteolin treatment.

Rahimuddin et al. (2007) researched the protective effect against UVA radiation for luteolin, luteolin-7-*O*-glucoside, and luteolin-4′-*O*-glucoside. Both luteolin and luteolin-7-*O*-glucoside prevented a significant increase in lipid peroxides. Measurements of MDA levels showed that luteolin was clearly more effective than irradiating human skin fibroblasts at doses of 250 and 500 kJ/m^2.

4.3.1.1.4 Wound healing activity Mhimed et al. (2023) investigated the effect of luteolin nanoparticle cream on wound healing in rabbits. The study demonstrated that topical application of cream containing 1%, 2%, and 4% luteolin nanoparticles for 7 days significantly reduced the wound compared with the control group.

Ozay et al. (2018) investigated the effects of luteolin ointments on wound healing in rat models of excisional wounds with and without streptozotocin-induced diabetes. Studies have shown that luteolin ointments improve the wound healing process of skin tissue in both nondiabetic and diabetic wounds. The best wound healing activity was observed in cuts treated with 0.5% (w/w) luteolin ointment on day 14.

4.3.1.1.5 Antiinflammatory activity Xia et al. (2014) investigated the antiinflammatory effects of luteolin on human umbilical vein endothelial cells (HUVEC). HUVECs were treated with luteolin in the presence/absence of TNF-α. Luteolin suppressed the TNF-α-activated ROS generation, as well as the NADPH oxidase 4 (Nox4), ubiquitous protein p22phox, and intercellular adhesion molecule-1 (ICAM-1) and vascular cell adhesion protein-1(VCAM-1) expression. Luteolin inhibited the TNF-α-induced transcriptional activity of NF-κB and p38, and ERK1/2 phosphorylation. Luteolin decreased TNF-α-induced oxidative stress and inflammation via its effect on the Nox4/ROS-NF-κB and MAPK pathways.

4.3.1.1.6 Antipsoriatic activity Palombo et al. (2016) studied the effect of luteolin-7-glucoside in the treatment of psoriasis. In vitro studies were performed on HEKn cells (human epidermal keratinocytes, neonatal) and rats with psoriasiform-like skin inflammation. Studies have shown that luteolin-7-glucoside leads to modification of the cell cycle and induction of keratinocyte differentiation, with modification of energy, fatty acids, and redox metabolism. Luteolin-7-glucoside treatment also neutralizes the proliferative stimulus induced by the pro-inflammatory cytokines IL-22 and IL-6 in HEKn. In the imiquimod (IMQ) mouse model of psoriasis, topical administration of luteolin-7-glucoside leads to a significant reduction in acanthosis and re-expression of epidermal differentiation markers.

Weng et al. (2014) investigated the effect of luteolin on the activation of cultured human keratinocytes. Luteolin concentrations (10–100 μM) reduced phosphorylation, nuclear translocation and DNA binding of nuclear factor kappa B (NF-κB), which is usually involved in the transcription of inflammatory mediators. Furthermore, luteolin reduced the induced mRNA expression of NFKB1 and RELA, encoding the NF-κB p50 and p65 subunits, respectively. Moreover, luteolin reduced keratinocyte proliferation, which is characteristic of psoriatic skin.

Zhou et al. (2020) investigated the effect of luteolin on psoriasis-like lesions induced by imiquimod in BALB/c mice and its possible antiinflammatory mechanisms in lipopolysaccharide-stimulated macrophages (RAW 264.7) cells. In vivo studies, luteolin alleviated psoriasis-like skin lesions, reduced skin infiltration of macrophages, T lymphocytes, and neutrophils, and inhibited the expression of cytokines such as IL-6, IL-1β, TNF-α, IL-17A, and IL-23. In vitro studies have shown that luteolin reduces the level of pro-inflammatory cytokines associated with psoriasis, such as IL-17A, IL-6, TNF-α, and IL-23, and inflammatory mediators such as nitric oxide NO, inducible NOS, and COX-2 in RAW264.7 cells.

4.3.1.1.7 Atopic dermatitis activity Jegal et al. (2021) conducted a study of the effect on inflammation associated with atopic dermatitis luteolin-7-methyl-ether isolated from *Wikstroemia ganpi*. The antiatopic activities of the isolated compounds were determined using TNF-treated RBL-2H3 cells and HaCaT cells. Luteolin-7-methyl-ether reduced IL-4, IL-6, granulocyte macrophage colony-stimulating factor (GM-CSF), granulocyte colony-stimulating factor (G-CSF), and transient receptor potential vanilloid 1 (TRPV1) mRNA.

4.3.1.1.8 Antibacterial activity Xi et al. (2022) investigated the antibacterial effect of luteolin on *Staphylococcus aureus*. The results demonstrated that luteolin had excellent antibacterial activity against *S. aureus* (MIC = 312.5 μg/mL). The developed disinfectant with luteolin showed a better antibacterial effect.

4.3.1.1.9 Practical utilization in cosmetology Luteolin and its derivatives are widely used in the cosmetics industry. According to the CosIng database (2026), one form of luteolin is used in cosmetic products, where it function as skin conditioning agent. Despite its beneficial effects on the skin, luteolin is not yet widely used as cosmetic ingredient. It is found mainly in cosmetic products manufactured in Malaysia and the USA, including facial, body, and hair care, such as shampoos and creams. Examples of cosmetics containing luteolin are presented in Table 8.1.

4.4 Apigenin

4.4.1 Chemical structure and plant resources

Apigenin (APG, 4′,5,7-trihydroxyflavone) (Fig. 8.11) is a bioactive plant flavonoid that belongs to the flavone subclass. The richest natural sources of APG include dried flowers of chamomile, parsley, celery, thyme, oregano, olives, onions, kumquats, cherries, oranges, broccoli, tomatoes, barley, carrots, and legumes. Botanically, the name apigenin is derived from the genus Apium in the family *Apiaceae*. The richest source of APG is dried parsley (45,035 μg/g). The other sources that contain high APG amounts are chamomile (dried flowers) (3000–5000 μg/g), celery seed (786.5 μg/g), vine spinach (622 μg/g), and Chinese celery (240.2 μg/g). The molar mass of this compound is 270.0528 g/mol. Apigenin is soluble in organic solvents such as ethanol, DMSO, and DMF, which should be handled under an inert gas. Apigenin is sparingly soluble in aqueous buffers (Jin et al., 2009).

According to the CosIng database (2026), in addition to apigenin, dimethoxyapigenin can be used in cosmetics, where it functions as skin conditioning and skin-protecting agent.

4.4.1.1 Review of the latest scientific studies focused on the cosmetological applications

4.4.1.1.1 Antioxidant properties Zhang et al. (2020) investigated the effect of apigenin on oxidative stress in the PIG3V vitiligo perilesional melanocyte cell model. The results showed that cells treated with apigenin exhibited higher viability than the negative control cells. Similarly, apigenin increased the expression of the cellular antioxidants SOD, CAT, and GSH-Px, but inhibited the production of MDA, a biomarker of oxidative stress.

The studies conducted by Perez-Leal et al. (2017) of natural compounds on the translation-inducing Nrf2. Studies have shown that apigenin induced Nrf2 translation. The antioxidant response is activated via pharmacologically induced Nrf2, which defeats Kelch-like ECH-associated protein 1 (Keap 1) regulation. The above-mentioned mechanism of action of apigenin has also been confirmed by other studies (Paredes-Gonzalez et al., 2015; Wu et al., 2015; Xu et al., 2016).

Begum et al. (2012) conducted studies on the possible effectiveness of radiological protection of apigenin against whole-body gamma (γ) radiation, oxidative damage, and hematological changes in Swiss albino mice. In addition, apigenin

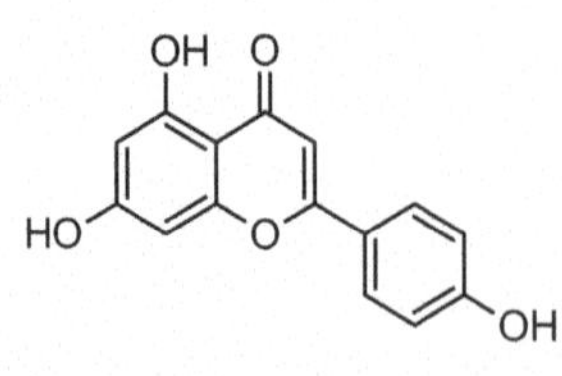

FIGURE 8.11

Chemical structure of apigenin.

increased antioxidant status and increased radiation-lowered levels of hemoglobin (Hb), white blood cells (WBC), and red blood cells (RBC).

4.4.1.1.2 Antiaging activity Bahadır Acıkara et al. (2019) investigated the antiaging potential of apigenin. Apigenin 7-*O*-β-glucoside and apigenin 7-*O*-β-rutinoside (100 μg/mL) extracted from *Podospermum canum* C.A. Meyer exerted an in vitro inhibitory effect on *Clostridium histolyticum* collagenase by 43.04% and 40.36%, and also on pancreatic porcine elastase by 26.48% and 24.48%.

Crascì et al. (2018) performed in vitro studies on the effect of apigenin on skin aging. Studies have shown a decreasing effect of apigenin (400 μg/mL) on advanced glycation end products in solutions of bovine serum albumin (BSA). A reduction in glycosaminoglycans and NO levels by treating normal human keratinocyte cells NCTC 2544 with apigenin at concentrations of 1–100 μM for 72 h.

4.4.1.1.3 Antiphotoaging activity Liu et al. (2023) investigated the effect of apigenin and doxycycline encapsulated in liposomes on photoaging of the skin. In vitro studies in HaCaT cells demonstrated that apigenin-loaded liposomes protected against oxidative stress-induced damage, reduced the production of inflammatory factors, and inhibited MMP activation. In vivo, these liposomal formulation attenuated skin damage and reduced collagen loss by suppressing MMP activation.

4.4.1.1.4 Photoprotective activity In a study by Abu-Yousif et al. (2008), apigenin showed a cytoprotective effect on UVB-induced apoptosis in HaCaT cells, neonatal foreskin and organotypic keratinocytes at a concentration of 10–50 μM.

Britto et al. (2017) investigated the protective effect of apigenin on UVB radiation. The model for the study was the cell line of human dermal fibroblasts (HDF). Apigenin administration reduced the formation of pyrimidine cyclobutane dimers (CPD). Apigenin treatment restored UVB-induced loss of NER proteins in HDF cells. In addition, apigenin showed strong UV absorption properties, and the SPF value was 10.08.

Choi et al. (2016) conducted in vitro studies of normal human dermal fibroblasts (nHDF) treated with apigenin (1–100 μM) after UVA irradiation. It has been shown that apigenin can increase cell viability and reduce UV-induced cytotoxicity, cell senescence, and MMP-1 expression. The study also assessed the effect of a cream with 1% apigenin on antiaging. The density of the dermis improved in those who used the cream with apigenin 1% for 4 weeks, while application of the control cream in the control group had no such effect. The use of the cream also improved skin hydration and reduced the visibility of crow's feet.

4.4.1.1.5 Skin lightening activity Yoshihisa et al. (2020) conducted animal studies on the effect of apigenin on skin discoloration. The model for the study was C57BL/6 mice with discoloration on the skin of the back. Apigenin was applied as a 0.1%–1% ointment 1 week before and 6 weeks after UVB exposure. The results showed that apigenin could significantly reduce UVB-induced hyperpigmentation and upregulation of macrophage migration inhibitory factor (MIF) by 60% suppression of the casein kinase II pathway. In vitro studies using mouse keratinocytes treated with apigenin (0.1–10 μg/mL) also showed a dose-dependent inhibitory effect of this compound on MIF-inducible protease-activated receptor (PAR-2) and stem cell factor (SCF) expression, two factors involved in skin pigmentation, due to 75% inhibition of casein kinase II.

4.4.1.1.6 Antipsoriatic activity Singh et al. (2023) conducted in vitro and in vivo studies of apigenin's antipsoriatic activity. Apigenin effectively reduced the expression and secretion of pro-inflammatory cytokines by the IL-23/IL-17/IL-22 axis. NF-κB nuclear translocation was suppressed by apigenin in RAW 264.7 cells induced by LPS. In HaCaT cells, the antiproliferative effect of apigenin was demonstrated. Apigenin was found to be safe in an acute dermal toxicity study.

4.4.1.1.7 Wound healing activity Cheng et al. (2018) investigated the potential of apigenin in the treatment of skin damage. Apigenin significantly reduces hemolysis of rabbit erythrocytes induced by *Staphylococcus aureus* and exhibited antiinflammatory effects. In a mouse model of a methicillin-resistant *S. aureus* (MRSA)-infected wound treated with apigenin, significantly reduced the average bacterial colony count.

Lopez-Jornet et al. (2014) investigated the effect of apigenin on wound healing. An in vivo studies were performed in the SKH-1/CRL mouse model of a cut wound. The animals were topically treated with an apigenin gel (5 mg apigenin per wound) for 2 weeks. Compared with the negative control, the apigenin-treated exhibited a faster rate of re-epithelialization and a lower degree of inflammation.

4.4.1.1.8 Antiallergic activity Yano et al. (2009) conducted in vitro studies on the antiallergic activity of apigenin. Mouse spleen cells were used as the in vitro model. The results showed that apigenin decreased IL-4-induced STAT6 tyrosine phosphorylation, followed by reduced levels of IgG1 and IgE. The study also confirmed the antiallergic effect of dietary apigenin on picryl chloride-induced atopic dermatitis-like skin lesion on the ears of NC/Nga (animal model of human atopic dermatitis) mice. Furthermore, apigenin reduced clinical skin signs, as well as expression of IgG1, IgE, and IFN-γ mRNA in animal spleen cells.

Che et al. (2019) investigated the effect of apigenin on atopic dermatitis in an animal model. In mice, apigenin significantly reduced pruritus and inflammatory cell infiltration in a dose-dependent manner. In addition, the expression of IL-31, an important cytokine involved in the symptoms of atopic dermatitis, was reduced compared with control animals. An in vitro study using human mast cells (HMC-1) revealed that inhibition of the phosphorylation of the MAPK and NF-κB/p65 complex is involved in reducing IL-13 mRNA and protein expression.

4.4.1.1.9 Practical utilization in cosmetology Apigenin and its derivatives are widely used in the cosmetics industry. According to the CosIng database (2026), two forms of apigenin are used in cosmetic products (Table 8.3). According to CosIng, these compounds exhibited antioxidant, skin conditioning, skin protecting, and hair conditioning properties. Apigenin is aingredient in cosmetic products marketed worldwide. It is used for facial, body, and hair care products, including creams, serums, ampoules, and shampoo. Examples of cosmetics containing apigenin are presented in Table 8.1.

Table 8.3 Apigenin and its derivative and their use in cosmetic preparations according to CosIng database.

The form	Activity
Apigenin	Antioxidant Hair conditioning
Dimethoxyapigenin	Skin conditioning Skin protecting

4.5 Kaempferol

4.5.1 Chemical structure and plant resources

Kaempferol (3,5,7-trihydroxy-2-(4-hydroxyphenyl)-4H-chromen-4one) (Fig. 8.12), is a flavonol, widely distributed in plants and foods such as onion leaves, broccoli, kale, tea, cabbage, apples, grapes, leeks, and tomatoes.It has also been isolated from *Delphinium* species, witch-hazel, grapefruit, and numerous other plant sources. The molar mass of kaempferol is 286.23 g/mol. It is a yellow crystalline solid with a melting point of 276–278 °C. Keampferol is slightly soluble in water and readily soluble in hot ethanol and diethyl ether (Chandekar et al., 2022; Shahbaz et al., 2023).

According to the CosIng database (2026), in addition to kaempferol, its glycosides—nicotiflorin and tiliroside, are also used in cosmetic products. Nicotiflorin has antioxidant, skin conditioning, and skin-protective properties, whereas tiliroside functions as a skin conditioning agent.

4.5.1.1 Review of the latest scientific studies focused on the cosmetological applications

4.5.1.1.1 Antioxidant properties Gabrielska et al. (2005) investigated the protective effect of kaempferol on the phosphatidylcholine liposomes membranes exposed to the pro-oxidative effects of diphenyltin dichloride (DPhT) and triphenyltin chloride (TPhT) induced by UV radiation (253.7 nm). Kaempferol inhibited the oxidation of UV-irradiated liposomes. The results also indicate the antiradical activity of kaempferol against the 2,2-diphenyl-1-picrylhydrazyl free radical (DPPH). Kaempferol is a scavenger of hydroxyl radicals, superoxide anions, and

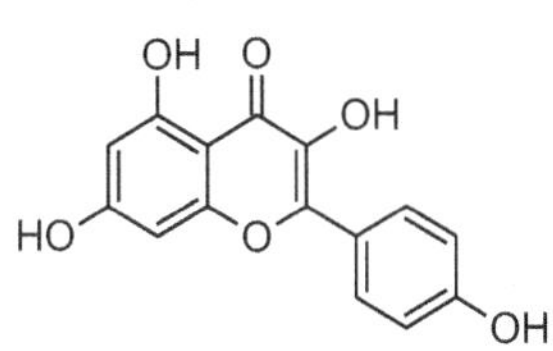

FIGURE 8.12

Chemical structure of kaempferol.

peroxynitrite. In addition, kaempferol also inhibits pro-oxidant-based enzymes such as xanthine oxidase, induces antioxidant enzymes such as heme oxygenase-1, superoxide dismutase, and catalase, and even suppresses hydroxyl radical generation by chelating of iron and copper ions.

Studies have shown the chemical structure of kaempferol contributes to its antioxidant activity. The hydroxyl groups at C3, C5, and C4', as well as the oxo group at C4 and the double bond between C2 and C3, are considered to contribute to its antioxidant properties. In addition, kaempferol inhibits the NF-κB signaling pathway and stimulates the Nrf2 transcriptional pathway. Kaempferol restoring cellular redox homeostasis. The anticancer effect of kaempferol is mainly attributed its antioxidant and anti-inflammatory effects (Chen et al., 2013).

4.5.1.1.2 Effect on the hypertrophic scar Li, Yang, et al. (2016) investigated the effect of kaempferol on the formation of hypertrophic scar in a mechanical load-induced mouse model. The results demonstrated that kaempferol significantly attenuated collagen synthesis, proliferation, and activation of fibroblasts in vitro and in vivo. The action of kaempferol was due to binding to the TGF-β type I receptor (TGFbRI).These findings suggest that kaempferol may be a promising therapeutic agent for the treatment of HPS or other fibroproliferative disorders.

4.5.1.1.3 Skin lightening activity Rho et al. (2011) investigated the depigmenting and antiinflammatory activity of kaempferol and its rhamnosides. The depigmenting activity was evaluated using B16 melanoma cells to evaluate the depigmenting activity and on LPS-stimulated RAW 264.7 cells to evaluate antiinflammatory activity. In all studies, kaempferol (1) and α-rhamnoisorobin (2) exhibited the strongest inhibitory effects. The order of inhibitory activity against melanogenesis and nitric oxide (NO) production was as follows: kaempferol (1) >α-rhamnoisorobin (2) > afzeline (3) > kaempferitrin (4).

4.5.1.1.4 Photoprotective activity Lee et al. (2010) investigated the effect of kaempferol on UVB-induced dermatitis. The result showed that kaempferol inhibited UVB-induced COX-2 protein expression in mice epidermal JB6 P+ cells and attenuated the transcriptional activity of COX-2 and AP-1.

4.5.1.1.5 Wound healing activity Özay et al. (2019) investigated the effect of kaempferol on wound healing. The study evaluated the effects of kaempferol of two concentrations (0.5% and 1%) on wound healing in diabetic and nondiabetic rats in cut and excised wound models. The best results were obtained after 14 days of treatment with a 1% kaempferol ointment. After 14 days, significantly higher hydroxyproline levels were observed in non-diabetic rats with excision and incision wounds treated with 1% kaempferol compared with the control groups. In addition, wound re-epithelialization was significantly enhanced after 7 and 14 days of kaempferol treatment compared with the control group.

4.5.1.1.6 Atopic dermatitis activity Lee and Jeong (2021) studied the effect of kaempferol on T cell activation and the underlying mechanisms of atopic dermatitis. Studies were performed on an immortalized T lymphocyte cell line (Jurkat cell line) and murine $CD4^+$ T cells, and in vivo studies were performed on a model of atopic dermatitis in mice. In vitro, kaempferol reduced type II C-type lectin receptor

(CD69) expression and production of inflammatory cytokines, including IL-2, from activated Jurkat cells and murine $CD4^+$ T cells without cytotoxicity. In an in vivo model, oral administration of kaempferol improved the symptoms of atopic dermatitis, a condition mediated by T lymphocytes. Mechanistic analyses demonstrated that kaempferol reduced JNK phosphorylation, which was associated with reduced multidrug resistance protein-1 (MRP-1) activity in vivo.

4.5.1.1.7 Antipsoriatic activity Li et al. (2023) investigated the effect of kaempferol in HaCaT cells, and in an imiquimod-induced psoriasis. In vitro, kaempferol reduced intracellular ROS production, inhibited rhIFN-γ-induced IFN-γR1 expression, and increased suppressor of cytokine signaling 1 (SOCS1) levels. In vivo, kaempferol alleviated imiquimod-induced psoriasis-like skin lesions in mice. In addition, kaempferol reduced the population of γδT17 cells in the lymph nodes of psoriatic mice, and also decreased the expression of several pro-inflammatory cytokines, including IL-23, IL-17A, TNF-α, IL-6, and IL-1β.

4.5.1.1.8 Effect on skin vitiligo Xie et al. (2023) conducted studies on the effect of kaempferol on melanogenesis in normal human PIG1 melanocytes and its response to oxidative stress. After kaempferol treatment, tyrosinase activity and melanin content in PIG1 cells increased, expression of TYR, TRP1, TRP2, and MITF mRNA and proteins increased, and ERK1/2 phosphorylation levels increased. After H_2O_2 stimulation, kaempferol decreased ROS production, decreased apoptosis, and increased HO-1 protein expression in PIG1 cells.

4.5.1.1.9 Antimicrobial activity Kaempferol has demonstrated antibacterial effect against Gram-positive bacteria: *Staphylococcus aureus* and *Staphylococcus epidermidis* (Kannanoor et al., 2021; Özçelik et al., 2006).

Studies have also shown the antifungal effect of kaempferol on such strains as *Candida albicans*, *Candida glabrata*, *Candida tropicalis*, and *Cryptococcus neoformans* (Tatsimo et al., 2012; Yordanov et al., 2008).

4.5.1.1.10 Practical utilization in cosmetology Kaempferol and its derivatives are widely used in the cosmetics industry. According to the CosIng database (2026), three forms of apigenin are used in cosmetic products (Table 8.4). According to CosIng, they exhibit antioxidant, skin conditioning, and skin protecting properties. Kaempferol is an ingredient in cosmetic products marketed worldwide. It is used in facial and body care products, including creams and serums. Examples of cosmetic products containing kaempferol are presented in Table 8.1.

4.6 Hesperidin

4.6.1 Chemical structure and plant resources

Hesperidin ((2S)-3′,5,7-trihydroxy-4′-methoxyflavan-4-one) (Fig. 8.13) is abundantly found as a flavanone glycoside (bioflavonoid) in tangerine (*Citrus reticulata*) and sweet orange (*Citrus sinensis*) peel. The name hesperidin is derived from the term “Hesperidium,” meaning citrus fruits. Furthermore hesperidin has been reported in some species of the family Rubiaceae and in the roots and aerial parts

Table 8.4 Forms of kaempferol and its derivatives and their use in cosmetic preparations according to CosIng database.

The form	Activity
Kaempferol	Antioxidant Skin conditioning—emollient
Nicotiflorin (kaempferol-3-rutinoside)	Antioxidant Skin conditioning—emollient Skin protecting
Tiliroside (kaempferol-3-O[-6-O-trans-p-coumaroyl-β-glucopyranoside)	Skin conditioning

of certain species belonging to the family Brassicaceae (Ahmad et al., 2023; Samota et al., 2023).

Pure hesperidin is a white crystalline compound that forms needle-shaped crystals upon recrystallization and has a melting point of 252–254°C. The molecular weight of hesperidin is 610.6 g/mol. Hesperidin is insoluble in water, soluble in DMF and DMSO (30 mg/mL), and slightly soluble in ethanol and methanol (~1 mg/mL) (Ahmad et al., 2023; Samota et al., 2023).

According to the CosIng database (2026), in addition to hesperidin, five of its derivatives are used in cosmetic products—glucosyl hesperidin acting as a humectant and nourishing the skin, hesperidin methyl chalcone with antioxidant properties, methyl hesperidin and neohesperidin with nutritional properties, as well as neohesperidin dihydrochalcone with perfuming, scent, and flavor properties.

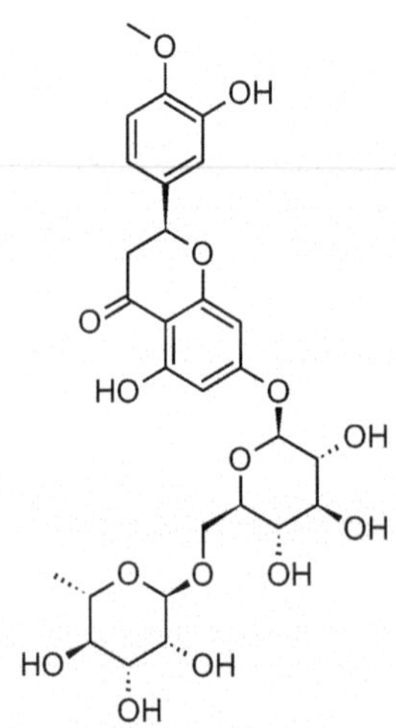

FIGURE 8.13

Chemical structure of hesperidin.

4.6.1.1 Review of the latest scientific studies focused on the cosmetological applications

4.6.1.1.1 Antioxidant properties Many studies focus on the antioxidant activities of hesperidin. One of the studies was aimed at evaluating the effect of hesperidin on erythrocyte damage caused by H_2O_2. Studies have shown that hesperidin inhibited damage to the erythrocyte membrane as well as free radical scavenging activity (Kalpana et al., 2009). Other studies have shown that hesperidin has a neutralizing effect on NO radicals, hydroxyl radicals, and superoxide anions (Wilmsen et al., 2005).

Studies have also shown the protective effect of hesperidin against damage to tissues, DNA, and proteins from internal and external factors. Hesperidin has been shown to protect against damage caused by nicotine, acrylonitrile, peroxynitrite, γ radiation, cadmium, hydrogen peroxide, lipopolysaccharide, and technetium. Hesperidin also increases the activity of antioxidant enzymes (Ahmad et al., 2012; Arafa et al., 2009; Balakrishnan & Menon, 2007; Chen, Gu, et al., 2010; Chen, Ye, et al., 2010; El-Sayed et al., 2008; Hosseinimehr et al., 2009; Kamaraj et al., 2011; Pradeep et al., 2012). Studies have also confirmed the inhibitory effect of hesperidin on the formation of advanced glycation products (Li, Cai, et al., 2012; Li, Mitsuhasi, et al., 2012).

4.6.1.1.2 Antiaging activity Novotná et al. (2023) conducted research on the antisenescence effect of hesperidin on physiologically aged normal human dermal fibroblasts (nHDF). Hesperidin inhibited the activity of elastase and hyaluronidase. The levels of MMP-1 and MMP-2 were reduced after the use of hesperidin.

Stanisic et al. (2020) conducted studies on the effect of hesperidin added to a cream applied to 3D artificial skin from cultured cells extracted from skin remnants after plastic surgery. Daily application of the cream reduced dark circles under the eyes, improve skin brightness, and alleviated UV-induced skin irritation.

4.6.1.1.3 Antiphotoaging activity Lee et al. (2018) conducted studies to evaluate the effectiveness of hesperidin against photoaging of the dorsal skin in hairless mice. Oral hesperidin reduced wrinkle depth, skin thickness, and epidermal hyperplasia. In addition, hesperidin reduced transepidermal water loss (TEWL) and pro-inflammatory cytokine expression, which was increased after UVB irradiation.

4.6.1.1.4 Photoprotective activity Santhanam et al. (2018) investigated the protective effects of hesperidin against UVB-induced cytotoxicity and its influence on the expression of pro-inflammatory cytokines (IL-6, IL-1β, and TNF-α), NF-κB, and matrix metalloproteinases (MMP1, 3, and 9) in human dermal fibroblasts (HDF). The results demonstrated that hesperidin prevented UVB-induced cytotoxicity in HDF cells in a dose-dependent manner. In addition, hesperidin suppressed the UVB-induced increase in expression of pro-inflammatory cytokines reducing their levels to values comparable to those of the non-irradiated control. It also inhibited the UVB-induced expression of NF κB, MMP 1, MMP 3, and MMP 9.

Li, Lin, et al. (2016) conducted research on the effect of hesperidin on skin oxidative stress and inflammation caused by UVA radiation. The studies were performed on UVA-irradiated HaCaT cells. Studies have shown that hesperidin significantly reduced oxidative stress and UVA-induced inflammatory response in *in vitro* studies.

Hou et al. (2012) investigated the effect of hesperidin on the epidermal permeability barrier function and its underlying mechanisms. Hairless mice were treated topically with 2% hesperidin twice daily for 6 days. The results demonstrated that hesperidin significantly accelerated the regeneration of the barrier. Hesperidin stimulated both epidermal proliferation and differentiation as well as increased lamellar body secretion.

Martinez et al. (2016) conducted studies to evaluate the effect of hesperidin on antioxidant activity in a mouse model of skin oxidative stress and UVB-induced inflammation. Hesperidin inhibited UVB-induced skin edema, reduction in antioxidant capacity (iron and ABTS-reducing capacity, and catalase activity), lipid peroxidation, superoxide anion production, and the mRNA expression of gp91phox (nicotinamide adenine dinucleotide phosphate [NADPH] oxidase subunit 2). Furthermore, hesperidin suppressed the UVB-induced production of cytokines (TNF-α, IL-1β, IL-6, and IL-10).

4.6.1.1.5 Antipsoriatic activity Li et al. (2019) investigated the effect of hesperidin on the imiquimod (IMQ) induced model of psoriasis and HaCaT stimulated with LPS. Studies have shown that hesperidin improved psoriatic skin lesions in mice treated with IMQ and inhibited LPS-induced proliferation of HaCaT cells. In addition, hesperidin reduced the thickness of the epidermis, proliferation, and differentiation of epidermal cells. Moreover, hesperidin inhibited mRNA expression of inflammatory factors, reduced local skin lesions, and reduced serum insulin and glucose levels.

4.6.1.1.6 Antiobesity activity Wang et al. (2011) investigated the protective effect against hypercholesterolemia and fatty liver in male Wistar rats fed a high-cholesterol diet. The results demonstrated that hesperidin reduced reduced hepatic lipid accumulation, adipose tissue, and liver weight. Hesperidin administration modulated the mRNA expression of proteins involved in lipid metabolism: RNA-binding protein (RBP), heart-type fatty acid-binding protein (H-FABP), and cutaneous fatty acid-binding protein (C-FABP) in liver and adipose tissue. The results showed that hesperidin has positive effects on hypercholesterolemia and fatty liver.

4.6.1.1.7 Practical utilization in cosmetology Hesperidin and its derivatives are widely used in the cosmetics industry. According to the CosIng database, six forms of hesperidin are used in cosmetic products (Table 8.5). According to the CosIng database, these compounds exhibit antioxidant, humectant, skin conditioning, flavoring, fragrance, and perfuming activity. Hesperidin is a component of cosmetics produced all over the world. It is used in wide range of facial and body care products, including creams and serums. Examples of cosmetic products containing hesperidin are presented in Table 8.1.

4.6.1.1.8 Summary The flavonoids discussed in this chapter exhibit a wide range of biological activities that support their application in cosmetology. They areincorporated into cosmetic products intended for facial, body, or hair care. The predominant biological activity of these compounds is their antioxidant effect, which has

Table 8.5 Forms of hesperidin and its derivatives and their use in cosmetic preparations according to CosIng database.

The form	Activity
Glucosyl hesperidin	Humectant Skin conditioning
Hesperidin	Skin conditioning—emollient
Hesperidin methyl chalcone	Antioxidant
Methyl hesperidin	Skin conditioning
Neohesperidin	Skin conditioning
Neohesperidin dihydrochalcone	Flavoring Fragrance Perfuming

been confirmed in numerous studies. In addition, flavonoids exhibit anti-inflammatory, photoprotective, and skin-lightening properties (Rho et al., 2011; Santhanam et al., 2018; Shahbaz et al., 2023; Xia et al., 2014). The CosIng database (2026) allows the use of the above flavonoids and their derivatives in cosmetic products. Beyond their established cosmetic applications, flavonoids also possess wound-healing properties, further highlighting their potential in skin care and dermatological formulations.

Rutin exhibit broad spectrum of biological activities including beneficial effects on cardiovascular and central nervous systems. In addition, rutin has shown anti-cancer activity. The results demonstrated that the administration of 50 and 100 mg/kg of rutin reduced the level of MDA and increased the activity of SOD and GSH in 5-fluorouracil-induced cardiotoxicity in male rats (Sengul et al., 2021). In addition, rutin demonstrated the cytotoxic effect on two melanoma cell lines: RPMI-7951 and SK-MEL-28 (Pinzaru et al., 2021). Studies have also confirmed the positive effect of rutin on disorders of the central nervous system (Arowoogun et al., 2021).

Quercetin demonstrated cytotoxic effects on two breast cell lines, MCF-7 and MDA-MB-231 (Ahmed et al., 2021). This compound has scientifically proven protective effects on hydrogen peroxide-induced neurodegeneration (Heo & Lee, 2004).

Luteolin inhibited the invasion of human A375 melanoma cells (Yao et al., 2019). The studies also confirmed the role of luteolin in liver apoptosis in an experimental model of liver ischemia-reperfusion in rats (Tsalkidou et al., 2014).

Apigenin clearly attenuated the inhibition of vasorelaxation. In addition, apigenin induced concentration-dependent relaxation in the aorta, increased superoxide anion inhibition, NO levels, and constitutive NOS activity (Jin et al., 2009). After apigenin administration, the pancreatic tissue architecture improved, inflammatory infiltration, edema, duct dilatation, and necrosis decreased (Lampropoulos et al., 2013).

Studies have shown that kaempferol suppressed the proliferation of MDA-MB-231 triple-negative breast cancer (TNBC) cells (Zhu & Xue, 2019). The studies conducted on the effect of kaempferol on the state of osteopenia in ovariectomized rats. The results demonstrated that kaempferol improved callus formation at the site of drill trauma in the femoral shaft after ingestion of kaempferol in animals with glucocorticoid-induced bone loss (Adhikary et al., 2018).

The conducted studies evaluated the effect of hesperidin on the viability and apoptosis of A549 (hypotriploid alveolar basal epithelial cell line) cells of human non-small cell lung cancer (NSCLC) (Xia et al., 2018). The hepatoprotective effect is related to counteracting hepatic oxidative stress and enhancing hepatic antioxidants, inhibition of NF-κB activation, and prevention of hepatic stellate cell activation (El-Sisi et al., 2017). Studies have shown that hesperidin can reduce the incidence and severity of multiple sclerosis (Haghmorad et al., 2017).

The use of the above-mentioned flavonoids in cosmetic preparations is truly significant. Cosmetics manufacturers focus primarily on their antioxidant properties, but also on their brightening and antiaging activity. The table below presents examples of cosmetic preparations using selected flavonoids (Table 8.1).

To sum up, flavonoids are widely used in cosmetology, mainly based on their antioxidant activity. The selected flavonoids presented above have been described in detail in terms of the mechanism of antioxidant action, as well as other activities that are used in cosmetology and cosmetic preparations.

5. Important phenolic acids for cosmetic applications

Phenolic acids are compounds that exhibit a wide spectrum of biological activity, which also includes activities used in the cosmetics industry (Călinoiu & Vodnar, 2018; Grace, 2007; Heleno et al., 2015; Robbins, 2003; Weng & Yen, 2012). Phenolic acids are characterized by proven antimicrobial activity. As low-molecular compounds, freely penetrating bacterial cell membranes, they exert a bactericidal effect by increasing the permeability of cell membranes as a result of the change in membrane potential caused by the dissociation of phenolic acid (Kampa et al., 2004; Rice-Evans et al., 1996; Rosa & Silva, 2016). Phenolic acids also have a strong antioxidant effect, which is associated with the neutralization of toxic compounds (such as mycotoxins, nitrosamines, polycyclic aromatic hydrocarbons) and protection of cells against oxidative stress (Anson et al., 2011; Gomes et al., 2003; Piazzon et al., 2012; Rice-Evans et al., 1996; Sánchez-Maldonado et al., 2011; Tarnawski et al., 2006).

Phenolic acids are mentioned as the strongest antioxidants in the plant world (Anson et al., 2011; Gomes et al., 2003; Piazzon et al., 2012; Rice-Evans et al., 1996; Sánchez-Maldonado et al., 2011; Tarnawski et al., 2006). Of particular importance here are, for example, acids: ellagic, ferulic, gallic, caffeic, and protocatechuic. Antioxidant activity is based on a multidirectional mechanism of action. The phenolic acids show, for example, the nature of compounds with reducing properties, compounds that bind free radicals, agents that chelate metal ions, enzymes that catalyze oxidation reactions, oxidase inhibitors, terminators that interrupt radical chain reactions, or stabilizers of free radicals formed in oxidative reactions by their hydrogenation or complexation. This property is particularly used in cosmetic preparations (Anson et al., 2011; Gomes et al., 2003). Phenolic acids are used not only as active ingredients of a given preparation, but also as compounds stabilizing the entire cosmetic formulation. Selected phenolic acids that are widely used in the cosmetics industry, the mechanism of antioxidant action, other biological activities, and examples of use in cosmetics are presented in detail below (Fig. 8.14).

5.1 Ferulic acid

5.1.1 Chemical structure and plant resources

Ferulic acid (4-hydroxy-3-methoxycinnamic acid) (Fig. 8.15) is chemically derived from cinnamic acid. It is poorly soluble in water, but better in ethanol and propylene glycol (Babbar et al., 2021; Damasceno et al., 2017; Piazzon et al., 2012). It occurs naturally in many plant species. The highest concentration of ferulic acid is found in the leaves, seeds, and bark of conifers, but also in grains: corn, rye, rice, and wheat. Representatives of the genus of plants where the accumulation of ferulic acid is the highest include *Salix*, *Ferula*, *Stellaria*, *Circaea*, *Geranium*, *Lemna,* and *Brassica* (Babbar et al., 2021). Ferulic acid and its derivatives are widely used in the cosmetics industry. According to the CosIng database (2026), thirteen forms of ferulic acid are used in cosmetics. They have an antioxidative effect primarily, resulting from the acid's biological activity profile. Ferulic acid is used both as a stabilizer of preparations, preventing the oxidation of other ingredients, and as an active substance that improves the skin condition.

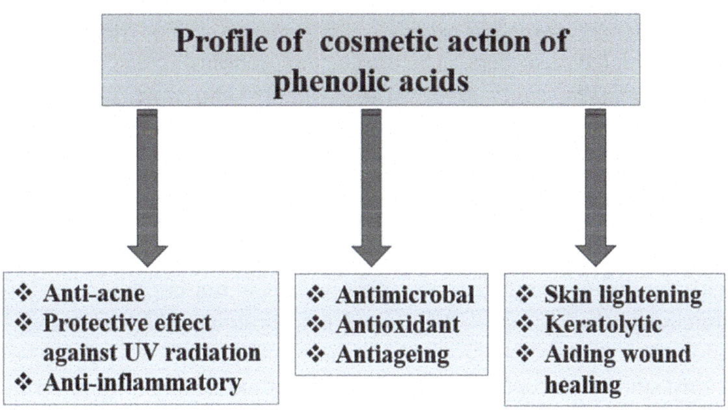

FIGURE 8.14

Profile of the action of phenolic acid.

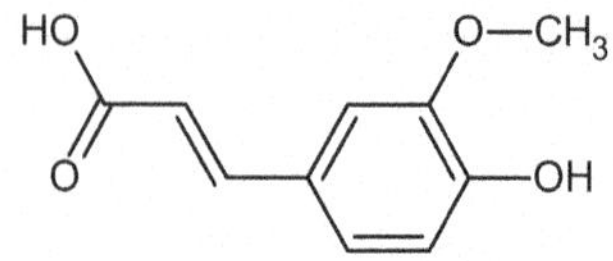

FIGURE 8.15

Chemical structure of ferulic acid.

5.1.1.1 Review of the latest scientific studies focused on the cosmetological applications

5.1.1.1.1 Antioxidant properties Ferulic acid is one of the strongest antioxidants among phenolic acids. The antioxidant mechanism is based primarily on the deactivation of active oxygen forms, the reduction in oxidative stress, and the prevention of the formation of thymidine dimers, which cause extensive damage to DNA and can lead to the initiation of the neoplastic process (Kim et al., 2020; Kyriakoudi et al., 2021; Murray et al., 2008; Oresajo et al., 2008; Piazzon et al., 2012).

A number of studies are focused on understanding the exact mechanism of the antioxidant action of ferulic acid.

Anis et al. (2020) conducted a study to assess the effect of ferulic acid as an antioxidant substance that may prevent neurodegenerative disorders during Parkinson's disease. The tests were based on a rat model damaged by hydroxydopamine, which was administered ferulic acid orally. It was shown that ferulic acid effectively reduced the level of oxidative stress caused by hydroxydopamine, inhibited morphological changes, and DNA fragmentation. It was also shown that ferulic acid effectively inhibited cell apoptosis. The mechanism of action consisted of reducing the expression of mitochondrial dynamin-related protein 1 (Drp1) and increasing the expression of genes and proteins peroxisome proliferator-activated receptor gamma coactivator 1-alpha (PGC1α), Cyt-c, p53, and cleaved poly(ADP-ribose) polymerase (PARP), thus restoring the proper dynamics of mitochondria.

Bumrungpert et al. (2018) conducted a study to evaluate the oral administration of ferulic acid on oxidative stress, inflammation, and lipid profiles in hyperlipidemia-related disorders. The study was double-blind, randomized, and placebo-controlled. The study involved 48 patients who were divided into two groups. One group was given ferulic acid at a dose of 1000 mg/day; the other group received a placebo for 6 weeks. Oral administration of ferulic acid was shown to effectively reduce total cholesterol (8.1%), triglycerides (12.1%), and low-density lipoprotein (LDL)-cholesterol (9.3%). High-density lipoprotein (HDL)-cholesterol levels also increased (4.3%) compared with placebo. The results also showed that ferulic acid significantly reduced the oxidative stress marker MDA (24.5%) and the inflammatory markers hs-CRP (C-reactive protein) (32.66%) and TNF-α (13.06%).

An in vitro study was conducted to assess the antioxidant activity of ferulic acid, catechin, hesperidin, and quercetin. The tests were based on the ability to inhibit copper-mediated oxidation of human lipoproteins. Compounds reacting with thiobarbituric acid were used in the test. Ferulic acid has been shown to have pro-oxidative activity (166.4%). However, the combination of all four compounds showed a high antioxidant capacity on human lipoproteins at all levels due to their synergistic effects (Cirico & Omaye, 2006).

A controlled, randomized, double-blind experiment was conducted to assess the effect of an antioxidant preparation containing various concentrations of ferulic acid, α-glycosylrutin and tocopheryl acetate on the inhibition of the symptoms of photodermatosis—polymorphic light eruption. The study was conducted on a group of 30 patients with polymorphic light eruption, who were irradiated after applying 3 preparations with different concentrations of antioxidants and one preparation of a placebo to the area of the extensors of the arms, which they applied 2 times/day for a period of 1 week. The results after the fourth day of photoprovocation showed that the preparation containing antioxidant compounds effectively reduced the development and severity of photodermatosis and reduced the intensity of pruritus, which is closely related to the antioxidant effect (Hadshiew et al., 1997).

5.1.1.1.2 Skin-brightening Kim et al. conducted a randomized, blinded study to evaluate the effect of a preparation containing ferulic acid, vitamins E and C as an adjuvant in the treatment of lentigines and chloasma with the Q-switched neodymium-doped yttrium aluminum garnet (Nd:YAG) laser. Eighteen subjects tested the preparation on one side of the face after using the laser. The studies were carried out using spectrophotometry and posttreatment photography to assess the melanin and erythema index. The side of the face treated with ferulic acid was shown to have a greater reduction in the melanin index, with no significant differences in terms of erythema. It was found that a greater clinical improvement was seen on the treated side compared with the untreated side (Kim et al., 2020).

A study was conducted focusing on the whitening activity of ferulic acid. The tests used ferulic acid, which was isolated from *Tetragonia tetragonioides*. The anti-pigmentation activity was tested on murine B16F10 melanoma cells stimulated with melanotropin α (α-MSH). The results showed that ferulic acid effectively inhibited melanin production, depending on the dose (5.10, 20 μg/mL) by 13.9%, 25.5%, and 43.6%. It was also found that the intracellular rate of tyrosinase synthesis using the same dosage was reduced by 4.7%, 11.8%, and 19.1%, respectively. The results suggest that the mechanism of action of ferulic acid is to inhibit melanin synthesis by inhibiting the expression of tyrosinase and MITF (Park et al., 2018).

5.1.1.1.3 Photoprotective activity Oresajo et al. (2008) conducted a study on 10 patients with II and III phototypes according to the Fitzpatrick scale, which was designed to assess the effect of a preparation containing ferulic acid, vitamin C, and phloretin in mitigating the adverse effects of UV radiation on healthy patients. The tests were performed using skin biomarkers in the lower back for 4 days. The preparation was randomly assigned. During day 3, the minimum erythema dose (MED) was determined for each subject. On day 4, UV irradiation was performed in UV doses of 1-5x MED at intervals of 1xMED. Then, immunohistochemistry and morphology were performed. It was shown that in patients treated with a preparation with ferulic acid, vitamin C, and phloretin, the combination of antioxidants blocked the formation of sunburn cells, thymine dimers, unblocking of p53 protein, and matrix metalloproteinase-9. The mixture effectively blocked the reduction in Langerhans cells in the skin.

Murray et al. (2008) performed in vivo tests of a topical preparation containing 15% vitamin C, 1% vitamin E, and 0.5% ferulic acid to assess the photoprotective effect. The mixtures were applied to areas of healthy human skin for 4 days. Areas where they were irradiated with UV radiation from 2 to 10 MED doses were separated at intervals of 2 MED. The following day, tests were performed to assess the skin areas for erythema index and the presence of sunburn cells. Immunohistochemical tests were also performed on the expression of p53 protein and thymine dimers. Polymerase chain reaction tests were also performed to assess the increase in the level of TNF-α, IL-1α, IL-6, IL-8, and IL-10. The results showed that the preparation with ferulic acid, vitamins C and E effectively protected the skin against harmful UV radiation. Assays showed a significant reduction in thymine dimer expression.

The study also focused on testing the antioxidant preparation containing ferulic acid, vitamin C, and vitamin E for photoprotective activity. Wu et al. tested the complex on 12 healthy women, on whom 4 unexposed spots were marked in the dorsal region. The formulation was applied over the next 4 days to 2 out of 4 sites. On day 4 of the experiment, 3 skin areas were treated with a 5-fold dose of MED. Photographic clinical documentation was carried out, and skin color was measured before and after the UV dose. A skin biopsy was then performed 24 h after UV exposure. Tests were performed to determine the level of hematoxylin and eosin, and immunohistochemical staining was performed. It was found that the complex of ferulic acid, vitamin C, and vitamin E effectively protected the skin against harmful UV radiation by inhibiting the formation of sunburn cells and thymine dimers, and also blocked the overexpression of the p53 protein. It was also shown that the preparation attenuated the depletion of CD1a-expressing Langerhans cells (LCs) (Wu et al., 2013).

Saija et al. (2000) also focused on the effect of ferulic acid as a photoprotective agent. Two phenolic acids were tested in the study—ferulic acid and caffeic acid. Aqueous solutions of acids at pH 3 or 7.2 were prepared, and their ability to penetrate through human cut skin placed in Franz cells was assessed. It has been shown that ferulic acid penetrates the epidermal layer much better, due to its higher lipophilicity compared with caffeic acid. Subsequently, in vivo tests were carried out, which showed that both ferulic acid and caffeic acid effectively protect the skin against harmful UV radiation.

5.1.1.1.4 Supporting the reconstruction of the skin barrier properties Suzuki et al. (2023) conducted a double-blind, placebo-controlled study to elucidate the effect of oral administration of ferulic acid on the skin barrier. The tests were carried out on a group of 16 patients, who were divided into 2 groups—one took a preparation containing 200 mg of ferulic acid, the other a placebo for 2 weeks. Measurements of skin barrier function, TEWL, and stratum corneum hydration were taken before and after 2 weeks. Studies have shown that oral administration of ferulic acid significantly reduced TEWL (from 6.1 to 4.8 $g/m^2/h$) and increased the level of stratum corneum hydration (from 30.1 to 32.3 a.u.). There was also a change in the amount of sympathetic nervous activity after ingestion of the preparation with ferulic acid (−0.7 vs. 1.1).

5.1.1.1.5 Wound healing properties Waibel et al. (2016) conducted a study to check whether a preparation containing ferulic acid, vitamin C, and vitamin E, after application to the skin of a fractional CO_2 ablative laser, has the effect of supporting and accelerating the healing of wounds after the procedure. For this purpose, a double-blind, single-center, randomized, prospective study was conducted with 15 volunteers. A complex containing ferulic acid, vitamin C, and vitamin E was applied to half of the face immediately after the laser treatment and every day for the next 7 days. Throughout the experiment, photographic documentation was made, and patients were assessed using molecular tests. The results showed that when applied to the face, the complex reduced inflammatory edema and erythema compared with the vehicle on days 3 and 5 after the treatment. Molecular tests showed that the preparation used increased the expression of the basic fibroblast growth factor (bFGF) on the fifth day compared with the test performed on the untreated side.

5.1.1.1.6 Antiaging effect Manosroi et al. (2012) conducted a study on the antiaging activity of *Oryza sativa* rice bran extract, which contains, among others, ferulic acid or phytic acid. A gel was prepared in which niosomes with bioactive compounds from the *Oryza stativa* extract were placed. In vitro tests were carried out, which showed that the preparation stimulated the production of human fibroblasts and showed an inhibitory effect on MMP-2. The in vivo study was carried out on shaved rabbit skin—it was shown that within 72 h the gel did not cause erythema or swelling of the skin. Then, an experiment was carried out on 30 volunteers who tested the gel for 28 days. It was shown that the level of hydration, elasticity, skin thickness increased, and skin roughness and discoloration decreased. The tests were carried out using the following apparatus: a viscometer, a mexameter, a corneometer, and a cutometer.

The use of an emulsion based on persimmon fruit extract (*Diospyros kaki*) containing the following acids: ferulic, gallic, p-coumaric, chlorogenic, cinnamic, catechin, and quercetin, the presence of which was confirmed by high-performance liquid chromatography (HPLC) for antiaging effects was investigated. The tests were carried out using apparatus for assessing facial skin parameters, such as the level of hydration, the presence of wrinkles, peeling, roughness, and elasticity. The results showed that the emulsion significantly increased the level of moisture and elasticity. The living skin surface assessment index (SELS) was used for the study, the level of which was reduced for the tested formulation, which contributed to the reduction in skin wrinkles (−12.40%), roughness (−11.76%), and peeling (−18.59%), which suggests that the tested emulsion can be used in antiaging preparations (Kashif & Akhtar, 2019).

5.1.1.1.7 Antiinflammatory activity A study was conducted to evaluate the antiinflammatory effect of ferulic acid in the treatment of inflammation in sciatica. The tests were carried out on 32 rats, which were divided into four groups—sham surgery, with chronic stenosis injury, and a group in which mecobalamin and ferulic acid were administered. The results were based on ELISA, polymerase chain reaction (PCR), behavioral, immunofluorescence, and Western blot analysis. Thermal hyperalgesia was also induced after surgery, which was effectively relieved by ferulic acid. In addition, ferulic acid has been shown to significantly reduce the levels of interleukins—IL1β, IL1β, IL6, CRP and TNF-α proteins as well as Rho-associated coiled-coil containing protein kinase 1 (Rock1), transient receptor potential vanilloid 1 (TRPV1), Transient receptor potential ankyrin 1 (TRPA1) and p-p38MAPK Ras homolog family member A (RhoA), RhoA-guanosine triphosphate (GTP), COX-2, ionized calcium binding adaptor molecule 1 (IBA-1), iNOS in root ganglia and was effective in lowering serum substance P, LPS, prostaglandin E2 and CRP levels. The mechanism of action of ferulic acid was to attenuate the M1 polarization of GMI-R1 cells through the RhoA/Rock pathway, which was induced by LPS (Zhang et al., 2022).

A study was conducted in which the effect of a gel based on essential oil containing menthol and thymol and antioxidant substances—ferulic acid and phloretin. Thirty patients with gingivitis took part in the study. They were randomly assigned to control groups treated with a placebo and the preparation. They applied the gel 2 times a day after brushing their teeth. Each patient had 3 orthodontic visits during which a periodontal examination was performed, bleeding on probing, plaque, and gingival index were assessed. Gel or placebo began to be applied between the first and second visit. After the second visit, by the third visit, they were asked to stop applying the preparation. The results showed that the use of the gel was effective in reducing gingival bleeding with the probe (−13.6%) and the rate of gingivitis (−0.14%), compared with the placebo group, suggesting that the administered preparation significantly reduced gingival inflammation (Martin et al., 2016).

5.1.1.1.8 Practical utilization in cosmetology In treatments performed in beauty salons, ferulic acid is most often found in mixtures of acids for exfoliation, that is, exfoliating treatments using chemical acids. Such preparations usually consist of a mixture of 3-4 acids (e.g., mandelic acid, salicylic acid, shikimic acid) selected according to the needs of the skin. In this case, ferulic acid acts as an antioxidant. Ferulic acid itself is also used as an exfoliating treatment, but it does not have such a strong exfoliating effect as other acids used in cosmetology (Apis-sklep). The advantage, however, is the short recovery time after the procedure; the skin is not severely irritated, and there is no strong exfoliation of the stratum corneum (Sesderma) (Table 8.6).

Banquet treatments with the use of ferulic acid and vitamin C are also very popular. They are aimed at brightening the skin, oxygenating it, without irritation and exfoliation (Sesderma).

5.2 Rosmarinic acid

5.2.1 Chemical structure and plant resources

Rosmarinic acid (rosemary acid, labiatenic acid, (R,E)-3-(3,4-dihydroxyphenyl)-2-((3-(3,4-dihydroxyphenyl)acryloyl)oxy)propanoic acid) (Fig. 8.16) is chemically an

Table 8.6 The examples of studies on the production of the phenolic compounds using plant in vitro cultures.

Compound	Species	Type of in vitro culture	Content	References
Flavonoids				
Rutin	*Aronia arbutifolia*	Agitated shoot cultures	9.08 mg/100 g DW	Szopa, Kubica, Snoch, and Ekiert (2018)
	Aronia × prunifolia	Agitated shoot cultures	13.44 mg/100 g DW	Szopa, Kubica, and Ekiert (2018)
	Centella asiatica	Agitated shoot cultures	910 mg/100 g DW	Skrzypczak-Pietraszek et al. (2019)
	Codonopsis pilosula	Hairy root cultures	13.44 mg/kg DW	Milentyeva et al. (2021)
	Eleutherococcus senticosus	Hairy root cultures	63.08 mg/kg DW	Milentyeva et al. (2021)
	Eryngium campestre	Shoot cultures	29.51 mg/100 g DW	Kikowska et al. (2022)
	Eryngium maritimum	Root cultures	14.32 mg/100 g DW	Kikowska et al. (2022)
	Hypericum perforatum cv. Elixir	Agitated shoot cultures	97.56 mg/100 g DW	Kwiecień et al. (2023)
	Nasturtium officinale	Microshoot cultures grown in a plant-form temporary immersion system bioreactor	16.03 mg/100 g DW	Klimek-Szczykutowicz et al. (2021)
	Ocimum basilicum	Cell suspension cultures	6.54 mg/g DW	Açıkgöz (2020)
	Platanthera bifolia	Hairy root cultures	53.55 mg/kg DW	Milentyeva et al. (2021)
	Polygonum multiflorum	Hairy root cultures	352.50 μg/g DW	Thiruvengadam et al. (2014)
	Pontechium maculatum	Shoot cultures grown in a plant-form temporary immersion system bioreactor	167.67 mg/100 g DW	Makowski et al. (2023)
	Rumex vesicarius	Microshoot cultures	np*	Khateeb et al. (2017)
	Ruta chalepensis	Callus cultures	21.60 μg/mg DW	Qahtan et al. (2022)
	Saposhnikovia divaricata	Hairy root cultures	30.65 mg/kg DW	Milentyeva et al. (2021)
	Schisandra henryi	Callus cultures	5.20 mg/100 g DW	Jafernik et al. (2020)
	Vitex agnus castus	Agitated shoot cultures	1085 mg/100 g DW	Skrzypczak-Pietraszek et al. (2018)

Continued

Table 8.6 The examples of studies on the production of the phenolic compounds using plant in vitro cultures.—*cont'd*

Compound	Species	Type of in vitro culture	Content	References
Flavonoids				
Quercetin	*Aronia arbutifolia*	Agitated shoot cultures	18.72 mg/100 g DW	Szopa, Kubica, Snoch, and Ekiert (2018)
	Aronia × *prunifolia*	Agitated shoot cultures	17.03 mg/100 g DW	Szopa, Kubica, Snoch, and Ekiert (2018)
	Codonopsis pilosula	Hairy root cultures	14.65 mg/kg DW	Milentyeva et al. (2021)
	Eleutherococcus senticosus	Hairy root cultures	12.17 mg/kg DW	Milentyeva et al. (2021)
	Eryngium campestre	Root cultures	22.96 mg/100 g DW	Kikowska et al. (2022)
	Platanthera bifolia	Hairy root cultures	17.12 mg/kg DW	Milentyeva et al. (2021)
	Polygonum multiflorum	Hairy root cultures	287.50 μg/g DW	Thiruvengadam et al. (2014)
	Rumex vesicarius	Microshoot cultures	np*	Khateeb et al. (2017)
	Saposhnikovia divaricata	Hairy root cultures	12.78 mg/kg DW	Milentyeva et al. (2021)
	Schisandra henryi	Callus cultures	9.55 mg/100 g DW	Jafernik et al. (2020)
Luteolin	*Hypericum perforatum* var. *angustifolium*	Callus cultures	97.00 μg/g DW	Dias et al. (1998)
	Rumex vesicarius	Microshoot cultures	np*	Khateeb et al. (2017)
Apigenin	*Centella asiatica*	Agitated shoot cultures	27.00 mg/100 g DW	Skrzypczak-Pietraszek et al. (2019)
	Codonopsis pilosula	Hairy root cultures	14.71 mg/kg DW	Milentyeva et al. (2021)
	Eleutherococcus senticosus	Hairy root cultures	32.74 mg/kg DW	Milentyeva et al. (2021)
	Platanthera bifolia	Hairy root cultures	75.23 mg/kg DW	Milentyeva et al. (2021)
	Saposhnikovia divaricata	Hairy root cultures	12.12 mg/kg DW	Milentyeva et al. (2021)

Kaempferol	*Centella asiatica*	Agitated shoot cultures	48.00 mg/100 g DW	Skrzypczak-Pietraszek et al. (2019)
	Eryngium campestre	Shoot cultures	14.53 mg/100 g DW	Kikowska et al. (2022)
	Eryngium maritimum	Shoot cultures	13.10 mg/100 g DW	Kikowska et al. (2022)
	Eryngium maritimum	Root cultures	8.82 mg/100 g DW	Kikowska et al. (2022)
	Polygonum multiflorum	Hairy root cultures	108.00 μg/g DW	Thiruvengadam et al. (2014)
	Pontechium maculatum	Callus cultures liquid medium	90.36 mg/100 g DW	Makowski et al. (2023)
	Schisandra chinensis cv sadova No.1	Agitated shoot cultures	30.81 mg/100 g DW	Szopa, Klimek-Szczykutowicz, et al. (2019)
	Schisandra henryi	Callus cultures	21.60 mg/100 g DW	Jafernik et al. (2020)
Hesperidin	*Cyclopia genistoides*	Callus cultures	0.33% DW	Kokotkiewicz et al. (2009)
	Cyclopia intermedia	Shoot cultures	0.899% DW	Kokotkiewicz et al. (2009)
	Cyclopia subternata	Shoot cultures	0.873% DW	Kokotkiewicz et al. (2009)
	Polygonum multiflorum	Hairy root cultures	136.00 μg/g DW	Thiruvengadam et al. (2014)
	Ruta chalepensis	Callus cultures	59.11 μg/mg DW	Qahtan et al. (2022)
	Rumex vesicarius	Microshoot cultures	np*	Khateeb et al. (2017)

Continued

Table 8.6 The examples of studies on the production of the phenolic compounds using plant in vitro cultures.—*cont'd*

Compound	Species	Type of in vitro culture	Content	References
Flavonoids				
Phenolic acids				
Ferulic acid	*Plectranthus scutellarioides*	Shoot cultures	np*	Kowalczyk et al. (2024)
	Opuntia genotype "Tapón"	Callus cultures	863,32 mg/100 g	Espino et al. (2023)
	Gardenia jasmonide variegata	Shoot cultures	2708 μg/g	Ashry et al. (2023)
	Angelica acutiloba	Suspension cultures	47.2 mg g g^{-1} DW	Hwang et al. (2022)
	Ruta graveolens	Shoot cultures	4.21 mg/100 g DW	Szopa et al. (2012)
	Verbena officinalis	Suspension cultures	10.44 mg/100 g DW	Kubica et al. (2020)
	Hyoscyamus niger	Root cultures	9.61 μg g^{-1}	Albayrak et al. (2024)
Rosmarinic acid	*Eryngium alpinum*	Shoot liquid cultures	179.20 mg/100 g DW	Kikowska et al. (2020)
	Verbena officinalis	Suspension cultures	26.34 mg/100 g DW	Kubica et al. (2020)
	Aronia melanocarpa	Shoot cultures	27.91 mg/100 g DW	Szopa et al. (2020)
	Aronia arbutifolia	Callus cultures	20.24 mg/100 g DW	Szopa, Kubica, Snoch, and Ekiert (2018)
	Agastache rugosa	Hairy root cultures	4.6 mg/g DW	Kozłowska et al. (2024)
	Origanum dictamnus	Root liquid cultures	1.8 mg/g DW	Sarropoulou et al. (2023)
	Hyoscyamus niger	Root cultures	31.72 μg g^{-1}	Albayrak et al. (2024)
	Basilicum polystachyon	Shoot cultures	45.09 μg/g DW	Das et al. (2024)
Ellagic acid	*Basilicum polystachyon*	Shoot cultures	7.17 μg/g DW	Das et al. (2024)
	Lagestromia tomentosa	Hairy root cultures	np*	Amer et al. (2024)
	Antigonon leptopus	Shoot cultures	np*	Taha et al. (2024)

Salicylic acid	*Schisandra chinensis*	Shoot cultures	21.17 mg/100 g DW	Szopa and Ekiert (2012)
Caffeic acid	*Eryngium alpinum*	Agitated cultures	3.46 mg/100 g DW	Kikowska et al. (2020)
	Verbena officinalis	Suspension cultures	4.80 mg/100 g DW	Kubica et al. (2020)
	Aronia melanocarpa	Shoot cultures	21.05 mg/100 g DW	Szopa et al. (2020)
	Hyoscyamus niger	Root cultures	1.67 μg g^{-1}	Albayrak et al. (2024)
	Basilicum polystachyon	Shoot cultures	18.51 μg/g DW	Das et al. (2024)
	Eryngium planum	Shoot cultures	16.51 mg/100 g DW	Kikowska et al. (2022)

The form	Activity
Ferulic acid	Antimicrobial Antioxidant
Cycloartenyl ferulate	Skin conditioning Emollient
Feruloyl oligopeptide-33	Skin conditioning
Feruloyl pentapeptide-71	Antioxidant
Feruloyl tetrapeptide-80	Antioxidant Emollient
Guanidine ferulamide	Skin conditioning
Hydrolyzed chitosan ferulyl linoleate	Antisebum Antimicrobial Antioxidant Emulsion stabilizing Skin conditioning
Isostearyl ferulate	Skin conditioning
Methyl polysilsesquioxane ferulate	Antioxidant Skin conditioning—miscellaneous
Methyl polysilsesquioxane ferulate hyaluronate	Antioxidant Humectant
Sodium hyaluronate ferulate	Colorant Skin conditioning Emollient Skin protecting
Soybean oil/ferulate esters	Skin conditioning
Tocopheryl ferulate	Antioxidant

np*, *not provided.*

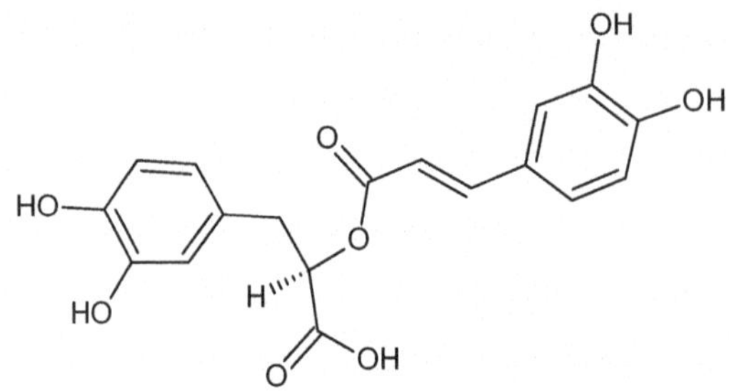

FIGURE 8.16

Chemical structure of rosmarinic acid.

ester of caffeic acid combined with α-hydroxydihydrocaffeic acid. Chemically, it is a red-orange powder, slightly soluble in water, but highly soluble in organic solvents. It belongs to the group of depsides/phenolic acid derivatives, which are a group of compounds similar in terms of chemical structure to tannins. Currently, rosmarinic acid has been confirmed in over 100 plant species, the most common of which is the *Lamiaceae* family. When it comes to plant species, the highest amount of rosmarinic acid was found in *Rosmarinus officinalis*, *Origanum vulgare*, *Melissa officinalis,* and *Salvia officinalis*. They are considered to be plant raw materials from which rosmarinic acid can be obtained. Rosmarinic acid is a relatively popular active substance used in cosmetology. The CosIng database (2026) contains three forms of rosmarinic acid approved for use in cosmetic preparations. They play the role of an antioxidant, a substance that soothes and conditions the skin.

5.2.1.1 Review of the latest scientific studies focused on the cosmetological applications

5.2.1.1.1 Antioxidant properties The mechanism of the antioxidant action of rosmarinic acid is based mainly on the direct capture of free radicals. Studies have confirmed that it affects the mitogen-activated protein kinase (MAPK) and modulates the signaling kinase1/2, which is closely related to the ERK1/2. It was also found that rosmarinic acid effectively scavenges reactive forms of peroxynitrites, nitrogen, and reactive oxygen species. Rosmarinic acid also affects the mechanisms related to the reduction in lipid and protein peroxidation. It has also been shown that rosmarinic acid reduces oxidative stress in cells, which further enhances its antioxidant activity. Studies have also found that rosmarinic acid has chelating properties.

Tsai et al. (2019) tested the effect of rosmarinic acid as an agent against oxidative stress. The study was conducted on red Sprague–Dawley puppies with selenite-induced cataracts. It has been shown that rosmarinic acid effectively reduces oxidative stress, and at the same time improves parameters related to oxidative damage by increasing the level of antioxidants, such as heme oxygenase-1 (HO1), NAD(P)H: quinone oxidoreductase 1 (NQO1), Nrf2, SOD, and antioxidant enzymes: catalase,

GSH-px, and GSH-Rd. In addition, rosmarinic acid reduced lipid peroxidation and inflammation.

A study was also conducted to determine the mechanism of antioxidant action of rosmarinic acid. Rosmarinic acid was tested in LPS-induced oxidative damage and inflammation in peripheral blood mononuclear cells. It has been shown that rosmarinic acid is involved in processes that reduce lipid peroxidation and NO levels. It also mediates in the processes that restore the constancy between the antioxidant and the oxidant (Sadeghi et al., 2020).

The effect of rosmarinic acid in combination with another polyphenolic compound—sinapic acid was also tested. It has been shown that their synergistic effect increases the level of other antioxidant compounds (e.g., superoxide dismutase) and reduces the peroxidation of proteins and lipids. This significantly reduced oxidative changes in the lenses of rats that had abnormally low levels of estrogen (Zych et al., 2019).

Zhang et al. (2015) tested the antioxidant activity of rosmarinic acid in mouse organs—kidneys and liver. It was shown that oral administration of rosmarinic acid (200 mg/kg) significantly increased the level of antioxidant enzymes—catalase, glutathione peroxidase and superoxide dismutase. Malondialdehyde levels were also found to have decreased.

Researchers also used the antioxidant potential of rosmarinic acid as an active substance encapsulated in ZnO/chitosan nanoparticles. The effects of the complex on oxidative stress damage and changes in caspase-3a mRNA levels induced by methamphetamine were tested. The tests were performed on the zebrafish brain. The rosmarinic acid/chitosan/ZnO nanoparticle complex was shown to be effective in lowering methamphetamine levels and combating changes caused by oxidative stress, including those at the mRNA level (Shahrestani et al., 2021).

It has also been shown that the mechanism of the antioxidant action of rosmarinic acid is mediated by the receptor for activated C kinase 1 (RACK1)/TNF-α pathway. Researchers confirmed this by experimenting with mice in which acute hepatitis was induced with acetaminophen (Yu et al., 2021).

A study was also conducted that showed that rosmarinic acid attenuated chromium-induced oxidative damage to the liver and kidneys and inhibited the formation of precancerous lesions in rats. The mechanism of action of rosmarinic acid has also been investigated. It has been shown to mediate the regulation of the Nrf2 signaling pathway and act as an antioxidant (Khalaf et al., 2020).

5.2.1.1.2 Antiinflammatory activity A study was conducted to assess the effect of rosmarinic acid on atopic dermatitis. The study involved 21 patients with mild atopic dermatitis. An emulsion preparation with 0.3% rosmarinic acid was prepared and applied to the elbows of volunteers twice a day. The skin condition was assessed before the administration of rosmarinic acid and after 4 and 8 weeks after the start of treatment. The results showed that atopic dermatitis severity and erythema were significantly less visible after 4 and 8 weeks. In addition, the rate of transepidermal water loss from the elbow joint was measured, which was visibly reduced 8 weeks after the start of therapy. Questionnaires were also conducted, where patients positively assessed the reduction in itching, dryness, and general symptoms associated with the disease (Lee et al., 2008).

Studies have been conducted on the antiinflammatory activity of mint extract with a high content of rosmarinic acid. In vitro tests were performed to show whether rosmarinic acid has an inhibitory effect on inflammation caused by LPS. It has been shown that rosmarinic acid effectively inhibits LPS-induced inflammation, as well as the release of nitric oxide and prostaglandin E2. Subsequently, the focus was on in vivo testing. For this purpose, a study was conducted on horses in which an extract rich in rosmarinic acid was added to the feed for 24 days in a blank trial. On day 21 of the study, LPS was injected into the left or right intercarpal joint. Tests were performed on synovial fluid samples that were collected the day before supplementation and after injection. The samples were tested for prostaglandin E2, glycosaminoglycans, and nitric oxide. The amount of protein was measured. The results indicated that horses given the rosmarinic acid extract had reduced levels of inflammatory biomarkers compared with the placebo group, suggesting a strong antiinflammatory activity of rosmarinic acid against LPS (Pearson et al., 2012).

5.2.1.1.3 Practical utilization in cosmetology In the preparations for professional use, rosmarinic acid is used as an antioxidant, exfoliant, antimicrobial, reducing inflammatory changes and reducing sebum secretion (Aruba.rzeszow). In the CosIng database (2026), as previously mentioned, there are three forms of rosmarinic acid approved for use in cosmetic preparations (Table 8.7).

Rosmarinic acid is not used in preparations for everyday use. Rosemary leaf extract and essential oil are most often used.

5.3 Ellagic acid

5.3.1 Chemical structure and plant resources

Ellagic acid (benzoaric acid, elagostasine, 4,4′,5,5′,6,6′-hexahydroxydiphenic acid 2,6,2′,6′-dilactone) is a dilactone of hexahydroxydiphenic acid (Fig. 8.17). It is one of the derivatives of gallic acid. It is formed through the hydrolysis of ellagitannins. It is in the form of a powder. Ellagic acid is found in many plant species, mainly

Table 8.7 Rosmarinic acid and its derivatives and their use in cosmetic preparations according to CosIng database.

The form	Activity
Rosmarinic acid	Antioxidant Soothing
Rosmarinoyl tetrapeptide-75	Antioxidant Skin conditioning
Rosmarinyl glucoside	Antioxidant Skin conditioning

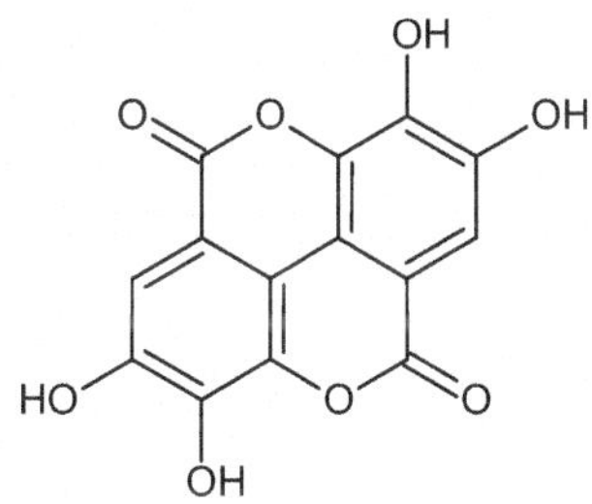

FIGURE 8.17

Chemical structure of ellagic acid.

dicotyledonous plants. Examples of plant species where the accumulation of ellagic acid is high include *Punica granatum*, *Terminalia catappa*, or *Prosopis humilis*. The largest accumulation of ellagic acid in plants is usually found in fruits and nuts. As an active substance, ellagic acid is used only in one form in cosmetic preparations according to the CosIng database (2026). It has a skin conditioning effect.

5.3.1.1 Review of the latest scientific studies focused on the cosmetological applications

5.3.1.1.1 Antioxidant properties The mechanism of the antioxidant effect of ellagic acid is mainly based on the capture of free radicals. Research focuses primarily on the effective action of ellagic acid on hydroxyl, peroxyl, peroxynitrite, and nitrogen dioxide radicals. The role of ellagic acid as a factor protecting against oxidative stress was also focused on. It has been shown that ellagic acid effectively protects against oxidative stress, even at low concentrations. Ellagic acid also has a chelating effect confirmed by research (Ahmed et al., 1999; Böyük et al., 2011; Diao et al., 2022; Mehrzadi et al., 2019).

Ahmed et al. (1999) conducted studies to determine the effect of ellagic acid on female Wistar rats after administration of nickel (250 mmol/kg b.w.), which caused changes in serum, liver and kidneys. Animals were supplemented with ellagic acid, which facilitated the excretion of heavy metals from cells and tissues while reducing toxicity. It was also shown that the mechanism of action of ellagic acid was based on reducing oxidative stress by lowering the biochemical parameters of lactate dehydrogenase, glutamate-oxaloacetic transaminase, and glutamate-pyruvate transaminase in the serum of rats.

Böyük et al. (2011) conducted studies in rat models after ischemic reperfusion. The animals were divided into two groups, in which one of them was administered ellagic acid (85 mg/kg b.w.). It has been shown that ellagic acid effectively lowered the level of malondialdehyde and thus effectively prevented the initiation and increase in lipid peroxidation.

Ellagic acid has also been shown to significantly inhibit the level of 8-oxo-2-deoxyguanosine, which is produced after oxidative DNA damage (Aiyer et al., 2008; Sharifi-Rad et al., 2024).

Diao et al. focused on examining the activity of the ellagic acid complex with the protein α-lactalbumin to increase the antioxidant potential. It was shown that the combination of compounds in molecular docking and simulations of the dynamics of both compounds significantly improved their antioxidant activity (Diao et al., 2022).

Meharzadi et al. also studied the protective effect of ellagic acid on the reduction in oxidative stress in methotrexate-induced hepatotoxicity. It was shown that in the tested Wistar rat models, the group to which ellagic acid (10 mg/kg b.w.) was administered significantly decreased the activity of aminotransferases, alkaline phosphatase (ALP) enzyme, and the level of parameters related to oxidative stress. In addition, the activity of antioxidant enzymes increased. The beneficial effect of ellagic acid was also confirmed in the histological examination of the liver (Mehrzadi et al., 2019).

A significant study by Hseu et al. (2012) showed that ellagic acid has a protective effect on HaCaT cells, protecting them from oxidative stress and UV-induced apoptosis. Ellagic acid was shown to increase the expression of the parameters Nrf-2, SOD, and HO-1.

5.3.1.1.2 Antiinflammatory activity The study focused on reducing inflammation in the chronic skin disease—atopic dermatitis. It was found that ellagic acid effectively suppressed the TNF-α/IFN-γ-stimulated inflammatory response in HaCaT keratinocytes by regulating the levels of mitogen-activated protein kinases, transcription activators, and signal transducers. The effect of ellagic acid on skin lesions caused by house dust mites in NC/Nga mice was also tested, and their etiology was similar to skin lesions caused by atopic dermatitis. In vivo tests on mice showed that the application of ellagic acid to skin lesions resulted in a reduction in skin lesions and their course by reducing the cascade of histological inflammatory reactions (Gil et al., 2021).

Another study confirmed antiinflammatory activity in the course of cardiac fibrosis, which involves increased accumulation of extracellular matrix and increased synthesis of fibrous collagen in the structures of the cardiac interstitium. In vitro tests were performed on immortalized human cardiac fibroblasts (IM-HCF) that were stimulated with transforming growth factor β1 (TGF-β1) at a concentration of 10 ng/mL for 24 h. It was shown that ellagic acid effectively inhibited the NF-κB pathway, consequently inhibiting the levels of TNF-α, IL-1β and IL-6, confirming its antiinflammatory effects at the molecular level (Mannino et al., 2023).

5.3.1.1.3 Practical utilization in cosmetology In cosmetic preparations, especially for the face in the form of serums or creams, ellagic acid is often combined with other active ingredients, which increases the effectiveness of their action. Cosmetics containing ellagic acid most often have brightening, illuminating, antiaging, and moisturizing properties. Only ellagic acid is listed in the CosIng database (2026), which functions as a skin conditioning agent.

5.4 Salicylic acid

5.4.1 Chemical structure and plant resources

Salicylic acid (2-hydroxybenzoic acid) (Fig. 8.18) is chemically a derivative of benzoic acid. It is in the form of a white crystalline powder. It is poorly soluble in water but highly soluble in ethanol. It is widely distributed in plants, but its greatest amounts are found in the bark of *Salix alba*. Salicylic acid is widely used in cosmetic preparations. CosIng database (2026) contains as many as twenty-four forms of salicylic acid. They function primarily as a skin conditioning agent, but also as a preservative, antimicrobial, and as an ingredient that absorbs UV radiation. Particularly noteworthy are salicylic acid compounds with peptides, which are very popular in antiaging cosmetics. In addition, salicylic acid itself has a keratolytic function, conditioning both the skin and hair, and has an antiseborrheic effect.

5.4.1.1 Review of the latest scientific studies focused on the cosmetological applications

5.4.1.1.1 Antioxidant properties Much research has been conducted on the antioxidant potential of salicylic acid. In addition to confirming the antioxidant activity, the focus was on examining the mechanism of action of this compound. Tests have shown that salicylic acid has mainly antiradical effects, reducing the accumulation of hydrogen peroxide while increasing the production of antioxidant enzymes. The second important mechanism that has been studied regarding salicylic acid is increasing the total amount of polyphenols. It was also found that salicylic acid derivatives significantly reduce oxidative stress by preventing the oxidation of carbonylated proteins as a result of an endogenous defense mechanism. In addition, salicylic acid had a beneficial effect on inhibiting lipid peroxidation caused by lipoxygenase (Blanch et al., 2020; Lapenna et al., 2009; Lavigne et al., 2022; Yang et al., 2022).

Yang et al. (2022) tested the effect of salicylic acid on the accumulation of H_2O_2 and O_2. It was found that salicylic acid effectively reduced the accumulation of free radicals by stimulating the increase in the level of antioxidant enzymes, such as catalase, ascorbic peroxidase, peroxidase, and superoxide dismutase. In addition, salicylic acid increased the level of antioxidant substances such as ascorbic acid and glutathione.

Blanch and other researchers (Blanch et al., 2020) focused on examining the effects of salicylic acid in grapes to enhance their health-promoting properties. It was

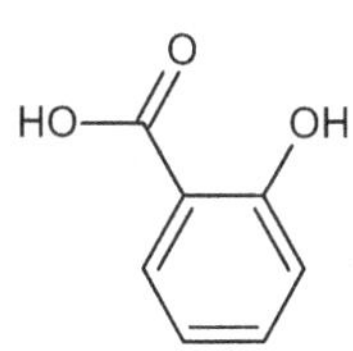

FIGURE 8.18

Chemical structure of salicylic acid.

found that the addition of salicylic acid effectively increases the level of total phenol content and, consequently, increases the free radical scavenging activity.

Lavigne et al. (2022) investigated the effect of a salicylic acid derivative on the aging of the nervous system and skin through in vitro tests on reconstituted human epidermis innervated with human sensory neurons and ex vivo human skin models. 5-Octanoyl salicylic acid was found to prevent the accumulation of carbonylated proteins in both cell lines and skin models. The mechanism of action was the process of protein autophagy and modulation of the NRF_2 antioxidant response genes, which prevented oxidative stress.

Khalil Alyahya et al. (2021) researched aqueous extracts from coffee peels, from which they determined the total content of polyphenols, including salicylic acid, on human mesenchymal stem cells in vitro. It was found that the extract, including salicylic acid, effectively reduced oxidative stress in mitochondria by reducing the expression of Nox-4, LPO mRNA, and increasing the expression of antioxidant compounds GPX, glycogen synthase kinase-3 beta (GSK-3β), GSH, and mammalian cytochrome P450 1A (CYP1A). Reduced activity of pro-inflammatory factors such as NF-κβ, IL-1β, toll-like receptor 4 (TLR-4), and TNF-α was also observed.

Another study that was performed concerned the inhibitory effect of salicylic acid on lipid peroxidation. Lapenna et al. found that salicylic acid effectively inhibited the peroxidation of fatty acids such as arachidonic acid, linoleic acid, docosahexaenoic acid, and human proteins caused by lipoxygenase (Lapenna et al., 2009).

5.4.1.1.2 Antiinflammatory activity Lu et al. (2019) focused on the use of the anti-inflammatory effect of salicylic acid in the treatment of acne vulgaris. The first stage of the research was an antiacne activity test on the human sebaceous cell (SEB-1) line. It was found that salicylic acid effectively reduced sebocyte lipogenesis by decreasing the activity of the AMPK-activated protein kinase pathway and reduced inflammation by suppressing the NF-κβ pathway. In addition, salicylic acid increased SEB-1 cell apoptosis *via* the death signal receptor pathway. Further analysis showed that after 3 weeks of using salicylic acid, it caused a decrease in the level of pro-inflammatory cytokines and pathogenic proteins around acne vulgaris lesions.

5.4.1.1.3 Practical utilization in cosmetology In cosmetics, salicylic acid is used as an active ingredient in tonics, serums, and creams. It has antibacterial and antiinflammatory properties, which are used for problematic skin, especially acne-prone skin. It is also often an ingredient that is intended to exfoliate dead skin cells. Salicylic acid also has an affinity for hair follicles, which has a beneficial effect on reducing inflammation caused by the pathogenic bacterial strain *Cutibacterium acnes*. It also has keratolytic and keratoplastic effects, depending on the concentration used.

A number of biological activities exhibited by salicylic acid determine its use in cosmetic preparations. The CosIng database (2026) includes as many as twenty-four forms of salicylic acid that are approved for use in cosmetic preparations (Table 8.8).

Table 8.8 Forms of salicylic acid and its derivatives and their use in cosmetic preparations according to CosIng database.

The form	Activity
Salicylic acid	Antiseborrheic Hair conditioning Keratolytic Fragrance Preservative Skin conditioning
Capryloyl salicylic acid	Skin conditioning
Glucosamine salicylate	Skin conditioning
Silver salicylate	Skin protecting
Titanium salicylate	Antimicrobial
Arginine salicylate	Skin conditioning—miscellaneous
Galactosyl salicylate	Antisebum Skin conditioning
Magnesium salicylate	Preservative
Sodium salicylate	Denaturant Preservative
Potassium salicylate	Antimicrobial Preservative
Mea-salicylate	Preservative
Calcium salicylate	Preservative
Methyl salicylate	Denaturant Oral care Soothing Perfuming Flavoring
Benzyl salicylate	UV absorber Perfuming
Tea-salicylate	Preservative Light stabilizer UV filter
Ethylhexyl salicylate	Light stabilizer UV absorber UV filter
Salicylic beeswax	Skin conditioning—emollient Humectant
Acetaminosalol	Antimicrobial UV absorber
Salicyloyl sh-nonapeptide-12	Antimicrobial Antioxidant Antisebum Skin protecting
Salicyloyl octapeptide-32	Skin conditioning
Salicyloyl decapeptide-10	Skin conditioning

Continued

Table 8.8 Forms of salicylic acid and its derivatives and their use in cosmetic preparations according to CosIng database.—*cont'd*

The form	Activity
Salicyloyl oligopeptide-143	Skin conditioning
Salicyloyl octapeptide-9	Antimicrobial Antioxidant Antisebum Skin protecting
Salicyloyl pentapeptide-33	Antioxidant Skin protecting

5.5 Caffeic acid

5.5.1 Chemical structure and plant resources

Caffeic acid (3,4-dihydroxycinnamic acid) belongs to the group of primary hydroxycinnamic acids (Fig. 8.19). It is a light yellow powder, slightly soluble in water, and freely soluble in ethanol. It occurs in leaves, grains, and fruits. Examples of plants where caffeic acid occurs are *Artocarpus heterophyllus*, *Coffea arabica*, *Coffea canephora*, *Daucus carota*, propolis, and *Olea europea* (Aryal et al., 2023). Caffeic acid is widely used in cosmetic preparations. There are as many as twenty-four forms of caffeic acid available in the CosIng database (2026). They mainly have antioxidant and skin conditioning properties. Most compounds also have UV-absorbing properties and antimicrobial activity. The increasing number of compounds is complexes of peptides with caffeic acid, which primarily act as an emollient in preparations, but also have a protective and chelating effect on the skin.

5.5.1.1 Review of the latest scientific studies focused on the cosmetological applications

5.5.1.1.1 Antioxidant properties Caffeic acid is a widely known antioxidant in the world of science. A large amount of research focuses on understanding the mechanism of the antioxidant action of this compound. One of them is to prevent damage caused by free radicals and other reactive compounds. In addition to scavenging free

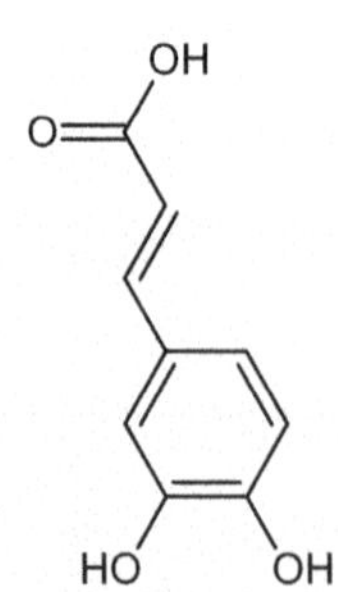

FIGURE 8.19

Chemical structure of caffeic acid.

radicals, caffeic acid also has chemical stability, which is extremely important for its antioxidant effect. Caffeic acid significantly minimizes oxidative stress, which has been confirmed by numerous studies. The mechanism of the antioxidant action of caffeic acid is also based on inhibiting the peroxidation of membrane lipids and performing chelating functions, which prevent the harmful effects of free radicals.

Jamali et al. (2021) conducted a study on the role of oxidative stress in the progression of endometriosis and the effect of caffeic acid on the activity of enzymes of ectopic and eutopic endometrial cells. It was found that caffeic acid significantly increased the expression level of the Nrf-2 gene and increased the activity of NQO1 and HO-1 enzymes. In addition, a significant decrease in ROS was observed in ectopic endometrial cells treated with caffeic acid.

Chavariia et al. (2022) developed substances based on caffeic acid that would counteract neurodegeneration occurring in Parkinson's disease. The inhibitory activity of caffeic acid against monoamine oxidases and catechol O-methyltransferases was tested. It was shown that caffeic acid derivatives significantly inhibited the action of both enzymes, with low half-maximal inhibitory concentration (IC50) μM values, suggesting that the compounds did not show cytotoxicity and effectively prevented oxidative damage, which may affect the neurotransmission of nigrostriatal dopamine.

Purushothaman et al. (2022) focused on investigating the mechanism of action of caffeic acid used to alleviate oxidative stress. It was found that the effectiveness of caffeic acid is based directly on the capture of free radicals and the production of new metabolites as a result of oxidative degradation. Tests were performed that confirmed the mechanism of the antioxidant action of caffeic acid. It is based on hydrogen atom transfer (HAT) and the formation of radical adducts.

Kassa et al. (2021) based their research on the antioxidant activity of caffeic acid. Previous research has confirmed that caffeic acid is involved in processes that inhibit the oxidation of myoglobin and hemoglobin. Further tests were conducted based on the interaction of caffeic acid with sickle hemoglobin, which tends to oxidize more quickly. The tests were carried out in the presence of hydrogen peroxide. Caffeic acid has been found to effectively prevent the oxidation of sickle hemoglobin.

Spagnol et al. (2019) aimed to determine the antioxidant potential of caffeic acid in vitro. Tests were carried out on the scavenging of radical anions, peroxide, hydrogen peroxide, hypochlorous acid, crocin bleaching test, and ABTS •+/DPPH • scavenging capacity. It was found that caffeic acid had a strong antiradical effect in all tests and had a greater antioxidant effect than the reference substances: ascorbic acid and Trolox.

Genaro-Mattos et al. (2015) focused on explaining the mechanisms of the antioxidant action of caffeic acid in chemical terms. It was found that caffeic acid effectively prevents the formation of hydroxyl radicals, which are determined by the oxidative degradation of 2-deoxyribose and hydroxylation of DMPO. It was also found that caffeic acid has a strong inhibitory effect on membrane lipid peroxidation. Further studies on the mechanism of action confirmed that the antioxidant potential of caffeic acid results from the formation of iron-caffeic acid complexes, which confirms its chelating effect.

5.5.1.1.2 Antiinflammatory activity Zielińska et al. (2021) investigated the properties of caffeic acid in modulating mechanisms related to intestinal inflammation. Tests were performed to determine the effect of caffeic acid at concentrations of 50–10 μM on the expression of COX-2, cytokines and chemokines (IL-8,6 and monocyte chemoattractant protein-1 (MCP-1)) and prostaglandin E2 (PGE2) in human colon myofibroblasts (CCD-18Co) treated with IL-1β. The focus was also on examining the inhibitory activity of caffeic acid on the angiotensin-converting enzyme. Caffeic acid was found to be effective against the expression of COX-2 and its product—PGE2, and also inhibited the biosynthesis of IL-8 in cells treated with IL-1β. In addition, it inhibits the production of the angiotensin-converting enzyme, which effectively inhibits inflammatory processes in the intestines.

Zaitone et al. (2019) conducted a study on the antiinflammatory activity of caffeic acid in a mouse model with rotenone-induced Parkinson's disease. In the tests, mice were given protective doses of caffeic acid (2.5, 5 or 10 mg/kg body weight) and nine injections of rotenone every 48 h. After completing the experiment, histopathological evaluation of substantia nigra neurons was performed. PCR analysis showed an increase in the expression of genes encoding microglial surface antigen (CD11 b), COX-2, inducible nitric oxide synthase, and nuclear factor κB. It was found that treatment with caffeic acid reduced the expression of microglia and inflammatory mediators.

5.5.1.1.3 Antimicrobial activity Kępa et al. (2018) assessed the antibacterial potential of caffeic acid against reference and clinical strains of *Staphylococcus aureus* obtained from infected wounds. It was found that caffeic acid had varied effects on *S. aureus* strains. The minimum inhibitory concentration (MIC) ranged from 256 μg/mL to 1024 μg/mL. Caffeic acid was also added to the Mueller-Hinton agar (MHA) 1/4 MIC medium, which resulted in a significant increase in the antibacterial activity of the antibiotics erythromycin, clindamycin, and cefoxitin, and to a lesser extent vancomycin.

Saivish et al. (2023) tested the antiviral properties of caffeic acid against Ilhéus virus in vitro. Ilhéus virus (ILHV) is transmitted by mosquitoes and can lead to Ilhéus fever, a febrile disease that may progress to meningitis and encephalitis. Caffeic acid was tested in vitro using Vero and A549 cell lines. It was found that caffeic acid at a concentration of 500 μM was virucidal against ILHV by interacting with an allosteric binding site on the coat protein.

De Vita et al. (2014) conducted research on the antifungal activity of caffeic acid and its ester derivatives on the biofilm of the *Candida albicans* strain. It was found that ester derivatives have higher antifungal activity compared with caffeic acid.

5.5.1.1.4 Practical utilization in cosmetology Caffeic acid is widely used in cosmetological treatments, primarily due to its effect on destroying free radicals and reducing the visibility of hyperpigmentation spots. Often used in the care of mature skin because it reduces the degradation of collagen and elastin fibers. In addition, it reduces inflammation. The CosIng database allows for the use of as many as 24 forms of caffeic acid, which suggests that it is widely used in cosmetic preparations (Table 8.9).

Table 8.9 Forms of caffeic acid and its derivatives and their use in cosmetic preparations according to CosIng database.

Form	Activity
Caffeic acid	Antioxidant Fragrance
Caffeoyl glucoside	Antioxidant Skin conditioning
Caffeoyl tripeptide-1	Antioxidant Skin protecting
Caffeoyl hexapeptide-56	Antioxidant Hair conditioning Skin protecting
Caffeoyl tetrapeptide-80	Antioxidant Skin conditioning—emollient
Caffeoyl hexapeptide-50	Antioxidant Chelating Skin protecting UV filter
Caffeoyl hexapeptide-82	Antioxidant Skin conditioning—emollient
Caffeoyl hexapeptide-43	Antioxidant Skin protecting
Caffeoyl hexapeptide-48	Antioxidant Skin protecting
Caffeoyl oligopeptide-77	Skin conditioning
Caffeoyl pentapeptide-27	Antioxidant Chelating Skin protecting UV absorber
Caffeoyl dipeptide-30	Skin conditioning
Caffeoyl tripeptide-35	Antioxidant Chelating Skin protecting UV filter
Caffeoyl hexapeptide-65	Antimicrobial Antioxidant Skin conditioning Skin protecting
Caffeoyl pentapeptide-4	Skin conditioning
Caffeoyl tripeptide-34	Skin conditioning
Caffeoyl tripeptide-7	Antioxidant Skin protecting
Caffeic/coco/coumaric/mangosteen pericarp esters	Antimicrobial Antioxidant Hair conditioning Skin conditioning

Continued

Table 8.9 Forms of caffeic acid and its derivatives and their use in cosmetic preparations according to CosIng database.—*cont'd*

Form	Activity
Caffeoyl sh-octapeptide-4	Antioxidant Skin protecting
Caffeoyl pentapeptide-20	Antioxidant Hair conditioning Skin protecting
Caffeoyl sh-decapeptide-9	Skin conditioning
Caffeoyl tetrapeptide-19 caffeamide	Antioxidant Chelating Skin protecting UV filter
Caffeoyl dipeptide-25 amide	Skin conditioning—emollient Skin conditioning Skin protecting
Methyl polysilsesquioxane caffeate	Antioxidant Skin conditioning miscellaneous

6. Summary

The selected phenolic acids presented have a number of biological activities that are used in cosmetology, both in preparations for everyday use and in professional salon care. The main action profile of all the compounds discussed is antioxidant activity. This area is of great interest to researchers, and a number of studies have been conducted to confirm the antioxidant effects of phenolic acids. In addition, phenolic acids have antiinflammatory, photoprotective, and skin-whitening properties. The CosIng database allows the use of the above acids and their derivatives in cosmetic preparations.

Focusing on the antioxidant effect, the phenolic acids described above actually use mechanisms of action based on, among others, fighting free radicals or chelating activity. In addition to their antioxidant properties and a number of activities that have been used in cosmetology, phenolic acids also have healing properties.

Research on the activity of ferulic acid used in medicine and pharmacy confirms its antiatherosclerotic effect and beneficial effect on the cardiovascular system, which is to some extent related to its antioxidant effect. Many studies focus on the use of the antioxidant potential in combinations of ferulic acid with vitamin C and vitamin E—the complex of these three antioxidants, which work synergistically, gives very good effects, both as an antioxidant, but also as an antiaging agent, brightening discolorations and photoprotective, where the antioxidant effect is used (Kim et al., 2020; Oresajo et al., 2008; Park et al., 2018).

Rosmarinic acid also has a broad spectrum of action. In medicine, it is used primarily for its antioxidant and antiinflammatory potential. In fact, these two properties are used to study other biological activities of rosmarinic acid. Below is an overview of studies focusing on antiinflammatory and antiallergic effects, but also neurodegenerative and anticancer effects (Böyük et al., 2011; Diao et al., 2022; Ekiert et al., 2013; Mehrzadi et al., 2019).

Salicylic acid is also widely used, both in cosmetology and medicine. In medicine, a derivative of salicylic acid is particularly well known, used as an antiinflammatory and analgesic drug—acetylsalicylic acid. Salicylic acid is also used in medicine to treat genetic dermatoses—psoriasis, where its keratolytic effect is used. In addition, salicylic acid is used to treat viral warts or combat dandruff due to its antimicrobial activity. Studies also confirm the anticancer activity of salicylic acid and immunomodulatory properties (Ausina et al., 2020; Blanch et al., 2020; Hassanpour et al., 2021; Lapenna et al., 2009; Lavigne et al., 2022; Lu et al., 2019; Radde et al., 2016; Yang et al., 2022).

Research on ellagic acid is mainly based on its antiinflammatory and antioxidant effects. Recent research also focuses on the anticancer potential of ellagic acid, in particular, using its antiproliferative and antimutagenic effects. In addition, researchers also focus on cardioprotective, antiallergic, hepato- and nephroprotective, antiatherosclerotic, and protective effects in neurodegenerative diseases (Aryal et al., 2023; Jamali et al., 2021; Kassa et al., 2021; Lu et al., 2019; Purushothaman et al., 2022; Yang et al., 2022).

Caffeic acid is also widely used in both cosmetology and medicine. First of all, it has very strong antioxidant and antiinflammatory properties. Studies also confirm anticancer and neuroprotective activity in Parkinson's disease. The antimicrobial activity of caffeic acid is also widely used (Chong et al., 2011; De Vita et al., 2014; Kassa et al., 2021; Kepa et al., 2018; Saija et al., 2000; Stasiłowicz-Krzemień et al., 2023; Zeng et al., 2018; Zhang et al., 2019; Zielińska et al., 2021).

The use of the above-mentioned acids in cosmetic preparations is truly enormous. Cosmetics manufacturers focus primarily on their antioxidant properties, but also on their brightening and beneficial effects on problematic skin, especially acne-prone skin. The table below presents examples of cosmetic preparations using selected phenolic acids (Table 8.10).

To sum up, phenolic acids are widely used in cosmetology, mainly based on their antioxidant activity, which is enhanced by other properties characteristic of a given acid. The selected phenolic acids presented above have been described in detail in terms of the mechanism of antioxidant action, as well as other activities that are used in cosmetology and cosmetic preparations.

7. Plant cell in vitro cultures as a new alternative source of flavonoids and phenolic acids for cosmetic use

Plant in vitro cultures, like organ cultures (shoots, roots) as well as callus and cell suspension culture techniques, can be a valuable and innovative source of secondary

Table 8.10 Examples of cosmetic preparations containing phenolic acids in their composition.

Producer	Country of origin	Trade name	Cosmetic form	The form of ferulic acid in the composition of the cosmetics (INCI)	Properties of the cosmetics according to the producer
Ferulic acid					
URGO aesthetics (www.retixc.com)	France	RETIX.C serum	Serum	Ferulic acid	Instant illumination, Protection against the harmful effects of external factors, Protection against free radicals, Smoothing fine wrinkles, Improves the skin's natural repair processes
Apis natural cosmetics (www.apis-sklep.pl/)	Poland	Ferulic terapis Ferulic acid 40%	Peeling	Ferulic acid	Stimulates the synthesis of elastin and collagen fibers Antiinflammatory properties and eliminates erythema and redness, Prevents the formation of discolorations and reduces the existing ones, Optimally moisturizes the skin
Apis natural cosmetics (www.apis-sklep.pl/)	Poland	Exfoliating home care Intensive skin renewal in 20 days I Step (10% emulsion)	Peeling	Ferulic acid	Signs of aging, Discoloration, oily, Acne and enlarged pores
Peel mission (www.peel-mission.pl)	Poland	Ferul tonic	Tonic	Ferulic acid	Soothes inflammation, Moisturizes the skin and stimulates the synthesis of collagen and elastin fibers, The ability to absorb UVA and UVB rays, thanks to which it acts as a natural sunscreen

Bielenda cosmetics (www/bielenda.com)	Poland	Ferul-X Antioxidant moisturizing and soothing cream	Cream	Ferulic acid	Brightening and illumination, Soothing irritations, Regenerating, Antioxidant effect, Reduction in the impact of negative environmental factors
Dermomedica cosmeceuticals (www.dermomedica.pl)	Poland, USA	Active serum with vitamin C and E, and ferulic acid	Serum	Ferulic acid	Antiwrinkle Antiaging Improves skin firmness Strong brightening Antiinflammatory
Sesderma (www.sesderma.com/es_es/home)	Spain	Ferulac, liposomal serum	Serum	Ferulic acid	Protects skin cells from environmental pollutants Adds shine Deeply moisturizes
Rosmarinic acid					
DECLEOR (www.decleor.co.uk)	France	Rosmarinic active essence	Serum	Rosmarinus officinalis leaf oil/rosmarinic leaf oil	Matting, cleansing, softening
Gen factor (www.genfactor.pl/gen-factor)	Poland	Ph rosmarinic	Peeling	Rosmarinic acid 5%	Cleansing, exfoliation, and removal of calloused epidermis, regulation of skin cell renewal, reduction in inflammatory changes (antiinflammatory and antibacterial), reduction in sebum secretion, skin lightening
LoveSkinFood (www.loveskinfood.pl)	Poland	Organic rosmarinic face mist	Tonic	Rosmarinic hydrosol	Stimulating to the skin, Improves circulation, prevents skin damage caused by environmental factors Soothe the skin, antiaging properties

Continued

Table 8.10 Examples of cosmetic preparations containing phenolic acids in their composition.—*cont'd*

Producer	Country of origin	Trade name	Cosmetic form	The form of ferulic acid in the composition of the cosmetics (INCI)	Properties of the cosmetics according to the producer
Ferulic acid					
Naturopathica (www.naturopathica.com)	USA	Rosmarinic & salicylic acid acne-clearing moisturizer—hydrate face & neck	Cream	Rosmarinus officinalis (rosmarinic) leaf oil, Rosmarinus officinalis (rosmarinic) leaf extract	Moisturizing, prevention of inflammation, matting the skin
Ellagic acid					
Cellex-C (www.cellex-c.com)	Canada	Skin hydration complex	Serum	Ellagic acid	Smoothes and brightens the skin, deeply moisturizes, and gives elasticity to the skin
Yves saint laurent (www.yslbeauty.com)	France	Blanc pur couture dark spot corrector	Serum	Ellagic acid	Against dark spots, brightens discolorations, and evens skin tone, skin whitening and brightening effect
Teinei (www.teinei-japan.com)	Japan	Teinei deep hydrating cream	Cream	Ellagic acid	Deep hydration and improvement of the functioning and reconstruction of the skin barrier, skin protection against irritants, leaving the skin moisturized for a long time
Transparent laboratory (www.nichebeautylab.com/en/collections/transparent-lab)	Spain	Gentle rejuvenation serum	Serum	Ellagic acid	Antiaging effect, smoothing the skin structure, brightening discolorations, strengthening the skin barrier, antioxidant effect
DermDoc (www.buydermdoc.com)	USA	Niacinamide and sodium hyaluronate serum	Serum	Ellagic acid	Antiacne effect, antioxidant effect, brightening the skin, moisturizing

Salicylic acid and its derivatives					
Creightons (www.creightons.com)	England	Intense serum 2% salicylic	Serum	Salicylic acid	Exfoliant, antiacne, soothing
Creightons (www.creightons.com)	England	Salicylic acid exfoliant tonic	Tonic	Salicylic acid	Antiacne, antioxidant, Exfoliant, skin brightening, soothing
Lacura (www.amazon.com/Skin-Care-Products-Lacura-Beauty-Personal/s?rh=n%3A11060451%2Cp_89%3ALacura)	Germany	Healthy Glow salicylic acid toner	Tonic	Salicylic acid	Antiacne, exfoliant, Scothing
Accufix cosmetics (www.accufixcosmetics.com)	Pakistan	Salicylic acid moisturizer	Cream	Salicylic acid	Antiacne, exfoliant, soothing, sunscreen, antioxidant
Peel mission (www.peel-mission.pl)	Poland	Antiacne 1, 10% capryloyl salicylic acid	Peeling	Capryloyl salicylic acid	Antiacne, antibacterial Antiinflammatory, regulates the functioning of sebaceous glands, and eliminates blackheads
Nacomi (www.nacomi-shop.pl)	Poland	Salicylic acid 2%	Serum	Salicylic acid	Antiacne, exfoliant, skin brightening, soothing
NeoGenesis (www.neogenesis.com)	USA	Salicylic acid gel	Cleasner	Salicylic acid	Antiacne, exfoliant, soothing
Caffeic acid					
Dr. ForHair (www.drforhair.com)	South Korea	Folligen Original shampoo	Shampoo	Caffeic acid	Exfoliant, soothing, antioxidant, inhibiting hair loss, strengthening hair bulbs
Holika Holika (www.holikaholika.ee/en)	South Korea	Penta madeca cream	Cream	Caffeic acid	Antiacne, antioxidant, exfoliant, soothing
Creightons (www.creightons.com)	Spain	Antiaging C light anti-ox moisturizing gel	Gel	Caffeic acid	Antioxidant, skin-brightening, soothing
Aqua + series (www.aquaplus.sg)	Thailand	Enriched-C serum	Serum	Caffeic acid	Antioxidant, skin-brightening, soothing

metabolites, such as phenolic acids and flavonoids. In recent years, there has been a growing interest in harnessing the potential of plant biotechnology to produce valuable bioactive compounds with various applications in the cosmetic, pharmaceutical, and food industries. There are numerous potential advantages, such as increased biomass yield and sustainability, associated with the utilization of plant biotechnology methods (Karuppusamy, 2009; Krasteva et al., 2021; Verpoorte et al., 2002). Biotechnological approaches, such as genetic and metabolic engineering, enable the manipulation of key enzymes and regulatory genes involved in the biosynthesis of bioactive compounds. However, in practical applications, techniques that do not involve genetic manipulation are often preferred because of their safety and regulatory acceptability. There are many biotechnological techniques that allow for increasing the production of metabolites as well as achieving high growth of in vitro cells, including optimization of the media composition, physical factors, and growth dynamics, application of precursors or elicitors, as well as expansion of biomass growth in bioreactors (Kirakosyan et al., 2009; Vasconsuelo & Boland, 2007). The results of plant culture optimization could lead to higher levels of flavonoids and phenolic acids compared with conventionally grown counterparts. In addition, bioreactor systems facilitate the scalability of production, making it feasible for large-scale commercial applications. This exploration contributes to the evolving landscape of biotechnological solutions in addressing the demand for natural compounds, paving the way for a sustainable and efficient source of flavonoids and phenolic acids (Chandran et al., 2020; Karuppusamy, 2009; Szopa, Kokotkiewicz, et al., 2019).

Scientific works focused on developing effective protocols for the production of phenolic compounds in plant in vitro cultures are carried out in numerous research centers around the world. Table 8.6 presents examples of amounts of compounds from the group of flavonoids and phenolic acids that were obtained in the cells of various plant species.

Overall, the integration of biotechnological approaches in plant biomass production has the potential to revolutionize the availability and sustainability of flavonoids and phenolic acids, offering new opportunities for industries ranging from pharmaceuticals to functional foods and cosmetics.

8. The delivery systems enhancing the stability and percutaneous absorption of plant polyphenols

The utilization of innovative delivery systems has emerged as a promising avenue for enhancing the stability and percutaneous absorption of plant polyphenols in various applications. Plant polyphenols, renowned for their potent antioxidative and health-promoting properties, often face challenges related to their stability during storage and their efficient delivery through the skin barrier (Kyriakoudi et al., 2021).

The stability of polyphenols is influenced by a complex interplay of factors that can impact their structural integrity, bioavailability, and overall efficacy. The key factors that contribute to the stability of polyphenols are their oxidation, instability in various pH ranges, high temperature, UV light, the presence of metal ions, as well as enzymes that can inactivate polyphenols. Exposure to oxygen, light, and heat can lead to oxidative degradation, resulting in the loss of bioactivity (Zillich et al., 2015). Antioxidants and proper storage conditions can help mitigate oxidation-induced instability. Moreover, the matrix in which polyphenols are present, such as cosmetic formulation ingredients, can influence their stability. The presence of protective compounds or the effects of matrix-related factors (e.g., water content) can impact stability. Interactions with other components in a formulation, such as proteins, sugars, or lipids, can impact polyphenol stability and bioavailability. Formulation optimization is crucial to prevent unwanted interactions (Zillich et al., 2015).

The chemical structure of compounds such as polyphenolic acids and flavonoids also affects the stability of individual compounds, which translates into their effectiveness. Structural diversity means that individual polyphenols differ in their resistance to external conditions. The number and position of phenolic hydroxyl groups, as well as the presence of conjugated double bonds, can influence susceptibility to degradation. Chemical modifications of polyphenol structure, such as hydroxylation, glycosylation, or acylation, can change molecular properties, including resistance to oxidation, temperature, and UV light. Hydroxylation may reduce polyphenol stability, whereas modifications such as glycosylation and acylation may improve it (Cao et al., 2020). Polyphenols in the trans form showed higher thermal stability than the cis form (Kosović et al., 2020). Moreover, many polyphenols may act as antioxidants themselves by scavenging reactive oxygen species. However, this antioxidative capacity can also affect polyphenol stability, because the compounds may undergo structural changes during antioxidative reactions (Stamatopoulos et al., 2014). Considering these multifaceted factors and their interactions is crucial when designing strategies to enhance polyphenol stability in topical formulations.

Encapsulation systems, such as liposomes, micro- and nanoparticles, emulsions, and solid lipid nanoparticles, have shown remarkable potential in protecting polyphenols from degradation and facilitating their controlled release (Sikora et al., 2020). These delivery systems offer the advantage of improved stability by shielding polyphenols from external factors, while also enabling controlled release kinetics that enhance percutaneous absorption (Tolun et al., 2016). In addition, strategies like iontophoresis, microneedle arrays, and nanocarriers can further enhance the transdermal delivery of polyphenols by overcoming the skin's inherent impermeability (Oliveira et al., 2022; Yanat & Schroën, 2021) (Table 8.11).

Table 8.11 Forms of ferulic acid and its derivatives and their use in cosmetic preparations according to CosIng database.

The form	Activity
Ferulic acid	Antimicrobial Antioxidant
Cycloartenyl ferulate	Skin conditioning Emollient
Feruloyl oligopeptide-33	Skin conditioning
Feruloyl pentapeptide-71	Antioxidant
Feruloyl tetrapeptide-80	Antioxidant Emollient
Guanidine ferulamide	Skin conditioning
Hydrolyzed chitosan ferulyl linoleate	Anti-sebum Antimicrobial Antioxidant Emulsion stabilizing Skin conditioning
Isostearyl ferulate	Skin conditioning
Methyl polysilsesquioxane ferulate	Antioxidant Skin conditioning - miscellaneous
Methyl polysilsesquioxane ferulate hyaluronate	Antioxidant Humectant
Sodium hyaluronate ferulate	Colorant Skin conditioning Emollient Skin protecting
Soybean oil/ferulate esters	Skin conditioning
Tocopheryl ferulate	Antioxidant

9. Conclusions

The diverse array of flavonoids and phenolic acids derived from plant sources offers immense potential as natural antioxidants for cosmetic applications. Their multifaceted antioxidative mechanisms, ranging from scavenging reactive oxygen species to modulating cellular defense systems, position them as valuable ingredients for addressing skin health concerns and promoting antiaging effects.

The most important flavonoids and phenolic acids described in this chapter are among the compounds frequently used in cosmetic and dermatological formulations. Their popularity is undoubtedly due to their valuable biological properties, which have been confirmed by numerous scientific studies. Ingredient databases for cosmetic use, such as CosIng, also indicate the wide use of these natural compounds and their derivatives. This is reflected by the presence of flavonoids and phenolic acids in cosmetics produced worldwide, as shown in Tables 8.1 and

8.10. An innovative aspect of the search for new sources of polyphenols is the development of plant in vitro cultures, which are less exposed to environmental pollution and toxic substances than field-grown plant material. The possibility of cultivating them in bioreactors may increase their feasibility for use in formulations.

The extensive body of research outlined in this article underscores their effectiveness in mitigating oxidative stress-induced damage, reducing inflammation, and enhancing collagen synthesis. The synergy between these natural compounds and the skin's innate defense mechanisms has been established as a cornerstone for improved skin health and radiance. However, challenges related to formulation stability, potential interactions, and variable bioavailability necessitate careful consideration during product development. As the demand for safe and sustainable cosmetic solutions grows, the application of phenolic acids and flavonoids is poised to flourish, offering both consumers and the cosmetic industry an opportunity to harness the power of nature for healthier, more vibrant skin. Continued exploration of their synergistic interactions, optimized formulation strategies, and clinical validation will undoubtedly pave the way for an exciting future where the potent antioxidative properties of flavonoids and phenolic acids shape the next generation of cosmetics.

Abbreviations

4-HNE	4-hydroxynonenal
A549	hypotriploid alveolar basal epithelial cell line
ABTS	2,2′-azino-bis-(3-ethylbenzothiazoline-6-sulfonic acid)
AChE	acetylcholinesterase
ALP	alkaline phosphatase
AP-1	activator protein-1
AQP	aquaporin
BACE1	β-site amyloid precursor cleaving enzyme 1
BALB/c	Bagg's albino, laboratory-bred strain
BChE	butyrylcholinesterase
Beclin 1	autophagy-related protein
bFGF	basic fibroblast growth factor
BSA	bovine serum albumin
C-FABP	cutaneous fatty acid-binding protein
C/EBPα	CCAAT-enhancer-binding protein α
CASP3	caspase-3
CB1/2	cannabinoid receptor 1/2
CCD-18Co	human colon myofibroblasts
CD11b	microglia surface antigen
CD69	type II C-lectin receptor
COX-2	cyclooxygenase-2
CPD	pyrimidine cyclobutane dimers
CRP	C-reactive protein
CYP1A	mammalian cytochrome P450 1A

Cyt-c	cytochrome c
DMF	dimethyl formamide
DMSO	dimethyl sulfoxide
DPhT	diphenyltin dichloride
DPPH	1,1-diphenyl-2-picrylhydrazyl
Drp1	dynamin-related protein 1
EGFR	epidermal growth factor receptor
ERK	extracellular regulated kinase
G-CSF	granulocyte colony-stimulating factor
GM-CSF	granulocyte macrophage colony-stimulating factor
GPR55	G-protein coupled receptor 55
GSH	glutathione
GSH-Px	glutathione peroxidase
GSK-3β	glycogen synthase kinase-3 beta
GTP	guanosine triphosphate
H-FABP	cardiac fatty acid binding protein
HaCaT	human keratinocyte line
HAT	hydrogen atom transfer
HDF	human dermal fibroblasts
HDL	high-density lipoprotein
HEKn cells	human epidermal keratinocytes, neonatal
HepG2	hepatoblastoma cell line
HIF-1α	hypoxia-inducible factor 1-alpha
HMC-1	human mast cells
HMGB	high-mobility group box
HMVII	melanin content in melanoma cells
HO1	heme oxygenase-1
HPLC	high-performance liquid chromatography
HSF	human skin fibroblasts
HSV-1/2	herpes virus 1/2
hUC-MSC	human umbilical cord mesenchymal stem cells
HUVEC	human umbilical vein endothelial cells
IBA-1	ionized calcium binding adaptor molecule-1
IC50	half-maximal inhibitory concentration
ICAM-1	intercellular adhesion molecule-1
IgE	immunoglobulin E
IgG2a	immunoglobulin G2a
IKK	IkappaB kinase
IL	interleukin
ILHV	Ilhéus virus
IM-HCF	immortalized human cardiac fibroblasts
IMQ	imiquimod
INF-g	interferon gamma
iNOS	inducible nitric oxide synthase
JB6P+	mouse epidermal cell line
JNK	c-Jun-N-terminal kinase
Keap 1	Kelch-like ECH-associated protein 1
LC3	microtubule-associated protein 1A/1B-light chain 3

LCs	CD1a-expressing Langerhans cells
LDL	low-density lipoprotein
LPS	lipopolysaccharide
MAPK	mitogen activated protein kinase
MCF-7	human breast cancer cell line
MCP-1	monocyte chemoattractant protein-1
MDA	malondialdehyde
MDA-MB-231	breast adenocarcinoma
MDM2	mouse double minute 2
MED	minimum erythema dose
MEK	mitogen-activated kinase kinase
MIF	macrophage migration inhibitory factor
MKK	mitogen-activated protein kinase
MMP-1	metalloproteinase-1
MRP-1	multidrug resistance protein-1
MRSA	methicillin-resistant *Staphylococcus aureus*
NADPH	nicotinamide adenine dinucleotide phosphate
NC/Nga	animal model of human atopic dermatitis
NCTC 2544	normal human keratinocyte cells
Nd	YAGNeodymium-doped Yttrium Aluminum garnet
NF-κB	nuclear factor kappa B
NFKB1	nuclear factor kappa B subunit 1
nHDF	normal human dermal fibroblasts
NHEM	normal human epidermal melanocytes
NIK	NF-κB-inducing kinase
Nox-4	NADPH oxidase 4
NQO1NAD(P)H	quinone oxidoreductase 1
Nrf2	erythroid 2-related factor 2
NSCLC	human non-small cell lung cancer
ONOO−	peroxynitrite
p22phox	ubiquitous protein
PAR-2	protease-activated receptor 2
PARP	poly(ADP-ribose) polymerase
PCR	polymerase chain reaction
PGC1α	peroxisome proliferator-activated receptor gamma coactivator 1-alpha
PGE2	prostaglandin E2
PPARγ	peroxisome proliferator-activated receptor γ
PTP1B	protein tyrosine phosphatase 1B
RACK1	receptor for activated C kinase 1
RAW264.7	lipopolysaccharide-stimulated macrophages
RBC	red blood cells
RBP	RNA-binding protein
RELA	proto-oncogene
RelB	transcription factor
RhoA	Ras homolog family member A
Rock1	Rho-associated coiled-coil containing protein kinase 1
ROS	reactive oxygen species
RPMI-7951	metastatic melanoma cell line

SCF	stem cell factor
SEB-1	human sebaceous cell line
SELS	skin surface assessment index
SIRT3	sirtuin 3
SK-MEL-28	human melanoma cell line
SOCS1	suppressor of cytokine signaling 1
SOD	superoxide dismutase
STAT3	signal transducer and activator of transcription-3
TEWL	transepidermal water loss
TGF-β1	transforming growth factor β1
TGFbRI	TGF-b type I receptor
TLR-4	toll-like receptor 4
TNBC	triple-negative breast cancer
TNF-α	tumor necrosis factor-α
TPhT	triphenyltin chloride
TRAF3	tumor necrosis factor receptor associated-factor 3
TRP-1/2	human tyrosinase-related protein 1/2
TRPA1	transient receptor potential ankyrin 1
TRPV1	transient receptor potential vanilloid 1
Trx	thioredoxin
UV	ultraviolet
V2R	vasopressin V2 receptor
VCAM-1	vascular cell adhesion protein-1
Vero	African green monkey kidney cell line
VIP	vasoactive intestinal peptide
VR1	vanilloid receptor 1
WBC	white blood cells
α-MSH	melanotropin α

References

Abarikwu, P. D. O., Lawrence, C. J., Wekere, F. C., Ochulor, A. C., & Barikuma, A. M. (2017). Rutin, an antioxidant flavonoid, induces glutathione and glutathione peroxidase activities to protect against ethanol effects in cadmium-induced oxidative stress in the testis of adult rats. *Andrologia, 49*(7), 1–12.

Abbas, H., El, S., Ali, N. S., & ME Elsheikh, M. A. (2022). Integrated lecithin–bile salt nanovesicles as a promising approach for effective skin delivery of luteolin to improve UV-induced skin damage in Wistar Albino rats. *Colloids and Surfaces, B: Biointerfaces, 211*, 112299.

Abramovič, H. (2015). *Antioxidant properties of hydroxycinnamic acid derivatives: A focus on biochemistry, physicochemical parameters, reactive species, and biomolecular interactions coffee in health and disease prevention* (pp. 843–852). Slovenia: Elsevier Inc.. https://doi.org/10.1016/B978-0-12-409517-5.00093-0

Abu-Yousif, A. O., Smith, K. A., Getsios, S., Green, K. J., Van Dross, R. T., & Pelling, J. C. (2008). Enhancement of UVB-induced apoptosis by apigenin in human keratinocytes and organotypic keratinocyte cultures. *Cancer Research, 68*(8), 3057–3065. https://doi.org/10.1158/0008-5472.CAN-07-2763

Açıkgöz, M. A. (2020). Establishment of cell suspension cultures of Ocimum basilicum L. and enhanced production of pharmaceutical active ingredients. *Industrial Crops and Products, 148*, 112278. https://doi.org/10.1016/j.indcrop.2020.112278

Adedara, I. A., Ego, V. C., Subair, T. I., Oyediran, O., & Farombi, E. O. (2017). Quercetin improves neurobehavioral performance through restoration of brain antioxidant status and acetylcholinesterase activity in manganese-treated rats. *Neurochemical Research, 42*(4), 1219–1229. https://doi.org/10.1007/s11064-016-2162-z

Ademosun, A. O., Oboh, G., Bello, F., & Ayeni, P. O. (2016). Antioxidative properties and effect of quercetin and its glycosylated form (Rutin) on acetylcholinesterase and butyrylcholinesterase activities. *Journal of Evidence-Based Complementary and Alternative Medicine, 21*(4), NP11. https://doi.org/10.1177/2156587215610032

Adhikary, S., Choudhary, D., Ahmad, N., Karvande, A., Kumar, A., Banala, V. T., Mishra, P. R., & Trivedi, R. (2018). Dietary flavonoid kaempferol inhibits glucocorticoid-induced bone loss by promoting osteoblast survival. *Nutrition, 53*, 64–76. https://doi.org/10.1016/j.nut.2017.12.003

Ahmad, A., Afzaal, M., Saeed, F., Ali, S. W., Imran, A., Zaidi, Y. R., Saleem, A., Hussain, M., & Jbawi, Al (2023). A comprehensive review of the therapeutic potential of citrus bioflavonoid hesperidin against lifestyle-related disorders. *Cogent Food & Agriculture, 9*(1), 2226427.

Ahmad, S. T., Arjumand, W., Nafees, S., Seth, A., Ali, N., Rashid, S., & Sultana, S. (2012). Hesperidin alleviates acetaminophen induced toxicity in wistar rats by abrogation of oxidative stress, apoptosis and inflammation. *Toxicology Letters, 208*(2), 149–161. https://doi.org/10.1016/j.toxlet.2011.10.023

Ahmadi, Z., Mohammadinejad, R., Roomiani, S., Ghasemipour Afshar, E., & Ashrafizadeh, M. (2021). Biological and therapeutic effects of troxerutin: Molecular signaling pathways come into view. *Journal of Pharmacopuncture, 24*(1), 1–13. https://doi.org/10.3831/kpi.2021.24.1.1

Ahmed, H. H., Aglan, H. A., Elsayed, G. H., Hafez, H. G., & Eskander, E. F. (2021). Quercetin offers chemopreventive potential against breast cancer by targeting a network of signalling pathways. *Research Journal of Pharmacy and Technology, 14*(5), 2829–2839. https://doi.org/10.52711/0974-360X.2021.00499

Ahmed, S., Rahman, A., Saleem, M., Athar, M., & Sultana, S. (1999). Ellagic acid ameliorates nickel induced biochemical alterations: Diminution of oxidative stress. *Human & Experimental Toxicology, 18*(11), 691–698. https://doi.org/10.1191/096032799678839563

Aiyer, H. S., Vadhanam, M. V., Stoyanova, R., Caprio, G. D., Clapper, M. L., & Gupta, R. C. (2008). Dietary berries and ellagic acid prevent oxidative DNA damage and modulate expression of DNA repair genes. *International Journal of Molecular Sciences, 9*(3), 327–341. https://doi.org/10.3390/ijms9030327

Akkoyun, D. C., Akyuz, A., Dogan, M., Erboga, M., Aktas, C., Caglar, V., Uygur, R., Topcu, B., Yilmaz, A., & Gurel, A. (2016). Quercetin inhibits heart injury in lipopolysaccharide-induced endotoxemic model by suppressing the effects of reactive oxygen species. *Analytical and Quantitative Cytology and Histology, 38*(3), 183–188.

Al Mamary, H. H. (2016). Phenolic compounds classification, chemistry, and updated techniques of analysis and synthesis. In: Badria F.A., Surguchov A. Phenolic compounds - Chemistry, synthesis, diversity, non-conventional industrial, pharmaceutical and therapeutic applications. *IntechOpen*, 1–21.

Albayrak, İ., Demirci, T., & Baydar, N. G. (2024). Enhancement of in vitro production of tropane alkaloids and phenolic compounds in Hyoscyamus niger by culture types and elicitor treatments. *Plant Cell, Tissue and Organ Culture, 156*(3). https://doi.org/10.1007/s11240-024-02692-x

Almeida, A. F., Borge, G. I. A., Piskula, M., Tudose, A., Tudoreanu, L., Valentová, K., Williamson, G., & Santos, C. N. (2018). Bioavailability of quercetin in humans with a focus on interindividual variation. *Comprehensive Reviews in Food Science and Food Safety, 17*(3), 714–731. https://doi.org/10.1111/1541-4337.12342

Amer, asmaa, Elgohary, R., Ibrahim, F., & Taha, H. (2024). Implement of biotechnology techniques for the production of anticoagulants from Lagestromia tomentosa L. *Egyptian Journal of Chemistry.* https://doi.org/10.21608/ejchem.2024.273296.9398

Anand David, A. V., Arulmoli, R., & Parasuraman, S. (2016). Overviews of biological importance of quercetin: A bioactive flavonoid. *Pharmacognosy Reviews, 10*(20), 84–89. https://doi.org/10.4103/0973-7847.194044

Anis, E., Zafeer, M. F., Firdaus, F., Islam, S. N., Khan, A. A., Ali, A., & Mobarak Hossain, M. (2020). Ferulic acid reinstates mitochondrial dynamics through PGC1α expression modulation in 6-hydroxydopamine lesioned rats. *Phytotherapy Research, 34*(1), 214–226. https://doi.org/10.1002/ptr.6523

Anson, N. M., Aura, A. M., Selinheimo, E., Mattila, I., Poutanen, K., Van Den Berg, R., Havenaar, R., Bast, A., & Haenen, G. R. M. M. (2011). Bioprocessing of wheat bran in whole wheat bread increases the bioavailability of phenolic acids in men and exerts anti-inflammatory effects ex vivo1-3. *Journal of Nutrition, 141*(1), 137–143. https://doi.org/10.3945/jn.110.127720Netherlands

Arafa, H. M. M., Aly, H. A. A., Abd-Ellah, M. F., & El-Refaey, H. M. (2009). Hesperidin attenuates benzo[α] pyrene-induced testicular toxicity in rats via regulation of oxidant/antioxidant balance. *Toxicology and Industrial Health, 25*(6), 417–427. https://doi.org/10.1177/0748233709106624

Arct, J., & Pytkowska, K. (2008). Flavonoids as components of biologically active cosmeceuticals. *Clinics in Dermatology, 26*(4), 347–357. https://doi.org/10.1016/j.clindermatol.2008.01.004

Arowoogun, J., Akanni, O. O., Adefisan, A. O., Owumi, S. E., Tijani, A. S., & Adaramoye, O. A. (2021). Rutin ameliorates copper sulfate-induced brain damage via antioxidative and anti-inflammatory activities in rats. *Journal of Biochemical and Molecular Toxicology, 35*(1), 1–8.

Aruwa, C. E., Amoo, S. O., Koorbanally, N., & Kudanga, T. (2021). Enzymatic dimerization of luteolin enhances antioxidant and antimicrobial activities. *Biocatalysis and Agricultural Biotechnology, 35*, 102105. https://doi.org/10.1016/j.bcab.2021.102105

Aryal, D., Joshi, S., Nabin, T. K., Chaudhary, Sirjana, B., Joshi, U., Bhandari, D., Rogers, H. M., Bhattarai, S., Sharma, K. R., & Regmi, N. P. (2023). Dietary phenolic compounds as promising therapeutic agents for diabetes and its complications: A comprehensive review. *Food Science and Nutrition, 12*, 3025–3045.

Ashry, H. S. E., Assaly, R. M. B., Bahr, & Gabr, A. M. M. (2023). Accumulation of ferulic acid under drought stress in in vitro culture of Gardenia jasmonide variegata. *Minia J. Agric. Res. & Develop., 1*(4), 727–750.

Askari, G., Ghiasvand, R., Feizi, A., Ghanadian, S. M., & Karimian, J. (2012). The effect of quercetin supplementation on selected markers of inflammation and oxidative stress. *Journal of Research in Medical Sciences, 17*(7), 637–641.

Auchter, E. C. (1945). *Report of the Administrator of Agricultural research 1943 United States of Agriculture* (pp. 32–33). Washington: Agricultural Research Administration.

Ausina, P., Branco, J. R., Demaria, E., Leandro, J. G. B., Ochioni, A. C., Mendonça, A. P. M., Palhano, F. L., Oliveira, M. F., Abou-Kheir, W., Sola-Penna, M., & Zancan. (2020). Acetylsalicylic acid and salicylic acid present anticancer properties against melanoma by promoting nitric oxide-dependent endoplasmic reticulum stress and apoptosis. *Scientific Reports, 10*(1), 1–15.

Babbar, R., Dhiman, S., Grover, R., Kaur, A., & Arora, S. (2021). A comprehensive review on therapeutic applications of ferulic acid and its novel analogues: A brief literature. *Mini-Reviews in Medicinal Chemistry, 21*(12), 1578–1593. https://doi.org/10.2174/1389557521666210120111702

Baghel, S. S., Shrivastava, N., Baghel, R. S., Agrawai, P., & Rajput, S. (2012). Review of quercetin: Antioxidant and anticancer properties. *World Journal of Pharmacy and Pharmaceutical Sciences, 1*(1), 146–160.

Bahadır Acıkara, Ö., Ilhan, M., Kurtul, E., Šmejkal, K., & Akkol, E. K. (2019). Inhibitory activity of Podospermum canum and its active components on collagenase, elastase and hyaluronidase enzymes. *Bioorganic Chemistry, 93*, 103330. https://doi.org/10.1016/j.bioorg.2019.103330

Balakrishnan, A., & Menon, V. P. (2007). Effect of hesperidin on nicotine toxicity and histopathological studies. *Toxicology Mechanisms and Methods, 17*(4), 233–239. https://doi.org/10.1080/15376510600970430

Begum, N., Prasad, N. R., & Thayalan, K. (2012). Apigenin protects gamma-radiation induced oxidative stress, hematological changes and animal survival in whole body irradiated Swiss albino mice. *International Journal of Nutrition, Pharmacology, Neurological Diseases, 2*(1).

Bhandary, B., Piao, C. S., Kim, D. S., Lee, G. H., Chae, S. W., Kim, H. R., & Chae, H. J. (2012). The protective effect of rutin against ischemia/reperfusion-associated hemodynamic alteration through antioxidant activity. *Archives of Pharmacal Research, 35*(6), 1091–1097. https://doi.org/10.1007/s12272-012-0617-6

Blanch, G. P., Gómez-Jiménez, M. C., & del Castillo, M. L. R. (2020). Exogenous Salicylic acid improves phenolic content and antioxidant activity in table grapes. *Plant Foods for Human Nutrition, 75*(2), 177–183. https://doi.org/10.1007/s11130-019-00793-z

Böyük, A., Önder, A., Kapan, M., Gümüs, M., Firat, U., Basarali, M. K., & Alp, H. (2011). Ellagic acid ameliorates lung injury after intestinal ischemia-reperfusion. *Pharmacognosy Magazine, 7*(27), 224–228. https://doi.org/10.4103/0973-1296.84236

Britto, S. M., Shanthakumari, D., Agilan, B., Radhiga, T., Kanimozhi, G., & Prasad, N. R. (2017). Apigenin prevents ultraviolet-B radiation induced cyclobutane pyrimidine dimers formation in human dermal fibroblasts. *Mutation Research: Genetic Toxicology and Environmental Mutagenesis, 821*, 28–35. https://doi.org/10.1016/j.mrgentox.2017.06.002

Bumrungpert, A., Lilitchan, S., Tuntipopipat, S., Tirawanchai, N., & Komindr, S. (2018). Ferulic acid supplementation improves lipid profiles, oxidative stress, and inflammatory status in hyperlipidemic subjects: A randomized, double-blind, placebo-controlled clinical trial. *Nutrients, 10*(6), 713. https://doi.org/10.3390/nu10060713

Călinoiu, L. F., & Vodnar, D. C. (2018). Whole grains and phenolic acids: A review on bioactivity, functionality, health benefits and bioavailability. *Nutrients, 10*(11). https://doi.org/10.3390/nu10111615

Calzada, F., Solares-Pascasio, J. I., Ordoñez-Razo, R. M., Velazquez, C., Barbosa, E., García-Hernández, N., Mendez-Luna, D., & Correa-Basurto, J. (2017). Antihyperglycemic activity of the leaves from Annona cherimola miller and rutin on alloxan-induced diabetic rats. *Pharmacognosy Research, 9*(1), 1–6. https://doi.org/10.4103/0974-8490.199781

Cao, H., Yi, L., Zhong, J., Högger, P., Wang, M., Prieto, M. A., Simal-Gandara, J., & Xiao, J. (2020). Investigation of new products and reaction kinetics for myricetin in DMEM via an in situ UPLC–MS–MS analysis. *Food Frontiers, 1*(3), 243–252. https://doi.org/10.1002/fft2.19

Carelli, S., Hebda, D. M., Traversa, M. V., Messaggio, F., Giuliani, G., Marzani, B., Benedusi, A., Di Giulio, A. M., & Gorio, A. (2012). A specific combination of zeaxanthin, spermidine and rutin prevents apoptosis in human dermal papilla cells. *Experimental Dermatology, 21*(12), 953–955. https://doi.org/10.1111/exd.12029

Casagrande, R., Georgetti, S. R., Verri, W. A., Dorta, D. J., dos Santos, A. C., & Fonseca, M. J. V. (2006). Protective effect of topical formulations containing quercetin against UVB-induced oxidative stress in hairless mice. *Journal of Photochemistry and Photobiology B: Biology, 84*(1), 21–27. https://doi.org/10.1016/j.jphotobiol.2006.01.006

Chandekar, L., Katgeri, R., & Takke, A. (2022). The potential clinical uses and nanoformulation strategies of kaempferol, a dietary flavonoid. *Revista Brasileira de Farmacognosia, 32*(5), 693–707. https://doi.org/10.1007/s43450-022-00290-6

Chandran, Meena, M., Barupal, & Sharma, K. (2020). Plant tissue culture as a perpetual source for production of industrially important bioactive compounds. *Biotechnology Reports, 26*, e00450.

Chavarria, D., Benfeito, S., Soares, P., Lima, C., Garrido, J., Serrão, P., Soares-da-Silva, P., Remião, F., Oliveira, P. J., & Borges. (2022). Boosting caffeic acid performance as antioxidant and monoamine oxidase B/catechol-O-methyltransferase inhibitor. *European Journal of Medicinal Chemistry, 243*, 114740.

Che, D. N., Cho, Shin, m Y., Kang, h H., Kim, J. S., & Jang, S. I. (2019). Apigenin inhibits IL-31 cytokine in human mast cell and mouse skin tissues. *Molecules, 24*(7), 1290.

Chen, M., Gu, H., Ye, Y., Lin, B., Sun, L., Deng, W., Zhang, J., & Liu, J. (2010). Protective effects of hesperidin against oxidative stress of tert-butyl hydroperoxide in human hepatocytes. *Food and Chemical Toxicology, 48*(10), 2980–2987. https://doi.org/10.1016/j.fct.2010.07.037

Chen, H. J., Lin, C. M., Lee, C. Y., Shih, N. C., Peng, S. F., Tsuzuki, M., Amagaya, S., Huang, W. W., & Yang, J. S. (2013). Kaempferol suppresses cell metastasis via inhibition of the ERK-p38-JNK and AP-1 signaling pathways in U-2 OS human osteosarcoma cells. *Oncology Reports, 30*(2), 925–932. https://doi.org/10.3892/or.2013.2490Taiwan

Chen, H., Lu, C., Liu, H., Wang, M., Zhao, H., Yan, Y., & Han, L. (2017). Quercetin ameliorates imiquimod-induced psoriasis-like skin inflammation in mice via the NF-κB pathway. *International Immunopharmacology, 48*, 110–117. https://doi.org/10.1016/j.intimp.2017.04.022

Chen, B. H., Park, J. H., Ahn, J. H., Cho, J. H., Kim, I. H., Lee, J. C., Won, M. H., Lee, C. H., Hwang, I. K., Kim, J. D., Kang, I. J., Cho, J. H., Shin, B. N., Kim, Y. H., Lee, Y. L., & Park, S. M. (2017). Pretreated quercetin protects gerbil hippocampal CA1 pyramidal neurons from transient cerebral ischemic injury by increasing the expression of antioxidant enzymes. *Neural Regeneration Research, 12*(2), 220–227. https://doi.org/10.4103/1673-5374.200805

Chen, M.-cang, Ye, Y.-yi, Ji, G., & Liu, J.-wen (2010). Hesperidin upregulates heme oxygenase-1 to attenuate hydrogen peroxide-induced cell damage in hepatic L02 cells. *Journal of Agricultural and Food Chemistry, 58*(6), 3330–3335. https://doi.org/10.1021/jf904549s

Cheng, M., Zhan, L., Zhang, H., Li, X., Wang, Y., Xia, F., Wang, B., Cai, R., Guo, Z., Zhang, Y., Ji, Y., Sun, C., Feng, X., Lei, L., Yang, Y., Han, W., & Gu, J. (2018). An ointment consisting of the phage lysin LysGH15 and apigenin for decolonization of methicillin-resistant Staphylococcus aureus from skin wounds. *Viruses, 10*(5). https://doi.org/10.3390/v10050244

Chiu, A., & Kimball, A. B. (2003). Topical vitamins, minerals and botanical ingredients as modulators of environmental and chronological skin damage. *British Journal of Dermatology, 149*(4), 681–691. https://doi.org/10.1046/j.1365-2133.2003.05540.x

Choi, J. S., Islam, M. N., Ali, M. Y., Kim, Y. M., Park, H. J., Sohn, H. S., & Jung, H. A. (2014). The effects of C-glycosylation of luteolin on its antioxidant, anti-Alzheimer's disease, anti-diabetic, and anti-inflammatory activities. *Archives of Pharmacal Research, 37* (10), 1354–1363. https://doi.org/10.1007/s12272-014-0351-3

Choi, J. K., & Kim, S. H. (2013). Rutin suppresses atopic dermatitis and allergic contact dermatitis. *Experimental Biology and Medicine, 238*(4), 410–417. https://doi.org/10.1177/1535370213477975

Choi, K. S., Kundu, J. K., Chun, K. S., Na, H. K., & Surh, Y. J. (2014). Rutin inhibits UVB radiation-induced expression of COX-2 and iNOS in hairless mouse skin: P38 MAP kinase and JNK as potential targets. *Archives of Biochemistry and Biophysics, 559*, 38–45. https://doi.org/10.1016/j.abb.2014.05.016

Choi, S., Lim, T. G., Hwang, M. K., Kim, Y. A., Kim, J., Kang, N. J., Jang, T. S., Park, J. S., Yeom, M. H., & Lee, K. W. (2013). Rutin inhibits B[a]PDE-induced cyclooxygenase-2 expression by targeting EGFR kinase activity. *Biochemical Pharmacology, 86*(10), 1468–1475. https://doi.org/10.1016/j.bcp.2013.08.066

Choi, S., Youn, J., Kim, K., Joo, D. H., Shin, S., Lee, J., Lee, H. K., An, I. S., Kwon, S., Youn, H. J., Ahn, K. J., An, S., & Cha, H. J. (2016). WaveApigenin inhibits UVA-induced cytotoxicity in vitro and prevents signs of skin aging in vivo. *International Journal of Molecular Medicine, 38*(2), 627–634. https://doi.org/10.3892/ijmm.2016.2626

Chong, K. P., Rossall, S., & Atong, M. (2011). HPC fingerprints and in vitro antimicrobial activity of syringic acid, caffeic acid and 4-hydroxybenzoic acid against Ganoderma boninense. *Journal of Applied Sciences, 11*(13), 2284–2291. https://doi.org/10.3923/jas.2011.2284.2291Malaysia

Cosmetic Ingredient Database (CosIng). Available online www.ec.europe.eu (Accessed 20 July 2026)

Cirico, T. L., & Omaye, S. T. (2006). Additive or synergetic effects of phenolic compounds on human low density lipoprotein oxidation. *Food and Chemical Toxicology, 44*(4), 510–516. https://doi.org/10.1016/j.fct.2005.08.025

Crascì, L., Cardile, V., Longhitano, G., Nanfitò, F., & Panico, A. (2018). Anti-degenerative effect of Apigenin, Luteolin and Quercetin on human keratinocyte and chondrocyte cultures: SAR evaluation. *Drug Research, 68*(3), 132–138. https://doi.org/10.1055/s-0043-120662

Dal Belo, S. E., Gaspar, L. R., Maia Campos, P. M. B. G., & Marty, J. P. (2009). Skin penetration of epigallocatechin-3-gallate and quercetin from green tea and Ginkgo biloba extracts vehiculated in cosmetic formulations. *Skin Pharmacology and Physiology, 22*(6), 299–304. https://doi.org/10.1159/000241299

Damasceno, S. S., Dantas, B. B., Ribeiro-Filho, J., Araújo, D. A. M., & da Costa, J. G. M. (2017). Chemical properties of caffeic and ferulic acids in biological system: Implications in cancer therapy. A review. *Current Pharmaceutical Design, 23*(20). https://doi.org/10.2174/1381612822666161208145508

Das, S., Sultana, K. W., Mondal, M., Chandra, I., & Ndhlala, A. R. (2024). Unveiling the dual nature of heavy metals: Stressors and promoters of phenolic compound biosynthesis in Basilicum polystachyon (L.) moench in vitro. *Plants, 13*(1).

De Vita, D., Friggeri, L., D'Auria, F. D., Pandolfi, F., Piccoli, F., Panella, S., Teresa Palamara, A., Simonetti, G., Scipione, L., Santo, R. Di, Costi, R., & Tortorella, S. (2014). Activity of caffeic acid derivatives against Candida albicans biofilm. *Bioorganic & Medicinal Chemistry Letters, 24*(6), 1502–1505. https://doi.org/10.1016/j.bmcl.2014.02.005

Diao, M., Liang, Y., Zhao, J., Zhang, J., & Zhang, T. (2022). Complexation of ellagic acid with α-lactalbumin and its antioxidant property. *Food Chemistry, 372*, 131307. https://doi.org/10.1016/j.foodchem.2021.131307

Dias, M. C., Pinto, D. C. G. A., & Silva, A. M. S. (2021). Plant flavonoids: Chemical characteristics and biological activity. *Molecules, 26*(17), 1–16.

Dias, A. C. P., Tomás-Barberán, F. A., Fernandes-Ferreira, M., & Ferreres, F. (1998). Unusual flavonoids produced by callus of Hypericum perforatum. *Phytochemistry, 48*(7), 1165–1168. https://doi.org/10.1016/S0031-9422(97)00963-1

Ding, B., Ma, G., Wang, Z., Liang, W., Gao, W., & Rehman, K. (2021). Mechanisms of kidney cell pyroptosis in chronic kidney disease and the effects of traditional Chinese medicine. *Evidence-Based Complementary and Alternative Medicine, 2021*, 1–10. https://doi.org/10.1155/2021/1173324

Doostkam, A., Mirkhani, H., Iravani, K., Karbalay-Doust, S., & Zarei, K. (2021). Effect of rutin on diabetic auditory neuropathy in an experimental rat model. *Clinical and Experimental Otorhinolaryngology, 14*(3), 259–267. https://doi.org/10.21053/ceo.2019.02068

Ebrahimzadeh, M. A., Pourmorad, F., & Bekhradnia, A. R. (2008). Iron chelating activity, phenol and flavonoid content of some medicinal plants from Iran. *African Journal of Biotechnology, 7*(18), 3188–3192.

Ekiert, H., Kwiecień, I., & Szopa, A. (2013). Rosmarinic acid production in plant in vitro cultures. *Polish Journal of Cosmetology, 16*(1), 49–58.

El-Sayed, E. S. M., Abo-Salem, O. M., Abd-Ellah, M. F., & Abd-Alla, G. M. (2008). Hesperidin, an antioxidant flavonoid, prevents acrylonitrile-induced oxidative stress in rat brain. *Journal of Biochemical and Molecular Toxicology, 22*(4), 268–273. https://doi.org/10.1002/jbt.20237

El-Sisi, A. E. D. E. S., Sokar, S. S., Shebl, A. M., & Mohamed, D. Z. (2017). Antifibrotic effect of diethylcarbamazine combined with hesperidin against ethanol induced liver fibrosis in rats. *Biomedicine & Pharmacotherapy, 89*, 1196–1206. https://doi.org/10.1016/j.biopha.2017.03.013

Espino, S., García, E., González, M., Valera-Montero, D., Benítez, L. F., Morales, O., Maldonado, G., Escalante-Garcia, H. S., & Olvera-Gonzalez, N. (2023). Interaction of culture medium and artificial light type on pigmentation of micro-propagated opuntia plants. *Horticulturae, 9*, 1348.

Falcone Ferreyra, M. L., Rius, S. P., & Casati, P. (2012). Flavonoids: Biosynthesis, biological functions, and biotechnological applications. *Frontiers in Plant Science, 3*. https://doi.org/10.3389/fpls.2012.00222

Farha, A. K., Gan, R. Y., Li, H. B., Wu, D. T., Atanasov, A. G., Gul, K., Zhang, J. R., Yang, Q. Q., & Corke, H. (2022). The anticancer potential of the dietary polyphenol rutin: Current status, challenges, and perspectives. *Critical Reviews in Food Science and Nutrition, 62*(3), 832–859. https://doi.org/10.1080/10408398.2020.1829541

Ferrier, M., Billet, K., Drouet, S., Tungmunnithum, D., Malinowska, M. A., Marchal, C., Dedet, S., Giglioli-Guivarc'h, N., Hano, C., & Lanoue, A. (2022). Identifying major drivers of antioxidant activities in complex polyphenol mixtures from grape canes. *Molecules, 27*(13), 1–13.

Ferulic terapis Kwas Ferulowy 40%/30ML. Available online: https://apis-sklep.pl53315-kwas-ferulowy-40-3-ml-p-398.html (Accessed 20 July 2026).

Francenia Santos-Sánchez, N., Salas-Coronado, R., Hernández-Carlos, B., & Villanueva-Cañongo, C. (2019). *Shikimic acid pathway in biosynthesis of phenolic compounds*. IntechOpen. https://doi.org/10.5772/intechopen.83815

Gabrielska, J., Soczyńska-Kordala, M., & Przestalski, S. (2005). Antioxidative effect of kaempferol and its equimolar mixture with phenyltin compounds on UV-irradiated liposome membranes. *Journal of Agricultural and Food Chemistry, 53*(1), 76–83. https://doi.org/10.1021/jf0401120

Ganesan, D., Albert, A., Paul, E., Ananthapadmanabhan, K., Andiappan, R., & Sadasivam, S. G. (2020). Rutin ameliorates metabolic acidosis and fibrosis in alloxan induced diabetic nephropathy and cardiomyopathy in experimental rats. *Molecular and Cellular Biochemistry, 471*(1–2), 41–50. https://doi.org/10.1007/s11010-020-03758-y

García-Mediavilla, V., Crespo, I., Collado, P. S., Esteller, A., Sánchez-Campos, S., Tuñón, M. J., & González-Gallego, J. (2007). The anti-inflammatory flavones quercetin and kaempferol cause inhibition of inducible nitric oxide synthase, cyclooxygenase-2 and reactive C-protein, and down-regulation of the nuclear factor kappaB pathway in Chang Liver cells. *European Journal of Pharmacology, 557*(2–3), 221–229. https://doi.org/10.1016/j.ejphar.2006.11.014

Gegotek, A., Rybałtowska-Kawałko, P., & Skrzydlewska, E. (2017). Rutin as a mediator of lipid metabolism and cellular signaling pathways interactions in fibroblasts altered by UVA and UVB radiation. *Oxidative Medicine and Cellular Longevity, 2017*. https://doi.org/10.1155/2017/4721352

Gen factor pH rosmarinic - kwas rozmarynowy 5%. Available online: https://www.aruba.rzeszow/pl/pl/Gen-Factor-pH-Rosmarinic-Kwas-Rozmarynowy5/1226?gclid=EAIaIQobChM (Accessed 20 July 2026).

Genaro-Mattos, T. C., Maurício, Â. Q., Rettori, D., Alonso, A., & Hermes-Lima, M. (2015). Antioxidant activity of Caffeic acid against iron-induced free radical generation-A chemical approach. *PLoS One, 10*(6). https://doi.org/10.1371/journal.pone.0129963

Georgiev, V., Ananga, A., & Tsolova, V. (2014). Recent advances and uses of grape flavonoids as nutraceuticals. *Nutrients, 6*(1), 391–415. https://doi.org/10.3390/nu6010391

Gil, T. Y., Hong, C. H., & An, H. J. (2021). Anti-inflammatory effects of ellagic acid on keratinocytes via mapk and stat pathways. *International Journal of Molecular Sciences, 22*(3), 1–14. https://doi.org/10.3390/ijms22031277

Goleniowski, M., Bonfill, M., Cusido, R., & Palazón, J. (2013). *Phenolic acids natural products: Phytochemistry, botany and metabolism of alkaloids, phenolics and terpenes* (pp. 1951–1973). Argentina: Springer Berlin Heidelberg. https://doi.org/10.1007/978-3-642-22144-6_64

Gomes, C. A., Girão Da Cruz, T. G., Andrade, J. L., Milhazes, N., Borges, F., & Marques, M. P. M. (2003). Anticancer activity of phenolic acids of natural or synthetic origin: A structure-activity study. *Journal of Medicinal Chemistry, 46*(25), 5395–5401. https://doi.org/10.1021/jm030956v

Grace, S. C. (2007). *Phenolics as antioxidants antioxidants and reactive oxygen species in plants* (pp. 141–168). United States: Blackwell Publishing Ltd. https://doi.org/10.1002/9780470988565.ch6

Granado-Serrano, A. B., Martín, M. A., Bravo, L., Goya, L., & Ramos, S. (2012). Quercetin modulates Nrf2 and glutathione-related defenses in HepG2 cells: Involvement of p38. *Chemico-Biological Interactions, 195*(2), 154–164. https://doi.org/10.1016/j.cbi.2011.12.005

Guardia, T., Rotelli, A. E., Juarez, A. O., & Pelzer, L. E. (2001). Anti-inflammatory properties of plant flavonoids. Effects of rutin, quercetin and hesperidin on adjuvant arthritis in rat. *Farmaco, 56*(9), 683–687. https://doi.org/10.1016/S0014-827X(01)01111-9

Hadshiew, I., Stäb, F., Untiedt, S., Bohnsack, K., Rippke, F., & Hölzle, E. (1997). Effects of topically applied antioxidants in experimentally provoked polymorphous light eruption. *Dermatology, 195*(4), 362–368. https://doi.org/10.1159/000245989

Haghmorad, D., Mahmoudi, M. B., Salehipour, Z., Jalayer, Z., Momtazi brojeni, A. A., Rastin, M., Kokhaei, P., & Mahmoudi, M. (2017). Hesperidin ameliorates immunological outcome and reduces neuroinflammation in the mouse model of multiple sclerosis. *Journal of Neuroimmunology, 302*, 23–33. https://doi.org/10.1016/j.jneuroim.2016.11.009

Hassanpour, M., Jafari, H., Sharifi, S., Rezaie, J., Mehri Lighvan, Z., Reza Mahdavinia, G., Gohari, G., & Akbari, A. (2021). Salicylic acid-loaded chitosan nanoparticles (SA/CTS NPs) for breast cancer targeting: Synthesis, characterization and controlled release kinetics. *Journal of Molecular Structure, 1245*, 131040. https://doi.org/10.1016/j.molstruc.2021.131040

Heim, K. E., Tagliaferro, A. R., & Bobilya, D. J. (2002). Flavonoid antioxidants: Chemistry, metabolism and structure-activity relationships. *Journal of Nutritional Biochemistry, 13*(10), 572–584. https://doi.org/10.1016/S0955-2863(02)00208-5

Heleno, S. A., Martins, A., Queiroz, M. J. R. P., & Ferreira, I. C. F. R. (2015). Bioactivity of phenolic acids: Metabolites versus parent compounds: A review. *Food Chemistry, 173*, 501–513. https://doi.org/10.1016/j.foodchem.2014.10.057

Heo, H. J., & Lee, C. Y. (2004). Protective effects of quercetin and vitamin C against oxidative stress-induced neurodegeneration. *Journal of Agricultural and Food Chemistry, 52*(25), 7514–7517. https://doi.org/10.1021/jf049243r

Hoang, H. T., Moon, J.-Y., & Lee, Y.-C. (2021). Natural antioxidants from plant extracts in skincare cosmetics: Recent applications, challenges and perspectives. *Cosmetics, 8*, 106.

Hosseinimehr, S. J., Ahmadi, A., Beiki, D., Habibi, E., & Mahmoudzadeh, A. (2009). Protective effects of hesperidin against genotoxicity induced by 99mTc-MIBI in human cultured lymphocyte cells. *Nuclear Medicine and Biology, 36*(7), 863–867. https://doi.org/10.1016/j.nucmedbio.2009.06.002

Hou, M., Man, M., Man, W., Zhu, W., Hupe, M., Park, K., Crumrine, D., Elias, P. M., & Man, M. Q. (2012). Topical hesperidin improves epidermal permeability barrier function and epidermal differentiation in normal murine skin. *Experimental Dermatology, 21*(5), 337–340. https://doi.org/10.1111/j.1600-0625.2012.01455.x

Hseu, Y. C., Chou, C. W., Senthil Kumar, K. J., Fu, K. T., Wang, H. M., Hsu, L. S., Kuo, Y. H., Wu, C. R., Chen, S. C., & Yang, H. L. (2012). Ellagic acid protects human keratinocyte (HaCaT) cells against UVA-induced oxidative stress and apoptosis through the upregulation of the HO-1 and Nrf-2 antioxidant genes. *Food and Chemical Toxicology, 50*(5), 1245–1255. https://doi.org/10.1016/j.fct.2012.02.020

Hwang, H. D., Han, J. E., Murthy, H. N., Kwon, H. J., Lee, G. M., Shin, J. H., & Park, S. Y. (2022). Establishment of bioreactor cultures for the production of chlorogenic acid and ferulic acid from adventitious roots by optimization of culture conditions in Angelica acutiloba (Siebold & Zucc. *Plant Biotechnology Reports, 16*(2), 173–182. https://doi.org/10.1007/s11816-022-00744-2

Irfan, F., Jameel, F., Khan, I., Aslam, R., Faizi, S., & Salim, A. (2022). Role of quercetin and rutin in enhancing the therapeutic potential of mesenchymal stem cells for cold induced burn wound. *Regenerative Therapy, 21*, 225–238. https://doi.org/10.1016/j.reth.2022.07.011

IfedibaluChukwu Ejiofor, I., Chikodili Igbokwe, M.-G. (2021). Flavonoids: Understanding their biosynthesis pathways in plants and health benefits. In: Zepka L.Q., Casagrande Do Nascimento, T., Jacob-Lopes, E.(Eds.) Bioactive compounds - biosynthesis, characterization and application. IntechOpen, pp-1-18.

Jafernik, K., Szopa, A., Barnaś, M., Dziurka, M., Ekiert, H., & Schisandra, C. B. (2020). Clarke in vitro cultures: A promising tool for the production of lignans and phenolic compounds. *Plant Cell, Tissue and Organ Culture, 143*(1), 45–60.

Jaisinghani, R. N. K. (2017). Antibacterial properties of quercetin. *Microbiology Research, 8* (1), 6877.

Jamali, N., Mostafavi-Pour, Z., Zal, F., Kasraeian, M., Poordast, T., & Nejabat, N. (2021). Antioxidant ameliorative effect of caffeic acid on the ectopic endometrial cells separated from patients with endometriosis. *Taiwanese Journal of Obstetrics & Gynecology, 60*(2), 216–220. https://doi.org/10.1016/j.tjog.2020.12.003

Jegal, J., Kim, T. Y., Park, Jo, o G., Choi, H. S., Kim, S. N., & Yang, M. H. (2021). Inhibitory effects of luteolin 7-methyl ether isolated from Wikstroemia ganpi on Tnf-A/Ifn-Γ mixture-induced inflammation in human keratinocyte. *Nutrients, 13*(12), 4387.

Jena, S., Bhol, N. K., & Duttaroy, A. K. (2023). Cellular red-ox system in health and disease: The latest update. *Biomedicine & Pharmacotherapy, 162*, 114606.

Jeong, H. J., Yoou, M. S., Han, N. R., Hwang, S. Y., Yoon, K. W., & Kim, H. M. (2018). The new therapeutic herbal drug HM0601 and its bioactive compound rutin exert potent antiproliferative activities in mast cells. *Fundamental and Clinical Pharmacology, 32*(3), 279–287. https://doi.org/10.1111/fcp.12350

Jin, B.h., Qian, L.b., Chen, S., Li, J., Wang, H.p., Bruce, I. C., Lin, J., & Xia, Q. (2009). Apigenin protects endothelium-dependent relaxation of rat aorta against oxidative stress. *European Journal of Pharmacology, 616*(1–3), 200–205. https://doi.org/10.1016/j.ejphar.2009.06.020

Jo, Y. J., Yoo, D. H., Lee, I. C., & Lee, J. (2022). Antioxidat and skin whitening activities of sub- and super-critical water treated rutin. *Molecules, 27*, 5441. https://doi.org/10.3390/molecules27175441

Johnson, E. L., & Schmidt, W. F. (2004). Flavonoids as chemotaxonomic markers for Erythroxylum australe. *Zeitschrift fur Naturforschung - Section C Journal of Biosciences, 59* (11–12), 769–776. https://doi.org/10.1515/znc-2004-11-1201

Kalil, A. C., Patterson, T. F., Mehta, A. K., Tomashek, K. M., Wolfe, C. R., Ghazaryan, V., Marconi, V. C., Ruiz-Palacios, G. M., Hsieh, L., Kline, S., Tapson, V., Iovine, N. M., Jain, M. K., Sweeney, D. A., El Sahly, H. M., Branche, A. R., Pineda, J. R., Lye, D. C., Sandkovsky, U., … Beigel, J. H. (2021). Baricitinib plus remdesivir for hospitalized adults with Covid-19. *New England Journal of Medicine, 384*(9), 795–807. https://doi.org/10.1056/NEJMoa2031994

Kalinowska, M., Bielawska, A., Lewandowska-Siwkiewicz, H., Priebe, W., & Lewandowski, W. (2014). Apples: Content of phenolic compounds vs. variety, part of apple and cultivation model, extraction of phenolic compounds, biological properties. *Plant Physiology and Biochemistry, 84*, 169–188. https://doi.org/10.1016/j.plaphy.2014.09.006

Kalita, B., & Das, M. K. (2018). Rutin–phospholipid complex in polymer matrix for long-term delivery of rutin via skin for the treatment of inflammatory diseases. *Artificial Cells, Nanomedicine and Biotechnology, 46*(1), 41–56. https://doi.org/10.1080/21691401.2017.1411931

Kalpana, K. B., Srinivasan, M., & Menon, V. P. (2009). Evaluation of antioxidant activity of hesperidin and its protective effect on H2O2 induced oxidative damage on pBR322 DNA and RBC cellular membrane. *Molecular and Cellular Biochemistry, 323*(1–2), 21–29. https://doi.org/10.1007/s11010-008-9960-9

Kamaraj, S., Anandakumar, P., Jagan, S., Ramakrishnan, G., & Devaki, T. (2011). Hesperidin attenuates mitochondrial dysfunction during benzo(a)pyrene-induced lung carcinogenesis in mice. *Fundamental and Clinical Pharmacology, 25*(1), 91–98. https://doi.org/10.1111/j.1472-8206.2010.00812.x

Kamel, R., & Mostafa, D. M. (2015). Rutin nanostructured lipid cosmeceutical preparation with sun protective potential. *Journal of Photochemistry and Photobiology B: Biology, 153*, 59–66. https://doi.org/10.1016/j.jphotobiol.2015.09.002

Kampa, M., Alexaki, V. I., Notas, G., Nifli, A. P., Nistikaki, A., Hatzoglou, A., Bakogeorgou, E., Kouimtzoglou, E., Blekas, G., Boskou, D., Gravanis, A., & Castanas, E. (2004). Antiproliferative and apoptotic effects of selective phenolic acids on T47D human breast cancer cells: Potential mechanisms of action. *Breast Cancer Research, 6*(2).

Kannanoor, M., Lakshmi, B. A., & Kim, S. (2021). Synthesis of silver nanoparticles conjugated with kaempferol and hydrocortisone and an evaluation of their antibacterial effects. *3 Biotech, 11*(7). https://doi.org/10.1007/s13205-021-02880-y

Karakas, F. P., Cingoz, G. S., & Turker, A. U. (2016). The effects of oxidative stress on phenolic composition and antioxidant metabolism in callus culture of common daisy. *African Journal of Traditional, Complementary and Alternative Medicines, 13*(4), 34–41. https://doi.org/10.21010/ajtcam.v13i4.6

Karuppagounder, V., Arumugam, S., Thandavarayan, R. A., Pitchaimani, V., Sreedhar, R., Afrin, R., Harima, M., Suzuki, H., Nomoto, M., Miyashita, S., Suzuki, K., Nakamura, M., & Watanabe, K. (2015). Modulation of HMGB 1 translocation and RAGE/NF κ B cascade by quercetin treatment mitigates atopic dermatitis in NC/Nga transgenic mice. *Experimental Dermatology, 24*(6), 418–423. https://doi.org/10.1111/exd.12685

Karuppusamy, S. (2009). A review on trends in production of secondary metabolites from higher plants by in vitro tissue, organ and cell cultures. *Journal of Medicinal Plants Research, 3*(13), 1222–1239.

Kashif, M., & Akhtar, N. (2019). Dermocosmetic emulgels for anti-aging effects: Evidence from chromatographic and non-invasive biophysical techniques. *Pakistan journal of pharmaceutical sciences, 32*(2), 845–852.

Kassa, T., Whalin, J. G., Richards, M. P., & Alayash, A. I. (2021). Caffeic acid: An antioxidant with novel antisickling properties. *FEBS Open Bio, 11*(12), 3293–3303. https://doi.org/10.1002/2211-5463.13295

Kepa, M., Miklasińska-Majdanik, M., Wojtyczka, R. D., Idzik, D., Korzeniowski, K., Smoleń-Dzirba, J., & Wasik, T. J. (2018). Antimicrobial potential of caffeic acid against staphylococcus aureus clinical strains. *BioMed Research International, 2018*. https://doi.org/10.1155/2018/7413504

Khalaf, A. A., Hassanen, Ibrahim, M. A., Tohamy, A. F., Aboseada, Hassan, H. M., & Zaki, A. R. (2020). Rosmarinic acid attenuates chromium-induced hepatic and renal oxidative damage and DNA damage in rats. *Journal of Biochemical and Molecular Toxicology, 34*(11), 1–12.

Khalil Alyahya, H., Subash-Babu, P., Mohammad Salamatullah, A., Hayat, K., Albader, N., Alkaltham, M. S., Ahmed, M. A., Arzoo, S., & Bourhia, M. (2021). Quantification of chlorogenic acid and vanillin from coffee peel extract and its effect on α-amylase activity, immunoregulation, mitochondrial oxidative stress, and tumor suppressor gene expression levels in H2O2-induced human mesenchymal stem cells. *Frontiers in Pharmacology, 12.* https://doi.org/10.3389/fphar.2021.760242

Khan, A., Iqtadar, S., Mumtaz, S. U., Heinrich, M., Pascual-Figal, D. A., & Livingstone, A. (2022). Oral co-supplementation of curcumin, quercetin, and vitamin D3 as an adjuvant therapy for mild to moderate symptoms of COVID-19—Results from a pilot open-label. *Frontiers in Pharmacology, 13*, 1–12.

Khateeb, Al, Aludatt, W., Zghoul, Al, Kanaan, H., El-Oqlah, R., & Lahham, A. (2017). Enhancement of phenolic compounds production in in vitro grown Rumex cyprius Murb. *Acta Physiologiae Plantarum, 39*(1), 14.

Kikowska, M., Chanaj-Kaczmarek, J., Derda, M., Budzianowska, A., Thiem, B., Ekiert, H., & Szopa. (2022). The evaluation of phenolic acids and flavonoids content and antiprotozoal activity of eryngium species biomass produced by biotechnological methods. *Molecules, 27*, 363.

Kikowska, M., Thiem, B., Szopa, A., & Ekiert, H. (2020). Accumulation of valuable secondary metabolites: Phenolic acids and flavonoids in different in vitro systems of shoot cultures of the endangered plant species—Eryngium alpinum L. *Plant Cell, Tissue and Organ Culture, 141*(2), 381–391. https://doi.org/10.1007/s11240-020-01795-5

Kim, J., Kim, J., Lee, Y. I., Almurayshid, A., Jung, J. Y., & Lee, J. H. (2020). Effect of a topical antioxidant serum containing vitamin C, vitamin E, and ferulic acid after Q-switched 1064-nm Nd:YAG laser for treatment of environment-induced skin pigmentation. *Journal of Cosmetic Dermatology, 19*(10), 2576–2582. https://doi.org/10.1111/jocd.13323

Kim, J. H., Kim, B. B., Lee, K. W., & Kang, N. J. (2013). Protective effect of rutin against ultraviolet b-induced cyclooxygenase-2 expression in mouse epidermal cells. *Food Science and Biotechnology, 22*(5), 1–6. https://doi.org/10.1007/s10068-013-0233-3

Kinaci, M. K., Erkasap, N., Kucuk, A., Koken, T., & Tosun, M. (2012). Effects of quercetin on apoptosis, NF-κB and NOS gene expression in renal ischemia/reperfusion injury. *Experimental and Therapeutic Medicine, 3*(2), 249–254. https://doi.org/10.3892/etm.2011.382

Kiokias, S., & Oreopoulou, V. (2021). A review of the health protective effects of phenolic acids against a range of severe pathologic conditions (including coronavirus-based infections). *Molecules, 26*(17), 5405.

Kirakosyan, A, Cseke, L. J., & Kaufman, P. B. (2009). *The use of plant cell biotechnology for the production of phytochemicals* (pp. 15–33).

Klimek-Szczykutowicz, M., Dziurka, M., Blažević, I., Đulović, A., Miazga-Karska, M., Klimek, K., Ekiert, H., & Szopa, A. (2021). Precursor-boosted production of metabolites in nasturtium officinale microshoots grown in plantform bioreactors, and antioxidant and antimicrobial activities of biomass extracts. *Molecules, 26*(15), 4660. https://doi.org/10.3390/molecules26154660

Kobori, M., Takahashi, Y., Akimoto, Y., Sakurai, M., Matsunaga, I., Nishimuro, H., Ippoushi, K., Oike, H., & Ohnishi-Kameyama, M. (2015). Chronic high intake of quercetin reduces oxidative stress and induces expression of the antioxidant enzymes in the liver and visceral adipose tissues in mice. *Journal of Functional Foods, 15*, 551–560. https://doi.org/10.1016/j.jff.2015.04.006

Kokotkiewicz, A., Wnuk, M., Bucinski, A., & Luczkiewicz, M. (2009). In vitro cultures of Cyclopia plants (Honeybush) as a source of bioactive xanthones and flavanones. *Zeitschrift fur Naturforschung - Section C Journal of Biosciences, 64*(7–8), 533–540. https://doi.org/10.1515/znc-2009-7-812

Kosović, E., Topiař, M., Cuřínová, P., & Sajfrtov. (2020). Stability testing of resveratrol and viniferin obtained from Vitis vinifera L. by various extraction methods considering the industrial viewpoint. *Scientific Reports, 10*(1), 1–9.

Kowalczyk, T., Sikora, J., Merecz-Sadowska, A., Kukula-Koch, W., Synowiec, E., Majda, A., Juda, D., Śliwiński, T., & Sitarek, P. (2024). Biological properties of extracts obtained from in vitro culture of plectranthus scutellarioides in a cell model. *International Journal of Molecular Sciences, 25*(2), 1043.

Kozłowska, W., Piątczak, E., Kolniak-Ostek, J., Kochan, E., Pencakowski, B., Stafiniak, M., Bielecka, M., Płachno, B. J., Strzemski, M., Matkowski, A., & Zielińska, S. (2024). Upscaling biomass production of rosmarinic acid-rich hairy root cultures of Agastache rugosa (Fisch. & C.A.Mey.) Kuntze. *Plant Cell, Tissue and Organ Culture, 156*(2). https://doi.org/10.1007/s11240-023-02626-z

Krasteva, G., Georgiev, V., & Pavlov, A. (2021). Recent applications of plant cell culture technology in cosmetics and foods. *Engineering in Life Sciences, 21*(3–4), 68–76. https://doi.org/10.1002/elsc.202000078

Kubica, P., Kokotkiewicz, A., Malinowska, M. A., Synowiec, Gniewosz, M., Hussain, S., Yaqoob, M., Bonn, G. K., Jakschitz, T., Mahmoud, El-Abedin, Z., Elansary, T. K., Luczkiewicz, H. O., Ekiert, M., & Szopa, H. (2022). Phenylpropanoid glycoside and phenolic acid profiles and biological activities of biomass extracts from different types of verbena officinalis microshoot cultures and soil-grown plant. *Antioxidants, 11*(2), 409.

Kubica, P., Szopa, A., & Ekiert, H. (2017). Production of verbascoside and phenolic acids in biomass of Verbena officinalis L. (vervain) cultured under different in vitro conditions. *Natural Product Research, 31*(14), 1663–1668. https://doi.org/10.1080/14786419.2017.1286477

Kubica, P., Szopa, A., Kokotkiewicz, A., Miceli, N., Gniewosz, M., Elansary, H. O., Mahmoud, El-Ansary, D. O., Nasif, O., Luczkiewicz, M., & Ekiert, H. (2020). Production of verbascoside, isoverbascoside and phenolic acids in callus, suspension, and bioreactor cultures of verbena offcinalis and biological properties of biomass extracts. *Molecules, 25* (5609), 1–27.

Kumar, N., & Goel, N. (2019). Phenolic acids: Natural versatile molecules with promising therapeutic applications. *Biotechnology Reports, 24*, e00370. https://doi.org/10.1016/j.btre.2019.e00370

Kumar, P., Sharma, S., Khanna, M., & Raj, H. G. (2003). Effect of quercetin on lipid peroxidation and changes in lung morphology in experimental influenza virus infection. *International Journal of Experimental Pathology, 84*(3), 127–134. https://doi.org/10.1046/j.1365-2613.2003.00344.x

Kwiecień, I., Miceli, N., Kędzia, E., Cavò, E., Taviano, M. F., Beerhues, L., & Ekiert, H. (2023). Different types of Hypericum perforatum cvs. (Elixir, Helos, Topas) in vitro cultures: A rich source of bioactive metabolites and biological activities of biomass extracts. *Molecules, 28*(5), 2376.

Kyriakoudi, A., Spanidi, E., Mourtzinos, I., & Gardikis, K. (2021). Innovative delivery systems loaded with plant bioactive ingredients: Formulation approaches and applications. *Plants, 10*(6), 1238. https://doi.org/10.3390/plants10061238

Lampropoulos, P., Lambropoulou, M., Papalois, A., Basios, N., Manousi, M., Simopoulos, C., & Tsaroucha, A. K. (2013). The role of apigenin in an experimental model of acute pancreatitis. *Journal of Surgical Research, 183*(1), 129–137. https://doi.org/10.1016/j.jss.2012.11.053

Lapenna, D., Ciofani, G., Pierdomenico, S. D., Neri, M., Cuccurullo, C., Giamberardino, M. A., & Cuccurullo, F. (2009). Inhibitory activity of salicylic acid on lipoxygenase-dependent lipid peroxidation. *Biochimica et Biophysica Acta, General Subjects, 1790*(1), 25–30. https://doi.org/10.1016/j.bbagen.2008.09.007

Lavigne, E. G., Cavagnino, A., Steinschneider, R., Breton, L., Baraibar, M. A., & Jäger, S. (2022). Oxidative damage prevention in human skin and sensory neurons by a salicylic acid derivative. *Free Radical Biology and Medicine, 181*, 98–104. https://doi.org/10.1016/j.freeradbiomed.2022.01.029

Lee, H. J., Im, A. R., Kim, S. M., Kang, H. S., Lee, J. D., & Chae, S. (2018). The flavonoid hesperidin exerts anti-photoaging effect by downregulating matrix metalloproteinase (MMP)-9 expression via mitogen activated protein kinase (MAPK)-dependent signaling pathways. *BMC Complementary and Alternative Medicine, 18*(1). https://doi.org/10.1186/s12906-017-2058-8

Lee, H. S., & Jeong, G. S. (2021). Therapeutic effect of kaempferol on atopic dermatitis by attenuation of T cell activity via interaction with multidrug resistance-associated protein 1. *Journal of Pharmacology, 178*(8), 1772–1788. https://doi.org/10.1111/bph.15396

Lee, J., Jung, E., Koh, J., Kim, Y. S., & Park, D. (2008). Effect of rosmarinic acid on atopic dermatitis. *Journal of Dermatology, 35*(12), 768–771. https://doi.org/10.1111/j.1346-8138.2008.00565.x

Lee, K. M., Lee, K. W., Jung, S. K., Lee, E. J., Heo, Y. S., Bode, A. M., Lubet, R. A., Lee, H. J., & Dong, Z. (2010). Kaempferol inhibits UVB-induced COX-2 expression by suppressing Src kinase activity. *Biochemical Pharmacology, 80*(12), 2042–2049. https://doi.org/10.1016/j.bcp.2010.06.042

Lee, S., Lee, H. H., Shin, Y. S., Kang, H., & Cho, H. (2017). The anti-HSV-1 effect of quercetin is dependent on the suppression of TLR-3 in Raw 264.7 cells. *Archives of Pharmacal Research, 40*(5), 623–630. https://doi.org/10.1007/s12272-017-0898-x

Leyva-López, N., Gutierrez-Grijalva, E. P., Ambriz-Perez, D. L., & Basilio Heredia, J. (2016). Flavonoids as cytokine modulators: A possible therapy for inflammation-related diseases. *International Journal of Molecular Sciences, 17*(6). https://doi.org/10.3390/ijms17060921

Li, R., Cai, L., Ren, D. Y., Xie, X. F., Hu, C. M., & Li, J. (2012). Therapeutic effect of 7, 3′-dimethoxy hesperetin on adjuvant arthritis in rats through inhibiting JAK2-STAT3 signal pathway. *International Immunopharmacology, 14*(2), 157–163. https://doi.org/10.1016/j.intimp.2012.07.001

Li, Y., Cui, H., Li, S., Li, X., Guo, H., Nandakumar, S., & Li, Z. (2023). International immunopharmacology kaempferol modulates IFN-γ induced JAK-STAT signaling pathway and ameliorates imiquimod-induced psoriasis-like skin lesions. *International Immunopharmacology, 114*, 109585.

Li, S., Li, J., Pan, R., Cheng, J., Cui, Q., Chen, J., & Yuan, Z. (2022). Sodium rutin extends lifespan and health span in mice including positive impacts on liver health. *British Journal of Pharmacology, 179*(9), 1825–1838. https://doi.org/10.1111/bph.15410

Li, M., Lin, X.f., Lu, J., Zhou, B.r., & Luo, D. (2016). Hesperidin ameliorates UV radiation-induced skin damage by abrogation of oxidative stress and inflammatory in HaCaT cells.

Journal of Photochemistry and Photobiology B: Biology, 165, 240–245. https://doi.org/10.1016/j.jphotobiol.2016.10.037

Li, D., Mitsuhashi, S., & Ubukata, M. (2012). Protective effects of hesperidin derivatives and their stereoisomers against advanced glycation end-products formation. *Pharmaceutical Biology, 50*(12), 1531–1535. https://doi.org/10.3109/13880209.2012.694106

Li, J., Ni, W., Aisha, M., Zhang, J., & Sun, M. (2021). A rutin nanocrystal gel as an effective dermal delivery system for enhanced anti-photoaging application. *Drug Development and Industrial Pharmacy, 47*(3), 429–439. https://doi.org/10.1080/03639045.2021.1890113

Li, X., Xie, X., Zhang, L., Meng, Y., Li, N., Wang, M., Zhai, C., Liu, Z., Di, T., Zhang, L., & Li, P. (2019). Hesperidin inhibits keratinocyte proliferation and imiquimod-induced psoriasis-like dermatitis via the IRS-1/ERK1/2 pathway. *Life Sciences, 219*, 311–321. https://doi.org/10.1016/j.lfs.2019.01.019

Li, H., Yang, L., Zhang, Y., & Gao, Z. (2016). Kaempferol inhibits fibroblast collagen synthesis, proliferation and activation in hypertrophic scar via targeting TGF-β receptor type I. *Biomedicine & Pharmacotherapy, 83*, 967–974. https://doi.org/10.1016/j.biopha.2016.08.011

Liga, S., Paul, C., & Péter, F. (2023). Flavonoids: Overview of biosynthesis, biological activity, and current extraction techniques. *Plants, 12*, 2732.

Lim, S. H., Jung, S. K., Byun, S., Lee, E. J., Hwang, J. A., Seo, S. G., Kim, Y. A., Yu, J. G., Lee, K. W., & Lee, H. J. (2013). Luteolin suppresses UVB-induced photoageing by targeting JNK1 and p90RSK2. *Journal of Cellular and Molecular Medicine, 17*(5), 672–680. https://doi.org/10.1111/jcmm.12050

Liu, C., Guo, X., Chen, Y., Zhao, M., Shi, S., Luo, Z., Song, J., Zhang, Z., Yang, W., & Liu, K. (2023). Anti-photoaging effect and mechanism of flexible liposomes co-loaded with apigenin and doxycycline. *Biomedicine & Pharmacotherapy, 164*, 114998.

Liu-Smith, F., & Meyskens, F. L. (2016). Molecular mechanisms of flavonoids in melanin synthesis and the potential for the prevention and treatment of melanoma. *Molecular Nutrition & Food Research, 60*(6), 1264–1274. https://doi.org/10.1002/mnfr.201500822

Lo, S., Leung, E., Fedrizzi, B., & Barker, D. (2021). Syntheses of mono-acylated luteolin derivatives, evaluation of their antiproliferative and radical scavenging activities and implications on their oral bioavailability. *Scientific Reports, 11*(1). https://doi.org/10.1038/s41598-021-92135-w

Lobo, V., Patil, A., Phatak, A., & Chandra, N. (2010). Free radicals, antioxidants and functional foods: Impact on human health. *Pharmacognosy Reviews, 4*(8), 118–126. https://doi.org/10.4103/0973-7847.70902

Lopez-Jornet, P., Camacho-Alonso, F., Gómez-Garcia, F., Molina Miñano, F., Cañas, X., Serafín, A., Castillo, J., & Vicente-Ortega, V. (2014). Effects of potassium apigenin and verbena extract on the wound healing process of SKH-1 mouse skin. *International Wound Journal, 11*(5), 489–495. https://doi.org/10.1111/j.1742-481X.2012.01114.xSpain

Lorigooini, Z., Jamshidi-Kia, F., & Hosseini, Z. (2020). Analysis of aromatic acids (phenolic acids and hydroxycinnamic acids). In *Recent advances in natural products analysis* (pp. 199–219). Iran: Elsevier. https://doi.org/10.1016/B978-0-12-816455-6.00004-4

Lu, J., Cong, T., Wen, X., Li, X., Du, D., He, G., & Jiang, X. (2019). Salicylic acid treats acne vulgaris by suppressing AMPK/SREBP 1 pathway in sebocytes. *Experimental Dermatology, 28*(7), 786–794. https://doi.org/10.1111/exd.13934

Lyu, S. Y., Rhim, J. Y., & Park, W. B. (2005). Antiherpetic activities of flavonoids against herpes simplex virus type 1 (HSV-1) and type 2 (HSV-2) in vitro. *Archives of Pharmacal Research, 28*(11), 1293–1301. https://doi.org/10.1007/BF02978215

Majeed, M., Majeed, S., Jain, R., Mundkur, L., Rajalakshmi, H. R., Lad, P. S., & Neupane, P. (2020). An open-label single-arm, monocentric study assessing the efficacy and safety of natural pterostilbene (Pterocarpus marsupium) for skin brightening and antiaging effects. *Clinical, Cosmetic and Investigational Dermatology, 13*, 105–116. https://doi.org/10.2147/CCID.S238358

Makowski, W., Królicka, A., Tokarz, B., Szopa, A., Ekiert, H., & Tokarz, K. M. (2023). Temporary immersion bioreactors as a useful tool for obtaining high productivity of phenolic compounds with strong antioxidant properties from Pontechium maculatum. *Plant Cell, Tissue and Organ Culture, 153*(3), 525–537. https://doi.org/10.1007/s11240-023-02487-6

Malacarne, G., Vrhovsek, U., Zulini, L., Cestaro, A., Stefanini, M., Mattivi, F., Delledonne, M., Velasco, R., & Moser, C. (2011). Resistance to Plasmopara viticola in a grapevine segregating population is associated with stilbenoid accumulation and with specific host transcriptional responses. *BMC Plant Biology, 11*. https://doi.org/10.1186/1471-2229-11-114Italy

Malinowska, M. A., Billet, K., Drouet, S., Munsch, T., Unlubayir, M., Tungmunnithum, D., Giglioli-Guivarc'h, N., Hano, C., & Lanoue, A. (2020). Grape cane extracts as multifunctional rejuvenating cosmetic ingredient: Evaluation of sirtuin activity, tyrosinase inhibition and bioavailability potential. *Molecules, 25*(9), 2203.

Mamani-Matsuda, M., Kauss, T., Al-Kharrat, A., Rambert, J., Fawaz, F., Thiolat, D., Moynet, D., Coves, S., Malvy, D., & Mossalayi, M. D. (2006). Therapeutic and preventive properties of quercetin in experimental arthritis correlate with decreased macrophage inflammatory mediators. *Biochemical Pharmacology, 72*(10), 1304–1310. https://doi.org/10.1016/j.bcp.2006.08.001

Manca, M. L., Castangia, I., Caddeo, C., Pando, D., Escribano, E., Valenti, D., Lampis, S., Zaru, M., Fadda, A. M., & Manconi, M. (2014). Improvement of quercetin protective effect against oxidative stress skin damages by incorporation in nanovesicles. *Colloids and Surfaces B: Biointerfaces, 123*, 566–574. https://doi.org/10.1016/j.colsurfb.2014.09.059

Mandrone, M., Coqueiro, A., Poli, F., Antognoni, F., & Choi, Y. H. (2018). Identification of a collagenase-inhibiting flavonoid from alchemilla vulgaris using NMR-based metabolomics. *Planta Medica, 84*(12–13), 941–946. https://doi.org/10.1055/a-0630-2079

Mannino, F., Imbesi, C., Bitto, A., Minutoli, L., Squadrito, F., D'Angelo, T., Booz, C., Pallio, G., & Irrera, N. (2023). Anti-oxidant and anti-inflammatory effects of ellagic and punicic acid in an in vitro model of cardiac fibrosis. *Biomedicine & Pharmacotherapy, 162*, 114666. https://doi.org/10.1016/j.biopha.2023.114666

Manosroi, A., Chutoprapat, R., Abe, M., Manosroi, W., & Manosroi, J. (2012). Anti-aging efficacy of topical formulations containing niosomes entrapped with rice bran bioactive compounds. *Pharmaceutical Biology, 50*(2), 208–224. https://doi.org/10.3109/13880209.2011.596206

Martin, B. J., Campbell, P. M., Rees, T. D., & Buschang, P. H. (2016). A randomized controlled trial evaluating antioxidant-essential oil gel as a treatment for gingivitis in orthodontic patients. *The Angle Orthodontist, 86*(3), 407–412. https://doi.org/10.2319/041515-251.1

Martinez, R. M., Pinho-Ribeiro, F. A., Steffen, V. S., Caviglione, C. V., Pala, D., Baracat, M. M., Georgetti, S. R., Verri, W. A., & Casagrande, R. (2016). Topical formulation containing hesperidin methyl chalcone inhibits skin oxidative stress and inflammation induced by ultraviolet B irradiation. *Photochemical and Photobiological Sciences, 15*(4), 554–563. https://doi.org/10.1039/c5pp00467e

Mehrzadi, S., Mehrabani, M., Malayeri, A. R., Bakhshayesh, M., Kalantari, H., & Goudarzi, M. (2019). Ellagic acid as a potential antioxidant, alleviates methotrexate-induced hepatotoxicity in male rats. *Acta Chirurgica Belgica, 119*(2), 69–77. https://doi.org/10.1080/00015458.2018.1455419

Memar, M. Y., Yekani, Sharifi, S., & Dizaj, S. M. (2023). Antibacterial and biofilm inhibitory effects of rutin nanocrystals. *Biointerface Research in Applied Chemistry, 13*(2), 132.

Messaadia, L., Bekkar, Y., Benamira, M., & Lahmar, H. (2020). Predicting the antioxidant activity of some flavonoids of Arbutus plant: A theoretical approach. *Chemical Physics Impact, 1*, 100007.

Mhimed, O. T., Al-Alsadoon, L. H., & Taqa, G. A. (2023). Evaluating the effect of topical flavonoid luteolin on skin wound Healing in rabbits. *Egyptian Journal of Veterinary Science, 54*(2), 199–209. https://doi.org/10.21608/EJVS.2022.165585.1401

Miceli, N., Filocamo, A., Ragusa, S., Cacciola, F., Dugo, P., Mondello, L., Celano, M., Maggisano, V., & Taviano, M. F. (2017). Chemical characterization and biological activities of phenolic-rich fraction from cauline leaves of isatis tinctoria L. (Brassicaceae) growing in Sicily, Italy. *Chemistry and Biodiversity, 14*(8). https://doi.org/10.1002/cbdv.201700073

Michalak, M. (2022). Plant-derived antioxidants: Significance in skin health and the ageing process. *International Journal of Molecular Sciences, 23*(2), 8–12.

Milentyeva, I. S., Le, V. M., Kozlova, O. V., Velichkovich, N. S., Fedorova, A. M., Loseva, A. I., & Yustratov, V. P. (2021). Secondary metabolites in in vitro cultures of Siberian medicinal plants: Content, antioxidant properties, and antimicrobial characteristics. *Foods and Raw Materials, 9*(1), 153–163.

Moraes, C. A. P., Arêas, E. P. G., & Velasco, M. V. R. (2017). Assessment of functional stability of photoprotective formulations containing rutin succinate. *Cosmetics, 4*(3), 1–14.

Morel, I., Lescoat, G., Cogrel, P., Sergent, O., Pasdeloup, N., Brissot, P., Cillard, P., & Cillard, J. (1993). Antioxidant and iron-chelating activities of the flavonoids catechin, quercetin and diosmetin on iron-loaded rat hepatocyte cultures. *Biochemical Pharmacology, 45*(1), 13–19. https://doi.org/10.1016/0006-2952(93)90371-3

Mu, J., Ma, H., Chen, H., Zhang, X., & Ye, M. (2021). Luteolin prevents UVB-induced skin photoaging damage by modulating SIRT3/ROS/MAPK signaling: An in vitro and in vivo studies. *Frontiers in Pharmacology, 12*. https://doi.org/10.3389/fphar.2021.728261

Murray, J. C., Burch, J. A., Streilein, R. D., Iannacchione, M. A., Hall, R. P., & Pinnell, S. R. (2008). A topical antioxidant solution containing vitamins C and E stabilized by ferulic acid provides protection for human skin against damage caused by ultraviolet irradiation. *Journal of the American Academy of Dermatology, 59*(3), 418–425. https://doi.org/10.1016/j.jaad.2008.05.004

Nagata, H., Takekoshi, S., Takeyama, R., Homma, T., & Osamura, R. Y. (2004). Quercetin enhances melanogenesis by increasing the activity and synthesis of tyrosinase in human melanoma cells and in normal human melanocytes. *Pigment Cell Research, 17*(1), 66–73. https://doi.org/10.1046/j.1600-0749.2003.00113.x

Novotná, R., Škařupová, D., Hanyk, J., Ulrichová, Křen, V., Bojarová, P., Brodsky, K., Vostálová, J., & Franková. (2023). Hesperidin, hesperetin, rutinose, and rhamnose act as skin anti-aging agents. *Molecules, 28*(4), 1728.

Oliveira, C., Coelho, C., Teixeira, J. A., Ferreira-Santos, P., & Botelho, C. M. (2022). Nanocarriers as active ingredients enhancers in the cosmetic industry—The European and North America regulation challenges. *Molecules, 27*(5), 1669.

Oliveira, C. A. D., Peres, D. D. A., Graziola, F., Chacra, N. A. B., Araújo, G. L. B. D., Flórido, A. C., Mota, J., Rosado, C., Velasco, M. V. R., Rodrigues, L. M., Fernandes, A. S., & Baby, A. R. (2016). Cutaneous biocompatible rutin-loaded gelatin-based nanoparticles increase the SPF of the association of UVA and UVB filters. *European Journal of Pharmaceutical Sciences, 81*, 1–9. https://doi.org/10.1016/j.ejps.2015.09.016

Oresajo, C., Stephens, T., Hino, P. D., Law, R. M., Yatskayer, M., Foltis, P., Pillai, S., & Pinnell, S. R. (2008). Protective effects of a topical antioxidant mixture containing vitamin C. ferulic acid, and phloretin against ultraviolet-induced photodamage in human skin. *Journal of Cosmetic Dermatology, 7*(4), 290–297. https://doi.org/10.1111/j.1473-2165.2008.00408.x

Ozay, Y., Guzel, S., Erdogdu, I. H., Yildirim, Z., Pehlivanoglu, B., Turk, B. A., & Darcan, S. (2018). Evaluation of the wound healing properties of luteolin ointments on excision and incision wound models in diabetic and non-diabetic rats. *Records of Natural Products, 12* (4), 350–366. https://doi.org/10.25135/rnp.38.17.08.135

Özay, Y., Güzel, S., Yumrutas, Ö., Pehlivanoğlu, B., Erdoğdu, İ. H., Yildirim, Z., Türk, B. A., & Darcan, S. (2019). Wound healing effect of kaempferol in diabetic and nondiabetic rats. *Journal of Surgical Research, 233*, 284–296. https://doi.org/10.1016/j.jss.2018.08.009

Özçelik, B., Orhan, I., & Toker, G. (2006). Antiviral and antimicrobial assessment of some selected flavonoids. *Zeitschrift fur Naturforschung - Section C Journal of Biosciences, 61*(9–10), 632–638. https://doi.org/10.1515/znc-2006-9-1003

Palombo, R., Savini, I., Avigliano, L., Madonna, S., Cavani, A., Albanesi, C., Mauriello, A., Melino, G., & Terrinoni, A. (2016). Luteolin-7-glucoside inhibits IL-22/STAT3 pathway, reducing proliferation, acanthosis, and inflammation in keratinocytes and in mouse psoriatic model. *Cell Death and Disease, 7*(8). https://doi.org/10.1038/cddis.2016.201

Panche, A. N., Diwan, A. D., & Chandra, S. R. (2016). Flavonoids: An overview. *Journal of Nutrition Sciences, 5*, e47. https://doi.org/10.1017/jns.2016.41

Pandey, P., Khan, F., Ramniwas, S., Saeed, M., & Ahmad, I. (2024). A mechanistic review of the pharmacological potential of narirutin: A dietary flavonoid. *Naunyn-Schmiedeberg's Archives of Pharmacology*. https://doi.org/10.1007/s00210-024-03022-w

Panzella, L., & Napolitano, A. (2019). Natural and bioinspired phenolic compounds as tyrosinase inhibitors for the treatment of skin hyperpigmentation: Recent advances. *Cosmetics, 6*(4), 57. https://doi.org/10.3390/cosmetics6040057

Parashar, D., Meena, A. K., & Sharma, M. (2023). Flavonoids - A review on natural antioxidants. *Research Journal of Pharmacy and Technology, 16*(10), 4952–4958. https://doi.org/10.52711/0974-360X.2023.00802

Paredes-Gonzalez, X., Fuentes, F., Jeffery, S., Saw, C. L. L., Shu, L., Su, Z. Y., & Kong, A. N. T. (2015). Induction of NRF2-mediated gene expression by dietary phytochemical flavones apigenin and luteolin. *Biopharmaceutics and Drug Disposition, 36* (7), 440–451. https://doi.org/10.1002/bdd.1956

Park, H. J., Cho, J. H., Hong, S. H., Kim, D. H., Jung, H. Y., Kang, I. K., & Cho, Y. J. (2018). Whitening and anti-wrinkle activities of ferulic acid isolated from Tetragonia tetragonioides in B16F10 melanoma and CCD-986sk fibroblast cells. *Journal of Natural Medicines, 72*(1), 127–135. https://doi.org/10.1007/s11418-017-1120-7

Patil, S. L., Mallaiah, S. H., & Patil, R. K. (2013). Antioxidative and radioprotective potential of rutin and quercetin in Swiss albino mice exposed to gamma radiation. *Journal of Medical Physics, 38*(2), 87–92. https://doi.org/10.4103/0971-6203.111321

Pearson, W., Fletcher, R. S., & Kott, L. S. (2012). Oral rosmarinic acid-enhanced Mentha spicata modulates synovial fluid biomarkers of inflammation in horses challenged with intra-articular LPS. *Journal of Veterinary Pharmacology and Therapeutics, 35*(5), 495–502. https://doi.org/10.1111/j.1365-2885.2011.01343.x

Peres, D. A., de Oliveira, C. A., da Costa, M. S., Tokunaga, V. K., Mota, J. P., Rosado, C., Consiglieri, V. O., Kaneko, T. M., Velasco, M. V. R., & Baby, A. R. (2016). Rutin increases critical wavelength of systems containing a single UV filter and with good skin compatibility. *Skin Research and Technology, 22*(3), 325–333. https://doi.org/10.1111/srt.12265

Perez-Leal, O., Barrero, C. A., & Merali, S. (2017). Pharmacological stimulation of nuclear factor (erythroidderived 2)-like 2 translation activates antioxidant responses. *Journal of Biological Chemistry, 292*(34), 14108–14121. https://doi.org/10.1074/jbc.M116.770925

Piazzon, A., Vrhovsek, U., Masuero, D., Mattivi, F., Mandoj, F., & Nardini, M. (2012). Antioxidant activity of phenolic acids and their metabolites: Synthesis and antioxidant properties of the sulfate derivatives of ferulic and caffeic acids and of the acyl glucuronide of ferulic acid. *Journal of Agricultural and Food Chemistry, 60*(50), 12312–12323. https://doi.org/10.1021/jf304076z

Pinto, D., Lameirão, F., Delerue-Matos, C., Rodrigues, F., & Costa, P. (2021). Characterization and stability of a formulation containing antioxidants-enriched Castanea sativa shells extract. *Cosmetics, 8*(2), 1–18.

Pinzaru, I., Chioibas, R., Marcovici, I., Coricovac, D., Susan, R., Predut, D., Georgescu, D., & Dehelean, C. (2021). Rutin exerts cytotoxic and senescence-inducing properties in human melanoma cells. *Toxics, 9*(9), 226.

Pizzino, G., Irrera, N., Cucinotta, M., Pallio, G., Mannino, F., Arcoraci, V., Squadrito, F., Altavilla, D., & Bitto, A. (2017). Oxidative stress: Harms and benefits for human health. *Oxidative Medicine and Cellular Longevity, 2017*. https://doi.org/10.1155/2017/8416763

Potts, R. H. (1941). Fats and their derivatives. *Oil & Soap, 18*(10), 199–202. https://doi.org/10.1007/BF02544182

Pradeep, K., Ko, K. C., Choi, M. H., Kang, J. A., Chung, Y. J., & Park, S. H. (2012). Protective effect of hesperidin, a citrus flavanoglycone, against γ-radiation-induced tissue damage in Sprague-Dawley rats. *Journal of Medicinal Food, 15*(5), 419–427. https://doi.org/10.1089/jmf.2011.1737

Purushothaman, A., Babu, S. S., Naroth, S., & Janardanan, D. (2022). Antioxidant activity of caffeic acid: Thermodynamic and kinetic aspects on the oxidative degradation pathway. *Free Radical Research, 56*(9–10), 617–630. https://doi.org/10.1080/10715762.2022.2161379

Qahtan, A. A., Faisal, M., Alatar, A. A., & Abdel- Salam, E. M. (2022). Callus-mediated high-frequency plant regeneration, phytochemical profiling, antioxidant activity and genetic stability in Ruta chalepensis L. *Plants, 11*(12), 1614.

Radde, B. N., Alizadeh-Rad, N., Price, S. M., Schultz, D. J., & Klinge, C. M. (2016). Anacardic acid, salicylic acid, and oleic acid differentially alter cellular bioenergetic function in breast cancer cells. *Journal of Cellular Biochemistry, 117*, 2521–2532. https://doi.org/10.1002/jcb.25544

Rahimuddin, S. A., Khoja, S. M., Zuhair, M. M., Howell, N. K., & Brown, J. E. (2007). Inhibition of lipid peroxidation in UVA-treated skin fibroblasts by luteolin and its glucosides. *European Journal of Lipid Science and Technology, 109*(7). https://doi.org/10.1002/ejlt.200700012, 647-55.

Rho, H. S., Ghimeray, A. K., Yoo, D. S., Ahn, S. M., Kwon, S. S., Lee, K. H., Cho, D. H., & Cho, J. Y. (2011). Kaempferol and kaempferol rhamnosides with depigmenting and antiinflammatory properties. *Molecules, 16*(4), 3338–3344. https://doi.org/10.3390/molecules16043338

Rice-Evans, C. A., Miller, N. J., & Paganga, G. (1996). Structure-antioxidant activity relationships of flavonoids and phenolic acids. *Free Radical Biology and Medicine, 20*(7), 933–956. https://doi.org/10.1016/0891-5849(95)02227-9

Robbins, R. J. (2003). Phenolic acids in foods: An overview of analytical methodology. *Journal of Agricultural and Food Chemistry, 51*(10), 2866–2887. https://doi.org/10.1021/jf026182t

Rosa, L. S., & Silva, N. J. A. (2016). Anticancer properties of phenolic acids in colon cancer – A review. *Journal of Nutrition & Food Sciences, 6*(2). https://doi.org/10.4172/2155-9600.1000468

Rosiak, N., Cielecka-Piontek, J., Skibiński, R., Lewandowska, K., Bednarski, W., & Zalewski, P. (2022). Antioxidant potential of resveratrol as the result of radiation exposition. *Antioxidants, 11*(11), 2097.

Sadeghi, A., Bastin, A. R., Ghahremani, H., & Doustimotlagh, A. H. (2020). The effects of rosmarinic acid on oxidative stress parameters and inflammatory cytokines in lipopolysaccharide-induced peripheral blood mononuclear cells. *Molecular Biology Reports, 47*(5), 3557–3566. https://doi.org/10.1007/s11033-020-05447-x

Sagit, M., Polat, H., Gurgen, S. G., Berk, E., Guler, S., & Yasar, M. (2017). Effectiveness of quercetin in an experimental rat model of allergic rhinitis. *European Archives of Oto-Rhino-Laryngology, 274*(8), 3087–3095. https://doi.org/10.1007/s00405-017-4602-z

Sahu, B. D., Kuncha, M., Sindhura, G. J., & Sistla, R. (2013). Hesperidin attenuates cisplatin-induced acute renal injury by decreasing oxidative stress, inflammation and DNA damage. *Phytomedicine, 20*(5), 453–460. https://doi.org/10.1016/j.phymed.2012.12.001

Saija, A., Tomaino, A., Trombetta, D., De Pasquale, A., Uccella, N., Barbuzzi, T., Paolino, D., & Bonina, F. (2000). In vitro and in vivo evaluation of caffeic and ferulic acids as topical photoprotective agents. *International Journal of Pharmaceutics, 199*(1), 39–47. https://doi.org/10.1016/S0378-5173(00)00358-6

Saivish, M. V., Pacca, C. C., Costa, G. M., Silva, Nebo, L., Silva, M., Teixeira, H. T. M., Hernandes, N. F. B., Zini, V. M., Carvalho, N., Fontoura, M. A. R., Sacchetto, P., Marques, R. E., & Nogueira, M. L. (2023). Caffeic acid has antiviral activity against Ilhéus Virus in vitro. *Viruses, 15*(2), 494.

Šamec, D., Karalija, E., Šola, I., Vujčić Bok, V., & Salopek-Sondi, B. (2021). The role of polyphenols in abiotic stress response: The influence of molecular structure. *Plants, 10*(1), 1–24. https://doi.org/10.3390/plants10010118

Samota, M. K., Kaur, M., Sharma, M., Sarita, V. K., Thakur, J., Rawat, M., Phogat, B., & Guru, P. N. (2023). Hesperidin from citrus peel waste: Extraction and its health implications. *Quality Assurance and Safety of Crops & Foods, 15*(2), 71–99. https://doi.org/10.15586/qas.v15i2.1256

Sánchez-Maldonado, A. F., Schieber, A., & Gänzle, M. G. (2011). Structure-function relationships of the antibacterial activity of phenolic acids and their metabolism by lactic acid bacteria. *Journal of Applied Microbiology, 111*(5), 1176–1184. https://doi.org/10.1111/j.1365-2672.2011.05141.xCanada

Santhanam, R. K., Fakurazi, S., Ahmad, S., Abas, F., Ismail, I. S., Rukayadi, Y., Akhtar, M. T., & Shaari, K. (2018). Inhibition of UVB-induced pro-inflammatory cytokines and MMP expression by Zanthoxylum rhetsa bark extract and its active constituent hesperidin. *Phytotherapy Research, 32*(8), 1608–1616. https://doi.org/10.1002/ptr.6092

Saraf, S., & Kaur, C. (2010). Phytoconstituents as photoprotective novel cosmetic formulations. *Pharmacognosy Reviews, 4*(7), 1–11. https://doi.org/10.4103/0973-7847.65319

Sarropoulou, V., Paloukopoulou, C., Karioti, A., Maloupa, E., & Grigoriadou, K. (2023). Rosmarinic acid production from Origanum dictamnus L. root liquid cultures in vitro. *Plants, 12*, 299.

Sesderma Ferulac Line. Available online: https://www.sesderma.pl/linie/ferulac (Accessed 20 July 2026).

Seleem, D., Pardi, V., & Murata, R. M. (2017). Review of flavonoids: A diverse group of natural compounds with anti-candida albicans activity in vitro. *Archives of Oral Biology, 76*, 76–83. https://doi.org/10.1016/j.archoralbio.2016.08.030

Semwal, R., Joshi, S. K., Semwal, R. B., & Semwal, D. K. (2021). Health benefits and limitations of rutin - A natural flavonoid with high nutraceutical value. *Phytochemistry Letters, 46*, 119–128. https://doi.org/10.1016/j.phytol.2021.10.006

Sengul, E., Gelen, V., & Gedikli, S. (2021). Cardioprotective activities of quercetin and rutin in sprague dawley rats treated with 5-fluorouracil. *Journal of Animal and Plant Sciences, 31*(2), 423–431. https://doi.org/10.36899/JAPS.2021.2.0231

Shahbaz, M., Imran, M., Alsagaby, S. A., Naeem, H., Al Abdulmonem, W., Hussain, M., Abdelgawad, M. A., El-Ghorab, A. H., Ghoneim, M. M., El-Sherbiny, M., Atoki, A. V., & Awuchi, C. G. (2023). Anticancer, antioxidant, ameliorative and therapeutic properties of kaempferol. *International Journal of Food Properties, 26*(1), 1140–1166. https://doi.org/10.1080/10942912.2023.2205040

Shahidi, F., & Yeo, J. D. (2018). Bioactivities of phenolics by focusing on suppression of chronic diseases: A review. *International Journal of Molecular Sciences, 19*(6). https://doi.org/10.3390/ijms19061573

Shahrestani, V. N., Haddadi, M., & Kermani, A. R. S. (2021). Behavioral and molecular analysis of antioxidative potential of rosmarinic acid against methamphetamine-induced augmentation of casp3a mrna in the zebrafish brain. *Basic and Clinical Neuroscience, 12*(2), 243–253. https://doi.org/10.32598/BCN.12.2.1777.1

Sharifi-Rad, J., Quispe, C., Castillo, C. M. S., Caroca, R., Lazo-Vélez, M. A., Antonyak, H., Polishchuk, A., Lysiuk, R., Oliinyk, P., Masi, L De, Bontempo, P., Martorell, M., Dastan, S. D., Rigano, D., Wink, M., & Cho, W. C. (2024). E. acid: A review on its natural sources, chemical stability, and therapeutic potential. *Oxidative Medicine and Cellular Longevity, 2024*.

Shi, G., Kang, Z., Liu, H., Ren, F., & Zhou, Y. (2021). The effects of quercetin combined with nucleopolyhedrovirus on the growth and immune response in the silkworm (Bombyx mori). *Archives of Insect Biochemistry and Physiology, 108*(2), 1–18.

Shu, Y., Liu, Y., Li, L., Feng, J., Lou, B., Zhou, X., & Wu, H. (2011). Antibacterial activity of quercetin on oral infectious pathogens. *African Journal of Microbiology Research, 5*(30), 5358–5361.

Sikora, E., Miastkowska, M., & Lasoń. (2020). *Selected skin delivery systems*. Wydawnictwo PK.

Simmler, C., Pauli, G. F., & Chen, S. N. (2013). Phytochemistry and biological properties of glabridin. *Fitoterapia, 90*, 160–184. https://doi.org/10.1016/j.fitote.2013.07.003

Simos, Y. V., Verginadis, I. I., Toliopoulos, I. K., Velalopoulou, A. P., Karagounis, I. V., Karkabounas, S. C., & Evangelou, A. M. (2012). Effects of catechin and epicatechin on superoxide dismutase and glutathione peroxidase activity, in vivo. *Redox Report, 17*(5), 181–186. https://doi.org/10.1179/1351000212Y.0000000020Greece

Singh, V. K., Sahoo, D., Agrahari, K., Khan, A., Mukhopadhyay, D., Chanda, N. P., & Yadav. (2023). Anti-inflammatory, anti-proliferative and anti-psoriatic potential of apigenin in RAW 264.7 cells, HaCaT cells and psoriasis like dermatitis in BALB/c mice. *Life Sciences, 328*, 12909.

Skrzypczak-Pietraszek, E., Piska, K., & Pietraszek, J. (2018). Enhanced production of the pharmaceutically important polyphenolic compounds in Vitex agnus castus L. shoot cultures by precursor feeding strategy. *Engineering in Life Sciences, 18*(5), 287–297. https://doi.org/10.1002/elsc.201800003

Skrzypczak-Pietraszek, E., Urbańska, A., PawełŻmudzki, & Pietraszek, J. (2019). Elicitation with methyl jasmonate combined with cultivation in the plantform™ temporary immersion bioreactor highly increases the accumulation of selected centellosides and phenolics in Centella asiatica (L.) urban shoot culture. *Engineering in Life Sciences, 19*(12), 931–943. https://doi.org/10.1002/elsc.201900051

Spagnol, C. M., Assis, R. P., Brunetti, I. L., Isaac, V. L. B., Salgado, H. R. N., & Corrêa, M. A. (2019). In vitro methods to determine the antioxidant activity of caffeic acid. *Spectrochimica Acta Part A. Molecular and Biomolecular Spectroscopy, 219*, 358–366. https://doi.org/10.1016/j.saa.2019.04.025

Stamatopoulos, K., Katsoyannos, E., & Chatzilazarou, A. (2014). Antioxidant activity and thermal stability of oleuropein and related phenolic compounds of olive leaf extract after separation and concentration by salting-out-assisted cloud point extraction. *Antioxidants, 3*(2), 229–244. https://doi.org/10.3390/antiox3020229

Stanisic, D., Liu, L. H. B., Santos, D., Costa, R. V., & Durán, N. T. (2020). New sustainable process for hesperidin isolation and anti-ageing effects of hesperidin nanocrystals. *Molecules, 25*(19), 1–18.

Stasiłowicz-Krzemień, A., Rosiak, N., Miklaszewski, A., & Cielecka-Piontek, J. (2023). Screening of the anti-neurodegenerative activity of caffeic acid after introduction into inorganic metal delivery systems to increase its solubility as the result of a mechanosynthetic approach. *International Journal of Molecular Sciences, 24*(11), 9218.

Stipcevic, T., Piljac, J., & Berghe, D. V. (2006). Effect of different flavonoids on collagen synthesis in human fibroblasts. *Plant Foods for Human Nutrition, 61*(1), 29–34. https://doi.org/10.1007/s11130-006-0006-8

Suzuki, A., Nomura, T., Jokura, H., Kitamura, N., Fujii, A., & Hase, T. (2023). Beneficial effects of oral supplementation with ferulic acid, a plant phenolic compound, on the human skin barrier in healthy men. *International Journal for Vitamin and Nutrition Research, 93*(1), 54–60. https://doi.org/10.1024/0300-9831/a000699

Szopa, A., & Ekiert, H. (2012). In vitro cultures of Schisandra chinensis (Turcz.) Baill. (Chinese magnolia Vine) - A potential biotechnological rich source of therapeutically important phenolic acids. *Applied Biochemistry and Biotechnology, 166*(8), 1941–1948. https://doi.org/10.1007/s12010-012-9622-y

Szopa, A., Ekiert, H., Szewczyk, A., & Fugas, E. (2012). Production of bioactive phenolic acids and furanocoumarins in in vitro cultures of Ruta graveolens L. and Ruta graveolens ssp. divaricata (Tenore) Gams. under different light conditions. *Plant Cell, Tissue and Organ Culture, 110*(3), 329–336. https://doi.org/10.1007/s11240-012-0154-5

Szopa, A., Klimek-Szczykutowicz, M., Kokotkiewicz, A., Dziurka, M., Luczkiewicz, M., & Ekiert, H. (2019). Phenolic acid and flavonoid production in agar, agitated and bioreactor-grown microshoot cultures of Schisandra chinensis cv. Sadova No. 1 – A valuable medicinal plant. *Journal of Biotechnology, 305*, 61–70. https://doi.org/10.1016/j.jbiotec.2019.08.021

Szopa, A., Kokotkiewicz, A., Bednarz, M., Jafernik, K., Luczkiewicz, M., & Ekiert, H. (2019). Bioreactor type affects the accumulation of phenolic acids and flavonoids in microshoot cultures of Schisandra chinensis (Turcz.) Baill. *Plant Cell, Tissue and Organ Culture, 139*(1), 199–206. https://doi.org/10.1007/s11240-019-01676-6

Szopa, A., Kubica, P., & Ekiert, H. (2018). Agitated shoot cultures of Aronia arbutifolia and Aronia × prunifolia: Biotechnological studies on the accumulation of phenolic compounds and biotransformation capability. *Plant Cell, Tissue and Organ Culture, 134*(3), 467–479. https://doi.org/10.1007/s11240-018-1436-3

Szopa, A., Kubica, P., Komsta, Ł., Walkowicz-Bożek, A., & Ekiert, H. (2020). The effect of feeding culture media with biogenetic precursors on high production of depsides in agitated shoot cultures of black and red aronias. *Plant Cell, Tissue and Organ Culture, 142*(2), 379–399. https://doi.org/10.1007/s11240-020-01869-4

Szopa, A., Kubica, P., Snoch, A., & Ekiert, H. (2018). High production of bioactive depsides in shoot and callus cultures of aronia arbutifolia and aronia × prunifolia. *Acta Physiologiae Plantarum, 40*(3), 1–11. https://doi.org/10.1007/s11738-018-2623-x

Taha, L. S., Abdel-Magied, H. M., Murad, S. A., & Youssef, N. M. (2024). Improvement of morphological and biochemical traits of in vitro antigonon leptopus plant grown under drought conditions. *Egyptian Journal of Chemistry, 67*(4), 225–236. https://doi.org/10.21608/ejchem.2023.235217.8576

Takekoshi, S., Matsuzaki, K., & Kitatani, K. (2013). Quercetin stimulates melanogenesis in hair follicle melanocyte of the mouse. *Tokai Journal of Experimental and Clinical Medicine, 38*(4), 129–134.

Tarnawski, M., Depta, K., Grejciun, D., & Szelepin, B. (2006). HPLC determination of phenolic acids and antioxidant activity in concentrated peat extract—a natural immunomodulator. *Journal of Pharmaceutical and Biomedical Analysis, 41*(1), 182–188. https://doi.org/10.1016/j.jpba.2005.11.012

Tatsimo, S. J. N., Tamokou, J. D. D., Havyarimana, L., Csupor, D., Forgo, P., Hohmann, J., Kuiate, J. R., & Tane, P. (2012). Antimicrobial and antioxidant activity of kaempferol rhamnoside derivatives from Bryophyllum pinnatum. *BMC Research Notes, 5*. https://doi.org/10.1186/1756-0500-5-158

Thiruvengadam, M., Praveen, N., Kim, E. H., Kim, S. H., & Chung, I. M. (2014). Production of anthraquinones, phenolic compounds and biological activities from hairy root cultures of polygonum multiflorum thunb. *Protoplasma, 251*(3), 555–566. https://doi.org/10.1007/s00709-013-0554-3

Thring, T. S. A., Hili, P., & Naughton, D. P. (2009). Anti-collagenase, anti-elastase and antioxidant activities of extracts from 21 plants. *BMC Complementary and Alternative Medicine, 9*. https://doi.org/10.1186/1472-6882-9-27

Tian, C., Guo, Y., Chang, Y., Zhao, J., Cui, C., & Liu, M. (2019). Dose-effect relationship on anti-inflammatory activity on LPS induced RAW 264.7 cells and antioxidant activity of rutin in vitro. *Acta Poloniae Pharmaceutica - Drug Research, 76*(3), 511–522. https://doi.org/10.32383/appdr/102677

Toiu, A., Mocan, A., Vlase, L., Pârvu, A. E., Vodnar, D. C., Gheldiu, A. M., Moldovan, C., & Oniga, I. (2019). Comparative phytochemical profile, antioxidant, antimicrobial and in vivo anti-inflammatory activity of different extracts of traditionally used Romanian Ajuga genevensis L. and A. Reptans L. (Lamiaceae). *Molecules, 24*(8). https://doi.org/10.3390/molecules24081597

Toiu, A., Vlase, L., Gheldiu, A. M., Vodnar, D., & Oniga, I. (2017). Evaluation of the antioxidant and antibacterial potential of bioactive compounds from Ajuga Reptans extracts. *Farmacia, 65*(3), 351–355.

Tolun, A., Altintas, Z., & Artik, N. (2016). Microencapsulation of grape polyphenols using maltodextrin and gum arabic as two alternative coating materials: Development and characterization. *Journal of Biotechnology, 239*, 23–33. https://doi.org/10.1016/j.jbiotec.2016.10.001

Tomazelli, L. C., de Assis Ramos, M. M., Sauce, R., Cândido, T. M., Sarruf, F. D., de Oliveira Pinto, C. A. S., de Oliveira, C. A., Rosado, C., Velasco, M. V. R., & Baby, A. R. (2018). SPF enhancement provided by rutin in a multifunctional sunscreen. *International Journal of Pharmaceutics, 552*(1–2), 401–406. https://doi.org/10.1016/j.ijpharm.2018.10.015

Tsai, C. F., Wu, J. Y., & Hsu, Y. W. (2019). Protective effects of rosmarinic acid against selenite-induced cataract and oxidative damage in rats. *International Journal of Medical Sciences, 16*(5), 729–740. https://doi.org/10.7150/ijms.32222

Tsalkidou, E. G., Tsaroucha, A. K., Chatzaki, E., Lambropoulou, M., Papachristou, F., Trypsianis, G., Pitiakoudis, M., Vaos, G., & Simopoulos, C. (2014). The effects of apigenin on the expression of Fas/FasL apoptotic pathway in warm liver ischemia-reperfusion injury in rats. *BioMed Research International, 2014*. https://doi.org/10.1155/2014/157216

Vasconsuelo, A., & Boland, R. (2007). Molecular aspects of the early stages of elicitation of secondary metabolites in plants. *Plant Science, 172*(5), 861–875.

Vermerris, W., & Nicholson, R. (2006). Biosynthesis of phenolic compounds. In *Phenolic compound biochemistry* (pp. 63–149). United States: Springer Netherlands. https://doi.org/10.1007/978-1-4020-5164-7_3

Verpoorte, R., Contin, A., & Memelink, J. (2002). Biotechnology for the production of plant secondary metabolites. *Phytochemistry Reviews, 1*(1), 13–25. https://doi.org/10.1023/A:1015871916833

Waibel, J. S., Mi, Q. S., Ozog, D., Qu, L., Zhou, L., Rudnick, A., Al-Niaimi, F., Woodward, J., Campos, V., & Mordon, S. (2016). Laser-assisted delivery of Vitamin C, Vitamin E, and ferulic acid formula serum decreases fractional laser postoperative recovery by increased beta fibroblast growth factor expression. *Lasers in Surgery and Medicine, 48*(3), 238–244. https://doi.org/10.1002/lsm.22448

Wang, X., Hasegawa, J., Kitamura, Y., Wang, Z., Matsuda, A., Shinoda, W., Miura, N., & Kimura, K. (2011). Effects of hesperidin on the progression of hypercholesterolemia and fatty liver induced by high-cholesterol diet in rats. *Journal of Pharmacological Sciences, 117*(3), 129–138. https://doi.org/10.1254/jphs.11097FP

Wang, J., & Sun, Y. C. (2022). Revealing the pharmacological effect and mechanism of darutoside on gouty arthritis by liquid chromatography/mass spectrometry and metabolomics. *Frontiers in Molecular Biosciences, 9*, 942303.

Weng, Z., Patel, A. B., Vasiadi, M., Therianou, A., & Theoharides, T. C. (2014). Luteolin inhibits human keratinocyte activation and decreases NF-kB induction that is increased in psoriatic skin. *PLoS One, 9*(2), 1–8.

Weng, C. J., & Yen, G. C. (2012). Chemopreventive effects of dietary phytochemicals against cancer invasion and metastasis: Phenolic acids, monophenol, polyphenol, and their derivatives. *Cancer Treatment Reviews, 38*(1), 76–87. https://doi.org/10.1016/j.ctrv.2011.03.001

Wilmsen, P. K., Spada, D. S., & Salvador, M. (2005). Antioxidant activity of the flavonoid hesperidin in chemical and biological systems. *Journal of Agricultural and Food Chemistry, 53*(12), 4757–4761. https://doi.org/10.1021/jf0502000

Wolber, R., Stäb, F., Max, H., Wehmeyer, A., Hadshiew, I., Wenck, H., Rippke, F., & Wittern, K. P. (2004). Alpha-Glucosylrutin ein hochwirksames Flavonoid zum Schutz vor oxidativem Streß. *Journal der Deutschen Dermatologischen Gesellschaft, 2*, 563–652.

Wu, P. S., Yen, J. H., Kou, M. C., & Wu, M. J. (2015). Luteolin and apigenin attenuate 4-hydroxy- 2-nonenal-mediated cell death through modulation of UPR, Nrf2-ARE and MAPK pathways in PC12 cells. *PLoS One, 10*(6). https://doi.org/10.1371/journal.pone.0130599

Wu, Y., Zheng, X., Xu, X. G., Li, Y. H., Wang, B., Gao, X. H., Chen, H. D., Yatskayer, M., & Oresajo, C. (2013). Protective effects of a topical antioxidant complex containing vitamins C and E and ferulic acid against ultraviolet irradiation-induced photodamage in Chinese women. *Journal of Drugs in Dermatology, 12*(4), 464–468.

Xi, M., Hou, Y., Wang, R., Cai, J. M., Ao, Y., Shen, J., Li, H., Wang, M., & Luo, J. (2022). Potential application of luteolin as an active antibacterial composition in the development of hand sanitizer products. *Molecules, 27*(21), 7342.

Xia, R., Sheng, X., Xu, X., Yu, C., & Lu, H. (2018). Hesperidin induces apoptosis and GO/G1 arrest in human non-small cell lung cancer A549 cells. *International Journal of Molecular Medicine, 41*(1), 464–472. https://doi.org/10.3892/ijmm.2017.3250

Xia, F., Wang, C., Jin, Y., Liu, Q., Meng, Q., Liu, K., & Sun, H. (2014). Luteolin protects HUVECs from TNF-α-induced oxidative stress and inflammation via its effects on the Nox4/ROS-NF-κB and MAPK pathways. *Journal of Atherosclerosis and Thrombosis, 21*(8), 768–783. https://doi.org/10.5551/jat.23697

Xie, Y., Mei, X., & Shi, W. (2023). Kaempferol promotes melanogenesis and reduces oxidative stress in PIG1 normal human skin melanocytes. *Journal of Cellular and Molecular Medicine, 27*(7), 982–990. https://doi.org/10.1111/jcmm.17711

Xu, X., Li, M., Chen, W., Yu, H., Yang, Y., & Hang, L. (2016). Apigenin attenuates oxidative injury in ARPE-19 cells thorough activation of Nrf2 pathway. *Oxidative Medicine and Cellular Longevity, 2016*, 1–9. https://doi.org/10.1155/2016/4378461

Yan, M., Liu, Z., Yang, H., Li, C., Chen, H., Liu, Y., Zhao, M., & Zhu, Y. (2016). Luteolin decreases the UVA-induced autophagy of human skin fibroblasts by scavenging ROS. *Molecular Medicine Reports, 14*(3), 1986–1992. https://doi.org/10.3892/mmr.2016.5517

Yanat, M., & Schroën, K. (2021). Preparation methods and applications of chitosan nanoparticles; with an outlook toward reinforcement of biodegradable packaging. *Reactive and Functional Polymers, 161*, 104849. https://doi.org/10.1016/j.reactfunctpolym.2021.104849

Yang, W., Kang, J., Liu, Y., Guo, M., & Chen, G. (2022). Effect of salicylic acid treatment on antioxidant capacity and endogenous hormones in winter jujube during shelf life. *Food Chemistry, 397*, 133788. https://doi.org/10.1016/j.foodchem.2022.133788

Yano, S., Umeda, D., Yamashita, S., Yamada, K., & Tachibana, H. (2009). Dietary apigenin attenuates the development of atopic dermatitis-like skin lesions in NC/Nga mice. *Journal of Nutritional Biochemistry, 20*(11), 876–881. https://doi.org/10.1016/j.jnutbio.2008.08.002

Yao, X., Jiang, W., Yu, D., & Yan, Z. (2019). Luteolin inhibits proliferation and induces apoptosis of human melanoma cells in vivo and in vitro by suppressing MMP-2 and MMP-9 through the PI3K/AKT pathway. *Food & Function, 10*(2), 703–712. https://doi.org/10.1039/c8fo02013b

Yingchutrakul, Y., Sittisaree, W., Mahatnirunkul, T., Chomtong, T., Tatpong, T., & Krobthong, S. (2021). Cosmeceutical potentials of grammatophyllum speciosum extracts: Anti-inflammations and anti-collagenase activities with phytochemical profile analysis using an untargeted metabolomics approach. *Cosmetics, 8*(4), 116.

Yong, D. O. C., Saker, S. R., Chellappan, D. K., Madheswaran, T., Panneerselvam, J., Choudhury, H., Pandey, M., Chan, Y. L., Collet, T., Gupta, G., Oliver, B. G., Wark, P., Hansbro, N., Hsu, A., Hansbro, P. M., Dua, K., & Zeeshan, F. (2020). Molecular and immunological mechanisms underlying the various pharmacological properties of the potent bioflavonoid, rutin. *Endocrine, Metabolic & Immune Disorders - Drug Targets, 20*(10), 1590–1596. https://doi.org/10.2174/1871530320666200503053846

Yordanov, M., Dimitrova, P., Patkar, S., Saso, L., & Ivanovska, N. (2008). Inhibition of Candida albicans extracellular enzyme activity by selected natural substances and their application in Candida infection. *Canadian Journal of Microbiology, 54*(6), 435–440. https://doi.org/10.1139/w08-029

Yoshihisa, Y., Andoh, T., Rehman, M. U., & Shimizu, T. (2020). The regulation of protein kinase casein kinase II by apigenin is involved in the inhibition of ultraviolet B-induced macrophage migration inhibitory factor-mediated hyperpigmentation. *Phytotherapy Research, 34*(6), 1320–1328. https://doi.org/10.1002/ptr.6597

Yu, Y., Wu, Y., Yan, H.z., Xia, Z.r., Wen, W., Liu, D.y., & Wan, L.h. (2021). Rosmarinic acid ameliorates acetaminophen-induced acute liver injury in mice via RACK1/TNF α medi ated antioxidant effect. *Pharmaceutical Biology, 59*(1), 1286–1293. https://doi.org/10.1080/13880209.2021.1974059

Zaitone, S. A., Ahmed, E., Elsherbiny, N. M., Mehanna, E. T., El-Kherbetawy, M. K., ElSayed, M. H., Alshareef, D. M., & Moustafa, Y. M. (2019). Caffeic acid improves locomotor activity and lessens inflammatory burden in a mouse model of rotenone-induced nigral neurodegeneration: Relevance to Parkinson's disease therapy. *Pharmacological Reports, 71*(1), 32–41. https://doi.org/10.1016/j.pharep.2018.08.004

Zeb, A. (2020). Concept, mechanism, and applications of phenolic antioxidants in foods. *Journal of Food Biochemistry, 44*(9), 1–22.

Zeng, N., Hongbo, T., Xu, Y., Wu, M., & Wu, Y. (2018). Anticancer activity of caffeic acid n-butyl ester against A431 skin carcinoma cell line occurs via induction of apoptosis and inhibition of the mTOR/PI3K/AKT signaling pathway. *Molecular Medicine Reports, 17*(4), 5652–5657. https://doi.org/10.3892/mmr.2018.8599

Zhang, Y., Chen, X., Yang, L., Zu, Y., & Qi, L. (2015). Effects of rosmarinic acid on liver and kidney antioxidant enzymes, lipid peroxidation and tissue ultrastructure in aging mice. *Food & Function, 6*(3), 927–931. https://doi.org/10.1039/C4FO01051E

Zhang, D., Jing, B., Chen, Z., Li, X., Shi, H., Zheng, Y., Chang, S., & Zhao, G. (2022). Ferulic acid alleviates sciatica by inhibiting peripheral sensitization through the RhoA/P38MAPK signalling pathway. *Phytomedicine, 106*, 154420.

Zhang, Y., Wu, Q., Zhang, L., Wang, Q., Yang, Z., Liu, J., & Feng, L. (2019). Caffeic acid reduces A53T α-synuclein by activating JNK/Bcl-2-mediated autophagy in vitro and improves behaviour and protects dopaminergic neurons in a mouse model of Parkinson's disease. *Pharmacological Research, 150*, 104538. https://doi.org/10.1016/j.phrs.2019.104538

Zheng, Y. Z., Deng, G., & Zhang, Y. C. (2022). Multiple free radical scavenging reactions of flavonoids. *Dyes and Pigments, 198*, 109877.

Zhou, W., Hu, M., Zang, X., Liu, Q., Du, J., Hu, J., Zhang, L., Du, Z., & Xiang. (2020). Luteolin attenuates imiquimod – Induced psoriasis-like skin lesions in BALB/c mice via suppression of inflammation response. *Biomedicine & Pharmacotherapy, 131*, 110696.

Zhu, X., Li, N., Wang, Y., Ding, L., Chen, H., Yu, Y., & Shi, X. (2017). Protective effects of quercetin on UVB irradiation-induced cytotoxicity through ROS clearance in keratinocyte cells. *Oncology Reports, 37*(1), 209–218. https://doi.org/10.3892/or.2016.5217

Zhu, Q., Sui, S., Lei, X., Yang, Z., Lu, K., Liu, G., Liu, Y.-G., Li, M., & Han, Y. (2015). Ectopic expression of the coleus R2R3 MYB-type proanthocyanidin regulator gene SsMYB3 alters the flower color in transgenic tobacco. *PLoS One, 10*(10), e0139392. https://doi.org/10.1371/journal.pone.0139392

Zhu, L., & Xue, L. (2019). Kaempferol suppresses proliferation and induces cell cycle arrest, apoptosis, and DNA damage in breast cancer cells. *Oncology Research, 27*(6), 629–634. https://doi.org/10.3727/096504018X15228018559434

Zielińska, D., Zieliński, H., Laparra-Llopis, J. M., Szawara-Nowak, D., Honke, J., & Giménez-Bastida, J. A. (2021). Caffeic acid modulates processes associated with intestinal inflammation. *Nutrients, 13*(2), 1–15. https://doi.org/10.3390/nu13020554

Zillich, O. V., Schweiggert-Weisz, U., Eisner, P., & Kerscher, M. (2015). Polyphenols as active ingredients for cosmetic products. *International Journal of Cosmetic Science, 37* (5), 455–464. https://doi.org/10.1111/ics.12218

Zych, M., Wojnar, W., Dudek, S., & Kaczmarczyk-Sedlak, I. (2019). Rosmarinic and sinapic acids may increase the content of reduced glutathione in the lenses of estrogen-deficient rats. *Nutrients, 11*(4), 803.

Zhang, B., Wang, J., Zhao, G., Lin, M., Lang, Y., Zhang, D., Feng, D., & Tu, C. (2020). Apigenin protects human melanocytes against oxidative damage by activation of the Nrf2 pathway. *Cell Stress Chaperones.* 2020, *25(2),* 277-285. https://doi.org/10.1007/s12192-020-01071-7.

Index

'*Note:* Page numbers followed by "f" indicate figures and "t" indicate tables.'

D

E

F

G

H

I

Q

R

S

T

U

V

W

Y